T0192489

Recent Advances in Science, Engineering and Technology (RASET-2023)

First edition published 2024
by CRC Press
4 Park Square, Milton Park, Abingdon, Oxon, OX14 4RN

and by CRC Press
2385 NW Executive Center Drive, Suite 320, Boca Raton FL 33431

CRC Press is an imprint of Informa UK Limited

British Library Cataloguing-in-Publication Data
A catalogue record for this book is available from the British Library

ISBN: 9781032864181(pbk)
ISBN: 9781003527442(ebk)

DOI: 10.1201/9781003527442

Font in Sabon LT Std
Typeset by Ozone Publishing Services

Recent Advances in Science, Engineering and Technology (RASET-2023)

Proceedings of the International Conference on Recent Advances in Science, Engineering & Technology, 29-30 September 2023

Editors

Dr. Nishu Gupta
Dr. Sandeep S. Joshi
Dr. Milind Khanapurkar
Dr. Asha Gedam
Dr. Nikhil Bhave

Contents

D. Emerging Trends in Basic & Applied Sciences

List of Figures

Chapter 1

Chapter 2

Chapter 3

Chapter 4

Chapter 5

Chapter 6

Chapter 7

Chapter 8

Chapter 9

Chapter 10

Chapter 11

Chapter 12

Chapter 13

Chapter 14

Chapter 15

Chapter 22

Chapter 23

Chapter 24

Chapter 25

Chapter 26

Chapter 27

Chapter 28

Chapter 29

Chapter 30

Chapter 31

Chapter 32

Chapter 33

Chapter 34

Chapter 35

Chapter 36

Chapter 38

Chapter 39

Chapter 40

Chapter 41

Chapter 42

Chapter 43

Chapter 44

Chapter 45

Chapter 46

Chapter 47

Chapter 48

Chapter 49

Chapter 50

Chapter 51

Chapter 52

Chapter 53

Chapter 61

Chapter 62

List of Table

Chapter 9

Chapter 10

Chapter 12

Chapter 13

Chapter 16

Chapter 17

Chapter 19

Chapter 20

Chapter 21

Chapter 22

Chapter 24

Chapter 25

Chapter 29

Chapter 30

Chapter 33

Chapter 38

Chapter 39

Chapter 40

Chapter 42

Chapter 43

Chapter 44

Chapter 46

Chapter 47

Chapter 48

Chapter 50

Chapter 51

Chapter 52

Chapter 53

Chapter 54

Chapter 55

Chapter 56

Chapter 58

Chapter 61

Acknowledgements

The content of book provides an insights into the advances in science and technology. We are express our warm gratitude to all researchers, co-researchers from industry professionals and academia for research contribution. We are also thankful for our patrons, advisory board members, reviewers and organizing committee members.

Preface

The advances in technology, engineering and science are necessary for better and sustailable life. It is not only beneficial for human growth but equally important for all the living and non living matters on the planet. Hence it is imperative to come togather and share the knowledge, innovations and developments in the technology and science happening around. The objective of 1st International Conference on "Recent Advances in Science, Engineering & Technology" (ICRASET-2023) was to provide platform to share various hypotheses, conclusions, and discoveries from students, researchers, professors, and industry experts. The conference was associated with the knowledge partners like ASM International, IEEE, IETE, ISTE, CSI and IE.

Editors

About Editors

Dr. Nishu Gupta

Dr. Nishu Gupta is a Senior Member, IEEE having interest in Computer Communication and Networking, Autonomous Vehicles, Smart Cities, Security in Sensor Networks and Internet of Things. He has authored and edited several books with international publishers like Taylor & Francis, Springer, Wiley and Scrivener. Along with chairing several International Conferences, Dr. Nishu served as a reviewer and is on the Editorial board of various Internationally reputed Journals and Transactions. Dr. Gupta worked on various research projects in international universities like University of Fribourg, Switzerland, King Abdulaziz University in Rabigh, Saudi Arabia, University of Oviedo, Gijón, Spain. He had also worked on projects of Ministry of Electronics and Information Technology (MeitY), Government of India. He is receipnt of MDPI 2022 Travel Award and has published 5 patents and more than 60 research articles in IEEE Transactions, SCI and Scopus Indexed Journals. Currently he is holding a position of Postdoctoral Fellow in the, Department of Electronic Systems at Norwegian University of Science and Technology (NTNU) in Gjøvik, Norway.

Dr. Sandeep S. Joshi

Dr. Sandeep Joshi is an accomplished editor with extensive expertise in the field of Mechanical Engineering. Currently serving as an Assistant Professor at Shri Ramdeobaba College of Engineering and Management in Nagpur, India, he holds a Ph.D. in Mechanical Engineering. Dr. Joshi specializes in Thermal Engineering and has made significant contributions to the field, with over 50 research papers published in renowned journals and conferences in India and abroad. His innovation extends beyond academia, as he has applied for around 30 patents, five of which have been granted by the Indian Patent Office. Dr. Joshi's editorial prowess is also notable, as he serves as a reviewer for more than 25 reputed International journals and research agencies in India, Poland and Hong Kong.

Dr. Milind Khanapurkar

Dr. Milind Khanapurkar has his specialization in Electronics Engineering with Intelligent Transportation System as area of interest with special focus on Embedded systems, modeling and IoT. He has 1 book, 1 book chapter and 50+ publication in international journals and conferences of repute. He has one patents to his credit with few design patents and several research and project grants from various organizations.

Dr. Asha Gedam

Dr. Asha H. Gedam, currently serving as an Assistant Professor in the department of chemistry, Cummins College of Engineering for Women, Nagpur India. The research background of Dr. Gedam is waste water reatment technology. She has published her research work in reputed journals and served as a reviewer for reputed international Elsevie, Taylor and Francis journals.

Dr. Nikhil Aniruddha Bhave

Dr. Nikhil Aniruddha Bhave is working as an Assistant Professor in the School of Energy and Environmental Systems at the Defence Institute of Advanced Technology, DIAT-DU-DRDO, Pune, Maharashtra, India. He has 3 patents awarded by Indian Patent Office and has quality publication in international journals and conferences of repute with major focus on I C Engines and Renewable fuels. He has worked as a reviewer for SAE Technical Papers, Proceedings of the Institution of Mechanical Engineers, Part D: Journal of Automobile Engineering, Journal of the Brazilian Society of Mechanical Sciences and Engineering, etc. He has been awarded the 'Best Paper Award" three times and is a life member of many professional societies. Presently, at DIAT-DRDO, he is working on the area of Green Hydrogen for a Sustainable Future.

Organizing Committee

Chief Patron:	Shri. Ravindra Deo Shri. Jayant Inamdar Dr. Dhananjay Kulkarni	
Patron:	Dr. PVS Shastry Dr. Makarand Karkare Mr. Hemant Ambaselkar Mr. Shrikant Gadge	
Honorary Chair:	Dr. Milind Khanapurkar	
Convenor:	Dr. Sanjivani Shastri Dr. Mahesh R. Shukla	
Organizing Chair:	Dr. Jitesh Shinde	
Organizing Secretary:	Mr. Prasanna Mahankar	
Technical Program Chair:	Dr. Shubhangi Bompilwar Dr. Jaya Raut Dr. Kanchan Wagh Mrs. Abhilasha Borkar	Dr. Priyadarshini Ramteke Dr. Shailesh Khekale Dr. Asha Gedam
Registration & Finance Chair:	Mr. Yogesh Dandekar Mr. Pravin Gorantiwar	Mrs. Rashmi Deshpande Mr. Anand Deshkar
Publicity, Sponsorship, Social Media Chair:	Mrs. Sharayu Deote Mr. Harshwardhan Kharpate Mr. Aditya Kawdaskar	Mrs. Pallavi Tanksale Mr. Abhijeet Getme
Web Chair:	Mr. Shailesh Sahu Ms. Pinky Gangwani	

A. Pioneer Engineering in Electronics Engineering

Chapter 1

Circular Slotted UWB Antenna Design For Microwave Medical Imaging Application

Bhawna Khokher[1], *V. K. Sharma*[2], *Sanjeev Sharma*[3], *and Vinod Kumar Singh*[4]

[1]Research Scholar, Bhagwant University, Ajmer, India,
[2]Vice Chancellor, Bhagwant University, Ajmer, India,
[3]Dean, New Horizon College of Engineering, Bangalore, India
[4]Professor, CSJM University, Kanpur, India

Abstract: Ultra-wideband (UWB) systems have attracted a lot of attention in recent medical research areas, especially in the fields of microwave imaging (MI) application. The exceptional features of UWB, such as its wide bandwidth and low energy usage, are driving this increased interest in healthcare segment. In this research, a novel wide band microstrip antenna for medical imaging support WLAN 5G band is proposed. The antenna's overall dimensions are 42×32×1.6 mm^3, and it is constructed on an FR4 substrate with a loss tangent of 0.025 and a dielectric constant of 4.3, respectively. This proposed antenna consists of a circular ring antenna with a rectangular strip with tiny square cut, along with a partial ground plane using the Defected ground structure (DGS) technique. Impedance bandwidth (S11 ≤ –10 dB) is approximately 7 GHz (3.5–10.5) GHz and VSWR is 1.05 at 5.008GHz with gain of 3.68 dB, and its maximum effective gain of 4.7 dB at 9.82 GHz. This patch antenna is modelled utilising the description of modified grounds with various types of stage constructed using CST.

Keywords: Circular-shaped antenna, monopole, defective ground structure, UWB, medical imaging.

1. Introduction

Biomedical research has gained prominence in recent decades, particularly in the early diagnosis of tumors. It offers various benefits over current imaging methods including computed tomography (CT) scans and microwave imaging (MI) comparison with X-ray image, as demonstrated Figure 1 (Wörtge *et al.*, 2018). UWB refers to a range of electromagnetic frequencies with very low-power density and extremely short pules to take precision imaging of biological tissues and structures to detect abnormalities and diagnosing medical condition. UWB

DOI: 10.1201/9781003527442-1

sensors can penetrate through various materials, including skin and bone, without causing harm. The non-invasiveness is essential for monitoring vital signs, such as blood pressure, heart rate and even glucose levels. UWB antennas are essential components of medical imaging systems such as UWB radar and MI compared with x-ray image as shown in Figure 1(a–b). Optimising their performance is essential because antennas are important components of the MI system. Due to the growing demand for multifunctional wire-free systems with operational frequencies ranging suggested by FCC from 3.1 to 10.6 GHz especially for UWB application, recent research in wireless communication systems (WCS) has mostly concentrated on UWB antennas (Lakrit *et al.*, 2020).

Particularly, UWB antennas have the potential to be able to satisfy the needs for low- cost, low-profile and high-gain antenna designs for broadband operating systems including radar imaging, medical imaging and high-speed mobile communications for biomedical applications (Loktongbam *et al.*, 2023). UWB signals can only be delivered for a short period of time because of this, but they still offer high temporal precision and non-ionising radiation (Di Biase *et al.*, 2022). Using electrical characteristics, imaging methods can differentiate between healthy and unhealthy tissues (Alani *et al.*, 2021). Extensive studies have indicated that the permittivity of tumors might be 10–20% higher than that of the surrounding healthy tissues (Selvaraj *et al.*, 2020).

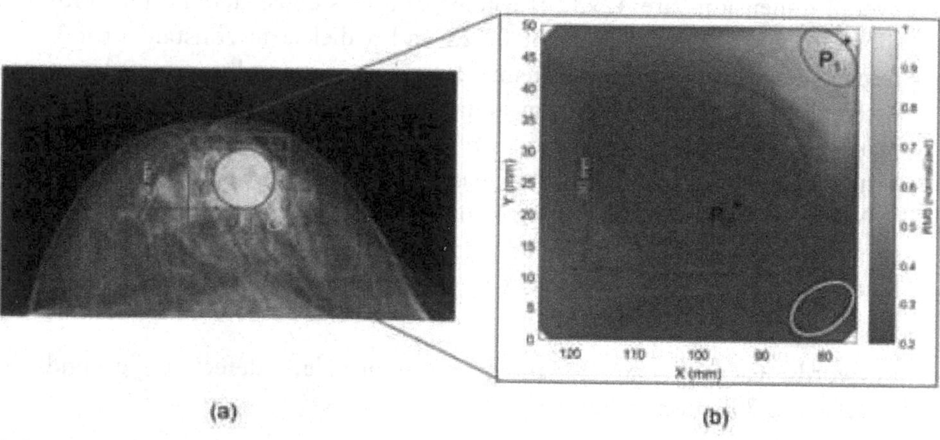

Figure 1: Comparison between (a) X-ray image and (b) microwave image (Dennis et al., 2018).

2. Literature Review

For MI, a unique patch antenna suggested by Ahadi *et al.* (2021), incorporating a circular form with a triangular cut and a low-cost FR4- substrate dielectric constant of 4.4. The antenna has an impedance bandwidth of 2.03 GHz, spanning from 1.22 to 3.45 GHz. As measured, the antenna exhibits a gain above 5 dBi and an efficiency over 85% by Samsuzzaman *et al.* (2023). Similarly, for UWB applications, disc-shaped antenna with a FR4 substrate and dimensions of 13×12×1.6 mm^3 with bandwidth of 11.2 GHz allows it to send and receive data at a wide range of frequencies. The antenna also has a modest directivity of 3.2 dBi and gains of 1.2 dBi by Markkandan *et al.* (2021). Diamond-slotted UWB antenna suggested by Kumar *et al.* (2022) for the frequency band from 1.71 to 12 GHz, material substrate FR4 discussed. Furthermore, it provides a brief analysis of different types of substrate materials affect the performance of patch antennas. The pattern is printed on an FR4 substrate with a size of 46×34×1.6 mm^3. Antenna impedance bandwidth measurements 11 GHz. At 2.1 GHz, it records the best radiation efficiency of 97%, while at 10.1 GHz it records the highest peak gain of 7.25 dBi. Multiband band antenna suggested is printed on a FR4 flat substrate and has a surface area of 33 mm^2. Its permittivity is 4.2 and its loss tangent is 0.02 for the whole frequency range. Wang *et al.* (2022) suggested configuration covers several bands and offers a wider impedance bandwidth of 2–10.7 GHz at a return loss of −10 dB.

UWB antennas are essential components for modern medical imaging, in general. They significantly contribute to the development of healthcare technology, which eventually leads to better patient care and diagnostic accuracy, by being able to provide high bandwidth, restrict interference, allow for novel modalities and promote patient safety.

2.1. Proposed Antenna Design

The antenna is examined using the CST programme to completely understand how its construction functions and determine the ideal parameters that show the physical characteristics of the proposed antenna. The proposed antenna's parameters are depecited in Table 1. To analyse the suggested antenna, one parameter is modified at a time while others remain constant. The front and back structures of the proposed antenna, a circular monopole made from a substrate with a dielectric constant for the FR4 of 4.3 and thickness 1.6 mm, are shown in Figure 2(a–b).

Table 1: Front parameters and back side parameters of antenna.

Front Side Parameter									
Parameters	**W**	**L**	**A**	**B**	**C**	**D**	**R1**	**R2**	**a**
Unit (mm)	**32**	**42**	**12.6**	**11.63**	**3.2**	**2.5**	**13.50**	**9.75**	**1.50**
Back Side Parameter									
Parameters	**W**	**x**	**Y**	**Z**	**p**	**q**			
Unit (mm)	**2.20**	**12.60**	**5.20**	**4.45**	**3.0**	**0.80**			

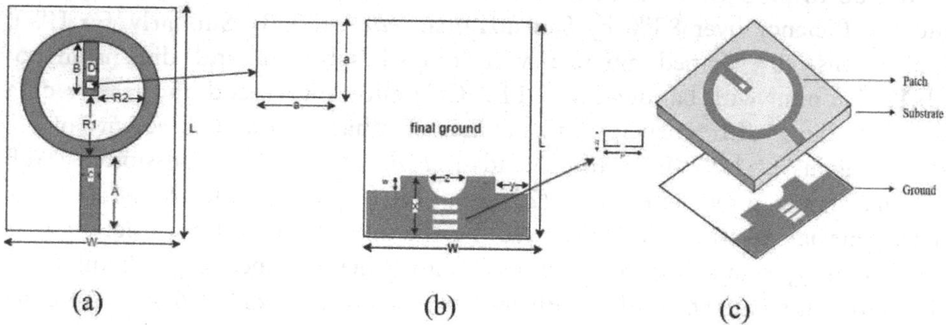

(a) (b) (c)

Figure 2: Antenna design – (a) Front side measurement, (b) Back side measurement and (c) Antenna design.

2.2. Defected Ground Structure (DGS) Evaluation for UWB

This article presents an intuitive circular ring antenna with DGS, wideband and low profile for use in biomedical applications, particularly medical imaging. The antenna further modified in stage 1–4 by cutting a rectangular slot from the lower side of the ground as shown in Figure 3.

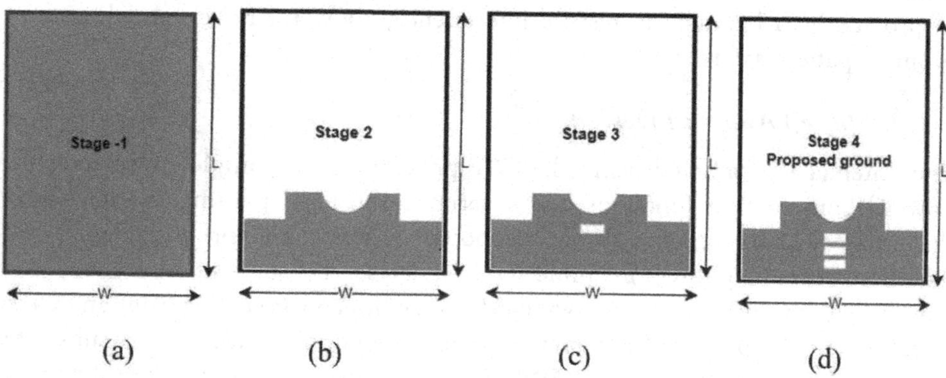

(a) (b) (c) (d)

Figure 3: Defected ground structure with (a) Stage 1 (b) Stage 2 (c) Stage 3 and (d) Stage 4.

3. Results and Discussion

The schematic of the suggested antenna with design evaluation processes for UWB antenna is shown in Figure 3(a–d), and their corresponding reflection coefficient (|S11|) graph is illustrated in Figure 4, respectively. The simple full-ground plane is used as the initial design step as stage 1. The antenna exhibits two resonances at approximately 9 GHz and 11 GHz in stage 1 with a minimum bandwidth of 0.4 GHz as shown in Figure 4. Stage 2 is depicted in Figure 3(b), the antenna resonant at 3.8 GHz and 5 GHz with corresponding reflection coefficient (|S11| < –51 dB) with VSWR factor is 1.05 with a gain of 4.71 dBi. In stage 3 as shown in Figure 3(c), the antenna resonated at round 3.8 GHz and 4.8 GHz with corresponding reflection coefficient (|S11| < –27 dB) with VSWR factor is 1.09 with a gain of 4.67 dBi. Finally, stage 4 as depicted in Figure 3(d) covered maximum bandwidth around 7 GHz with Figure 4 plots the comparison graph of S11 of the proposed antenna stages along with Figure 5 of the gain graph with a resonant frequency of 5 GHz and 3.86 dB gain.

Corresponding reflection coefficient (|S11| < –39.56 dB) with VSWR factor is 1.05 with a gain of 4.71 dB. In Figure 5 plot, the comparison graph of S11 of proposed antenna along with gain graph with resonant frequency 5 GHz with 3.86 dB gain.

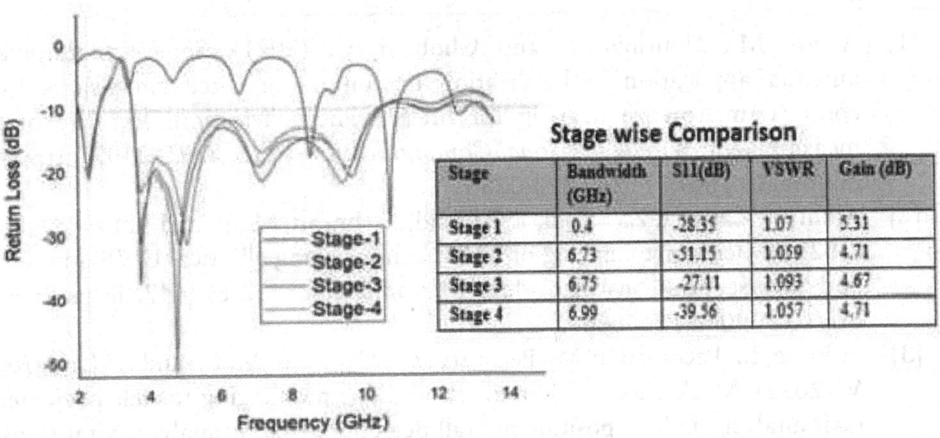

Stage wise Comparison

Stage	Bandwidth (GHz)	S11(dB)	VSWR	Gain (dB)
Stage 1	0.4	-28.35	1.07	5.31
Stage 2	6.73	-51.15	1.059	4.71
Stage 3	6.75	-27.11	1.093	4.67
Stage 4	6.99	-39.56	1.057	4.71

Figure 4: Simulation results of the defective ground structure of an antenna.

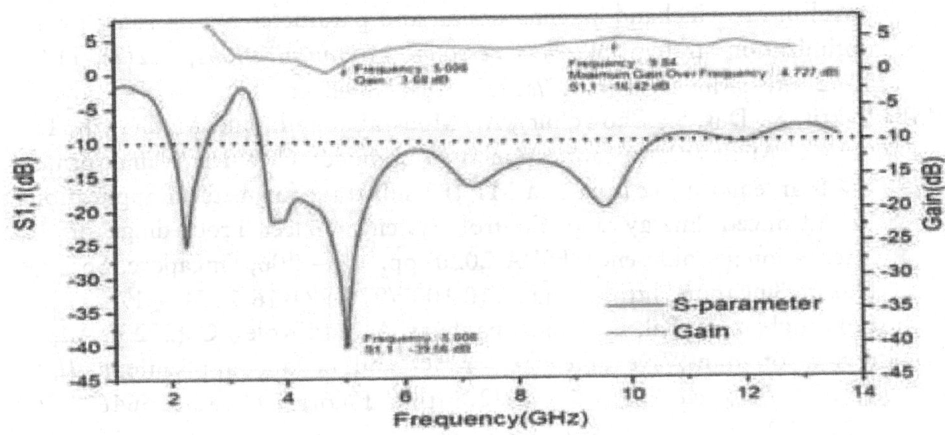

Figure 5: Comparison graph of return loss (S11) & gain of proposed antenna.

4. Conclusion

This paper reported a detailed design of UWB antenna with DGS which covers frequency from 3.5 to 10.5 GHz. The design of an UWB antenna with FR4 substrate circular patch slotted is covered in the article. Analysis is also done of the various ground structure deformation performance of size 42×32×1.6 mm^3 with return loss S11 ≤ –10 dB from 3.5 to 10.5 GHz with approx. 7 GHz frequency band. 4.74 dB maximum gain at 9.3 GHz recorded and 3.68 dB gain at 5.008 GHz which is suitable for the 5G band application. To achieve the needed range for the biomedical application, the antenna configuration is discussed stage-by-stage with bandwidth impedance.

References

[1] Ahadi, M., Nourinia, J., and Ghobadi, C. (2021). Square monopole antenna application in localization of tumors in three dimensions by confocal microwave imaging for breast cancer detection: Experimental measurement. *Wireless Personal Communications*, 116, 2391–2409. https://doi.org/10.1007/s11277-020-07801-5.

[2] Alani, S., Zakaria, Z., Saeidi, T., Ahmad, A., Imran, M. A., and Abbasi, Q. H. (2021). Microwave imaging of breast skin utilizing elliptical UWB antenna and reverse problems algorithm. *Micromachines*, 12(6), 647. https://doi.org/10.3390/mi12060647.

[3] di Biase, L., Pecoraro, P. M., Pecoraro, G., Caminiti, M. L., and Di Lazzaro, V. (2022). Markerless radio frequency indoor monitoring for telemedicine: Gait analysis, indoor positioning, fall detection, tremor analysis, vital signs and sleep monitoring. *Sensors*, 22(21), 8486. https://doi.org/10.3390/s22218486.

[4] Kumar, P., Shilpi, S., Kanungo, A., Gupta, V., and Gupta, N. K. (2022). A novel ultra- wideband antenna design and parameter tuning using hybrid optimization strategy. *Wireless Personal Communications*, 122(2), 1129–1152 https://doi.org/10.1007/s11277-021-08942-x.

[5] Lakrit, S., Das, S., Chowdhury, A., MoussaLabbadi, and Madhav, B. T. P. (2022). Analysis and fabrication of a compact CPW-fed planar printed UWB antenna using isola tera MT (R) substrate for medical applications, In Advanced Energy and Control Systems: Select Proceedings of 3rd International Conference, ESDA 2020, pp. 195– 206, Singapore: Springer Nature Singapore. https://doi.org/10.1007/978-981-16-7274-3_17.

[6] Loktongbam, P., Pal, D., Bandyopadhyay, A., and Koley, C. (2023). A brief review on mm-Wave antennas for 5G and beyond applications. *IETE Technical Review*, 40(3), 397–422. https://doi.org/10.1080/02564602.2022.2121771.

[7] Markkandan, S., Malarvizhi, C., Raja, L., Kalloor, J., Karthi, J., and Atla, R. (2021). Highly compact sized circular microstrip patch antenna with partial

ground for biomedical applications. *Materials Today: Proceedings*, 47, 318–320. https://doi.org/10.1016/j.matpr.2021.04.480.

[8] Samsuzzaman, M., Talukder, M. S., Alqahtani, A., Alharbi, A. G., Azim, R., Soliman,

[9] M. S., and Islam, M. T. (2023). Circular slotted patch with defected grounded monopole patch antenna for microwave-based head imaging applications. *Alexandria Engineering Journal*, 65, 41–57. https://doi.org/10.1016/j.aej.2022.10.034.

[10] Selvaraj, V., Sheela, J. B. J. J., Krishnan, R., Kandasamy, L., and Devarajulu, S. (2020). Detection of depth of the tumor in microwave imaging using ground penetrating radar algorithm. *Progress In Electromagnetics Research M*, 96, 191–202. https://doi.org/10.2528/PIERM20062201.

[11] Wang, M., Wang, H., Chen, P., Ding, T., Xiao, J., and Zhang, L. (2022). A butterfly-like slot UWB antenna with WLAN band-notch characteristics for MIMO applications. *IEICE Electronics Express*, 19(14), 20220233. https://doi.org/10.1587/elex.19.20220233.

[12] Wörtge, D., Moll, J., Krozer, V., Bazrafshan, B., Hübner, F., Park, C., and Vogl, T. J. (2018). Comparison of X-ray-mammography and planar UWB microwave imaging of the breast: First results from a patient study. *Diagnostics*, 8(3), 54. https://doi.org/10.3390/diagnostics8030054.

Chapter 2

Metamaterial Dual L-Slot Triangle Patch Antenna for Optimal X-Band Applications

Bhawna Khokher[1], Gurulakshmi A. B[2], Rajesh G.[3], Sanjeev Sharma[4], Y. Veni[4], S. Meghana[4], Y. R. Leele Vara Prasad Reddy[1], and S. Bhavishya[1]

[1]Assistant Professor, New Horizon College of Engineering, Bengaluru, India.
[2]Associate Professor, New Horizon College of Engineering, Bengaluru, India.
[3]Professor & Dean, New Horizon College of Engineering, Bengaluru, India.
[4]UG Student, New Horizon College of Engineering, Bengaluru, India.

Abstract: Recently, metamaterials have made an increasingly significant impact across several domains. They have emerged as a promising technology for enhancing the performance of various electromagnetic devices. Metamaterials are highly attractive for antenna design because they can control electromagnetic waves which cannot be from.

Keywords: Metamaterial, CPW feeding, microstrip feeding, 5G.

1. Introduction

Designing antennas that handle the broad frequency range needed for 5G services is one of the main issues facing the development of 5G networks (Chettri & Bera., 2019). One such band, from 7 to 12 GHz, can offer fast, high-bandwidth connections for a variety of uses, including video streaming, virtual reality and Internet of Things devices (Gurulakshmi *et al.*, 2023). But a significant problem is creating to design an effective wideband small-size antenna for certain applications. Patch antennas are low profile, lightweight and economical along with limited bandwidth are their biggest flaw (AL-Amoudi, 2021). Though, researchers used a variety of strategies to increase bandwidth in order to solve this issue, including expanding substrate size, making slots (Mahendran *et al.*, 2020), employing magneto-dielectric substrates and adding parasitic patches (Aktar *et al.*, 2011) using an artificial magnetic ground plane (Laabadli *et al.*, 2023).

To design efficient and effective antennas for a particular frequency range, designers introduced a new material known as metamaterial, which are artificially engineered composite materials that exhibit unique characteristics that are not present in natural materials. There are some other advanced techniques such as multi-beam antennas and frequency-selective surfaces (Abdelgwad, 2018). These techniques enable antennas to achieve high gain, high directivity and low loss,

DOI: 10.1201/9781003527442-2

which are essential for reliable and robust 5G communication. Furthermore, the use of this frequency range in 5G technology opens new opportunities for innovative applications such as smart healthcare, smart transportation and smart cities.

2. Literature Review

Left-handed materials (LHM) are a type of metamaterials with unique electromagnetic characteristics including negative refractive index, anomalous dispersion and reverse propagation. These characteristics result from the unique structure of LHM, which is made up of periodic arrays of subwavelength unit cells. The unit cells are designed to have a negative electric permittivity or magnetic permeability or both (Xie, *et al.*, 2022), resulting in a negative refractive index. LHM are widely used for invisibility cloaking, superlensing and antenna design. Viktor Veselago, a Russian scientist, presented theoretical research in 1967 (He, *et al.*, 2022) in which he hypothesised the potential of materials having negative permittivity and permeability. He demonstrated that such materials would have unusual electromagnetic characteristics, such as negative refraction, backward wave propagation and a reversed Doppler effect.

Rodger M. Walser and his colleagues at the University of Texas in Austin published "Experimental Verification of Negative Refraction" in 1999, demonstrating the first experimental demonstration of negative refractive index in a metamaterial. They achieved this by creating a metamaterial structure consisting of an array of split-ring resonators (SRSS) and wires on a printed circuit board substrate, which exhibited negative refractive index a microwave frequency.

2.1. Antenna Design

2.1.1. Geometry of Antenna Without Metamaterial Using Microstrip feed/CPW feed

Figure 1(a–b) depicts the geometry that comprises the proposed compact antenna with substrate FR-4 and Rogers RO3003 substrates with permittivity values of 4.3–3 and a thickness of 1.5 mm. To analyse the suggested antenna, one parameter is modified at a time while the others remain constant. The CST programme is used to evaluate an antenna to completely understand how its structure functions and to determine the optimal settings. Figure 1 displays the proposed antenna's physical parameters. The CST programme is used to analyse the antenna in order to fully comprehend how its structure behaves and identify the ideal parameters. Table 1 displays the proposed antenna's physical parameters.

Table 1: Microstrip feed antenna parameters without metamaterial.

Parameters	*a*	*b*	*c*	*d*	*e*	*f*	*g*	*H*	*I*	*j*	*k*	*L*	*M*	*n*	*O*
Units (mm)	6.3	1	2.5	1	2.6	0.95	0.3	0.4	0.5	0.85	0.4	0.3	0.4	12	15

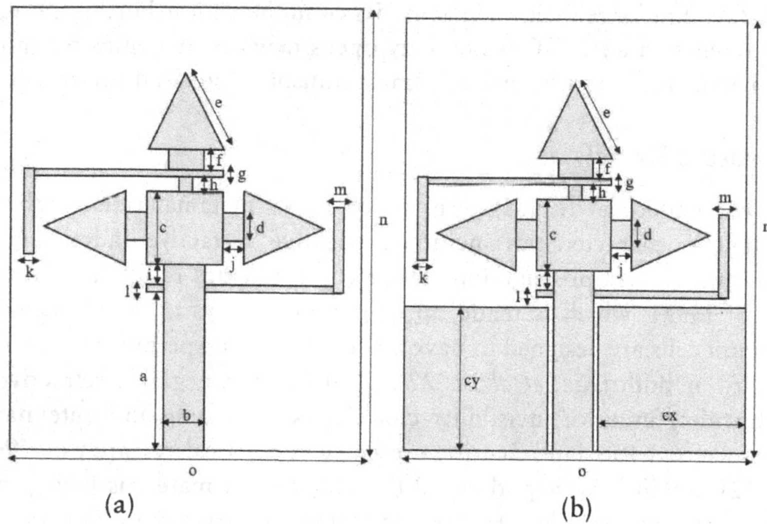

(a) (b)

***Figure1:** Geometry of antenna without metamaterial using (a) Microstrip feed and (b) CPW feed.*

2.1.2. Geometry of Antenna Metamaterial Using Microstrip feed/CPW feed

Figure 2(a–b) shows the geometry of the proposed minimalist antenna using metamaterials with substrate materials FR-4 and Rogers RO3003 permittivity values of 4.3–3 and a thickness of 1.5 mm. To achieve a 50Ω characteristic impedance, the microstrip feed line's width is fixed at 1mm. Table 2 displays the proposed antenna's physical parameters metamaterial.

Table 2: Microstrip feed antenna parameters with metamaterial.

Parameters	*a1*	*b1*	*c1*	*d1*	*e1*	*f1*	*g1*	*h1*	*i1*	*j1*	*k1*	*l1*	*m1*	*n1*
Units (mm)	6.3	1	2.5	1	2.6	0.95	0.3	0.4	0.5	0.85	0.4	0.3	0.4	12

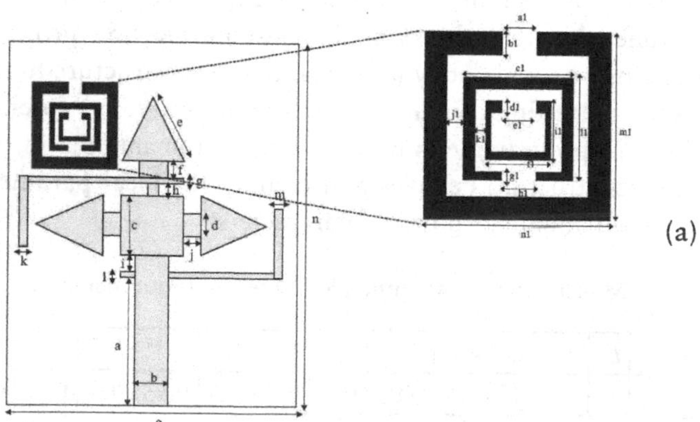
(a)

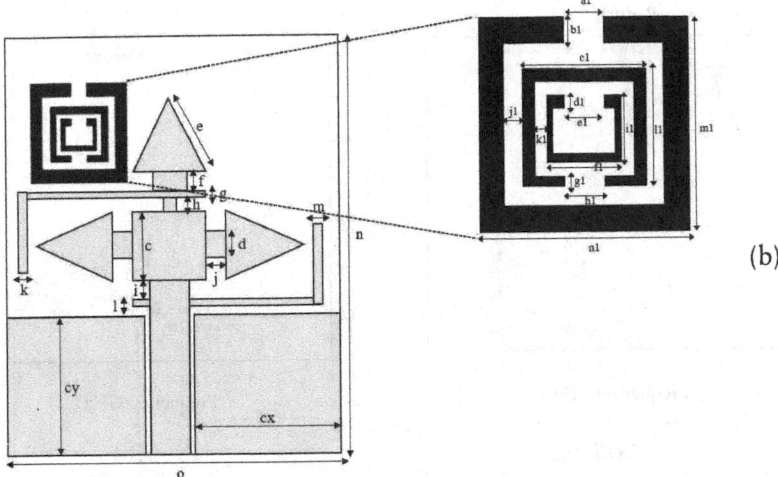

Figure 2: Geometry of antenna metamaterial using (a) Microstrip feed and (b) CPW feed.

3. Results and Discussion

3.1. S-Parameters Comparison

The graphs in Figure 3(a–b) below provide a comparison of the S-parameters of antennas that are developed using FR-4 and Rogers RO3003 as the substrate, Figure 3(a–b) compares a microstrip patch antenna with and without metamaterial. The S11 parameters of the antenna without metamaterial are –45.148 dB, whereas those of the antenna with metamaterial are –17.875. As a result, we may conclude that if metamaterials are forcibly introduced into a good-parameter antenna, it is possible to reduce its parameters. A microstrip patch antenna equipped with or without metamaterial on a Rogers RO3003 substrate is shown in Figure 3(a). An antenna's

|S11| parameters without and with metamaterial are –17.608 and –36.138, respectively. Its bandwidth expands in tandem with its return loss. We may conclude from this that metamaterial will be beneficial in terms of return loss and bandwidth. With FR-4 used as the substrate, Figure 3(b) compares the CPW feed antenna with and without metamaterial. The |S11| parameters of an antenna without metamaterial are –11.601 dB, whereas the |S11| parameters of an antenna with metamaterial are –12.789 dB. Additionally, its bandwidth has increased. Figure 3(b) gives a comparative analysis of CPW feed antenna using Rogers RO3003 as a substrate.

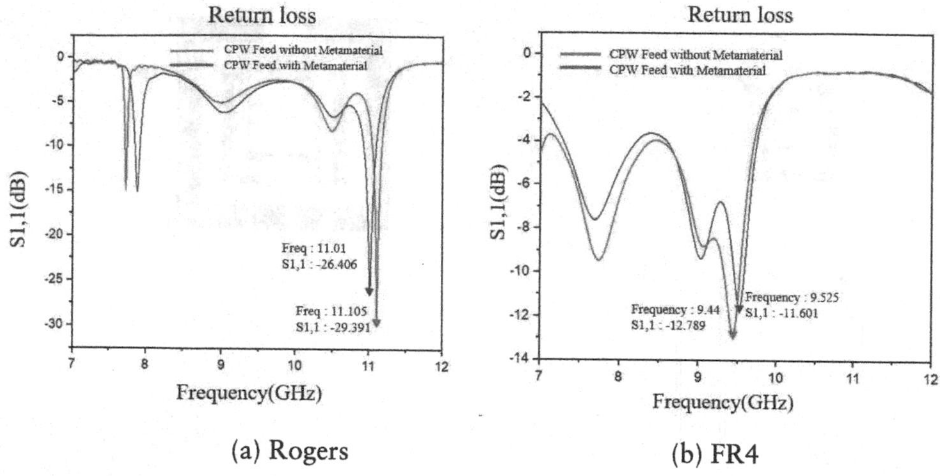

(a) Rogers (b) FR4

Figure 3: S-parameters graph comparison between microstrip feed antenna with and without metamaterial using (a) Rogers as substrate and (b) FR-4.

3.2. *VSWR Comparison*

Figure 4(a–b) displays a comparison graph between microstrip feed with and without metamaterial. Using Rogers RO3003 as the substrate, Figure 4(a) displays a comparison graph between microstrip feed with and without metamaterial. Using FR- 4 as the substrate, Figure 4(b) displays a comparison graph between CPW feed with and without metamaterial. Using Rogers RO3003 as the substrate, Figure 4(a) displays a comparison graph between CPW feed with and without metamaterial.

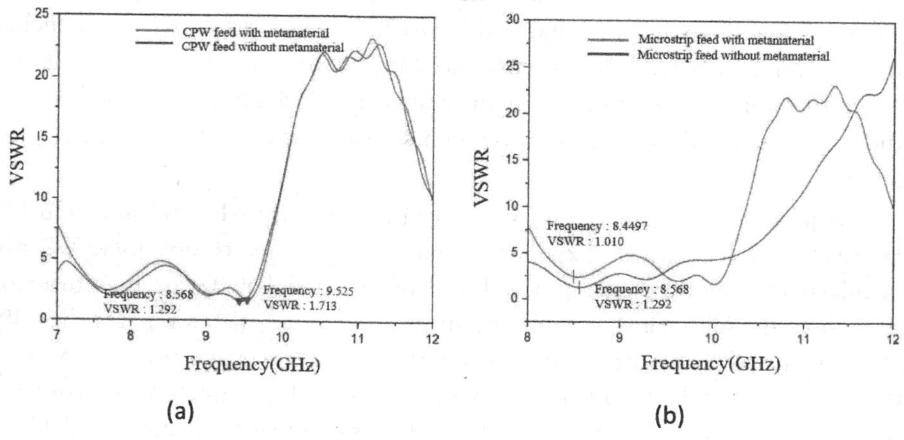

(a) (b)

Figure 4: Comparison of VSWR graphs for (a) microstrip feed antennas with and without metamaterial using Rogers as the substrate (b) CPW feed antenna with and without metamaterial using Rogers as substrate.

3.3. *Gain and S-parameter Comparison*

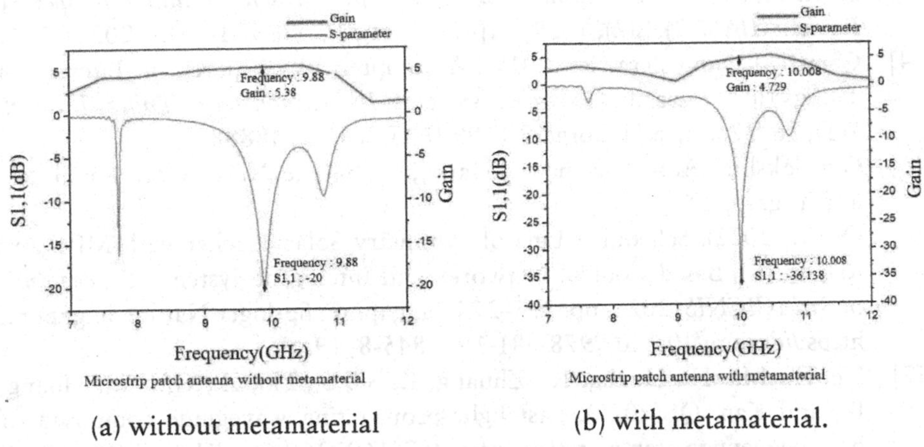

(a) without metamaterial (b) with metamaterial.

Figure 5: *Gain and S-parameter comparison of microstrip patch antenna using Roger's substrate.*

Figure 5(a–b) shows a comparison graph of microstrip feed antenna using Roger's substrate. The gain of antenna at resonant frequency 9.88 is 5.38. Figure 5(a) shows a microstrip feed antenna comparison graph utilising Roger's substrate. At resonant frequency 10.008, the antenna gain is 4.729.

4. Conclusion

Finally, the incorporation of metamaterials into patch antennas has proven to be a potential strategy for increasing antenna performance in 5G applications. Metamaterials have unique electromagnetic properties that can improve antenna attributes like gain, directivity, bandwidth and efficiency. The patch antenna design has been improved by including metamaterial structures such as SRRS, frequency-selective surfaces (FSSS) and metamaterial substrates. For instance, the use of metamaterial structures allows for the antenna to be miniaturised, making it more compact and ideal for integration into smaller devices. Furthermore, metamaterial-integrated patch antennas have shown increased bandwidth, allowing antennas to accommodate multiple frequency bands at the same time. This is critical for X-band applications that rely on a variety of frequency bands to enable high data speeds and reliability.

References

[1] Abdelgwad, A. H. (2018). Microstrip patch antenna enhancement techniques. *Int. J. Electr. Commun. Eng*, 12(10), 703–710. https://publications.waset.org/vol/142.

[2] Aktar, M. N., Uddin, M. S., Amin, M. R., and Ali, M. (2011). Enhanced gain and bandwidth of patch antenna using ebg substrates. *Int. J. Wireless Mobile Netw*, 3(1), 62–69. https://doi.org/10.5121/ijwmn.2011.3106.

[3] AL-Amoudi, M. A. (2021). Study, design, and simulation for microstrip patch antenna. *International Journal of Applied Science and Engineering Review* (*IJASER*), 2(2), 1–29. https://doi.org/10.52267/IJASER.2021.2201.

[4] Chettri, L., and Bera, R. (2019). A comprehensive survey on Internet of Things (IoT) toward 5G wireless systems. *IEEE Internet of Things Journal*, 7(1), 16–32. https://doi.org/10.1109/JIOT.2019.2948888.

[5] Gurulakshmi, A. B., Sharma, S., Manoj, N., Bhinge, N. A., Santhosh, H. M., and Yogesh,

[6] O. M. (2022). Scheduled Line of Symmetry Solar Tracker with MPT and IoT. In IoT Based Control Networks and Intelligent Systems. *Proceedings of 3rd ICICNIS,* 2022, pp. 265–274, Singapore: Springer Nature Singapore. https://doi.org/10.1007/978-981-19- 5845-8_19.

[7] He, Z., Ma, H., Huang, R., Zhuang, F., Su, S., Lin, Z., Qiu, W., Huang, B., and Kan, Q. (2022). Fast light propagating waveguide composed of heterogeneous metamaterials. *Optik*, 262, 169326. https://doi.org/10.1016/j.ijleo.2022.169326.

[8] Laabadli, A. A., Mejdoub, Y., El Amri, A., and Tarbouch, M. (2023). A miniaturized rectangular microstrip patch antenna with negative permeability unit cell metamaterial for the band 2.45 GHz. *In ITM Web of Conferences*, 2023, 52, 03002, EDP Sciences. https://doi.org/10.1051/itmconf/20235203002.

[9] Mahendran, K., Dhivya, K., and Prasanniya, V. (2020). Microstrip patch antenna enhancement techniques: a survey. *Int. J. Eng. Appl. Sci. Technol*, 4(11), 245–249. Xie, P., Shi, Z., Feng, M., Sun, K., Liu, Y., Yan, K.,... Guo, Z. (2022). Recent advances in radio-frequency negative dielectric metamaterials by designing heterogeneous composites. *Advanced Composites and Hybrid Materials*, 5(2), 679–695.

Chapter 3

A Multifunctional Robot for COVID-19 Hospital

Nikhil Pawar[1], *Umesh Kubade*[2], *Pranjali Jumle* [3], *and Pankaj Chandankhede*[3]

[1]M.Tech Scholar, G H Raisoni College of Engineering, Nagpur, Maharashtra, India
[2]Managing Director, LGPS Hybrid Energy Pvt Ltd, Maharashtra, India
[3]Assistant Professor, G H Raisoni College of Engineering, Nagpur, Maharashtra, India

Abstract: A technology known as a multifunctional robot allows the COVID-19 hospital to assist with the delivery of meals, healthcare, temperature monitoring and other tasks for COVID-positive patients without requiring additional labour from frontline staff or medical professionals. Everyone has a significant chance of contracting the virus. Therefore, a multifunctional robot can reduce the amount of time that patients and frontline employees or medical professionals spend interacting. Using a transmitter or remote, the user may operate the robot. The robot's main duties involve travelling, delivering meals and remotely tracking patients' vital signs, such as temperatures, without physically interacting with the patient. The robot also features an IP web camera that can send live footage of the patient. Additionally, it will make it easier for anyone to follow hospital guidelines and operate the robot. The authors used four motors with specific wheels connected to drive the robot from one location to another. A microcontroller can control these motors using an radio frequency (RF) wireless connection module. The robot can support a weight between 5 and 10 kg because it is strong enough to do so. Furthermore, as family members are not allowed to spend time with patients in the COVID ward, patients can communicate with their relatives. In this configuration, the transmitter component acts as the controller, while the robot itself acts as the slave. The adaptable robot offers several advantages because it utilises inexpensive innovation, requires minimal upkeep, is simple to use, can be quickly set up and has the ability to observe.

Keywords: Human assistance robot, healthcare robot, rehabilitation robots, robotic assistance, surgical robotic technologies.

1. Introduction

In the year 2020, the COVID-19 pandemic swept across the globe, subjecting individuals to some unusual and dire circumstances. This dangerous virus has been

DOI: 10.1201/9781003527442-3

detected in millions of people and has also claimed the lives of many. Currently, there are 1,22,714 active cases, which account for 0.36% of all cases, and 4,65,662 fatalities, comprising 1.36% of all deaths in India, according to official figures. Furthermore, the number of these cases continues to rise.

As a result, hospitals are becoming increasingly overcrowded with sick patients, making it very challenging to find available beds in both government and private facilities. Additionally, it is becoming more difficult for hospital staff to provide proper and effective care to hospitalised patients. This is because there are comparatively fewer staff members than there are patients. Therefore, to enable employees to adequately and continually manage COVID-19 positive individuals, a certain level of assistance is required. A multipurpose robot designed for COVID-19 situations proves highly valuable. Several such robots have been developed and tested in real-world scenarios to confirm their effectiveness (Meghana *et al.*, 2017).

They have proposed the development of a robot designed to assist medical personnel in administering medication to patients. The robot operates on a framework built upon the principles of line-follower robots. Each patient is provided with an RFID card containing their medical data. DC motors are utilised for the robot's mobility, enabling it to follow the black markings painted on the ground. These markings facilitate the robot's navigation within the facility and allow it to use the RFID reader to access relevant patient information. Furthermore, they have proposed that the robot also checks the patient's temperature, SpO2 levels and blood pressure. The individual operating the robot receives the patient's vital parameter values via NODEMCU suggested by Chandankhede *et al.* (2019) and Selvi *et al.* (2019).

For control, the surveillance robot employs RF communication modules. It utilises the accelerometer component to send commands based on x and y plane readings to the receiving end. Before transmission through the RF modules on the receiving end, these values undergo initial processing by the Uno programme on the transmitting end. Whichever signal is received by the RF recipient, the Uno will then employ the L298 motor control modules to operate the electric actuators by Harshita *et al.* (2018).

The technology utilises an IR line follower and IR sensors to dispense medication to COVID-19 patients. To detect obstacles in the robot's path, ultrasonic sensors have been incorporated, which will prompt the robot to stop upon detection. Moreover, the robot is equipped with a built-in real-time clock, ensuring timely administration of medications according to each patient's specific schedule. To identify the patient, a robot equipped with an RFID scanner scans the patient's RFID tags, containing all necessary information about the patient's medication and other relevant details. The name and dosage of the medication are displayed on the robot's LCD screen. Shinde *et al.* (2021) have also integrated an SMS alert system, which notifies the patient's family or caregiver when the medication is dispensed as prescribed. One limitation of this study, however, is the absence of a centralised platform for uploading all patient information to the system when a new patient is admitted to the hospital.

The main aim of this study is as follows:

- To design a multipurpose robot for COVID-19 patients by reducing close contact between medical staff and patients.
- The robot can remotely assess a patient's temperature levels.
- To design low-cost and low-maintenance robots doing a lot of duties in hospitals.

2. Proposed Methodology

The transmitter segment and receiver section are the two sections that make up the entire system. Someone else can provide input via the toggle switch at the transmitting area within the system when the operation starts. In that case, there are two choices on the toggle button: motor operation and temperature monitoring management. Currently, the actions of each of those two types may be described by Kale *et al.* (2022) separately. operating mode The Atmega328p, nRF24L01 wireless module, L298N drive motorist and motor (12 V, 100 rpm) are used in this mode of operation to do the task. First, once the individual chooses a driving method, the joystick is activated.

Minutes can be moved in one of the four unique ways using the joystick's gadget: left, right, upward and backward; every one of them is regulated via an x-axis or a y-axis (Hirak *et al.*, 2022). Each of those signals is detected by the processor by looking for a change in the resistance's spectrum, which could vary from low (for backward or forward) to elevated (for left or right). Following that, such controller input information is transmitted via the radio's component nRF24L01. In the 2.4 GHz ISM band, the transmitter's RF component can connect to the recipient's RF modules at a rate known as band of 115200 bps.

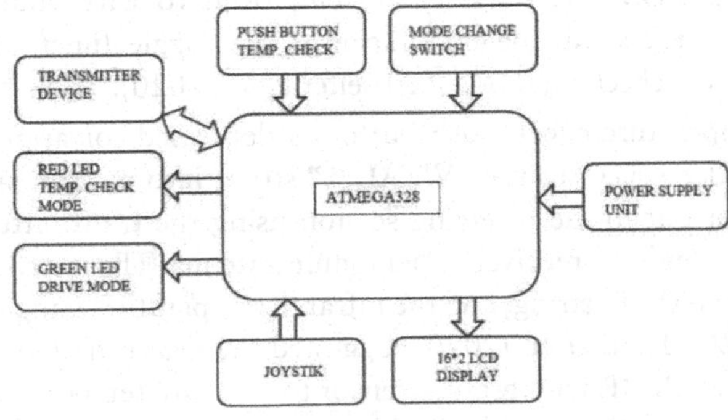

***Figure 1:** Modular schematic for a transmitter portion of a multifunctional robot.*

When the messages in controversy (digital amounts from 0 to 255) have been transmitted through the antennae of the getting radio component nRF24L01, which is then sent to the Arduino (receiver portion), the resulting 100 RPM DC-directed algorithms will begin to spin according to the inputs generated by a microcontroller by (Anandravisekar *et al.*, 2022). It goes through processing to give the motor control L298N modules their input controller impulses as a set of current levels. Bagul *et al.* (2020) suggested electric motor controller is needed to function as a connection between the electric motor and the computer circuitry to transform the lightweight into a high-power outlet that may

operate the drive mechanisms from the device's messages since its motors require a 12V supply to operate, whereas its Arduino only needs a maximum of 5V.

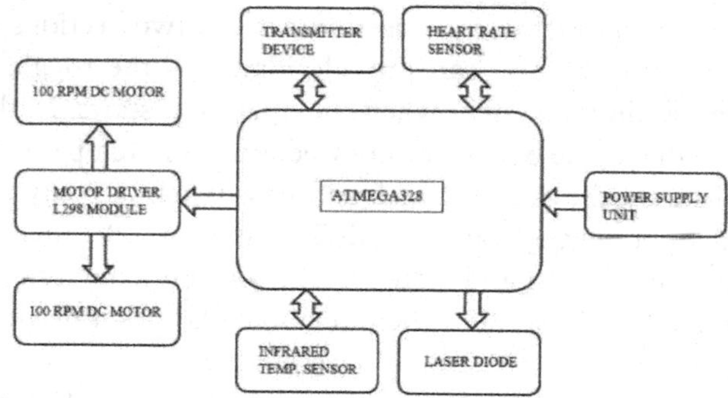

***Figure 2:** Multipurpose robot (receiver section).*

Motor controller L298N uses the h-bridge circuits to reverse the path of the electricity that is delivered to the load (Shu *et al.*, 2020). The engine is able to move in one of the four directions: left to right, upward or backward. Checking a key characteristic: the microprocessor Atmega328p, wireless antenna NRF24L01, IR thermometer MLX90614, cardiovascular and oxygenation of oxygen sensors MAX30100, and LCD screen (162) are employed to carry out tasks in this environment. The procedure begins whenever the toggle function is set to the essential parameters checking mode (Feil-seifer *et al.*, 2020).

While a temperature check push button is depressed for approximately 5 s, the microcontroller processes the "VITAL_1" string into a radio signal, which is subsequently sent out to the receiving section using the transmitter's RF section and picked up through the receiver's RF module antenna. The microcontroller then compares the "VITAL_1" string, and the infrared temperature gauge is then turned (Khan *et al.*, 2020). Fanti *et al.* (2020) suggested the heat radiated by the body of a person is used by the IR temperature sensor to measure temperatures. Following getting transferred to the broadcast's side antennae in the reverse orientation, a CPU on the receiver will analyse its temperature worth, translate it to Celsius units and degrees Fahrenheit and display identical temperatures on the colour LCD screen.

3. Experimental Result

The receiver component is only half developed, but the transmitter section has been fully implemented. The joystick module and Arduino have been connected by the authors in the transmitter section (Quaresima *et al.*, 2020). The joystick's X and Y axis analogue values are now transferred to its digital range, for example, 0 to 1,023 (analog range) values are mapped to 0 to 255 (digital range) values, which will cause the motors to rotate as necessary. Value mapping simply entails mapping the analogue and digital voltage ranges needed for ADC conversion. The 4-bit mode of operation is used to link the 16×2 LCD displays with the Arduino. VCC, GND, RS, RW, EN and DATA0 through DATA3 pins are connected to the Arduino suggested by Vogan *et al.* (2020). One can transition from mode of operation to humidity or temperature status or vice versa thanks to the connection between the toggle button and Uno. Connecting toggle switches to microcontroller wires A4 and A5 is a common practice. In this scenario, driving modes are going to be activated when a green light turns on if pin A4 triggers. However, the red LED will illuminate if pin A5 enters temperature tracking mode (Tegmark *et al.*, 2020).

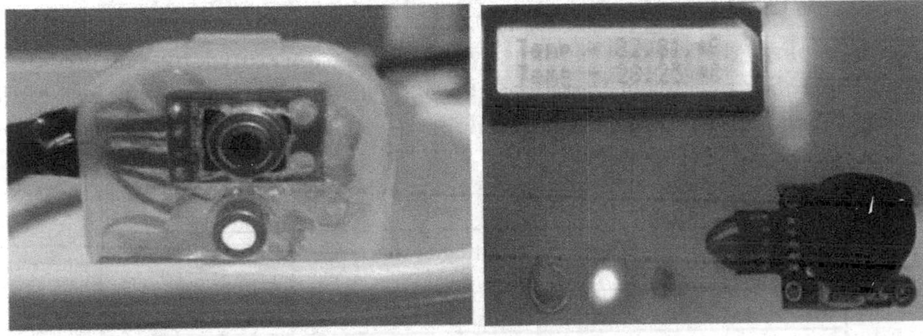

Figure 3: *IR temperature sensor at receiver section.*

Figure 4: *Temp measurement from the entire system's transmission portion.*

Figure 5: *Prototype of sanitisation section with top and side views.*

Someone can monitor a patient's temperature via an invisible machine via an offset of approximately 800 meters using this authors' unusual solution, which they carried out and examined in tandem with traditional equipment like monitoring robots. The person using the device can configure the disinfection area independently or via bluetooth. This allows the robot vehicles to drive about the rooms or other designated areas while spraying the liquid disinfectant, sanitising, cleaning the medical facility beds and eliminating the need for human involvement. Various angles, sides and top perspectives of the prototypes are shown in Figure 5.

Table 1: Various vital sign monitoring of different patients.

Patient	Age	SpO_2	Pulse	Temp inC
Patient-1	35	98	74	36
Patient-2	34	98	74	38
Patient-3	40	94	73	39
Patient-4	45	98	76	37
Patient-5	28	98	75	36

Table 1 displays the various parameters of patients measured by the proposed robot. The robot effectively measured the patients' parameters, which are quite accurate when compared to existing available devices.

Table 2: Comparative analysis of the proposed model with existing system.

Capabilities	Existing System	ProposedModel
Multifunctional	No	Yes
Medication	No	Yes
Vital Sign Monitoring	No	Yes
Data Transmission	No	Yes

Table 2 shows the comparative analysis of the proposed model with the existing (Manual) system that offers a promising solution for enhancing healthcare delivery during the pandemic. It combines various essential functions into a single, versatile system, potentially improving patient care and reducing infection risks. However, the implementation of such a system should consider factors such as cost, scalability and compatibility with existing healthcare infrastructure.

4. Conclusion

The multipurpose robot is actually the best approach to keeping people who have tested positive for COVID away from doctors and other medical professionals. With this low- cost, low-maintenance robot doing a lot of duties in hospitals' regular routines, the facilities will undoubtedly gain greater advantages. The robot can remotely assess a patient's temperature levels. It can also assist in delivering meals and medications to patients with chronic illnesses, which saves on labour costs. These have yet to be completed, connecting a 16×2 LCD section, an ON/OFF switch, LEDs, a push-to-press button, a controller section, the L298N motor racer, DC drives and an IR thermometer (MLX90614) for wireless interaction using an Uno. The same study will be published again in the future months with the ability for patients' heart rates, SpO2 levels and temperature to all be monitored wirelessly using the MAX30100 sensor. Additionally, an IP web camera for audio and video transmission will be put on the receiver part to facilitate contact with patients and their family members. The working model performed well by being controlled from a far to sanitise an area without the user having touched any other nearby things. The employment of the indicated tool will halt the virus' transmission and ensure that the existence of servicemen serving in the conflict's trenches is safe.

References

[1] Anandravisekar, G., Clinton, A., Raj, M. T., and Raveen L. (2018). IOT Based Surveillance Robot. *International Journal of Engineering Research & Technology (IJERT)*, 7, 03.

[2] Chandankhede, P., Titarmare, A. S., and Chauhvan, S. (2021). Voice recognition-based security system using convolutional neural network. *Proceedings-IEEE International Conference on Computing, Communication, and Intelligent Systems, ICCCIS*, pp. 738–743.

[3] Dehankar, V., Jumle, P. M., and Tadse, S. (2023). Design of Drowsiness and Yawning Detection System. *Proceedings of the 2nd International Conference on Electronics and Renewable Systems, ICEARS*, pp. 1585–1589.

[4] Fanti, M. P., Mangini, A. M., Roccotelli, M., and Silvestri, B. (2020). Hospital drug distribution with autonomous robot vehicles. *16th IEEE International Conference on Automation Science and Engineering (CASE)*.

[5] S. Meghana, T. V. Nikhil, R. Murali, S. Sanjana, R. Vidhya and K. J. Mohammed, "Design and implementation of surveillance robot for outdoor security", 2017 2nd IEEE International Conference on Recent Trends in Electronics,Information & Communication Technology (RTEICT), Bangalore.

[6] Feil-seifer, D., Haring, K., Rossi, S., Wagner, A., and Williams, Y. (2020). Where to next? The impact of COVID-19 on human-robot interaction Research. *ACM Trans. Hum.-Robot Interact*, 10, 1, 1–7.

[7] Kale, C., Khanapurkar, M., and Kubde, U. (2022). Design and implementation of multipurpose robot for Covid-19 ward. *10th International Conference on Emerging Trends in Engineering and Technology – Signal and Information Processing*, Nagpur, India, 2022, pp. 01–06.

[8] Kalbande, K., Choudhary, S., Singru, A., Mukherjee, I., and Bakshi, P. (2021). Multi- way controlled feedback oriented smart system for agricultural application using Internet of Things. *5th International Conference on Trends in Electronics and Informatics* (ICOEI), Tirunelveli, India, pp. 96–101.

[9] Kalkotwar, K., Ukey, M., Nanwatkar, S., Gulwade, S., Padole, Y., and Jumle, P. M. (2023). Internet of things (IoT) based smart helmet and intelligent bike

system. *7th International Conference on Trends in Electronics and Informatics*, ICOEI - *Proceedings*, pp. 1458–1461.

[10] Kolhe, P., Kalbande, K., and Deshmukh, A. (2022). Internet of thing and machine learning approach for agricultural application: a review. *International Conference on Emerging Trends in Engineering and Technology*, ICETET.

[11] Mandale, A., Jumle, P. M., Wanjari, M. M., and Biranje, D. (2023). Automated Parcel Sorting System. *International Conference on Emerging Trends in Engineering and Technology*, ICETET.

[12] Mathur, R., and Kalbande, K. (2020). Internet of Things (IoT) based energy tracking and bill estimation system. *Fourth International Conference on I-SMAC* (IoT in Social, Mobile, Analytics and Cloud) (I-SMAC), Palladam, India, pp. 80–85.

[13] Nikhate, P., Deshmukh, A. R., and Choudhari, S. (2021). Study and analysis in MIMO wireless channel for STBC and equalization techniques by using Matlab. *Proceedings of the 3rd International Conference on Inventive Research in Computing Applications*, ICIRCA, pp. 422–429.

[14] Pawar, N. N., Kubade, U., and Jumle, P. M. (2023). Application of multipurpose robot for Covid-19. *7th International Conference on Trends in Electronics and Informatics*, ICOEI 2023 - *Proceedings*, pp. 25–30.

[15] Rodke, S. J., Ghodmare, S., and Deshmukh, A. (2022). Critical analysis of saturation flow rate and affecting parameters. *International Conference on Emerging Trends in Engineering and Technology*, ICETET, 60.

Chapter 4

A Novel Approach for Compressing Electrocardiogram Signals

Abhijeet. Y. Men[1] *and A. S. Joshi*[2]

[1]Research Scholar, Sipna College of Engineering and Technology, Amravati, Maharashtra, India

[2]Professor, Department of Electronics and Telecommunication Engineering, Sipna College of Engineering and Technology, Amravati, Maharashtra, India

Abstract: Electrocardiogram (ECG) signals serve as vital diagnostic tools in modern healthcare. However, the increasing volume of ECG data necessitates efficient storage and transmission methods. ECG signal compression addresses this challenge by reducing data size while preserving clinical information. This paper explores various compression techniques applied to ECG signals, focusing on their effectiveness in maintaining diagnostic accuracy. We present a detailed analysis of compression ratios, peak signal-to-noise ratios (PSNR) and mean squared error (MSE) across two datasets. The proposed approach demonstrates promising compression ratios of 12.7% and 17% for dataset-1 and dataset-2, respectively, with PSNR values of 38.5 and 42.3. The final findings underline the potential of ECG signal compression in optimising resource utilisation and enhancing telemedicine and remote monitoring applications.

Keywords: Electrocardiogram, signal compression, compression ratio.

1. Introduction

In the field of healthcare, medical signal compression plays a vital role in efficiently managing and transmitting large volumes of medical data while preserving critical clinical information (Zhu *et al.*, 2022). Medical signal compression techniques aim to reduce the size of various types of medical signals, including physiological waveforms like electrocardiograms (ECGs) and electroencephalograms (EEGs), as well as medical imaging modalities such as computed tomography (CT) scans and magnetic resonance imaging (MRI) images (Chowdhury *et al.*, 2020).

The need for medical signal compression arises from the ever-increasing volume of medical data generated in clinical settings. Efficient compression enables more efficient storage, transmission and analysis of medical signals, contributing to improved healthcare delivery and diagnostic capabilities (Zhang *et al.*, 2017). The compressed medical data allows for seamless integration into electronic health record (EHR) systems, telemedicine applications and remote monitoring platforms.

DOI: 10.1201/9781003527442-4

Unique to medical signal compression is the requirement to maintain the clinical relevance and accuracy of the compressed data. Unlike general data compression, medical signals contain vital information that directly influences diagnoses, treatment decisions and patient care (Hla *et al.*, 2017). Therefore, compression techniques must strike a delicate balance between achieving high compression ratios and preserving the integrity and diagnostic value of the medical signals.

Various compression methods have been developed to address the unique characteristics of medical signals. These methods can be broadly classified into lossless and lossy compression approaches. Lossless compression ensures exact reconstruction of the original signal, while lossy compression allows for some loss of information in exchange for higher compression ratios suggested by Al-Senwi *et al.* (2017). Transform-based compression techniques, such as discrete wavelet transform (DWT) (Alsenwi *et al.*, 2016), exploit the inherent redundancies present in medical signals to achieve efficient compression. Other approaches include region of interest (ROI) compression, where specific areas of clinical significance within an image are given priority during compression, and compressed sensing (CS), which leverages the sparsity or compressibility of medical signals to obtain high compression ratios (Wu *et al.*, 2013).

Medical signal compression is a critical area of research that focuses on reducing the size of medical data without significantly compromising its clinical information. It plays a crucial role in healthcare systems by enabling efficient storage, transmission and analysis of medical signals, such as ECGs, EEGs and medical imaging modalities like CT and MRI.

Medical signal compression techniques aim to strike a balance between efficient storage and transmission of medical data while preserving clinical information. These methods continue to evolve as researchers explore new algorithms, standards and technologies to address the increasing demands of healthcare systems in terms of data volume, transmission bandwidth and storage capacity. The development of effective medical signal compression techniques has significant implications for telemedicine, remote monitoring, efficient data exchange and facilitating advanced data analysis in healthcare settings.

The organisation of the paper is as follows: Section 2 presents the related work of previous research. Section 3 presented the proposed methodology. Section 4 presented the evaluation parameters used to measure the performance of experimental result analysis. Section 4 and Section 5 presented the discussion and conclusion.

2. Related Work

Luo *et al.* (2016) examined hybrid compression techniques that combine multiple compression algorithms for medical signals. It discusses the integration of transform-based compression, wavelet-based compression and predictive coding to exploit different signal characteristics. The review evaluates the performance of hybrid compression approaches in terms of compression ratio, computational complexity and preservation of diagnostic information.

Hamza *et al.* (2021) focus on wavelet-based compression techniques for medical signals, such as ECG, EEG and medical images. It discusses various wavelet transforms, including discrete wavelet transform (DWT) and stationary wavelet transform (SWT). The review explores the performance of wavelet-based compression algorithms in terms of compression ratio, signal reconstruction quality and computational efficiency.

Pal *et al.* (2023) provide an in-depth analysis of lossless compression techniques for medical signals. It discusses traditional methods like Huffman coding, arithmetic coding and run-length encoding, as well as more advanced techniques like Burrows–Wheeler Transform (BWT) and its variants. The review evaluates the performance of these techniques in terms of compression ratio, entropy and computational complexity.

These literature reviews offer comprehensive insights into the advancements, challenges and applications of medical signal compression techniques, providing valuable resources for researchers and practitioners in the field of medical signal processing.

3. Proposed Methodology

The ECG signal analysis is given before going into further depth about the suggested compression approach. This section and the parts that follow go into great detail concerning the ECG signal. It discusses the creation of ECG, recording techniques, varieties of different frequency bands and ECG applications. The suggested model is explained, and the section below provides a brief overview of this study.

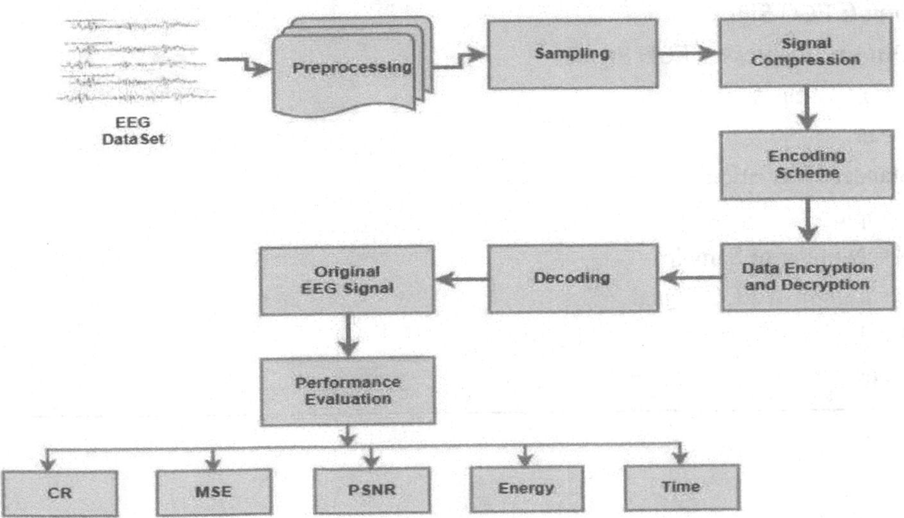

Figure 1: *Complete architecture of proposed system.*

The use to modify correlation with observed data is the biggest obstacle to modelling the data compressing technique. Since they depend on data preprocessing, the standard data compression techniques for ECG signal data are better suited for detecting high-correlation signal data (Jalaleddine *et al.*, 1990). The fundamental concept behind this coding method is to build a tree model and include the variance seen in ECG data. Additionally, the coding scheme is heavily dependent on the amount of tree development.

3.1. Preprocessing

Preparation before analysing data is important in all investigations because without organising the information correctly, the patterns in the ECG dataset are not clear. The brainwave data contains a lot of unwanted disturbances and some weaker ECG signals. It is necessary to separate the genuine signals from the collected brainwave signals during this preparation phase. The unwanted signals are eliminated or filtered out to prevent distortion in the signals. This preprocessing step involves three main tasks: reading the ECG data recordings, standardising the data and segmenting the signals. Standardisation ensures that the ECG data is evenly scaled and improves the compression rate. It changes the average value of the ECG data to make it centred around zero deviation. However, there is no widely accepted technique for processing ECG data, so we have flexibility in modifying the raw data. The signals recorded from the scalp may have incorrect representations that result in the loss of spatial information. Additionally, there could be noise, eye blinking and muscle activity that can contaminate the ECG signals.

Algorithm 1: Data Preprocessing

Input: ECG Signal

Output: Prepressed ECG Signal

S = ECG input

mean

Standard Deviation

Sample

N = Number of Samples

L = Length of ECG Signal

A = 1

Steps

```
Steps 1: while A <= N do
if A = 1 then
ECG Signal = (1:s)
        else
         Processed Signal = (A -1)*s + 1;
         Signal = A*s
        ECG Signal = (Initial ECG Signal: Signal Attained)
        end if
        if A = N then
         vector Map = ECG Signal (1:L + A.s)
Data = [vector map]
end if
A = A + s;
end while
```

3.2. *Sampling*

When signals are converted into a numerical format, the act of sampling is described as the procedure of selecting the right values. The reduction of labour hours and expenses is the main goal. To improve the effectiveness of the approach, sampling data must be evaluated using a variety of techniques. Since the information size is adaptable and provides lossless capability, it is employed to convey data effectively and without loss.

3.3. *Inverse Discrete Cosine Transforms* (IDCT)

The source of each transformation coefficient is converted into a time series. Each set of coefficients utilises this conversion to create a time series. Both IDCT and discrete cosine transformation (DCT) possess a pronounced statistical component. The reverse-compression unit approach is employed to retrieve the original ECG data. The initial ECG data is fully reconstructed using the inverse DCT by Nasim *et al.* (2019) and Pal *et al.* (2022).

Time-series data undergoes a transformation into initial frequencies through the DCT method. It is particularly effective for dataset reduction and feature retrieval. DCT's success stems from its notable energy compression and the correlation between signals.

Assume *f(a)* is the EEG input signal that is composed of *N* number of samples, *y(u)* represents the output signal, which is composed of *N* coefficients.

$$y(u) = \sqrt{\frac{2}{N}} \propto (u) \sum_{a=0}^{N-1} e(k) f(a) cos\left(\frac{\pi(2a+1)u}{2N}\right)$$

Where, $e(k)\begin{cases}\frac{1}{\sqrt{2}} if\ k=0\\ 1, otherwise\end{cases}$

3.4. Signal Compression

The compression of information greatly decreased the amount of storage required for signal storage and delivery latency while taking the loss of data reduction into consideration by Tripathi *et al.* (2017) and Surekha *et al.* (2014). The compressing procedure is carried out via Huffman coding. This technique uses concatenation to increase the compression ratio while reducing computing difficulty; as a result, it tackles multiple methods with straightforward and quick encoding and decoding techniques.

3.5. Lossless Compression

Although compression without loss shrinks files, it leaves behind unnecessary data. For long-term documentation conservation, conformity and preservation without sacrificing file effectiveness, compression without loss is excellent suggested by Mahajan *et al.* (2014). The size of files is greatly decreased with efficient compression while keeping important data to comply with laws. The key benefits of this compression method are listed below: 1. No information is lost and 2. The decompression procedure allows for the restoration of the original file.

3.6. Compression Unit

Lossy encryption and regenerative methods are also included in the compressing component. DCT is used in this situation as a compressing method. To increase the reliability of converted data, thresholds are set for noisy and lossy data. The numerical values of the converted data are set to 0. As a result, changes in a threshold affect how many zero coefficients there are. However, this threshold level controls how accurate the compressing mechanism is.

3.7. Huffman Coding

Huffman's variable-length source code method is identified as Huffman coding. Because of its greater effectiveness, it is widely employed in the encryption and communications fields. It is a prefix code that shortens the typical length of coding. Shorter codes are used to specify signals that have greater probabilities, while lengthy code words are used to specify symbols with lower probabilities suggested by Hamza *et al.* (2021). The above coding method, therefore, relies on the likelihood of sign recurrence. The typical encoding duration is longer whenever the likelihood of an image seems closer.

3.8. Huffman Encoding

The quantity of values for data is decreased using Huffman coding. Since the source is a list of representations, encryption is necessary. Every symbol location and the trace of each symbol at the basic level carry an index value. According to the letter rate, encryption created an unintelligible data structure. This approach to coding is employed for a certain sequence, and a few characters are described by Mahsa *et al.* (2016). Each letter is replaced with a variable-length code based on the comparative letter rate, which shortens the typical code duration.

Algorithm: Huffman Encoding Scheme

Input: List of probabilities {p1, p2, ..., pm} for ECG signal samples

Output: Compressed ECG signal using Huffman encoding

Huffman tree for the given probabilities, follow these steps:

Steps

1. Create initial nodes for each symbol with their probabilities:

Symbol: A B C D E
Probability: 0.45 0.40 0.30 0.20 0.10

2. Construct a priority queue (min-heap) based on the probabilities:
Queue: (E: 0.10) (D: 0.20) (C: 0.30) (B: 0.40) (A: 0.45)

3. Build the Huffman Tree:
Take the two nodes with the lowest probabilities (E and D) and combine them into a new internal node with a probability equal to the sum of their probabilities: E + D = 0.10 + 0.20 = 0.30.

4. Add the new node back to the queue:
Queue: (C: 0.30) (B: 0.40) (A: 0.45) (ED: 0.30)

5. Repeat the process:
Queue: (C: 0.30) (BED: 0.70) (A: 0.45)

6. Final Queue: (C: 0.30) (BAED: 1.15)

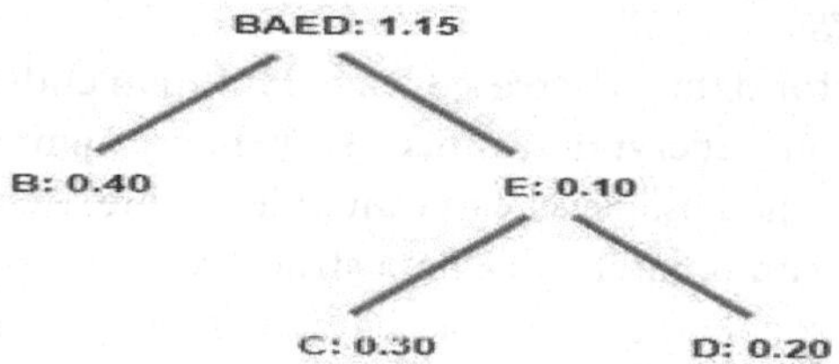

Figure 2: Huffman Tree.

Table 1: Final Huffman code.

Symbol	A	B	C	D	E
Probability	0.45	0.40	0.30	0.20	0.10
Huffman Code	1	00	01	10	001

4. Evaluation Parameters

To validate the findings, specific productivity criteria are taken into account. These include CR, PSNR, PRD, QS and MSE. The ECG datasets from (https://archive.ics.uci.edu/ml/datasets/eeg?database) are used to test the predictive compression methods. The recordings were from seizures and individuals, respectively. ECG information compression is used to address the security threats and constraints of traditional models in order to address these problems.

Compression Ratio

$$Compression\,Ratio = \frac{Uncompressed\,Size}{Compressed\,Size}$$

Peak Signal-to-Noise Ratio (PSNR)

$$PSNR = ASignal_{dB} - Anoise_{dB}$$

The percentage of the variation in the root-mean square: A deformation factor known as PRD is used as a bridge between the origin and alternate EEG waves. It serves as a quality indicator and is fairly straightforward. It is also employed to assess the dependability of the rebuilt signal.

Quality Score: The proportion of CR to PRD is used to characterise it. A crucial performance parameter called quality score helps choose the best compression method by taking into account how a mistake at reconstruction is compensated.

Mean Squared Error

$$MSE = \frac{1}{n}\sum_{i=1}^{n}(y_i - y_i)^2$$

5. Result and Analysis

This paper suggests a technique that utilises a DCT and an entropy calculation method based on Huffman coding. No other multimodal indication exists here. Approaches like independence or primary component analyses are employed to break down ECG data because they may skip sections of the information where artificial differentiation would be laborious and unreliable. The process of deconstructing ECG data enables the detection of flaws by calculating the entropy of each signal. For instance, when the inherent modal frequency range is obtained, it is substantially broader than when the IMF is created at low frequencies, which is not ideal. To tackle this problem, limited-bandwidth ECG data are acquired via DCT signal breakdown, which is then used to produce focused values for the required frequency range. The amount of entropy is then calculated, providing a different source to differentiate motion and improve the compression procedure.

Table 2: Overall performance of the proposed model.

Performance Metrics	Yielded Values for Dataset 1	Yielded Values for Dataset 2
CR	12.7	17
MSE	0.23	0.27
PSNR	38.5	42.3
PRD	8.36	10.85
QS	2.68	3.42
Construction Time	0.90 s	1.5 s

Table 2 shows the comparative result analysis of various evaluation parameters. It is clearly shown that PRD gives the maximum compression ratio on dataset-1 and dataset-2. The results confirm that employing IDCT-based compression is a cost-effective approach when applied on less expensive, compact devices, aiming to minimise the usage of wireless communication bandwidth.

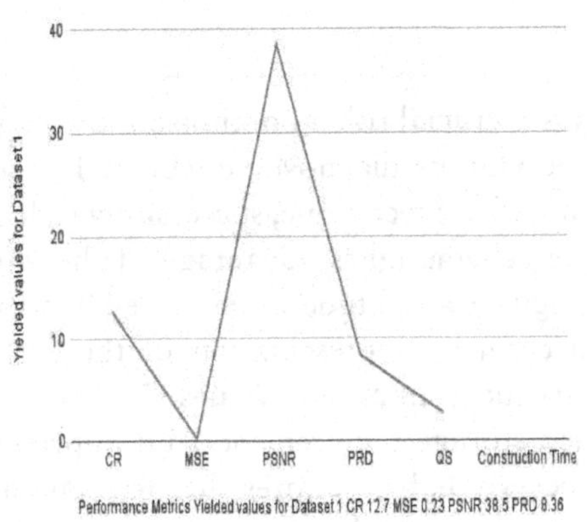

Figure 3: Performance evaluation for dataset-1.

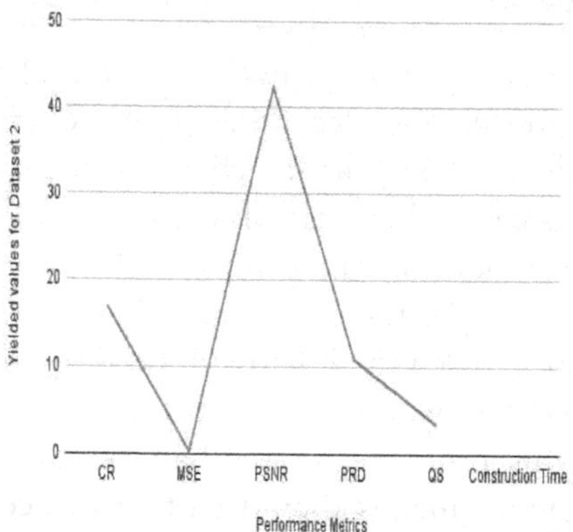

Figure 4: Performance evaluation for dataset-2.

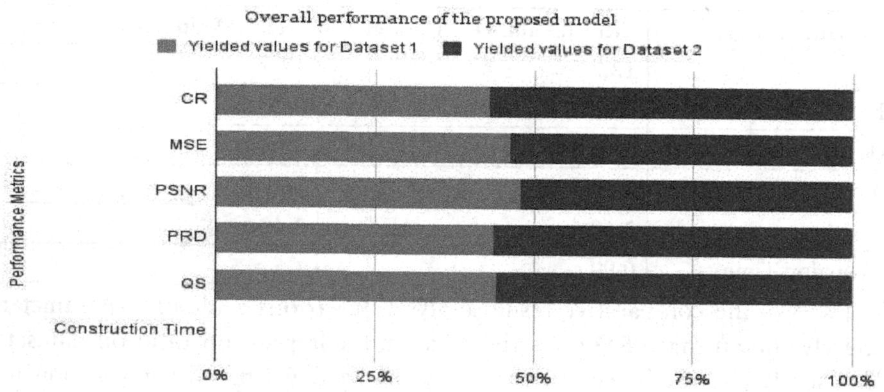

Figure 5: Overall performance of the proposed model.

6. Conclusion

Signal compression plays a crucial role in optimising the storage and transmission of ECG data while preserving its diagnostic integrity. The utilisation of Huffman encoding, a lossless compression technique, successfully reduced the data size of the signal. This compression scheme takes advantage of the variable probabilities of occurrence in the ECG signal's amplitude values. The Huffman encoding algorithm efficiently constructs a compact representation of the ECG signal by assigning shorter codes to more frequent amplitude values. This reduction in bitstream size contributes to minimised storage requirements and improved data transfer rates. The decompression process reliably restores the original signal without any loss of information. Huffman encoding demonstrates its effectiveness in compressing signals, making it suitable for applications in remote monitoring, medical databases and telemedicine, where efficient data handling is essential. The proposed approach achieves compression ratios of 12.7% and 17% for dataset-1 and dataset-2, respectively. The PSNR values are 38.5 for dataset-1 and 42.3 for dataset-2. The corresponding MSE values are 0.23 and 0.27, respectively.

References

[1] Alsenwi, M., Ismail, T., and Mostafa, H. (2016). Performance analysis of hybrid lossy/lossless compression techniques for EEG data. *28th International Conference on Microelectronics* (ICM), Giza, Egypt, pp. 1–4. doi: 10.1109/ICM.2016.7847849.

[2] Al-Senwi, M., Darweesh, M. S., and Ismail, T. (2017). Hybrid compression technique for EEG data based on lossy/lossless algorithms.

[3] Chowdhury, M., Poudel, K., and Hu, Y. (2020). Compression, denoising and classification of ECG signals using the discrete wavelet transform and deep convolutional neural networks. *2020 IEEE Signal Processing in Medicine and Biology Symposium* (SPMB), Philadelphia, PA, USA, pp. 1–4. doi: 10.1109/SPMB50085.2020.9353618.

[4] Hamza, R. K., Rijab, K. S., and Hussien, M. A. (2021). The ECG data compression by discrete wavelet transform and Huffman Encoding. *2021 7th International Conference on Contemporary Information Technology and Mathematics* (*ICCITM*), Mosul, Iraq, pp. 75–81. https://doi.org/10.1109/ICCITM53167.2021.9677704.

[5] Hla, T. (2017). Analysis on ECG data compression using wavelet transform technique. *International Journal of Psychological and Brain Sciences*, 2(6), 127. https://doi.org/10.11648/j.ijpbs.20170206.12.

[6] Jalaleddine, S. M. S., Hutchens, C. G., Strattan, R. D., and Coberly, W. A. (1990). ECG data compression techniques-a unified approach. *In IEEE Transactions on Biomedical Engineering*, 37(4), 329–343. https://doi.org/10.1109/10.52340.

[7] Luo, C. H., et al. (2016). An ECG acquisition system prototype design with flexible PDMS dry electrodes and variable transform length DCT-IV based compression algorithm. *IEEE Sensors Journal*, 16, 23, 8244–8254. https://doi.org/ 10.1109/JSEN.2016.2584648.

[8] Mahajan, R., and Bansal, D. (2014). Hybrid ECG signal compression system: a step towards efficient tele-cardiology. *2014 International Conference on Reliability Optimization and Information Technology* (ICROIT), Faridabad, India, pp. 437–442. https://doi.org/ 10.1109/ICROIT.2014.6798380.

[9] Mahsa, R., Quchani, S. R., Mahdi, M., and Kambiz, B. (2016). Compression and Encryption of ECG Signal Using Wavelet and Chaotically Huffman Code in Telemedicine Application. *Journal of Medical Systems*, 40, 10. https://doi.org/10.1007/s10916-016-0433-5.

[10] Nasim, Sbrollini, A., Marcantoni, I., Morettini, M., and Burattini, L. (2019). Compressed segmented beat modulation method using discrete cosine transform. *41st Annual International Conference of the IEEE Engineering in Medicine and Biology Society* (EMBC), Berlin, Germany, 2019, pp. 2273–2276. https://doi.org/ 10.1109/EMBC.2019.8857267.

[11] Pal, H. S., Kumar, A., Vishwakarma, A., and Balyan, L. K. (2022). A hybrid 2D ECG compression algorithm using DCT and embedded zero tree wavelet. *IEEE 6th Conference on Information and Communication Technology* (*CICT*), Gwalior, India, 2022, pp. 1–5. https://doi.org/10.1109/CICT56698.2022.9997915.

[12] Pal, Kumar, A., Vishwakarma, A., and Singh, (2023). Optimized tunable-Q wavelet transform-based 2-D ECG compression technique using DCT. *IEEE Transactions on Instrumentation and Measurement*, 72, pp. 1–13. https://doi.org/ 10.1109/TIM.2023.3279885.

[13] Surekha, and Patil, (2014). ECG signal compression using hybrid 1D and 2D wavelet transform. *Science and Information Conference*, London, UK, pp. 468–472. https://doi.org/10.1109/SAI.2014.6918229.

[14] Tripathi, and Mishra, (2017). Study of various data compression techniques used in lossless compression of ECG signals. *2017 International Conference on Computing, Communication and Automation (ICCCA)*, Greater Noida, India, pp. 1093–1097. https://doi.org/10.1109/CCAA.2017.8229958.

[15] Wu, T., Hung, K., Liu, J., and Liu, T. (2013). Wavelet-based ECG data compression optimization with genetic algorithm. *Journal of Biomedical Science and Engineering*, 06, 746–753. https://doi.org/10.4236/jbise.2013.67092.

[16] Zhang, B., Zhao, J., Chen, X., and Wu, J. (2017). ECG data compression using a neural network model based on multi-objective optimization. *PLoS ONE*, 12(10), e0182500. https://doi.org/10.1371/journal.pone.0182500.

[17] Zhu, Z., Chen, H., Xie, S., Hu, Y., and Chang, J. (2022). Classification and Reconstruction of Biomedical Signals Based on Convolutional Neural Network. *Computational Intelligence and Neuroscience*, 13. https://doi.org/10.1155/2022/6548811.

Chapter 5

A Compressive Approach for Cancer Classification using Text Data and Machine Learning

Punam Gulande[1] *and R. N. Awale*[2]

[1]PhD Scholar, VJTI, Matunga, Mumbai University

[2]Professor, VJTI, Matunga, Mumbai University

Abstract: Lung illness generated by a number of factors, which is having age, gender, diabetes, high blood pressure (BP), cholesterol, irregular rate of pulse and others. Many techniques of data prospecting and machine learning have been used to uncover the cause of lung illness. This disease's algorithms are classified using several approaches such as KNN algorithm, decision trees (DT), Naive Bayes (NB), logistic regression (LR) and support vector machine (SVM). Because the nature of lung disease is quite complex, the condition must be managed properly. In medical science and data mining, the opinion of lung cancer is used to uncover various metabolons disorders. Data prospecting with classification and clustering is essential for lung disease prediction and query data. Many issues harm the lungs and possibly cause early death.

Keywords: Lung disease, K-Nearest Neighbor (KNN), text data, machine learning algorithm.

1. Introduction

This Lung cancer is the leading cause of death in this generation and will persist in the next generation. If the symptoms of lung cancer are identified early, it can be treated. Current advances in computational intelligence can be used to create sustainable early models for lung cancer treatment without negatively influencing the environment. By reducing the amount of wasted resources and the efforts required to perform labour-intensive tasks, the resources in terms of time and money can be saved. A machine learning (ML) model based on SVM is proposed to optimise the detection method from the lung cancer dataset. Identified cancer patients based on symptoms using the SVM classifier and build models using the MatLab programming language. A variety of metrics have been used to access the efficiency of our SVM model. Big and Metro cities will be able to afford to superior healthcare to its resident because of the positive outcomes of this study. Real-time treatment can have received to the patients with lung cancer in the minimum time and effort. The

DOI: 10.1201/9781003527442-5

proposed model was pitted against the existing SVM. When compared to existing approaches, the proposed method has a 91.90 % accuracy rate.

Symptoms are classified based on the location and tumour size. It is difficult to analyse and diagnose in the early stages because it does not cause pain or symptoms. Symptoms like cough, pain in chest, breathing shortness, fainting, haemoptysis (blood coughing), Pancoast syndrome (joint and body pain), hoarseness, weight loss, paleness and weakness are all symptoms of lung cancer. Figure 1 depicts many types of lung cancer.

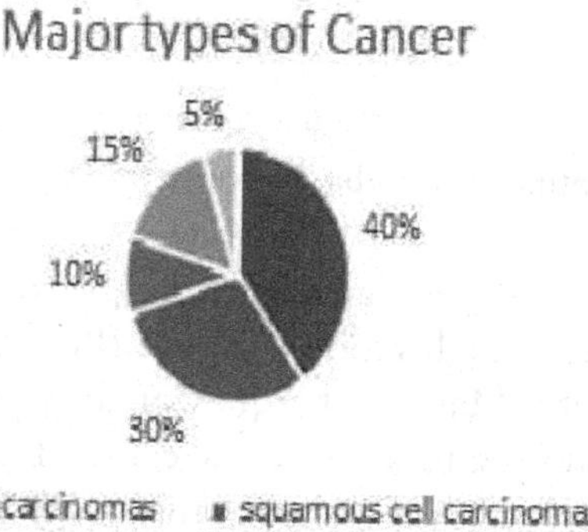

Figure 1: Different kinds of lung cancer.

Active smoking amounts to 90% of the lung disease. Tobacco smoke impregnation, commonly known as submissive smoking, causes lung cancer. One of the major risk factors of lung cancer is heredity factor and pollution caused by vehicles. In business and in productions units, the breathing of hazardous vapour's such as Radon is the second leading factor for fatality of lung cancer. Table 1 (Raoof *et al.*, 2020) shows the factors that cause lung cancer as well as the fatality rate. The 5-year relative survival rate in women is

33.00% for NSCLC in the United States, 23.00 % is a survival relative rate of 5 years for Men. The overall 5-year relative survival rate for those with localised NSCLC, which indicates the cancer has not migrated outside the lung, is 65% (Miller *et al.*, 2021).

Table 1: Lung cancer causing factors and mortality rate.

Reasons	Mortality (percentage)	Figures
Cigarette Smoking	90	70,000 (USA)
Gases like Radon	12	21,000
Passive Smoking	2–4	---

Cancer diagnosis can be detected and made with a variety of techniques. According to NSCLS, four stages to severity like X-Rays, CT/Bone/MRI Scans or PET Scan. Stage 1 is limited to the lungs, stage 2 is limited to the chest, stage 3 is restricted to the chest but includes bigger and more destructive tumours and stage 4 has cover extent to the entire body. SLCC is clinically divided into limited-stage and extensive stage (ES) SCLC. Deaths details are on the rise in India All over world record related to cases and deaths of Lung Cancer are shown in Table 2 (Raoof *et al.*, 2020). Table 3 depicts the death statistical information of lung cancer.

Table 2: Cases and deaths of Lung cancer percentage.

Region	Population (%age)	Cancer Cases (%age)	Cancer Deaths (%age)
Asia	60.00	66.00	57.30
Europe	9.00	23.40	20.30
USA	13.30	21.00	14.40
African Region	17.00	9.00–10.00	7.30

Table 3: Cases and deaths of Lung cancer in India.

Lung Cancer (category)	New Cases (%age)	Death Cases (%age)
Men	60.00	66.00
Women	9.00	23.40
Both cases	13.30	21.00

For evaluation and analysis, ML/AI employs complex and sophisticated algorithms. Traditional AI technologies are distinguished by the intelligence of algorithms in receiving information, analysing the same and handing it out by processing it, and finally determining the accurate result. AI in the field of healthcare is divided into two groups based on the type of data. For example, ML approaches are used to analyse structured and organised data as well as photos, genes and biomarkers. Natural language processing (NLP) technologies are used to analyse unstructured and amorphous data such as notes, publications and journals. First, the gathered data is transformed into binary representation with the help of NLP methods, and then evaluated with the help of ML techniques in order to yield precise output and judgements (Raoof *et al.*, 2020).

Cancer, neurology and cardiology remains the key areas of medical research where AI can be used to achieve maximum benefits. This disease has a higher fatality rate. Aside from these disorders, AI is being used in other medical fields for prediction, analysis and treatment. SVM, NN, logistic regression, discriminant analysis, decision trees, linear regression, K-nearest-neighbor, Naive Bayes and other well-known ML techniques are widely used in the healthcare sector (Maleki *et al.*, 2021). As an extension, Taher *et al.* (2021) compared with CNN approach. The various machine learning models using text dataset are the objectives of this paper. Study of related work and introduction of machine learning algorithm in the field of lung disease recognition are done in the first sector. The next section provides the intuitions design of machine learning model used and results shown in tabular as well as graphical format as the Accuracy (Ac), Precision (Pr), Recall (Re) and F1-Score (F1sc) gained by the model. Finally, task performance of an optimal model concludes. Many steps of AI algorithms that can now be implemented in healthcare such as algorithm that detects tumour variation and forecast heart attacks, image classification, AI-assisted skincare and breast cancer, risk associated with suicide and ICU AI system etc.

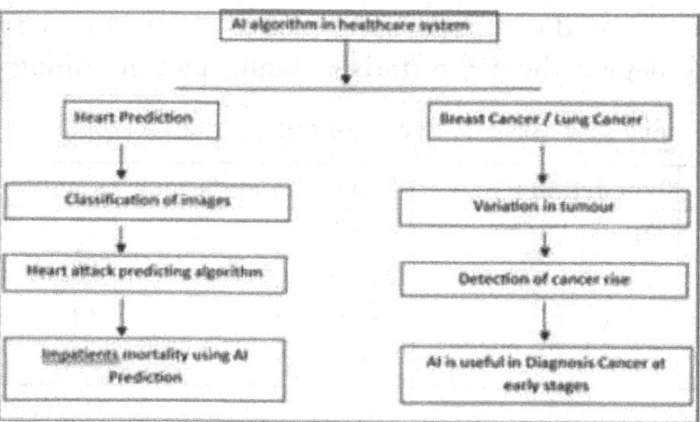

Figure 2: Steps for AI algorithm.

Image-based effective ML systems are established using SVM, linear regression, decision tree, K-nearest neighbor and so on. SVM uses a data-driven approach, which is practicable without having an imaginary system. In various healthcare fields, ANN application is used for diagnosis and adaptation, such as breast, lung cancer, and other perspective predictions, investigative systems and exploration of drug. The simplest form is K-nearest neighbor method which is centered solely on supervised learning techniques. The proposed algorithm has high similarities with the quantity of cases and the existing data and new data. DT algorithms typically belong to the family of supervised learning algorithms. It is used to train the model and predict either value or class of the target variable to learn simple preference data decisions.

2. Literature Review

Support vector regression is a type of machine that accepts real values in a binary form. To minimise the gap between observed and predicted data, to improve performance, create subclasses of training data called support vectors as mentioned by Raoof *et al.* (2020). Alam *et al.* (2018) pronounces an operative proposal using SVM to detect, diagnose and even predict the probability of lung cancer. The detection and feature extraction are performed using grey level co-occurrence method (GLCM). SVM is used for classification purposes. A UCI ML database containing 500 contaminated and 500 uncontaminated CT images. The suggested method identifies 126 out of 130 images as infected and predicted. 87 out of 100 predefined images as cancerous. Experimental analysis shows a discrimination accuracy of 97% and a prediction accuracy of 87%. Yakar *et al.* (2021) defined CAD model for the examination and investigation of details of cancer area in CT images. A basic step is to detect the area of investigational interest. Extraction of the lung area is followed by segmentation of the same area. Recognition of cancer modules is achieved using the fuzzy possibility CMean (FPCM) clustering algorithm. Maleki *et al.* (2021) suggested specialized CAD model aimed at sensing lung nodules on chest films. The proposal grants a CAD prototype with an image server and relevant software. Rule-based analysis and ANN are used for this purpose. For experiments, the developed model was analysed on a database consisting of 274 radiographs with 323 lung nodules. 315,235(75%) of the false-positive images were found to be normal autonomic structures and 155(49%) were found to be pulmonary vessels. SVM is widely used across many industries, including medicine. However, cancer prognostic models are still being built as mentioned by Jenipher and Radhika (2021). Precision of lung malignancy identification and classification of tumor stages to improves the decision-making process were proposed by Ashwini *et al.* (2021) and Vijh *et al.* (2021). WEKA Tool was used to investigate the accuracy by Dakhaz and Adnan (2021). Jaweed and Afsha (2021) surveyed that the accuracy has increased in the past years and also pointed disadvantages.

3. Methodology and Model Specifications

3.1. *Proposed Work*

A ML technique enabled to provide an accurate classification and prediction of lung cancer using technology. Poor performance of classification algorithms based on traditional methods is seen. At a starting stage, data need to be collected. After that, removal of inconsistencies and parameter tuning of machine learning model are done. Figure 3 illustrates utilisation technology for the classification and prediction of lung cancer that enables machine learning. The result analysis of which is depicted in Table 4 where Ac, Pr, Re and F1sc are metrics of evaluation in a tabular form and also depicted graphically in Figure 4. Region of convergence (ROC) is represented in Figure 5.

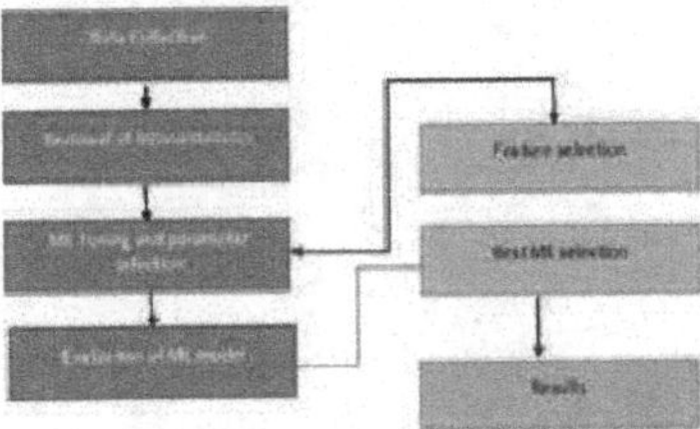

Figure 3: *Proposed classification Process.*

4. Empirical Results

4.1. *Measurement of Classification Performance*

In machine learning, various different parameters play a vital role in representing results. The performance revealed by the different classifiers is evaluated based on the evaluation metrics, like accuracy, sensitivity and specificity. Efficiency of the ML model performance can be determined by using Performance Metrics (PM).

4.2. *Dataset Availability*

Lung Data Set: https://www.dataworld.com link "sta427ceyin/survey-lung-cancer" No of samples-309 and software used- MatLab.

Tabular results for the performance of various classifier are shown in the Table 4 and graphical representation is shown in Figure 4 and ROC curve in Figure 5 below. We use different cancer datasets to evaluate the performance of the model. It is adopted to evaluate its performance. All cancer datasets were considered to limit the number of available features.

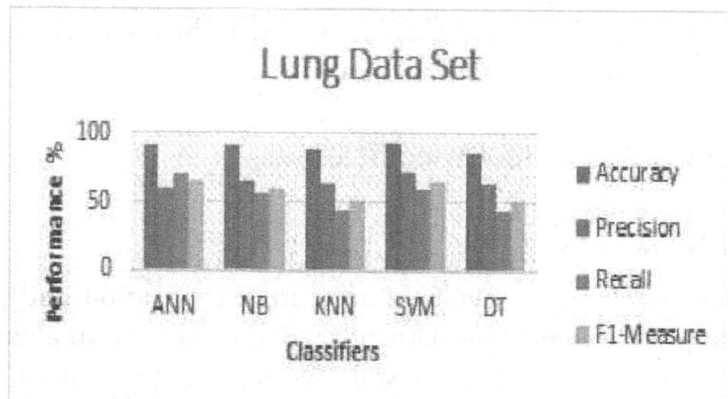

Figure 4: Performance analysis.

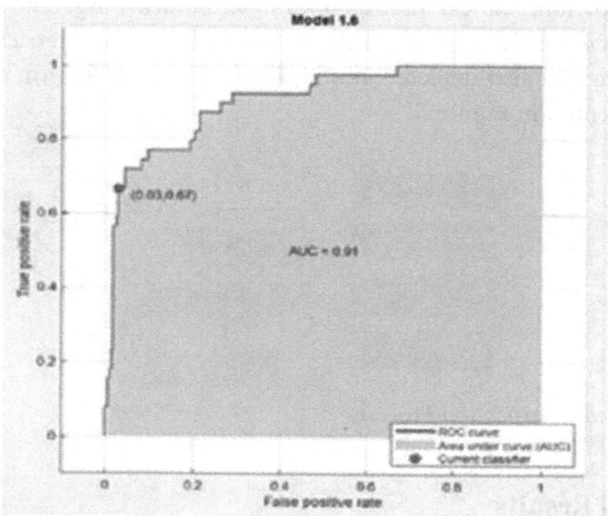

Figure 5: ROC curve.

Table 4: Performance analysis.

Methods/Parameter	ANN	NB	KNN	SVM	TREE
Accuracy	90.29	90.61	89.64	91.90	85.43
Precision	60	64.70	62.96	71.87	62.96
Recall	69.23	56.41	43.58	58.97	43.58
F1-Measure	64.28	60.27	51.50	64.78	51.50

Table 5: Comparative analysis of accuracy.

Method	Accuracy
Proposed Method	91.90%
Verma *et al.* (2022)	89%
Bharathy *et al.* (2022)	88.50%
Alam *et al.* (2018)	87%

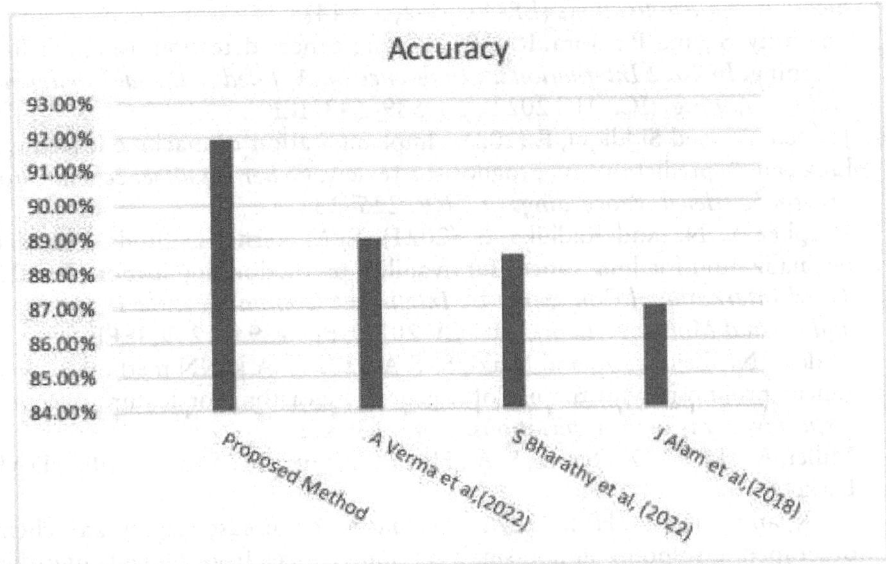

Figure 6: Accuracy analysis.

As can be seen from the above Table 5, the accuracy of the proposed method is 91.9% which is relatively higher than those of other approaches. Therefore, the suggested methodology as depicted in Figure 6 proves to be superior in predicting the target category and accomplishing elevated levels of accuracy.

5. Conclusion

An effective classification of lung cancer is one of the machine learning approach which is very essential for surgical treatment. This effort warrants the use of different classifiers and learning approaches for its classification. In this work, SVM is used to forecast the progress of lung cancer as it is beneficial for early detection, and treatment of patient can be cured. The basic aim of this system is to make available the early warning, which will save money and time. The proposed method performance evaluation yielded positive results that demonstrated that SVM can be effectively used by oncologists to help identify lung cancer. In the field of medical science if the prediction is correct, then it is possible that the doctor can diagnose the patient in a better way.

References

[1] Alam, J., Alam, S., and Hossan, A. (2018). Multi-stage lung cancer detection and prediction using multi-class SVM classifier. *In 2018 International Conference on Computer, Communication, Chemical, Material and Electronic Engineering* (*IC4ME2* 2018), pp. 1–4, IEEE.

[2] Ashwini, S. S., Kurian, M. Z., and Nagaraja, M. (2021). Lung cancer detection and prediction using customized selective segmentation technique with SVM classifier. *In Emerging Research in Computing, Information, Communication and Applications* (*ERCICA*), 2, 37–44.

[3] Bharathy, S., and Pavithra, R. (2022). Lung cancer detection using machine Learning. *In 2022 International Conference on Applied Artificial Intelligence and Computing*, (ICAAIC 2022), pp. 539–543. IEEE.

[4] Jaweed, A., and Siddiqui, F. (2021). Implementation of machine learning in lung cancer prediction and prognosis: a review. *Cyber Intelligence and Information Retrieval: Proceedings of CIIR*, 225–231.

[5] Jenipher, V. N., and Radhika, S. (2021). SVM kernel methods with data normalization for lung cancer survivability prediction application. *In 2021 Third International Conference on Intelligent Communication Technologies and Virtual Mobile Networks* (ICICV 2021), pp. 1294–1299, IEEE.

[6] Maleki, N., Zeinali, Y., and Niaki, S. T. A. (2021). A k-NN method for lung cancer prognosis with the use of a genetic algorithm for feature selection. *Expert Systems with Applications*, 164, 113981.

[7] Miller, A. H., Yin, X., Smith, S. A., Hu, X., Zhang, X., Yan, J., Miller, D.M., Berkel,

[8] V. H., and Frieboes, H. B. (2021). Evaluation of disease staging and chemotherapeutic response in non-small cell lung cancer from patient tumor-derived metabolomic data. *Lung cancer*, 156, 20–30.

[9] Mustafa, A. D., Abdulazeez A. M., and Sallow, A. B. (2021). Lung cancer prediction and classification based on correlation selection method using machine learning techniques. *Qubahan Academic Journal*, 1(2), 141–149.

[10] Raoof, S. S., Jabbar, M. A., and Fathima, S. A. (2020). Lung cancer prediction using machine learning: a comprehensive approach. *In 2020 2nd International Conference on Innovative Mechanisms for Industry Applications* (*ICIMIA*), pp. 108–115.

[11] Taher, F., Prakash, N., Shaffie, A., Soliman, A., and El-Baz, A. (2021). An overview of lung cancer classification algorithms and their performances. *IAENG International Journal of Computer Science*, 48(4).Verma, A., Shah, C. K., Kaur, V., Shah, S., and Kumar, P. (2022). Cancer detection and analysis using machine learning. *In 2022 Second International Conference on Computer Science, Engineering and Applications* (ICCSEA 2022), pp. 1–5, IEEE.

[12] Vijh, S., Sarma, R., and Kumar, S. (2021). Lung tumor segmentation using marker-controlled watershed and support vector machine. *International Journal of E-Health and Medical Communications* (*IJEHMC*), 12(2), 51–64.

[13] Yakar, M., Etiz, D., Metintas, M., Guntulu, A. K., and Celik, O. (2021). Prediction of radiation pneumonitis with machine learning in stage III lung cancer: a pilot study. *Technology in cancer research & treatment*, 20, 1–10.

Chapter 6

Solar-Based Real-Time Solid Waste Separation and Management System Using IOT

Suranjali Jagtap[1] *and Meenakshi Pawar*[2]

[1]PG Student, SVERI's College, Pandharpur, Solapur University

[2]Assistant Professor, SVERI's College, Pandharpur, Solapur University

Abstract: These days proper waste management is one of the most serious issues for the environment. The current problems of waste bin in cities are as overflowing of waste bin, manual operation and maintenance and create bad smell within the city. To overcome these problems, in this study we present a solution for the efficient separation and management of solid waste which is a valuable and complex aspect of waste management. To tackle this issue smart cities employ a device known as the smart bin. Our proposed system enhances traditional dustbins by incorporating embedded bins that enable real-time monitoring of waste levels and location detection. This proposed solar-based real-time solid waste separation and management system uses GPS, NodeMCU, GSM and ultrasonic sensors. The NodeMCU is utilised to control the dustbin's functionality. Additionally specialised IoT sensors are employed to separate wet and dry waste. Furthermore a solar panel is integrated to harness solar energy which is then stored in a battery for future use.

Keywords: NodeMCU, GPS, GSM, ultrasonic sensor, solar panel, rain sensor.

1. Introduction

Currently, we can observe that waste management system is a crucial thing when it comes to urban communities. The amount of waste produced by multiple sources is increasing with each passing day. Conventional dustbins often become unclean and overflow due to inadequate planning and inefficient management resulting in a detrimental environment within our society. To ensure the safety of the public health and to reduce pollution, waste management needs to be addressed. Waste management and awareness building are new concepts in developing nations. Waste management is critical to ensuring environmental sustainability and

DOI: 10.1201/9781003527442-6

future development. However in underdeveloped countries, waste management has become less efficient due to a lack of infrastructure and inefficient methods resulting in environmental pollution. To collect waste from open dumpsites it may causes health-problems like skin infection. The problem increases as a result of the dense population in slum areas. It is clear that incorrect waste management creates dangers to both health and the environment. Growing populations, growing urbanisation and industrialisation all contribute to increased waste production across the world. Figure

1 shows a bin for waste at different places. Urban regions do not implement environmentally friendly disposal of waste. Nearly all of the dustbins routinely have rubbish overflowing into them. As a result a large number of people, particularly youngsters, die away each year from various infectious diseases. This paper proposed an effective waste separation and management system based on the above issue raised. For separation of solid waste, the proposed system uses a rain sensor. For indicating the waste level in both wet and dry waste bins, an ultrasonic sensor is used in the proposed system. Continuous data regarding waste (dry and wet) level are available on the Blynk app. This smart wastebin provides an automated lid opening and closing system. If person is detected then lid open automatically. Separation mechanism separates the solid wet and dry waste. When the bin is filled up to 100% then it will send an alert message to the supervisor and waste collector with the level of waste present in a particular bin along with the location of waste bin. From the proposed smart waste management system, we developed an effective small-scale wastebin prototype. A solar panel is integrated to harness solar energy which is then stored in a battery for future use.

IoT is a technical advancement in which each device is given a unique IP address and is having the ability to automatically allocate data over a network. Kumar *et al.* (2016) developed IoT-based smart waste alert system for sending alert messages to the municipal web server. These alert messages depend on the amount of garbage in the dustbins. So that the waste bins may be cleaned immediately with proper verification. The entire system is supported with the help of combined module having a facility of IOT and RFID. Wijaya *et al.* (2017) developed waste management system to display the total amount of waste in the waste bin and measure the weight of the waste which is present in the waste bin. These systems consist of waste bins that are embedded with a sensor network. Pardini *et al.* (2018) developed smart trash can for waste management. To detect the location of waste bin this system used GPS. An ultrasonic sensor is used for the identification of person. GSM is used for the mobile communication and GPRS is used for the communication system. Malapur *et al.* (2017) developed trash management system based on IoT to sound alarm of bin when bin is full. Also he developed simple Android app to help in the efficient collection of waste. Monika *et al.* (2016) developed a management system for waste to collects waste from inaccessible urban locations by using a camera. Blynk mobile app is used to connect Arduino UNO and Wi-Fi module to the internet. Poddar *et al.* (2017) developed wireless IoT-based system to helps municipal corporations for observing the waste content of dustbin remotely with the help of web server. When waste bin is full then it sends alert messages to authority. The authority then sends vehicles to collect waste at a particular place.

So, in a summarised form, the paper's contributions are:

1. It gives eco-friendly waste separation system.
2. The designed system is simpler to operate as compared to traditional waste management system.
3. The system has lid opening and closing mechanism that operates automatically to avoid any direct interaction with the waste bin.
4. IOT applications help to prevent waste overflow and spreading.
5. It also saves the power. It also traces the location of waste bin easily.

Figure 1: *Waste problems at different places.*

Table 1: Summary of some of the existing solutions.

No. of Ref.	Similar Challenges with the Designed System	Technologies Used	Beneficial Impacts	Limitations
Aazam *et al.* (2016)	It is able to determine the status of the waste level.	There is simply mention of a sensor and router.	Road improvement to reduce travelling time and fuel usage.	Actual realisation of prototype is not detailed. The provision for a display and an automated lid opening and closing mechanism is not present.

Baby *et al.* (2017)	Informs authority by checking the level of waste bin.	LCD, Ultrasonic and IR sensor, Raspberry Pi, Arduino UNO, Power BI, IFTTT.	Improvement in waste management outcomes through the development of a well-structured plan.	Actual realisation of prototype is not detailed. The provision for a display and an automated lid opening and closing mechanism is not present.
Kumar *et al.* (2017)	Informs authority by checking the level of waste bin.	LCD screen, Ultrasonic sensor, android application, LED.	Provides nearby bin finding facility.	Real-time monitoring of bin is not happen. Not providing automatic lid mechanism.
Malapur *et al.* (2017	Sends alert SMS to the collector and use of ultrasonic sensor to monitor waste bin.	GSM, Ultrasonic sensor, Arduino, Genetic algorithm, GUI.	User friendly software with the waste bin location.	-Difficult to operate, -The prototype's practical application is not defined -Display and an automated lid system are not shown Without Function buzzer

2. System Design

The complete structure of proposed smart waste separation and management system is represented in Figure 2. This proposed system is made up with the help of an identification system, separation unit, power generation unit, an automated lid opening and closing unit, monitoring system and communication unit. Using the NodeMCU microcontroller, all of these systems are synchronised. Each system is explained below.

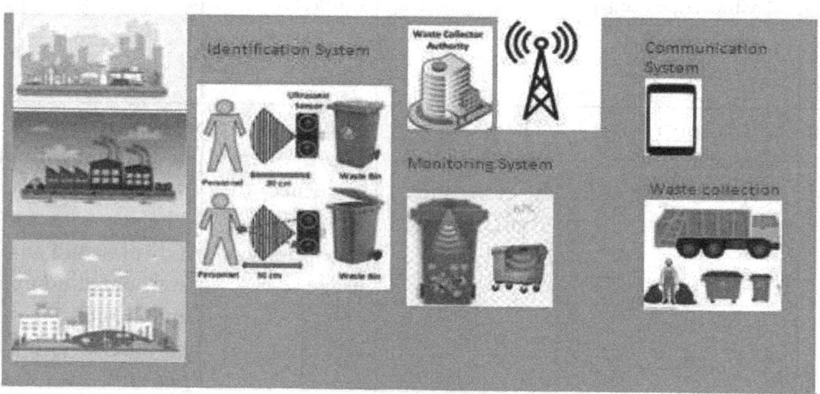

Figure 2: Overall system design of proposed system.

2.1. Identification System

Ultrasonic sensors operate well for measuring amplitude because the shape of the object being measured has little effect on how the sound wave is reflected [12]. As a result, sensors have been used to measure waste levels and identify persons. The distance is determined on time taken for the echo signal to return. The system can find the space occupied by the waste inside the waste bin. Figure 3 shows the interfacing of ultrasonic sensor with Node MCU and proposed system prototype.

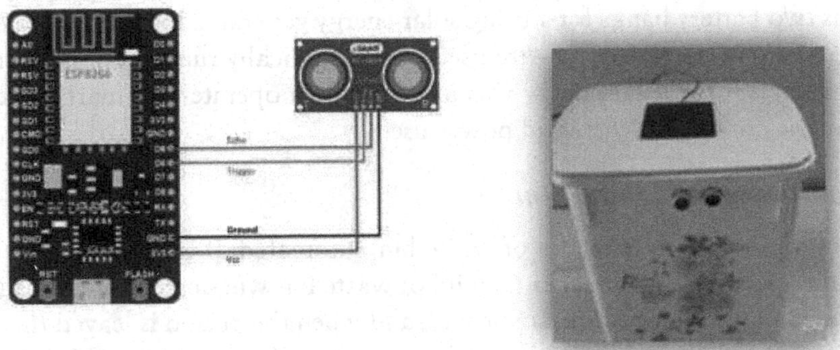

***Figure 3:** Interfacing of ultrasonic sensor with NodeMCU and proposed system prototype.*

The identifying system makes use of two HC-SR04 ultrasonic sensors. From that one sensor is used for detection of people within a specific range and another one is used for calculating how much waste is in the smart waste bin, which is mounted inside the waste bin.

2.2. Separation System

The separation systems consist of a servo motor and rain sensor. A device used for detecting the height of the Great Depression acts as the rain sensor's component. It is typically employed to measure the intensity of the rain when it comes down on the rainboard. The analogue outputs are utilised in rainfall measurement to determine rainfall. The LED is tastefully activated with a 5V power when the inductive board is not wet and the output voltage. When the output is low, a small amount of water is poured while activating the switch mark. Pass through the water drops, and then you're back to the basic formal productions at important side. Rain sensor is used for separating the dry /wet solid waste from the mixed waste. The separation of waste is depending upon the threshold value, which is set on the smart waste bin sensor. Firstly, the waste is put in the waste bin then separation mechanism separates the waste into solid dry and solid wet waste. If the solid dry waste is detected, then the tilt (separation mechanism) is moved automatically 90 degree (clockwise) inside the waste bin with the help of servo motor. If the waste is wet, then the tilt (separation mechanism) is moved automatically 90 degree (anticlockwise) inside waste bin by using servo motor. These separated wastages are supplied to industries for recycling process.

2.3. Power Generation Unit

2.3.1. Battery and Solar Panel

The battery used for a solar panel system is a lithium ion battery. Lithium battery is used for security; wind energy, solar energy and other uses are possible. The battery technology known as lithium-iron-phosphate (LiFePO4) is used in smart bins and other applications because of its high cycle count. To capture solar energy from the sun, photovoltaic panel of rated at 100 W and 24 V is used. Smart waste bin contains two battery banks for storing solar energy generated by photovoltaic cells. These renewable energy sources are used to automatically run the entire smart bin infrastructure. If the solar energy was insufficient to operate the smart bin due to atmospheric conditions, then grid power used.

2.4. An Automated Lid System

Servo motor is used to make lid of waste bin automated. If the person is detected within the specific detection area then lid of waste bin will open automatically and it will remain open until the person leaves, and when the person is leaved then after some time it will close automatically. A perfect angular or linear position can be perfectly controlled by a servo motor, which is a rotary or linear actuator. It can rotate clockwise and anticlockwise up to 90 degrees in both directions.

2.5. A Monitoring System

To monitoring level of waste inside the waste bin to the authorities, Blynk app is used. Thereby speedy action would be taken to empty the bins. To carry out this Arduino Integrated Development Environment (IDE) tool is used. Only solid waste should be thrown in the waste dumps. The NodeMCU is first interfaced to the Wi-Fi hotspot and then to the Blynk app. To support the Internet of Things, Blynk was developed. The hardware is closely monitored over here, and sensor data is shown here for the Blynk to store, interpret and do various wonderful things with.

2.6. A Communication System

To communicate with waste collector and supervisor, a GSM module is used. GSM module send message to the registered mobile number of respective authority, depending on the set level. The location of waste bin will be identified with the help of GSM

Module. GPS module provides coordinates of waste bin. Message signal also include coordinates of waste bin.

3. Methodology

To control the identification unit, separation unit, power generation unit, an automated lid unit, a monitoring unit and a communication unit, Arduino programme is developed. The waves of sound are reflected from solid objects are collected by the receiver of ultrasonic sensor. The ultrasonic sensor's transmitter emits ultrasonic waves that exceed the upper limit of human hearing. As a result, servo motor will automatically open the lid of waste bin if a person is identified within a specific range of the sensor. The planned identifying system was set up with a 30-cm range. Someone may throw solid waste into the waste bin after opening the lid, and lid of waste bin will remain open up to the user is inside the detecting range. The lid will automatically close after a predetermined interval if the user walks outside the sensing range. The solar panels and battery system produce the electricity. To maintain the system operating, a power supply is used. Here, a closed container is shown to prevent interference from outside factors like animals and abnormal atmospheric conditions that may lead to waste being left out as litter. Ultrasonic sensors installed inside the waste bin can constantly check waste level inside both solid dry waste bin and solid wet waste bin. Waste inside both solid dry waste bins and solid wet waste reflects the sound waves sent out by the ultrasonic sensor, which uses ultrasonic sound to convey information. Ultrasonic sound is transmitted, and then the reflected sound waves are picked up after a delay. This time interval is used to calculate the proportion of the trashcan that is filled. The Blynk Mobile app is used to display the calculated data. Below is a conversion formula for the proposed system's usage of distance to percentage of capacity. Flow chart of the proposed system is shown in fig.4 which shows the working flow of proposed system.

$$D = [(V \times t)/2] \tag{1}$$

where V is the sound's velocity and t is the sensor's response time.Percentage of capacity achieved = $[100 - (100/H) \times D]$ (2) where D is the calculated distance and H is the waste bin's height.

In the waste bin (dry or wet), the waste bin level is 80% then the alert message is send to the waste collector along with location and percentage filled status of waste bin. In waste bin (dry or wet), waste level is 100% then the alert message is send to the waste collector as well as supervisor or manager along with location and percentage filled status of the waste bin. After that waste collection truck will arrive to gather the segregated waste. Waste bin can be reused, and the automated lid and separation system can be turned again on.

Flowchart

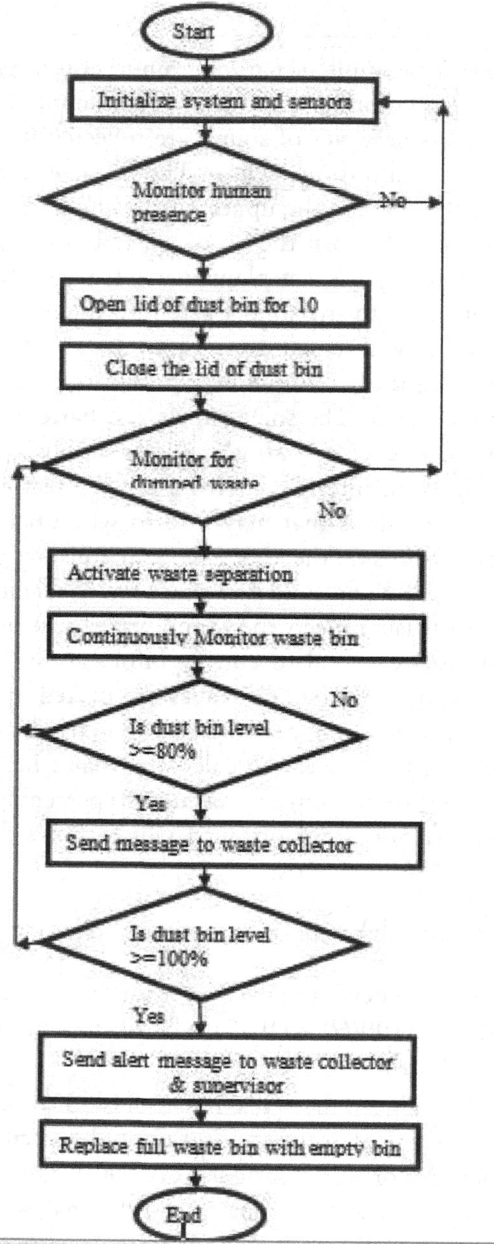

Figure 4: *Waste bin level monitoring on Blynk App.*

3.1. Software Used

An Arduino integrated development environment (IDE) is used to write (develop) and upload programme to Arduino compatible boards. The GNU General Public Agreement has been used to make the IDE's source code available. With the help of unique code organisation principles, Arduino IDE supports C and C++ programming languages. Arduino IDE has wiring project's product library. This library provides a wealth of standard data and production techniques.

4. Experiments and Result

A prototype of a smart waste separation and management system is practically developed. The smart bin is a combination of an identification unit, separation unit, power generation unit, automated lid opening and closing unit, monitoring unit and communication unit. The NodeMCU acts as microcontroller and has a programme uploaded to it that is used to automatically control the system as a whole. This publication previously addressed the waste bin's complete structure and operational procedure. Prototype of proposed system was verified for different conditions to verify the working of prototype is perfect. Table 2 shows summary of some experimental assessment of different smart waste bin scenarios. The filled level of waste bin is shown on Blynk App, for these practical observations. The morning of each day the corresponding authority collects waste and visit to every waste bin for collecting waste. The traditional system has this major advantage, as the amount of waste for every location is not the same. To remove the overflowing problem of waste in such location, we need to cleaned waste bin twice or thrice in a day. Some places required two or more days to be filled waste bin 100%. Due to such reason, traditional waste management system is whole waste of work and time. To overcoming this kind of problem, proposed waste separation and management system is efficient. The top and side view of proposed system is shown in fig.5

Figure 5: Side and top view of proposed system.

Incident 7 in the below Table 2 indicates that the bin is 100% filled up. If the waste disposal container is completely full, then it will alert the appropriate authority to collect waste with the location of the waste bin in the message.

Table 2: Summary of some experimental assessment of different smart waste bin scenarios.

Test No.	Waste level inside the waste bin	Identification Unit	Automated lid Unit	Power generation Unit	Separation Unit	Monitoring Unit	Communication Unit
1	Waste in Small quantity	Doing well	Working	Working	Working	9	-
2	Quantity of waste is increased	Doing well	Working	Working	Working	25	-
3	Quantity of waste is increased more	Doing well	Working	Working	Working	54	-
4	Quantity of waste is increased more than earlier	Doing well	Working	Working	Working	69	-
5	Quantity of waste is 80%	Doing well	Working	Working	Working	80	Successful
6	Quantity of waste is increased more than 80%	Doing well	Working	Working	Working	100	Successful
7	Quantity of waste is 100%	Doing well	Working	Working	Working	100	Successful

5. Conclusion

This proposed system provides valuable assistance to residents of smart cities in effectively utilising smart bins. The integration of solar panels allows the smart bins to harness renewable energy, enabling their overall functionality. Dry/wet segregation, human detection, level of monitoring and solar panels, these all features are embedded in the proposed smart waste bin. Smart bin is convenient for users due to their ability to automatically open and close waste bin lid. Additionally, it efficiently and accurately measures the bin level, sending alert messages based on the waste level. Looking ahead, this work has the potential to serve as an autonomous and mobile smart bin solution for cities worldwide. By leveraging these advancements, cities can make significant towards sustainable waste management and contribute to building smarter and cleaner urban environments.

References

[1] Aazam, M., St-Hilaire, M., Lung, C. H., and Lambadaris, I. (2016). Cloud-based smart waste management for smart cities. *In 2016 IEEE 21st international workshop on computer aided modelling and design of communication links and networks* (*CAMAD*), pp. 188–193, IEEE.

[2] Baby, C. J., Singh, H., Srivastava, A., Dhawan, R., and Mahalakshmi, P. (2017). Smart bin: an intelligent waste alert and prediction system using machine learning approach. *In 2017 international conference on wireless communications, signal processing and networking* (*WiSPNET*), pp. 771–774, IEEE.

[3] Kumar, S. V., Kumaran, T. S., Kumar, A. K., and Mathapati, M. (2017). Smart garbage monitoring and clearance system using internet of things. *In 2017 IEEE international conference on smart technologies and management for computing, communication, controls, energy and materials* (*ICSTM*), pp. 184–189, IEEE.

[4] Kumar, N. S., Vuayalakshmi, B., Prarthana, R. J., and Shankar, A. (2016). IOT based smart garbage alert system using Arduino UNO. *In 2016 IEEE region 10 conference* (*TENCON*), pp. 1028–1034, IEEE.

[5] Malapur, B. S., and Pattanshetti, V. R. (2017). IoT based waste management: an application to smart city. *In 2017 International Conference on Energy, Communication, Data Analytics and Soft Computing* (*ICECDS*), pp. 2476–2486, IEEE.

[6] Monika, K. A., Rao, N., Prapulla, S. B., and Shobha, G. (2016). Smart dustbin: an efficient garbage monitoring system, 6(6), 7113–7116.

[7] Pardini, K., Rodrigues, J. J., Hassan, S. A., Kumar, N., and Furtado, V. (2018). Smart waste bin: a new approach for waste management in large urban centers. *In 2018 IEEE 88th Vehicular Technology Conference* (*VTC-Fall*), pp. 1–8, IEEE.

[8] Poddar, H., Paul, R., Mukherjee, S., and Bhattacharyya, B. (2017). Design of smart bin for smarter cities. *In 2017 Innovations in Power and Advanced Computing Technologies* (*i-PACT*), pp. 1–6, IEEE.

[9] Rahman, M. M., Bappy, A. S., Komol, M. M. R., and Podder, A. K. (2019). Automatic product 3D parameters and volume detection and slide force separation with feedback control system. *In 2019 4th International Conference on Electrical Information and Communication Technology* (*EICT*), pp. 1–6, IEEE.

[10] Rozenberg, A. (2013). Municipal solid waste: Is it garbage or gold. *Journal of Management Studies*, 38(4), 489–513.

[11] Singh, P. & Agrawal, R. (2018). A customer centric best connected channel model for heterogeneous and IoT networks. *Journal of Organizational and End User Computing* (*JOEUC*), 30(4), 32–50.

[12] Wijaya, A. S., Zainuddin, Z., and Niswar, M. (2017). Design a smart waste bin for smart waste management. *In 2017 5th International Conference on Instrumentation, Control, and Automation* (*ICA*), pp. 62–66, IEEE.

Chapter 7

AES and SIMON Algorithms in 45-nm and 90-nm Technology Implemented on Spartan-6 FPGA

Anil Gopal Sawant[1], *Akhilesh Kumar Mishra*[2], *and Vilas N. Nitnaware*[3]

[1]Research Scholar, Shri Jagdishprasad Jhabarmal Tibrewala University, Rajasthan, India
[2]Professor, Shri Jagdishprasad Jhabarmal Tibrewala University, Rajasthan, India
[3]Professor, D. Y. Patil School of Engineering Academy, Pune, India

Abstract: Many new technologies have arisen in the 21st century, such as data science, the internet of things (IOT), blockchain technology, artificial intelligence and cyber security. In several engineering disciplines, these innovations are expanding increasing their commercial as well as research applications. These innovations extend to new horizons in many industries, exploring new possibilities at each stage with a goal to serve a better planet and the entire humanity. The security requirements of these technologies vary based on their deployment in various fields, services and applications. The proposed VHDL implementation of the AES algorithm was simulated and synthesised in the ISE Xilinx before being verified on the Spartan 6 FPGA. The suggested SIMON Cipher technique is created in Verilog HDL, synthesised and analysed in Xilinx and proved on Spartan 6 Device. The suggested Simon Cipher design was developed at a speed of 100 MHz using Cadence RC compiler and Synopsys Design Vision. AES and Simon used far less power than previously achieved designs, 4.81 mw and 0.86 mw in 45 nm accordingly as well as 9.77 mw and 0.0638 mw in 90-nm technology. The suggested Simon cipher algorithm is included in the class of lightweight block ciphers that operate well in embedded contexts with limited resources, little hardware and low-power consumption, which are common applications for IOT gadgets.

Keywords: Cryptography, internet of things, block Cipher, FPGA, artificial intelligence.

1. Introduction

With the advancement of technology in all fields of research and applications, global populations are now communicating with one another in a variety of ways (Bouillaguet *et al.*, 2012). Electronics-based communication and interaction link engage the entire planet (Beaulieu *et al.*, 2013). These electrical gadgets make

DOI: 10.1201/9781003527442-07

linkages and interact with one another incredibly quickly, in microseconds, and at an ever-increasing rate (Le *et al.*, 2019). Every time these electronic devices engage or share information, a lot of data is either created or transferred among them, therefore any exchange of information among them must be safe (Sawant *et al.*, 2020). That refers to offering protected data and connectivity (Dofe *et al.*, 2015). Different levels of protection must be used at different stages (Le *et al.*, 2019).

Fundamentally, cryptography is utilised to secure data according to their needs by using certain cryptographic techniques. (Beaulieu *et al.*, 2015). The cryptographic techniques work with two separate forms of data encryption and decryption, which include block cipher & stream cipher. (Amal *et al.*, 2020). Three alternative methods, including symmetric, asymmetric and hash types, may be used to implement these block ciphers and stream ciphers. In today's technological age, it has become very necessary to protect the interconnections and data of gadgets using symmetric block cipher methods (Alkamil *et al.*, 2020).

The 64/128-bit Simon block cipher technique is the proposed cryptographic algorithm under discussion. In real-time applications, resources are typically limited. In addition to these restrictions, increased security must be prioritised with speedy computation and less power consumption with compact size.**2.**

2. SIMON Block Cipher Structure

Designed to satisfy the needs of future IOT devices in environments with power or resource constraints, SIMON is a lightweight block cipher (Sawant *et al.*, 2021). The Simon encryption is a one-bit-based encryption that makes use of left shifts, logical XOR (⊕) and logical AND (∧) operations in a balanced Feistel network. A complicated key schedule is utilised by the cipher against the data over a number of rounds. (Dofe *et al.*, 2015). To further fortify the key, this key schedule consists of a series of constant bits. The Simon provides a wide range of key and data block combinations to accommodate security needs while minimising power, energy and physical space requirements. Here, Figure 1 shows a single-round function. 2n block divided into two pieces The rightmost component is Xi, while the leftmost

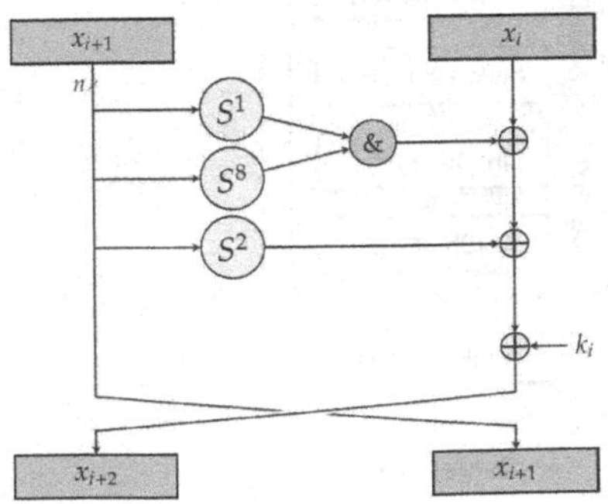

Figure 1: *SIMON one round Feistel Structure (Sawant et al., 2021).*

part is Xi+1. Key used in each round is denoted by Ki, where i stands for the round number. Shifting, anding and Xoring operations were performed by the round function. In this study, when all rounds are finished, Xi+1 logic is generated using the current block Xi, and then the blocks are exchanged. The number of rounds eventually gets determined by the block size and key size presented in Table 1. More focus is being placed on enhancing the features of confusion and diffusion to improve the security of the crypto that relies on rounds. The greater the number of rounds, the more secure the situation (Le *et al.*, 2019).

Table 1: SIMON parameters and rounds (Sawant *et al.*, 2021).

Block Size 2n	Key Size mn	Word size n	Key Word m	Rounds
32	64	16	4	32
48	72	24	3	36
48	96	24	4	36
64	96	32	3	42
64	128	32	4	44
96	96	48	2	52
96	144	48	3	54
128	128	64	2	68
128	256	64	4	72

2.1. *AES Block Cipher Structure*

A block cipher technique, known as the AES, is essentially what it is today. It is used for decryption as well as encryption purposes and supports a wide range of applications that operate in real time (Sawant *et al.*, 2020). Data chunks have a

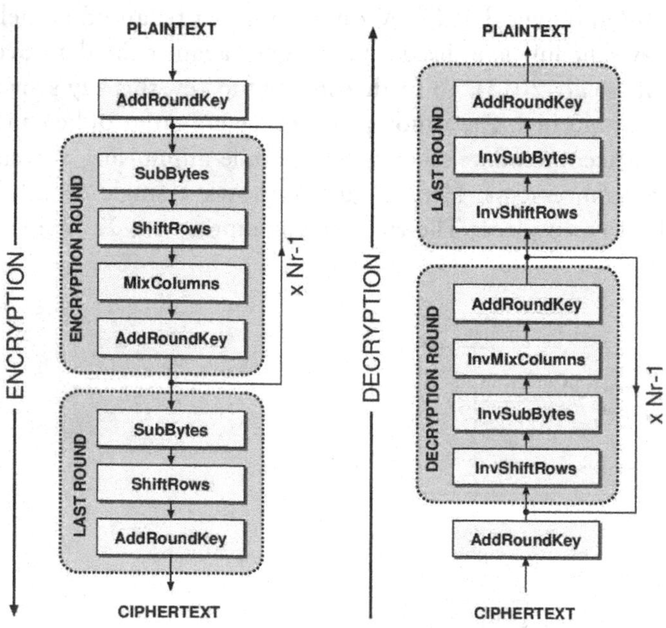

***Figure 2:** Encryption and decryption of AES Algorithm (Sawant et al., 2020).*

defined size of 128 bits, whereas key lengths can vary 10 rounds for data blocks of 128 bits presented in Table 2. It was discovered that AES offered sufficient security and operated much quicker compared to previous ciphers. (Amal *et al.*, 2020). The proposed paper is an alternative methodology that included various steps for the AES algorithm's encryption and decryption, as seen in Figure 2. The opposite of the encryption method's related procedure is the decryption algorithm. Utilised operations include Add round, Mix column, Shift row and Substitute byte. To provide proper security, these four steps are repeated numerous times.

3. Results with Discussion

The proposed algorithm for AES and Simon is simulated, synthesised and validated on hardware in the way of a Spartan 6 FPGA using Xilinx 14.2 Integrated Environment Synthesis software depicted in Figure 3, Figure 4, and Figure 5. The suggested AES and Simon cipher achieved overall power usage of 4.81 mw and 0.86 mw in 45 nm as well as 9.77 mw and 0.0638 mw in 90 nm are reported in Table 3 and Table 4 accordingly at 100 MHz frequency, which is significantly less than that of earlier systems. Hence these proposed designs are much superior to earlier systems.

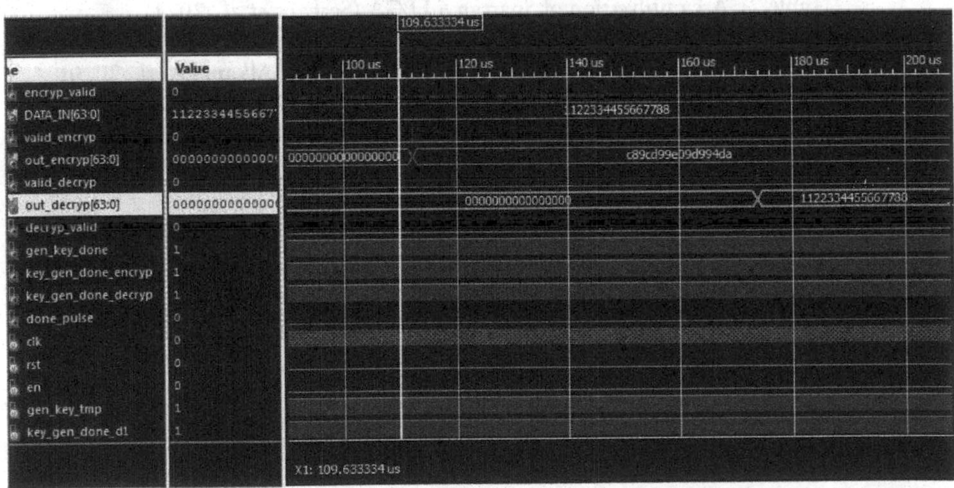

Figure 3: Simon simulation result.

Figure 4: AES simulation result.

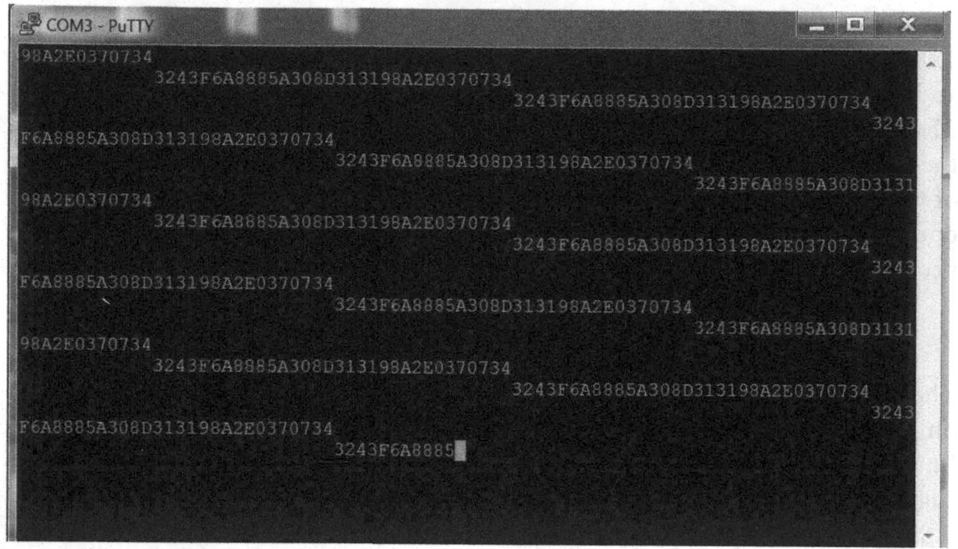

Figure 5: *AES output on Spartan 6 FPGA.*

Table 2: Area utilization of Spartan 6 FPGA (Sawant *et al.*, 2021).

Sr. No	Parameters	Utilization of Proposed Simon	Utilization of Proposed AES	(Alkamil *et al.*, 2020) Utilization of Previous Simon
1	Block Size	64	128	64
2	Key Size	128	128	128
3	Rounds	44	10	44
4	Slice Registers	2395	70	8222
5	Slice LUTs	3817	151	9509
6	Occupied Slices	3816	46	4000

Table 3: Power consumption for Simon (Sawant *et al.*, 2021).

Sr. No	Parameters	Proposed 45-nm Simon Cipher	Proposed 90-nm Simon Cipher	(Dofe *et al.*, 2015) Previously Implemented 65-nm Simon Cipher
1	Overall Power Usage	0.86 mw	0.0638mw	1.9 mw

Table 4: Power consumption for AES.

Sr. No	Parameters	Proposed 45-nm AES Algorithm	Proposed 90-nm AES Algorithm	(Amal *et al.*, 2020) Previously Implemented 90-nmAES algorithm
1	Overall Power Usage	4.81 mw	9.77 mw	151.68 mw

4. Conclusion

The AES and Simon algorithms being proposed were validated on the Spartan 6 FPGA after being simulated and synthesised in Xilinx Software. With 45-nm and 90-nm technology, the proposed design of the AES and Simon algorithm is also generated, synthesised reports in Synopsys Design Vision and Cadence RC compiler. The overall power usages for 45-nm technology at 100 MHz were 4.81 mw and 0.86 mw, while for 90 nm technology, it were 9.77 mw and 0.0638 mw respectively, which is far less power intensive than earlier designs. The proposed AES algorithm uses less FPGA hardware, a lesser amount of power and improves the performance of the cryptosystem.

The proposed Simon Block cipher performs better in sophisticated embedded systems built on FPGAs for IOT applications than previous block cipher algorithms. The Simon Cipher algorithm's suggested technique outperformed in hardware implementation for FPGA, displaying extremely low-power usage, low hardware usage with improved secured structure.

References

[1] Alkamil, A., et al. (2020). Efficient FPGA-based reconfigurable accelerators for SIMON cryptographic algorithm on embedded platforms. *International Conference on ReConFigurable* Computing and FPGAs (*ReConFig*). https://doi.org/10.1109/ReConFig48160.2019.8994803.

[2] Amal, H., et al. (2020). Serial-Serial AES as custom hardware interface using NIOS II Processor. *2020 IEEE International Conference on Design & Test of Integrated Micro & Nano-Systems* (DTS), (September). https://doi.org/10.1109/DTS48731.2020.9196191.

[3] Beaulieu, R., et al. (2013). The SIMON and SPECK families of lightweight block ciphers.

[4] Bouillaguet, C., et al. (2012). Low-data complexity attacks on AES. *IEEE transactions on information theory*, 58, 11, 7002–7017.

[5] Dofe, J., et al. (2015). Strengthening SIMON implementation against intelligent fault attacks. *IEEE Embedded Systems Letters*, 7, no 4, (December). https://doi.org/10.1109/LES.2015.2477273.

[6] Le, D., et al. (2019). Algebraic differential fault analysis on SIMON block Cipher. *IEEE Transactions on Computers*, 68, 11. https://doi.org/10.1109/TC.2019.2926081.

[7] Sawant, A., Nitnaware, V., and Deshpande, A. (2020). Spartan-6 FPGA Implementation of AES Algorithm. *Springer Nature, Lecture Notes in Electrical Engineering*, 570. https://doi.org/10.1007/978-981-13-8715-9_26.

[8] Sawant, A., Mishra, A., and Nitnaware, V. (2021). AES and SIMON Block Cipher Algorithms on 45 nm and 90 nm Technology Implemented on Spartan-6 FPGA. *International Journal of Innovative Research in Science, Engineering and Technology (IJIRSET)*, 10(8), 2023–12026. https://doi.org/10.15680/IJIRSET.2021. 1008001

Chapter 8

Design of Microstrip Patch Antenna for 5G Applications

Anand Deshkar[1], *Saroj Sahu*[1], *Sabyasachi Bhattacharya*[1], *Amit Fulsunge*[1], *Trupti Dipakwar*[2], *and Pallavi Gajghate*[2]

[1]Assistant Professor, Cummins College of Engineering for Women, Nagpur

[2]Engineering students, Cummins College of Engineering for Women, Nagpur

Abstract: This paper presents the design and simulation of a high-performance microstrip patch antenna tailored for operation at 28 GHz, a crucial frequency for 5G applications. The antenna, meticulously crafted and simulated using AWR software, was physically realised on an FR4 substrate. Impressively, the antenna exhibits exceptional characteristics, including a reflection coefficient of 12.43 dB, a low VSWR of 1.609, a commendable gain of 5.436 dB and substantial directivity at 7.35 dBi. These outstanding metrics position the antenna as an ideal choice for diverse 5G applications, ranging from satellite communications and smart cities to artificial intelligence and IoT deployments, where its ability to ensure robust signal transmission, low reflection and high gain significantly enhances performance and connectivity in these critical domains.

Keywords: Microstrip patch antenna, return loss, VSWR, substrate.

1. Introduction

Microstrip patch antennas are highly relevant in 5G communication due to their inherent characteristics, offering both high gain and directional radiation patterns, prerequisites for point-to-point communication and applications necessitating low latency and superior data throughput. Colaco *et al.* (2020) found that these antennas, characterised by their planar configuration, comprise a conductive patch affixed to a dielectric substrate, which, in turn, rests atop a ground plane. The patch interfaces with a transmission line, typically implemented as a microstrip trace etched onto the substrate. Darboe (2019) observed that these antennas feature numerous advantages, including a low-profile form factor conducive to seamless integration, adaptability across a spectrum of frequencies (including microwave and millimetre-wave), the ability to provide directional radiation patterns for elevated gain and directivity and cost-effective fabrication via standard printed circuit board (PCB) methodologies.

Microstrip patch antennas, often referred to as printed antennas, consist of three fundamental elements: the ground plane, the dielectric substrate and the conductive

DOI: 10.1201/9781003527442-8

patch. Garg *et al.* (2020) found that the metallic patch resides upon the dielectric material and finds support from the ground plane, with a transmission line serving as the interface to the patch. This antenna class finds extensive applications in mobile communication, satellite communication and global positioning systems (GPS). Design endeavours commonly involve specialised software tools like AWR. Crucial attributes encompass the choice of a small conductive patch (typically composed of copper) and the inclusion of a dielectric material with a low loss tangent (e.g., FR4 with a loss tangent of 0.018 square, or triangular). Different antennas for 5G applications have also been developed using this technology. A dual-band antenna is developed in Guo et al., (2021) for 5G applications. Two radiating patches are developed in this design; two radiate at two different frequencies required for 5G applications. Similarly Wang *et al.* (2022) described that four patches are developed to increase the bandwidth of antenna for 5G applications. A patch is developed with eight metal pins in Gao *et al.* (2022) which generates different transverse magnetic (TM) modes to develop a broadband antenna for 5G mobile applications. Ullah *et al.* (2021) described that microstrip line-fed aperture coupled patches are developed to achieve circular polarization.

The radiating patch and feedlines are photos etched on the dielectric substrate. For designing of antenna rectangular Microstrip patch is used and for conducting ground, copper is used as shown in figure 1. As per Puttaswamay (2014) the substrate is a dielectric material which should have low loss tangent. In the present work, FR4 is used which is having loss tangent 0.018.

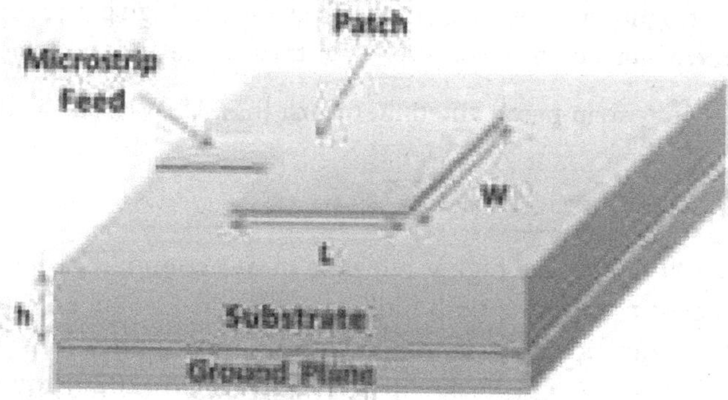

***Figure 1:** Structure of microstrip patch antenna.*

1.2. Microstrip Line Feed

The microstrip feedline in antenna design serves as the transmission line that connects the antenna's radiating patch to the external RF circuitry. It consists of a carefully engineered trace on the substrate, featuring specific dimensions and characteristics to ensure efficient power transfer. Wang *et al.* (2022) found that the feedline's width, dielectric properties and impedance matching are pivotal in guiding electromagnetic energy to and from the radiating element. This microstrip feedline's precision is critical for achieving optimal antenna performance, minimising signal loss and maximising radiation efficiency, particularly in applications demanding stringent impedance matching and signal fidelity, such as wireless communication systems and satellite links. The microstrip feedline is a transmission line utilised in microwave and RF circuits, comprising a narrow conducting strip, typically composed of copper, that is precisely etched onto a dielectric substrate. This conducting strip functions as the signal conductor, while the dielectric substrate provides electrical insulation. This arrangement is engineered to efficiently guide and propagate high-frequency electromagnetic signals, making it an essential component in microwave and RF circuitry for applications such as communication systems and radar.

1.3. Specifications of Antenna

Table 1: Specifications of proposed antenna.

Operating Frequency	**28 GHz**
Substrate Material	**FR4**
Loss Tangent	**0.018**
Gain	**5–6 dB**
Effective Dielectric Constant (ϵr)	**4.4**
Height of Substrate (h)	**1.6 mm**

2. Design Methodology

The design parameters for a microstrip patch antenna encompass critical dimensions like width (W), length (L) and thickness (H), in addition to the dielectric constant (ε_r) and effective height (H_{eff}). Width and length affect the resonant frequency and impedance, with ε_r impacting size and resonant frequency. The thickness influences the resonant frequency and the antenna's physical size. Effective height represents the distance from the ground plane to the patch's centre and is used in design calculations. Balancing these parameters is essential to optimise antenna performance, including impedance matching, radiation pattern, gain and bandwidth, according to specific application requirements.

2.1. Formulae for Design Parameter of Microstrip Patch Antenna

The resonant frequency (fr) of a patch antenna can be calculated using the formula (1), where fr represents the resonant frequency in Hertz, c stands for the speed of light in free space (approximately 3×10^8 metres per second), εeff represents the effective dielectric constant of the substrate, L represents the length of the patch in metres and W represents the width of the patch in metres.

$$\text{fr} \approx \frac{C}{2\times\sqrt{(seff \times L \times W)}} \tag{1}$$

The effective dielectric constant (εeff) is determined according to formula (2), which takes into account the relative dielectric constant (εr) and the substrate's thickness (h) and width (W).

$$\epsilon_{\text{effective}} = \frac{\epsilon_r + 1}{2} + \frac{\epsilon_r - 1}{2}\sqrt{(1 + 12(h/w))} \tag{2}$$

The effective length of the microstrip patch antenna ($Leffective$) is computed using formula (3), where c is the speed of light, $f0$ is the resonant frequency and εeff is the effective dielectric constant.

$$L_{\text{effective}} = \frac{c}{2f_0\sqrt{\epsilon_{\text{effective}}}} \tag{3}$$

For the patch width (W), formula (4) approximates its value based on the speed of light (c), resonant frequency ($f0$) and the relative dielectric constant (εr).

$$W \approx \frac{c}{2f_0\sqrt{\frac{sr + 1}{2}}} \tag{4}$$

The width of the feedline (wf) can be determined through formula (5), taking into account the substrate's thickness (h), characteristic impedance ($Z0$) and εeff.

$$W_f = \frac{hZ0}{60\sqrt{\epsilon_{\text{effective}}}} \tag{5}$$

The length extension (ΔL) is calculated using a complex formula that incorporates various factors such as ε_{eff}, patch dimensions (W and h) and constants.

$$W_f = \frac{\epsilon_W}{(\epsilon_{\text{effective}} - 0.258)\left(\frac{W}{h} + 0.8\right)} \tag{6}$$

Finally, the actual length of the patch (L) is obtained by subtracting twice the length extension ($2\Delta L$) from the effective length ($L_{\text{effective}}$). These equations provide a rigorous and scientific foundation for the design and analysis of microstrip patch antennas.

$$L = Leffective - 2\,\Delta L \tag{7}$$

3. Results and Discussions

The schematic diagram of microstrip patch antenna is shown in Figure 2 and the fabricated antenna is shown in Figure 7. The microstrip patch antenna exhibits exceptional performance across several key parameters. The return loss consistently falls below −10 dB, signifying efficient power reception with minimal reflection. Its broad radiation pattern makes it suitable for various applications. The S11 parameter, indicating the reflection coefficient, remains consistently below –10 dB, confirming excellent impedance matching and minimal power reflection. Moreover, the voltage standing wave ratio (VSWR) consistently stays below 2 dB, ensuring minimal power mismatch between the antenna and the feedline. These results collectively underscore the outstanding performance and impedance characteristics of the microstrip patch antenna design, rendering it highly suitable for a wide range of applications.

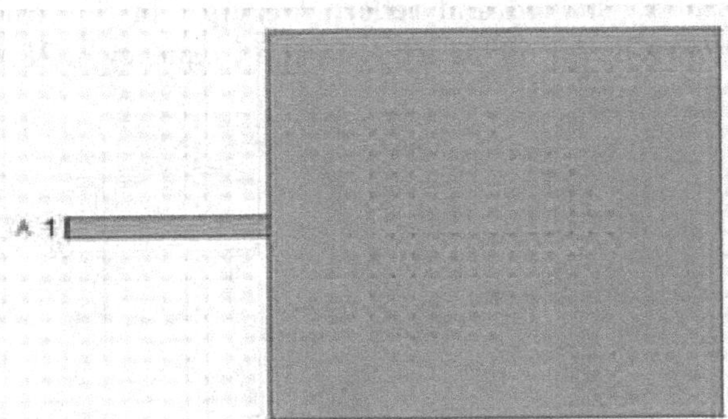

Figure 2: Design of patch antenna in AWR software.

3.1. Return Loss

To ensure optimal antenna performance, it is imperative to attain a return loss value exceeding –10 dB. This parameter signifies the ability of the antenna to preserve the maximum received power while minimising the power reflected back. Achieving this characteristic requires meticulous impedance matching at the antenna's input, wherein the impedance of the feedline is harmonised with that of the coaxial cable. This fine-tuning process is executed using specialised Microwave Office (MWO) software, resulting in an input return loss of –12.43 dB at a frequency of 28 GHz.

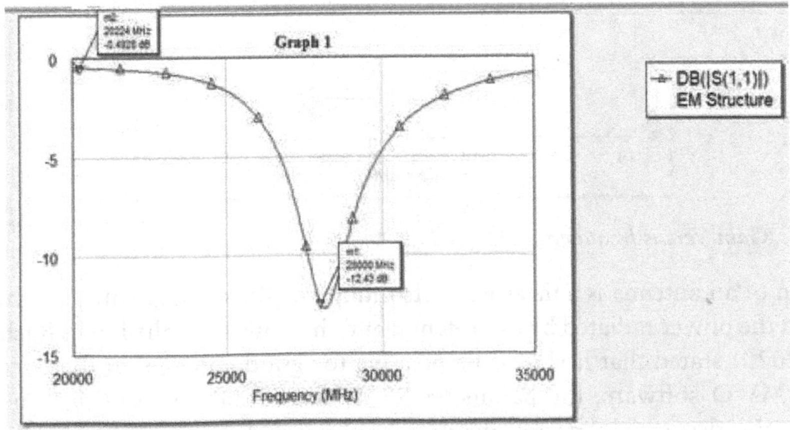

Figure 3: S11 parameter Versus frequency.

3.2. *VSWR*

For a patch antenna, it is imperative that the VSWR remains below 2 dB. VSWR is intricately linked to the reflection coefficient, forming a nonlinear relationship between the two parameters. Sharma *et al.* (2013) found that it should be less for good antenna. Typically, when the reflection coefficient falls below –10 dB, a VSWR below 2 dB is observed. This metric serves as a clear indicator of the power mismatch between the antenna and the feedline, offering a precise quantification of the ratio between transmitted and reflected energy. In this study, Figure 4 demonstrates that a VSWR of 1.609 is achieved at a frequency of 28.230 GHz.

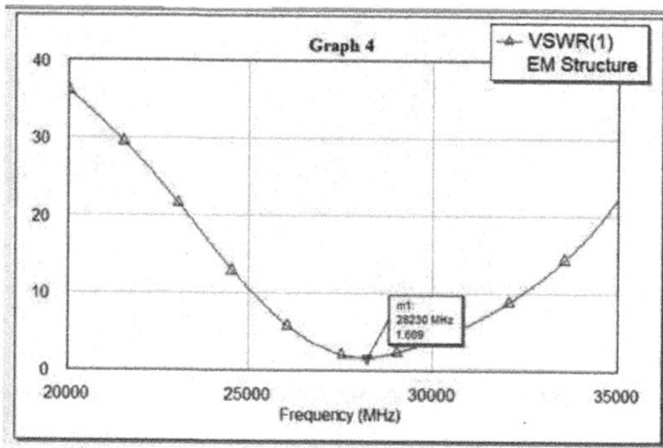

Figure 4: VSWR Versus frequency.

3.4 *Antenna Gain*

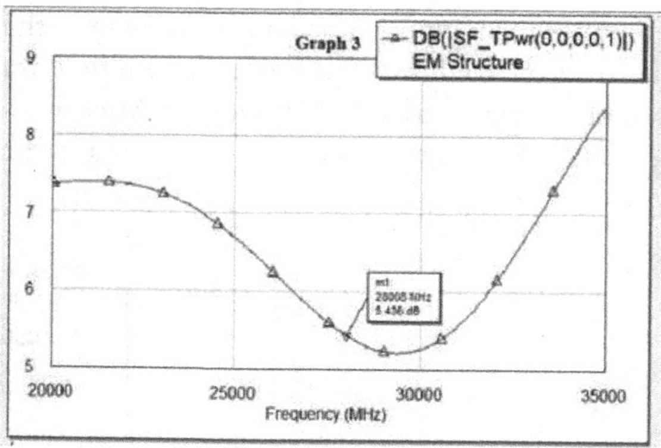

Figure 5: Gain Versus frequency.

The gain of an antenna is a measure of its radiative efficiency, quantifying the ratio between the power radiated by the antenna and the power supplied to it. Rashmitha *et al.* (2020) stated that it should be positive for good antenna. In the context of AWR's MWO software, the parameter SF_TPWR is indicative of antenna gain. Notably, this design exhibits an antenna gain of 5.436 dB at a frequency of 28 GHz, as visually represented in Figure 5.

3.5. Radiation Pattern

The antenna gain serves as a measure of the power radiated by the antenna in a specific direction, elucidating its directional efficiency. In the case of a patch antenna, a characteristic feature often observed is a broad radiation pattern. As depicted in Figure 6, this particular antenna design attains a directivity of 7.35 dB, signifying its ability to radiate power primarily in the vertical upward direction.

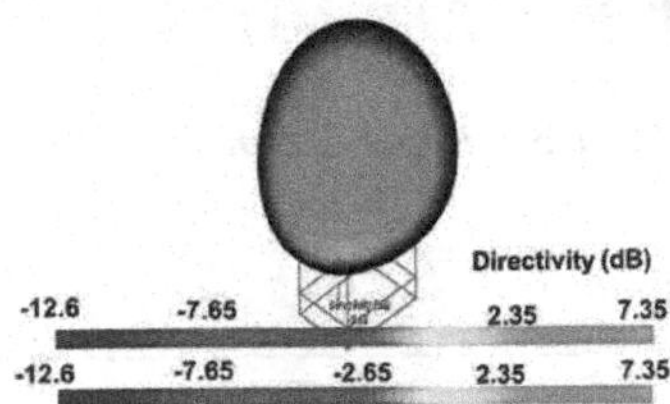

Figure 6: Radiation pattern.

4. Conclusion

In summary, microstrip patch antennas excel in 5G applications due to their compact size, ease of integration and high-frequency capabilities. They offer high gain and directional radiation patterns for efficient point-to- point communication, vital for low-latency, high-throughput scenarios. The antenna designed at 28 GHz using AWR software exhibited strong performance metrics, including a low S11 parameter (–12.43 dB), favourable VSWR (1.609), substantial gain (5.436 dB), small size (3.5 mm × 2 mm) and high directivity (7.36 dBi). These attributes make it suitable for satellite communication, smart cities, AI and IoT applications.

Table 2: Comparison of presented work.

Parameters	Omar Darboe et al. (2019)	Ubaid Ullah et al. (2021)	Presented paper
Input Return loss	-13.48 dB	Below -10 dB	-12.43 dB
Gain	6.63 dB	11.65 dB	5.436 dB
Frequency	27.954 GHz	28GHz	28 GHz
Material Used	RTduroid 5880	RO4003C	FR4
Size of antenna	6.285mm * 7.235mm	10mm * 27.7mm	2 mm * 3.5mm

Figure 7: *Fabricated antenna of size 3.5 mm × 2 mm.*

Acknowledgment

This work is performed in Cadense AWR supported RF Design Lab at CCOEW, Nagpur.

References

[1] Colaco, J., and Lohani, R. (2020). Design and implementation of microstrip patch antenna for 5G applications. *Conference Paper.*

[2] Darboe, O., and Konditi, D. B. O. (2019). *28GHz rectangular microstrip patch antenna for 5G Applications. Research Paper.*

[3] Gao, Y., Wang, Y., Wang, X., and Wei, M. (2022) A low profile broadband multimode patch antenna for 5G mobile applications. *IEEE Antennas and wireless propogation letters.*

[4] Garg, M., and Singh, N. (2020). Rectangular microstrip patch antenna.

[5] Guo, Y. Q., Mei, Y., and Zheng, S. Y. (2021) A singly fed dual microstrip antenna for microwave and millimetre wave applications in 5G wireless communication. *IEEE transaction on vehicular technology.*

[6] Mishra, R., Mishra, R. G., Chaurasia, R. K., and Shrivastava, A. K. (2019). Design and analysis of microstrip patch antenna for wireless communication. *International Journal of Innovative Technology and Exploring Engineering*, 8(7).

[7] Puttaswamay, P., Murti, P., and Thomas, B. (2014). Analysis of loss tangent effect on microstrip antenna gain.

[8] Rashmitha, R., Nataraj, N., Jugale, A., and Ahmed, M. R. (2019). Microstrip patch antenna design for fixed mobile and satellite 5G Communications. *Third International Conference on Computing and Network Communications.*

[9] Sharma, R., and Upadhyay, P. (2013). Design and study of inset feed square microstrip patch antenna. *International Journal of Application or Innovation in Engineering & Management.*

[10] Ullah, U., Al-Hasan, M., Koziel, S., and Mabrouk, I. B. (2021), A series inclined slot-fed circularly polarized antenna for 5G 28 GHz applications. *IEEE Antennas and wireless propogation letters.*

[11] Wang, J., Li, Y., and Wang, J. (2022). A low profile dual mode slot patch antenna for 5G Millimeter Wave applications. *IEEE Antennas and Wireless Propagation Lettersa.*

Chapter 9

Electric Vehicles: A Review on Technological Trends and Performance Improvement Strategies

Amit B. Jadhav[1]*and Dr. Mahadev S. Patil* [2]

[1]Department of Electrical Engineering, Annasaheb Dange College of Engineering & Technology, Ashta, Sangli, Maharashtra, India

[2]Department of Electronics and Telecommunication Engineering, Rajarambapu Institute of Technology, Skharale, Sangli, Maharashtra, India

Abstract: The study provides a summary of current technical developments as well as several techniques to improve performance of electric cars. The research explores the configurations of power electronic converters applied in charging stations, along with the corresponding charging procedures. Charging stations for electric vehicles encompass plug-in AC chargers, two-way converters and DC to DC converters. The many methods for recharging electric car batteries are examined. It is also explored how to control charging station energy utilising grid-to-vehicle (G2V) and vehicle-to-grid(V2G) modes to reduce power consumption. The research investigates the influence of harmonics produced while electric cars are being charged at charging stations. To address harmonic reduction at these stations, the utilisation of shunt active power filters is being suggested as a potential solution. For the design of the active power filter, reference current generating methods in the time and frequency domain are evaluated. Finally, several electric vehicle speed control techniques dependent on the kind of motor are explored. The advantages and uses of the algorithms used to regulate the speed of electric cars are examined.

Keywords: Electric vehicles charging stations, active filters, speed control of electric vehicles.

1. Introduction

As the demand for sustainable energy-based technologies rises, the adoption of electric vehicles (EVs) has also seen a significant increase. Modern technology is helping the traditional internal combustion-based systems to get better. However, the dwindling supply of fossil fuels and their negative long-term consequences on the environment have accelerated the design and advancement of electric

DOI: 10.1201/9781003527442-9

propulsion systems. The amount of automobiles on the road has led to a rise in the greenhouse gases that cause global warming. Governments around the world are implementing diverse measures to encourage the widespread adoption of EVs. Additionally, the time has seen a marked rise in energy demand. Hybrid electric systems and plug-in electric systems are two categories for the electric propulsion systems. A system with many facets, the electric vehicle, has thrust areas in the following areas:

- Electric drive system,
- Charging system,
- Speed control algorithms,
- Safety requirements,
- Battery management systems and
- Fault diagnosis and predictive maintenance.

Hybrid electric cars consist of both mechanical and electrical subsystems. The electric drive system, power electronic converters, batteries, sensors and microcontroller units are some of the electrical components employed (Chan, 2007). The internal combustion engine and gear system are examples of the mechanical component, also. In order to improve efficiency and power production, it is crucial to efficiently monitor, manage and regulate both of these systems. Internal combustion engine design already uses cutting edge technologies. Both fields benefit from the efficient fusion of mechanical and electrical subsystems. Better control, more energy effectiveness and reduced emissions are some of the benefits. The effectiveness of the electric drive system determines how well EVs function (Plesko *et al.*, 2008; Sarlioglu *et al.*, 2017). EV power systems need to provide both high torque and power density, a wide speed range, and the capability to function consistently at a constant power output. The purpose of this paper is to highlight the current research trends in the field of electric vehicle design. The paper explores the design of various DC–DC converters suitable for charging

stations, along with the application of bidirectional converters for vehicle-to-grid (V2G) and grid-to-vehicle (G2V) power transmission. Bidirectional converters can manage their output power in a way that aligns with the torque and power prerequisites of EVs and Hybrid Electric Vehicles (HEVs) (Plesko *et al.*, 2008), (Javadi and Haddad, 2015). The functionalities of V2G and G2V allow EV batteries to contribute power to the grid in times of power disruptions. The paper provides a broad review of different speed control strategies employed for torque and speed control of induction, PMSM and BLDC motors employed in EVs. The speed control methods that are discussed include control schemes using techniques like sliding mode controllers, and fuzzy controllers are employed to boost the steady-state performance of electric drive systems. The effectiveness of the speed controller was evaluated by considering factors such as resilience in speed regulation and adaptability to load changes.

Additionally, the study offers an overview of various power charger topologies for EVs and HEVs according to power needs. The several power electronic converters are discussed, along with any power quality problems they may have. The significance of non-linear components on the power system is considerable, given the requirement for high-power converters by EVs and HEVs (Emadi *et al.*, 2005). The design of electric vehicles must consider the multitude of electricity quality concerns, similarly harmonics and the demand for reactive power. Active power filters are utilised to demonstrate harmonic reduction. Selecting suitable methods for generating reference currents is a crucial factor in active power filter design. Both the time and frequency domains have classifications for the reference current generation approaches. By removing the harmonics from the charging current, harmonics are reduced. Then, by employing an inverter whose pulses are generated using the proper reference current production approach, these harmonics are removed selectively. The discussion in the study (Sarlioglu *et al.*, 2017) is mostly centred on the various EV and HEV charging techniques. The evaluation document lists the rankings of several wireless chargers as well as the appropriate motors chosen by various global corporations. In the summary, the effects of harmonics on the grid complications of electric car chargers are provided. The authors describe how electric cars produce harmonics and how to reduce such harmonics.

The article is organised into six sections. The initial segment introduces diverse EV types and outlines the paper's structure. The second section of the paper presents the category of EVs and configurations of EVs. In Section III, the design considerations of various power electronic converters for charging are discussed. Section IV discusses the impact of harmonics that are generated at the charging stations. It emphasises the utilisation of active power filters to mitigate the harmonics and enhance the quality of power. The fifth section of the paper outlines the different speed control strategies employed in EVs and HEVs. Finally, in section VI conclusions are given based on the analysis of the different papers.

2. Technology Behind Electric Vehicles

The main basic categories of EVs include plug-in EVs, HEVs and fuel Cell EVs (Chan, 2007) as seen in Figure 1. The electric driving system in battery-operated EVs is coupled to a battery or ultracapacitor. The electric grid serves as the EVs' energy supply during charging. The construction of plug-in electric cars is the most straightforward since the transmission connects the motor system directly to the wheels. These EVs employ PMSM, brushless DC and induction motors. Depending on the battery type, they have zero emissions and good energy efficiency. However, the initial cost is considerable, and they are unable to go long distances without a charging station (Abu-Rub *et al.*, 2014).

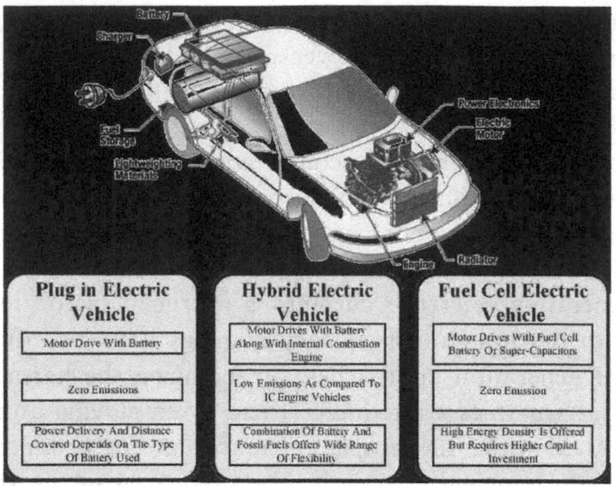

***Figure 1:** Classification of electric vehicles.*

In addition to their internal combustion engines, HEVs also include electric motor drives. Both conventional fossil fuels and electric grid charging are used as the energy source for HEVs. The HEVs combine IC engines with electrical motor drive systems, giving them the best features of both, including low emissions, cutting-edge IC engine technology and long range. They are categorised as series and parallel HEVs as a result. In parallel HEVs, one notable characteristic of this hybrid vehicle setup involves the integration of both an internal combustion engine (IC engine) and an electric motor to the drive train, unlike the series HEVs where only an electric motor is utilised. This enables the vehicle to operate in various modes, such as motor-only mode, IC engine mode or a combination of both, depending on the specific driving situation. The third form of EV is a fuel cell EV, which uses an ultracapacitor or a fuel cell as its energy source. These EVs are completely independent from fossil fuels and have extremely low emissions. However, this EV has a large upfront cost and is hardly utilised. Prominent automotive manufacturers, namely TESLA, Toyota, Mercedes and TATA, have been actively introducing vehicles in both the plug-in electric vehicle (PEV) and HEV segments.

There are numerous varieties of fuel cells, and two reactions take place between its three components. As a consequence, the fuel (often hydrogen) is used up, water is formed as a result of a reaction between the fuel and oxygen, and an electric current is generated that may be used to power the associated load. The type of electrolyte material used in fuel cells often defines these devices. These cells can primarily be categorised into the following types:

a. Alkaline
b. Molten Carbonate
c. Phosphoric Acid
d. Proton Exchange Membrane
e. Solid Oxide

When compared to alternative fuel cell varieties, within the range of different fuel cell types, the alkaline fuel cell stands out as one of the most well-established and advanced technologies, boasting the highest efficiency. Compared to alternative fuel cell variants, AFCs also possess notable advantages, such as reduced pollution compared to acidic fuel cells, lower production costs and the ability to use a wider range of readily accessible chemicals (EE&RE, U.S. Dept of Energy, 2022). For use in the military, molten carbonate fuel cells (MCFCs) were created. Because they don't require the use of precious metals as catalysts and because they work at high temperatures, fuels are transformed into hydrogen inside the cell itself, making MCFCs cost effective. However, the corrosion process is much expedited and cell life is decreased because of the high working temperatures (Reuters, 2007).
Compared to AFCs and proton-exchange membrane fuel cells (PEMFCs), phosphoric acid fuel cells (PAFC) are far less susceptible to CO poisoning, but they have a relatively poor power density. Additionally, handling large concentrations of phosphoric acid is quite hazardous (EE&RE, U.S. Dept of Energy, 2022). A solid oxide fuel cell (SOFC) is extremely effective, reliable, fuel-flexible, environmentally benign and substantially less expensive. The main drawback is the prolonged startup times caused by the high-operating temperature (EG&G Technical Services, Inc., 2004). Transport applications are where PEMFCs are most often employed. They offer the benefits of being lightweight, operating at low temperatures, and having increased efficiency. They can therefore take the place of AFCs. They are less versatile because they require high-purity oxygen, which is not the case with other fuel cell types (Badwal *et al.*, 2015; Lee *et al.*, 2006).

The comparison of various motor topologies has been carried out in publications (Rajashekara, 2013; Santiago *et al.*, 2012; Sarlioglu *et al.*, 2017; Welchko and Nagashima 2003) taking the specific demands of electric cars into consideration. This paper offers a comparison among switching reluctance motors, permanent magnet motors and induction motors concerning battery requirements, electronic systems, and stator and rotor design challenges. The energy density of batteries has been compared in order to estimate the range that each class of electric cars offers (Rajashekara, 2013). For the design of passenger cars (Ahrabi *et al.*, 2017; Chau *et al.*, 2008; Sarlioglu *et al.*, 2017) offer the ratings of several electric motors and the winding structure for the motor. As motor weight and volume decrease with speed, it can be seen that high-speed motors are typically favoured in EVs.

The many topologies for electric cars, including series, hybrid, parallel hybrid and fuel cell-based topologies, are described in the study (Emadi *et al.*, 2005) along with their operational characteristics. Due to its appropriateness at low speeds, intermittent braking and covering of all operating circumstances, the report names the series-parallel hybrid architecture of EV as the most common topology. Although the fuel cell-based architecture needs power comparable to an ultracapacitor, it is useful for low-energy and high-performance requirements (Estima and Cardoso, 2012) present the effectiveness of the EV's powertrain components. Two topologies are shown for the performance comparison between driving and regenerative braking modes. In the second configuration, a bidirectional DC–DC converter is connected to the inverter through a variable voltage control technique, while the first topology relies on a battery-powered inverter employing PWM. With effective dynamic features in both driving and regenerative braking modes, the variable voltage control technology aids in regulating the necessary system voltage. Auxiliary DC–DC converters for HEVs are accomplished in the study (Ahrabi *et al.*, 2017) by including DC–DC converter within traction drive system. subsequently, thc system as a whole is lighter and less expensive.

3. Impact of Harmonics and Power Quality Improvement

In the reference (Crosier & Shuo, 2013), enhancement of power quality for electrified transportation is achieved through the implementation of a series of active power filters. The article presents a sequence of active power filters that have been developed to counteract harmonics originating from the integration of vehicle loads with the grid. The design of polyvalent hybrid topology is implemented on a real-time controller. A linear closed-loop discrete-time model is discussed in (Karvekar & Patil, 2014) for active filters at the EV charging station. Instead of using the Zero Order Hold model for the EV charging station/APF, the paper develops an equivalent model known as the triangle hold model, which leads to a simpler power system model. The linear closed-loop model implemented using DQ reference frame theory is simulated to evaluate transient performance. The non-linear closed-loop model of APF is converted into a linear feedback system for power quality analysis using Simulink. The paper [25] investigates the use of EPL for the mitigation of harmonics under balanced and unbalanced conditions. The enhanced phase-locked loop (PLL) is utilised to identify the fundamental component of load current even when non-linear loads are present. The reference current generation technique makes use of this peak value and the loss of current value needed charging capacitor of the inverter. According to the simulation results, the utilisation of the EPLL-based reference current generation technique, as specified in the IEEE 519-2014 standard (IEEE P519/D5.1, January 2021), leads to a reduction in total harmonic distortion at the source side.In reference (Karvekar & Patil, 2013) the paper discusses the creation of an active power filter aimed at mitigating both harmonics and imbalanced load currents that result from non-linear loads. The approach suggests a two-level four-leg voltage source inverter with predictive control. The predictive control algorithm is accustomed to maintain the constant dc-link voltage supplied to the inverter. The proposed method demonstrates the mitigation of harmonics under both steady-state and transient conditions. The paper (Akin *et al.*, 2009) investigates the performance of electric motors employed in electric transportation systems under real-time conditions. The fault diagnosis algorithm for measurement of rotor asymmetry condition using variation in current harmonics during running mode is implemented using Texas Instruments TMS320F2812. The proposed method is model-independent and hence applicable to different powertrain components.

The reference current generation methods for active power filters using sliding mode control and the enhanced PLL have been elaborated upon in articles (Karvekar & Patil, 2013; Karvekar *et al.*, 2014). The suggested system enhances power quality by diminishing harmonic content and compensating for reactive power, thereby enhancing current and voltage quality. In the articles Karvekar & Kumbhojkar (2013) and Karvekar & Patil (2014), the authors examine and compare different reference current generation techniques in both the time domain and frequency domain. The time-domain approach encompasses the utilisation of both instantaneous reactive power theory and synchronous reference frame theory. In the other hand, in frequency domain, the discrete Fourier transform method is employed and enhance the overall power quality of the system as illustrated in Table 1 Top of Form. In Chen *et al.* (2019), the concept of selective harmonics elimination for non-linear loads is investigated. It explains the estimation of harmonics using the method of generalized trigonometric function delayed signal cancellation. The concept can be extended to the computation of specific harmonics at charging stations in minimum time for different load conditions.

The paper (Yeetum & Kinnares, 2019) proposes a parallel active power filter using a single harmonic detector offering robustness to parametric variations. The technique provides suppression of source current harmonics and enhances power quality. The Onboard charger described in (Nguyen *et al.*, 2018) explains the role of the charger as an active filter to erase low-frequency ripple power in dc links. A frequency adaptive pre-filter is designed using differential control for the estimation of frequency in (Mishra *et al.*, 2019). The control algorithm is employed to generate reference currents aiming to lower the total harmonic distortion of grid currents to less than 5%, in accordance with the IEEE 519-2014 standard. Congyin Shi *et al.* have proposed a hybrid continuous-time low pass filter and a discrete-time finite impulse response filter. The hybrid filter is useful in the cancellation of harmonics using three-tap harmonics cancellation architecture. The method described in reference (Li & Fei, 2018), suggests the utilisation of a second-order sliding mode controller and fuzzy controller for improving the power quality of the system. The adaptive laws are designed in line with the Lyapunov criterion to ensure system stability.

Table 1: Classification of reference current generation methods for power quality improvement at charging station.

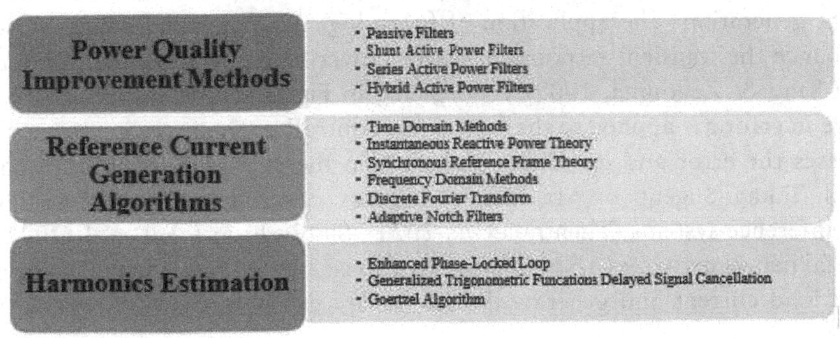

Power Quality Improvement Methods	• Passive Filters • Shunt Active Power Filters • Series Active Power Filters • Hybrid Active Power Filters
Reference Current Generation Algorithms	• Time Domain Methods • Instantaneous Reactive Power Theory • Synchronous Reference Frame Theory • Frequency Domain Methods • Discrete Fourier Transform • Adaptive Notch Filters
Harmonics Estimation	• Enhanced Phase-Locked Loop • Generalized Trigonometric Funcations Delayed Signal Cancellation • Goertzel Algorithm

The system effectively decreases the total harmonic distortion at the source side, and the second-order sliding mode controller successfully addresses chattering issue. In Charger *et al.* (2018), same inverter hardware is used as rectifier hardware through a new topology. It presents a solution for the elimination of second-order harmonics generated from quasi-Z-source rectifiers if used separately.

The studies in the paper (Basta & Morsi, 2010) evaluate the impact of higher order harmonics and lower order harmonics on the grid. The lower and higher order harmonics are quantified by IEEE 1459-2000 standards. Lower harmonics include frequencies below 2.4 KHz (from 2nd to 39th harmonic) while the higher-order harmonics include all the harmonics above 2.4 KHz (above 40th harmonic). The impact of harmonics due to ac–dc rectification phase at the charging station is measured at various time intervals and the measurement is quantified in terms of lower order harmonics and higher order harmonics also called supra-harmonics. In the paper (Yong *et al.*, 2015), the paper investigates the impact on the grid caused by EV charging. A model to study the impact on voltage profile is presented by simulating 6–8 EVs at a time. The study showed that voltage profile and stability deteriorated to a point below safe operating level even when six vehicles are charged simultaneously. A universal ac–dc converter based on the direct voltage control method is designed for compensating reactive power with measurements of only the three-phase grid voltage and dc-link voltage. Hardware-in-loop simulation using wavelet transform (WT) for the design of SAPF is implemented in the paper (Kumar *et al.*, 2019). The Haar wavelet transform is employed for the purpose of harmonics mitigation and power factor correction to unity. The hardware implementation was carried out using OPAL-RT real-time target platform. The use of wavelet transform extracts the features of current/voltage signals in time as well as frequency domain. The WT is used along with synchronous reference frame theory for the generation of control signals as depicted in Figure 2. The load current is sensed and processed using Haar WT to extract the information about fundamental current and harmonics.

The reference voltage $Vdcref$ is analysed with the actual dc-link voltage Vdc and PI controller is utilised to minimise the error and generate the reference current for loss reduction. This loss current is added to the output of the WT block to generate the current i_d^*. For synchronisation of the inverter output, the source frequency is extracted using PLL from the three-phase source voltage and used for reference current generation. The application of fuzzy logic and artificial neural networks to enhance the transient response of active power filters is presented (Iqbal *et al.*, 2021; Saad & Zellouma, 2009). As depicted in Figure 3, the values of error and change in error are applied to the fuzzy logic controller. The fuzzy inference system processes the error and its variations based on the rule base to produce control signals. Takagi-Sugeno or Mamdani-type fuzzy controllers are used to develop fuzzy inference systems (Thentral *et al.*, 2021). Similarly, in Iqbal *et al.* (2021), the artificial neural network (ANN) is employed to isolate the fundamental component of the load current and generate the reference signals. In paper Verma & Singh

(2021), a second-order frequency adaptive integrator is designed with dc offset rejection capability. The adaptive integrator is employed to deduce the magnitude and phase of the fundamental current, and a comprehensive controller is designed to enhance power quality. The unified controller is investigated for a charging station powered by solar and wind energy. The experimental results show that the supply side harmonics are efficiently mitigated using the second-order adaptive integrator-based reference current generation technique.

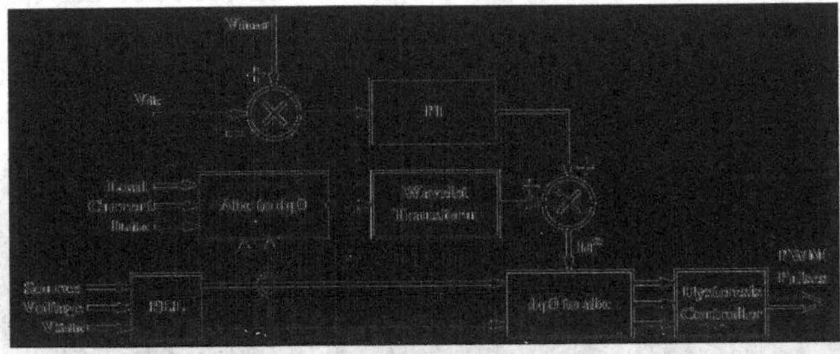

Figure 2: Sinusoidal battery charging method using series resonant converter.

Figure 3: A fuzzy logic controller is utilised to manage the voltage across the DC link capacitor.

4. Speed Control Strategies

The work done in this paper Syed *et al.* (2009) is mainly related to using of a fuzzy-based gain scheduling algorithm to control battery power and speed. The paper suggests the implementation of a fuzzy controller to regulate and control the engine power and speed. The fuzzy controller has been applied to determine the gain values of the PI controller, taking into account the system's operating conditions. The simulation has been carried out to demonstrate the improvement in transient parameters like settling time and overshoot. The results of the system are tested on the Ford Hybrid vehicle using a PI controller and fuzzy-based proposed algorithm. The paper Allaoua & Laoufi (2013) elucidates a method to control the torque of induction machines used in electrified transportation systems. In this method, a sliding mode fuzzy-based controller, in conjunction with the space vector modulation technique, is employed for torque control of the induction machine.

The stability of the sliding mode controller is guaranteed through the Lyapunov criterion. A comparison between the performance of the proposed system and a conventional PI controller is conducted to demonstrate the effectiveness of the proposed approach. Figure 4 shows the speed control of the induction motor using a fuzzy logic controller. Here, the technique of indirect speed estimation is used by sensing the motor currents. The Park's transformation block (ABC to DQ) converts the three- phase currents into a rotating reference frame. The speed and flux of the system are estimated and they are compared with the reference speed ω^* and reference flux ϕ^* for generating reference signals. These signals are then converted into the reference current signals using Clarke's transformation block (DQ to ABC) for generating PWM pulses to achieve the desired speed.

Figure 4: Speed control of induction machine using fuzzy logic controller.

In Bitar *et al.* (2015); Cheng *et al.* (2012) the testing equipment for performance analysis of induction motors used for Electric Cars is developed. The testing was carried out on a special induction motor whose construction was designed to satisfying the needs and specifications of EVs. The performance of the EV was carried out under the conditions of no load, running the motor with constant acceleration for a finite time under different resistive loads. The results for variation in speed and torque were studied for these cases. In this paper (Janpan & Boonyanant, 2012), the speed control of the BLDC motor has been examined using the combined mode. The combined model implies the generation of electrical and mechanical energy simultaneously using the proposed method. The generated electrical energy is stored in the batteries and is made available to the system. A BLDC motor with three coils is considered viz A, B and C. In a three-phase BLDC motor, only two windings are excited at a given time. The third inactive winding is used to generate electrical energy during motor operation. The paper (Kraa *et al.*, 2014) investigates the use of an adaptive fuzzy logic controller for speed control of motor drives used in EVs. The concept of maximum control structure is incorporated in the fuzzy logic controller to speed control and stability. The adaptive fuzzy control coupled with the maximum control structure provides additional robustness to the system under dynamic conditions.

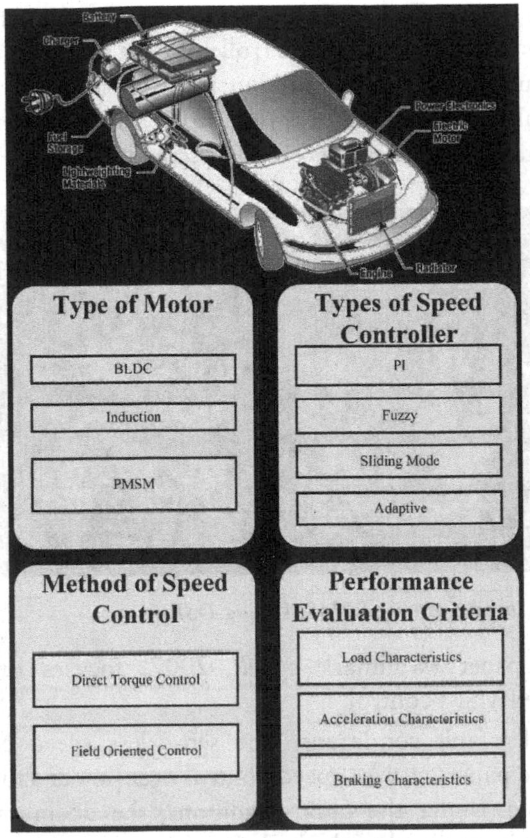

***Figure 5:** Speed control techniques for electric vehicles.*

The authors in (Kommuri *et al.*, 2016) have successfully implemented a speed observer based on a higher order sliding mode (HOSM) controller to estimate the speed of the PMSM motor. The main advantage of using a higher order sliding mode controller is robustness with minimum convergence time. The controller employs a super- twisting algorithm to minimise the chattering effect and ensure linear perturbation. Additionally, the HOSM controllers exhibit better performance under the conditions of load disturbances. The use of disturbance torque observer in sliding mode control is studied in Wang *et al.* (2019) where velocity control of reaching law is adaptively selected. The results show that for smaller gains in the sliding mode controller, the chattering is reduced but settling time increases whereas the increase in gain reduces the settling time but causes chattering problems. Therefore, the disturbance torques the observer adaptively selects these gain values to reduce the chattering effect due to external disturbances. In Lee & Nam (2021), the parameters of the EV are estimated using the recursive least squares algorithm. A two-speed gear system is implemented using torque control mode where the reference torque is calculated using gear ratio vehicle dynamics. Sensorless speed control of PMSM is validated in Batzel & Lee (2005) using DSP and the performance is analysed. The rotor position and speed are estimated using stator current and stator voltage. The difference between the estimated speed and the actual speed is applied to the PI controller to generate torque reference. This torque reference and the information about rotor position are further processed using speed control algorithms like field-oriented control to generate pulses for the inverter and maintain the required speed as depicted in Figure 5.

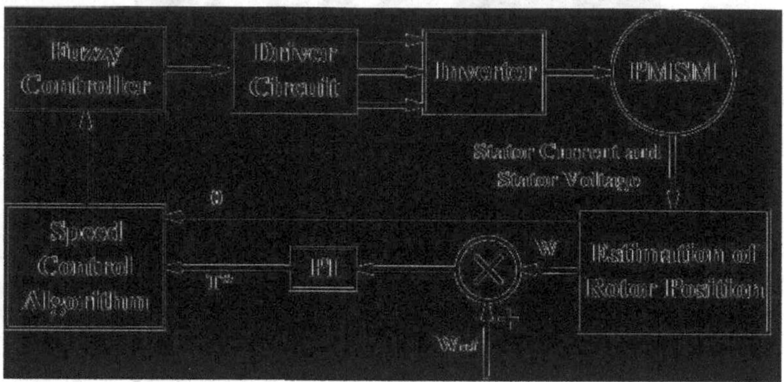

Figure 6: Sensorless speed control of PMSM using DSP.

The research paper (Wallmark *et al.*, 2007) focuses on examining the performance of the PMSM control

algorithm under fault conditions. The sensorless speed control algorithm implemented in the paper ensures speed control even under fault conditions like shorting of a phase. Under the fault conditions, the dc-link voltage reference applied to inverter is modified with the help of an extra inverter leg. For speed control of PMSM, a fuzzy control-based indirect field-oriented control approach is utilised (Alsayed *et al.*, 2019). As depicted in Figure 6 to achieve efficient control, the stator current's torque and flux components are decoupled. This decoupling process renders the control of PMSM akin to the speed control of DC machines The disparity between the reference speed and the actual speed is used as input to the fuzzy logic controller, which then generates the current reference. This current reference is compared with the current i^* and processed using a PI controller. The fuzzy controller avoids overshoot and improves the overall transient response. For implementing the algorithm, dSpace 1104 is used as a real-time target. Similarly, in (Li & Gu, 2012), the fuzzy adaptive internal model control is used while the Takagi-Sugeno control is designed by authors in the paper (Wang *et al.*, 2019). The modified internal model control avoids the saturation of control input and avoids anti-windup problems faced by controllers. The Takagi–Sugeno type of fuzzy controller is implemented in the discrete-time domain directly which ensures that the closed-loop response converges to zero. In paper (Li *et al.*, 2013), the algorithm represented in Figure 7 is modified by using a terminal sliding mode controller. The terminal sliding mode controller make sure faster convergence and improved disturbance rejection, and the use of a non-singular type of SMC reduces the chattering effect.

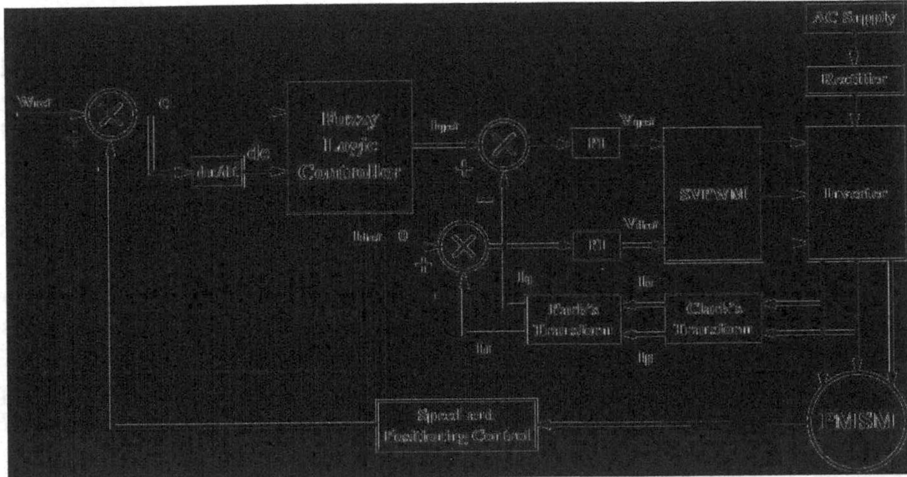

Figure 7: Sensorless speed control of PMSM using DSP using fuzzy control.

The use of the MRAS control technique for sensorless speed control of PMSM is investigated in Kivanc & Ozturk (2018); Chintawar *et al.* (2021) for feedforward estimation of stator voltage. MRAS controller updates the values of stator resistance and rotor flux linkage thereby making the controller adaptive. Moreover, a PLL-based speed estimator is implemented using TMS320F28335 and it is observed that the system offers improved speed measurement as compared with a sliding mode observer.

5. Conclusion

The report analyses the various EV models, together with their energy sources and uses, in its conclusion. Modern electric motor configurations utilised in EVs and HEVs are addressed, along with changes made to the motor to meet power density and torque requirements. The many power electronic converter topologies used in charging stations are described along with their benefits and drawbacks. Various topologies of DC to DC and DC to AC converters have been utilised to illustrate the design of power electronics converters, enabling unidirectional and bidirectional power flow and thereby enhancing power efficiency. The study explores the impact of harmonics on the power system during electric car charging, revealing that they significantly affect the power quality of the system. As a result, it is crucial that the harmonics at the charging stations are minimised. As described in the sections before, active power filters, bidirectional dc–dc converters and dc–ac converters have all been used to reduce harmonics. The many methods for regulating the speed of the electric motors utilised in EVs have finally been described. The enhancement of the system's steady-state and transient responsiveness is emphasised by conventional speed control methods as well as intelligent control approaches in both the driving and regenerative braking regions. The article also aids in pinpointing the areas that need investigation in the realm of electric car technology.

References

[1] Abeywardana, D. B. W., Acuna, P., Hredzak, B., Aguilera, R. P., and Agelidis, V. G. (2017). *Single-Phase Boost Inverter-Based Electric Vehicle Charger with Integrated Vehicle to Grid Reactive Power Compensation. In IEEE Transactions on Power Electronics*, 3462–71.

[2] Abu-Rub, K., Malinowski, M., and Al-Haddad, K. (2014). Power *Electronics for Renewable Energy Systems, Transportation and Industrial Applications. IEEE Publications*, Wiley.

[3] Acuna, P., Moran, L., Rivera, M., Dixon, J. J., and Rodriguez. (2014). Improved active power filter performance for renewable power generation systems. *IEEE Transactions on Power Electronics*, 687–94.

[4] Ahrabi, R. R., Ardi, H., Elmi, M., and Ajami, A. (2017). A novel step-up multiinput DC-DC converter for hybrid electric vehicles application. *IEEE Transactions on Power Electronics*, 3549–61.

[5] Akin, B., Ozturk, Toliyat, and Rayner, M. (2009). DSP-based sensorless electric motor fault diagnosis tools for electric and hybrid electric vehicle powertrain applications. *IEEE Transactions on Vehicular Technology*, 2679–88.

[6] Allaoua, B., and Laoufi, A. (2013). A novel sliding mode fuzzy control based on SVM for electric vehicles propulsion system. *Energy Procedia*, 36.

[7] Alsayed, Y. M., Maamoun, A., and Shaltout, A. (2019). High-performance control of PMSM drive system implementation based on DSP real-time controller. *International Conference on Innovative Trends in Computer Engineering* (ITCE), 225–30.

[8] Badwal, S. P. S., Giddey, S., Munnings, C., and Kulkarni, A. (2015). Review of progress in high-temperature solid oxide fuel cells. *Journal of the Australian Ceramics Society*.

[9] Basta, B., and Morsi, W. G. (2021). Low and high order harmonic distortion in the presence of fast charging stations. *International Journal of Electrical Power & Energy Systems*, 126.

[10] Batzel, T. D., and Lee, K. Y. (2005). Electric propulsion with sensorless permanent magnet synchronous motor: Implementation and performance. *In IEEE Transactions on Energy Conversion*, 575–83.

[11] Bitar, Z., Sandouk, A., and Al Jabi, S. A. (2015). Testing performances of a special ac induction motor used in electric car. *Energy Procedia*, 74, 160–71.

[12] Chan, C. C. (2007). The state of the art of electric, hybrid, and fuel cell vehicles. *Proceedings of the IEEE*, 704–18.

[13] Charger, Q. I. O., Na, T., Zhang, Q., Tang, J., and Wang, J. (2018). Active power filter for single-phase quasi- z- source integrated on-board charger, 197–201.

[14] Chen, M., Peng, L., Wang, B., and Wu, W. (2019). Accurate and fast harmonic detection based on the generalized trigonometric function delayed signal cancellation. *IEEE*, 3438–47.

[15] Cheng, S., Li, C., Chai, F., and Gong, H. (2012). Research on induction motor for mini electric vehicles. *Energy Procedia*, 17, 249–57.

[16] Chintawar, S., Ghodke, S., Khatavkar, V., Alset, U., and Mehta, H. (2021). Performance evaluation of speed behaviour of fuzzy-pi operated BLDC mo-

tor drive. *International Conference on Computational Performance Evaluation* (ComPE), 179–84.
[17] Choi, H. H., and Jung, J. (2013). Discrete-Time Fuzzy Speed Regulator Design for PM Synchronous Motor. *IEEE Transactions on Industrial Electronics*, 600–7.
[18] Crosier, R., and Shuo, W. (2013). DQ-frame modeling of an active power filter integrated with a grid- connected, multifunctional electric vehicle charging station. *Power Electron*, translated by *IEEE*, 28, 5702–16.
[19] EG&G Technical Services. (2004). Fuel Cell Handbook, Seventh Edition.
[20] Emadi, A., Rajashekara, K., Williamson, and Lukic,. (2005). Topological overview of hybrid electric and fuel cell vehicular power system architectures and configurations. *IEEE Transactions on Vehicular Technology*, 763–70.
[21] Estima, J., and Cardoso, A. J. M. (2012). Efficiency analysis of drive train topologies applied to electric/hybrid vehicles. *IEEE Transactions on Vehicular Technology*, 1021–31.
[22] Iqbal, M., Jawad, M., Jaffery, M. H., Akhtar, S., Rafiq, M. N., Qureshi, M. B., Ansari, A. R., and Nawaz, R. (2021). Neural networks based shunt hybrid active power filter for harmonic elimination. *IEEE* Access 9, 69913–25.
[23] Javadi, A., and Al-Haddad, K. (2015). A single-phase active device for power quality improvement of electrified transportation. *IEEE Transactions on Industrial Electronics*, 3033–41.
[24] Janpan, I., Chaisricharoen, R., and Boonyanant, P. (2012). Control of the *Brushless DC Motor in Combine Mode. Procedia Engineering*, 32, 279–85.
[25] Karvekar, S., and Patil, D. (2013). Implementation of shunt active power filter for dynamically distorted load conditions using goertzel algorithm. *International Conference on Circuits, Power and Computing Technologies* (ICCPCT), 6–11.
[26] Karvekar, S., and Patil, D. (2013). A novel technique for implementation of shunt active power filter under balanced and unbalanced load conditions. *International Conference on Circuits, Power and Computing Technologies* (ICCPCT), 1–5.
[27] Karvekar, S., and Kumbhojkar, A. (2013). Comparison of different methods of reference current generation for shunt active power filter under balanced and unbalanced load conditions. *Proceedings IEEE International Conference on Circuit, Power Computing Technologies.*
[28] Karvekar, S., Gadgune, S., and Patil, D. (2014). Implementation of shunt active power filter using sliding mode controller. *In International Conference on Circuits, Power and Computing Technologies.*
[29] Karvekar, S., and Patil, D. (2014). *Comparison of Reference Current Generation for Shunt Active Power Filter Using Goertzel Algorithm and Enhanced PLL.* In International Conference on Circuits, Power and Computing Technologies.
[30] Kesler, M., Kisacikoglu, M. C., and Tolbert, L. M. (2014). Vehicle-to-grid reactive power operation using plug-in electric vehicle bidirectional offboard charger. *In IEEE Transactions on Industrial Electronics*, 6778– 84.
[31] Khan, A., Memon, S., and Sattar, T. P. (2018). Analyzing integrated renewable energy and smart-grid systems to improve voltage quality and harmonic dis-

tortion losses at electric-vehicle charging stations. *IEEE Access*, 6, 26404–15.

[32] Kisacikoglu, M. C., Metin, K., and Leon, M. T. (2015). Single-phase on-board bidirectional PEV charger for v2g reactive power operation. *IEEE Transactions on Smart Grid*, 767–75.

[33] Kisacikoglu, M. C., Ozpineci, B., and Tolbert, L. M. (2010). Examination of a PHEV bidirectional charger system for V2G reactive power compensation. *Twenty-Fifth Annual IEEE Applied Power Electronics Conference and Exposition* (APEC), 458–65.

[34] Kivanc, O. C., and Ozturk, S. B. (2018). Sensorless PMSM drive based on stator feed forward voltage estimation improved with MRAS multiparameter estimation. *In IEEE/ASME Transactions on Mechatronics*, 1326–37.

[35] Kommuri, S. K., Defoort, M.., Karimi, H. R., and Veluvolu, K. C. (2016). A robust observer-based sensor fault- tolerant control for pmsm in electric vehicles. *In IEEE Transactions on Industrial Electronics*, 7671–81.

[36] Kraa, O., Becherif, M., Ayad, M. Y., Saadi, R., Bahri, M., Aboubou, A., and Tegani, I. (2014). A novel adaptive operation mode based on fuzzy logic control of the electric vehicle. *Energy Procedia*, 50, 194–201.

[37] Kuperman, A., Levy, U., Goren, J., Zafransky, A., and Savernin, A. (2013). Battery charger for electric vehicle traction battery switch station. *IEEE Transactions on Industrial Electronics*, 5391–9.

[38] Kwon, M., and Choi, S. (2017). An electrolytic capacitorless bidirectional ev charger for v2g and v2h applications. *In IEEE Transactions on Power Electronics*, 6792–9.

[39] Lee, T., and Nam, K. (2021). Torque control based speed synchronization for two-speed gear electric vehicle. *IEEE* Access 9, 153518–27.

[40] Lee, J. S. et al. (2006). Polymer electrolyte membranes for fuel cells. Journal of Industrial & Engineering Chemistry.

[41] Li, S., and Gu, H. (2012). Fuzzy adaptive internal model control schemes for pmsm speed-regulation system. *IEEE Transactions on Industrial Informatics*, 767–79.

[42] Li, S., Zhou, M., and Yu, X. (2013). Design and implementation of terminal sliding mode control method for pmsm speed regulation system. *In IEEE Transactions on Industrial Informatics*, 1879–91.

[43] Li, S., and Fei, J. (2018). Adaptive second-order sliding mode fuzzy control based on linear feedback strategy for three-phase active power filter. *IEEE Access*, 6, 72992–3000.

[44] Lulhe, A. M., and Date, T. N. (2015). A technology review paper for drives used in the electrical vehicle (EV) and hybrid electrical vehicles (HEV). *Int. Conf. Control. Instrum. Commun. Comput. Technol*, ICCICCT, 2015, 632–6.

[45] Madawala, U. K., and Thrimawithana, D. J. (2011). A bidirectional inductive power interface for electric vehicles in V2G systems. *In IEEE Transactions on Industrial Electronics*, 4789–96.

[46] Mishra, S., Hussain, I., Singh, B., Chandra, A., Al-Haddad, K., and Shah, P. (2019). Frequency adaptive prefiltering stage for differentiation-based control of shunt active filter under polluted grid conditions. *IEEE Transactions on Industry Applications*, 882–91.

[47] Monteiro, V., Pinto, J. G., and Afonso, J. L. (2016). Operation modes for the electric vehicle in smart grids and smart homes: present and proposed modes. *In IEEE Transactions on Vehicular Technology*, 1007–20.

[48] Nguyen, H. V., To, D. D., and Lee, D. C. (2018). Onboard battery chargers for plug-in electric vehicles with dual functional circuit for low-voltage battery charging and active power decoupling.

[49] Plesko, H., Biela, J., Luomi, J., and Kolar, (2008). Novel concepts for integrating the electric drive and auxiliary DC-DC converter for hybrid vehicles. *IEEE Transactions on Power Electronics*, 3025–34.

[50] Rajashekara, K. (2013). Present status and future trends in electric vehicle propulsion technologies. *IEEE Journal of Emerging & Selected Topics in Power Electronics*, 3–10.

[51] Ravinder, K., and Bansal, H. O. (2019). Hardware in the loop implementation of wavelet-based strategy in shunt active power filter to mitigate power quality issues. *Electric Power Systems Research*, 169, 92–104.

[52] Reuters [Editorial]. (2007). Platinum-free fuel cell developed in Japan.

[53] Rivera, S., and Wu, B. (2017). Electric vehicle charging station with an energy storage stage for split-dc bus voltage balancing. *IEEE Transactions on Power Electronics*, 2376–86.

[54] Saad, S., and Zellouma, L. (2009). Fuzzy logic controller for three-level shunt active filter compensating harmonics and reactive power. *Electric Power Systems Research*, 1337–41.

[55] Said, D., and H. T. Mouftah. Mar. (2020). A novel electric vehicles charging/discharging management protocol based on queuing model. *In IEEE Transactions on Intelligent Vehicles*, 100–11.

[56] Sahoo, S., Prakash, S., and Mishra, S. (2018). Power quality improvement of grid-connected dc microgrids using repetitive learning-based pll under abnormal grid conditions. *IEEE Transactions on Industry Applications*, 82–90.

[57] Santiago, J. D., Bernhoff, H., Ekergård, B., Eriksson, S., Ferhatovic, S., Waters, R., and Leijon. M. (2012). Electrical motor drivelines in commercial all-electric vehicles: a review. *IEEE Transactions on Vehicular Technology*, 475–84.

[58] Sarlioglu, B., Morris, C. T., Han, D., and Li, S. (2017). Driving toward accessibility: a review of technological improvements for electric machines, power electronics, and batteries for electric and hybrid vehicles. *IEEE Industry Applications Magazine*, 14–25.

[59] Saxena, N., Singh, B., and Vyas, (2017). Integration of Solar Photovoltaic with Battery to Single-Phase Grid. *IET Generation, Transmission & Distribution*, 2003–12.

[60] Shi, C., and Sánchez-Sinencio, E. (2019). An on-chip built-in linearity estimation methodology and hardware implementation. *In IEEE Transactions on Circuits & Systems*, 897–908.

[61] Syed, F. U., Kuang, Smith, M., Okubo, S., and Ying, H. (2009). Fuzzy gain-scheduling proportional-integral control for improving engine power and speed behavior in a hybrid electric vehicle. *IEEE Transactions on Vehicular Technology*, 69–84.

[62] Tang, Y., Lu, J., Wu, B., Zou, S., Ding, W., and Khaligh, A. (2018). An integrated dual-output isolated converter for plug-in electric vehicles. *In IEEE Transactions on Vehicular Technology*, 966–76.

[63] Thentral, T. M. T., Vijayakumar, K., Usha, S., Palanisamy, R., Babu, T. S., Alhelou, H. H., and Al-Hinai, A. (2021). Development of control techniques using modified fuzzy based sapf for power quality enhancement. *IEEE Access*, 9, 68396–413.

[64] Verma, A., and Singh, B. (2021). AFF-SOGI-DRC control of renewable energy based grid-interactive charging station for ev with power quality improvement. *In IEEE Transactions on Industry Applications*, 588–97.

[65] Wallmark, O., Harnefors, L., and Carlson, O. (2007). Control algorithms for a fault-tolerant pmsm drive. *IEEE Transactions on Industrial Electronics*, 1973–80.

[66] Wang, Q., Yu, H., Wang, M.., and Qi, X. (2019). An improved sliding mode control using disturbance torque observer for permanent magnet synchronous motor. *In IEEE Access*, 7, 36691–701.

[67] Wang, L., Madawala, U. K., and Wong, M. C. (2021). A wireless vehicle-to-grid-to-home power interface with an adaptive dc link. *In IEEE Journal of Emerging & Selected Topics in Power Electronics*, 2373–83.

[68] Welchko, B. A., and Nagashima, J. M. (2003). The influence of topology selection on the design of EV/HEV propulsion systems. *IEEE Power Electronics Letters*, 36–40.

[69] Xu, W., Chan, K. W., Or, S. W., Ho, S. L., and Liu, M. (2020). A low-harmonic control method of bidirectional three-phase z-source converters for vehicle-to- grid applications. *In IEEE Transactions on Transportation Electrification*, 464–77.

[70] Yeetum, W., and Kinnares, V. (2019). Parallel active power filter based on source current detection for antiparallel resonance with robustness to parameter variations in power systems. *IEEE Transactions on Industrial Electronics*, 876–86.

[71] Yong, J. Y., Ramachandaramurthy, V. K., Tan, K. M., and Mithulananthan, N. (2015). Bi-directional electric vehicle fast-charging station with novel reactive power compensation for voltage regulation. *International Journal of Electrical Power & Energy Systems*, 300–10.

[72] Zhang, D., Lin, H., Zhang, Q., Kang, S., and Lv, Z. (2018) Analysis, Design, and Implementation of a Single-Stage Multi-pulse Flexible-Topology Thyristor Rectifier for Battery Charging in Electric Vehicles. *IEEE Transactions on Energy Conversion.*

[73] Zhou, N., Wang, J., Wang, Q., and Wei, N. (2015). Measurement-based harmonic modeling of an electric vehicle charging station using a three-phase uncontrolled rectifier. *IEEE Transactions on Smart Grid*, 1332–40.

Chapter 10

Beamforming Algorithms for Massive MIMO: Developing and Evaluating Novel Beamforming Algorithms to Improve the Coverage, Capacity and Interference Management in Massive MIMO systems

Kanchan Wagh[1], *Amit Tripathi*[2], *and Vedant Adhau*[3]

[1]Assistant Professor, Cummins College of Engineering for Women, Nagpur
[2]Executive Engineer, Godrej Industries

Abstract: In the era of burgeoning wireless communication demands, the quest for enhancing the signal-to-interference-plus-noise ratio (SINR), channel capacity, throughput and user fairness has become paramount. Existing beamforming techniques, although proficient, exhibit limitations in balancing the trade-off between interference mitigation and signal amplification. This paper introduces a novel approach by synergistically integrating maximum ratio transmission (MRT) with regularised zero-forcing (RZF) beamforming, while employing the Elephant Herding Optimisation (EHO) for the astute selection of hyperparameters. The crux of our proposition lies in the adaptive capability of EHO, which aids in determining the optimal beamforming weights and regularisation parameters, thereby enhancing system performance. Benchmark results vividly indicate the supremacy of our model, recording improvements of 4.9% in SINR, 3.5% in channel capacity, 5.9% in throughput and 5.5% in user fairness when juxtaposed against the state-of-the-art techniques. Such advancements not only pave the way for more efficient and fair wireless systems but also spotlight the potential of leveraging bio-inspired algorithms in the realm of communication engineering for different scenarios.

Keywords: Massive MIMO beamforming, Elephant Herding optimisation, regularised zero-forcing, maximum ratio transmission, wireless communication optimisations.

1. Introduction

The burgeoning growth of wireless communications, underpinned by the digital transformation era and the Fourth Industrial Revolution, emphasises the

DOI: 10.1201/9781003527442-10

paramount importance of optimising wireless network performance. As users become increasingly reliant on such networks for a myriad of applications ranging from mundane data exchanges to mission-critical tasks, the pressure mounts on technological solutions to deliver more efficient, faster and fairer services. Beamforming, an essential technique in massive multiple input multiple output (Massive MIMO) systems, has emerged as a cornerstone to meet these demands by maximising the received signal power at the desired user's end while nullifying the interference from other undesired users (Le *et al.*, 2016). This is done via the distributed beamforming (DB) process.

Traditional beamforming strategies, such as maximum ratio transmission (MRT) and zero forcing (ZF), each possess their merits and demerits. While MRT leverages the channel state information to amplify the signal power of the desired user, it tends to be susceptible to interference. In contrast, ZF focuses on nullifying inter-user interference but often at the cost of signal power amplification. However, the selection of hyperparameters for RZF, which is pivotal for its performance, remains a challenging task for different scenarios (Zhang *et al.*, 2019). This can be achieved via threshold-based beamforming (TBB) operations. The challenge, then, lies not just in the methods themselves but in determining their optimal configuration. Enter the world of optimisation algorithms. While traditional optimisation techniques have their value, bio-inspired algorithms have recently gained traction owing to their adaptability, efficiency and capability to navigate complex solution spaces. Among these, the Elephant Herding Optimization (EHO) algorithm, inspired by the social behavior and memory of elephants, has shown potential in diverse fields, raising the question of its applicability in beamforming optimisations. In this paper, we delve into the amalgamation of the strengths of MRT and RZF, using the EHO algorithm for hyperparameter optimisation. By doing so, we aim to transcend the limitations of current models and usher in an era of heightened wireless communication performance levels.

1.1. *Motivation and Objectives*

The digital era, characterised by smart cities, Internet of Things (IoT), and ubiquitous connectivity, is creating a wireless data traffic surge. Such an interconnected landscape demands higher throughput, enhanced signal quality and fair resource allocation among users to ensure a seamless user experience. Beamforming, central to massive MIMO systems, plays a quintessential role in shaping this experience. However, the present beamforming strategies often fall short in managing the intricate trade-off between interference mitigation and optimal signal amplification. Their deterministic nature, influenced by predefined parameters, often confines them to suboptimal solutions. This realisation underscores the imperative to revisit, reimagine and revitalise these methods, ensuring they are primed for the challenges of contemporary wireless landscapes. Our motivation stems from this very need to augment the existing beamforming strategies.

2. Literature Review

Massive MIMO systems, with their potential to support numerous antennas at the base station, have emerged as a cornerstone in modern wireless communication. The principal advantage of MIMO antennas is their beamforming capability, which aims at directing the energy towards desired users, thereby amplifying the signal quality and minimising interference. Over the years, various models have been proposed to enhance this beamforming efficiency. This review provides an in-depth look into these models.

2.1. *Maximum Ratio Transmission* **(MRT):** MRT is one of the most straightforward and widely adopted beamforming techniques. It leverages the channel state information to maximise the signal power at the intended user's end for different scenarios. While MRT ensures that the transmitted signal aligns with the channel's conditions, it often does so at the expense of exacerbating interference to other users, especially in dense user scenarios (Liu *et al.*, 2020).

2.2. *Zero Forcing* **(ZF):** ZF, in stark contrast to MRT, focuses primarily on nullifying the inter-user interference levels. By designing the beamforming weights, ZF ensures that the signals intended for different users are orthogonal, effectively minimising interference levels (Geng *et al.*, 2018). Methods like adaptive radar beamforming (ARB) use ZF for internal operations. However, its primary limitation is the potential reduction in the signal power of the intended user, especially when the number of users approaches the number of antennas (Rihan *et al.*, 2018).

2.3. *Regularized Zero-Forcing* **(RZF):** Recognising the individual limitations of MRT and ZF, RZF was introduced. By adding a regularisation factor to the ZF, RZF smartly incorporates aspects of both the techniques, balancing interference mitigation and signal amplifications. However, the efficiency of RZF largely depends on the optimal selection of the regularization parameter, which is non-trivial for these scenarios (Le *et al.*, 2020).

2.4. *Eigen beamforming*: Building upon the mathematical framework of eigenvectors and eigenvalues, Eigen beamforming employs the channel's eigenmodes to direct the transmissions. While this method promises significant gains in specific scenarios, especially when channel conditions are favourable, its computational complexity, particularly for real-time applications, poses a challenge for different scenarios (Li *et al.*, 2021).

2.5. *Robust Beamforming*: Real-world scenarios often entail uncertainties in channel state information. Robust beamforming techniques are designed to be resilient against these uncertainties. By considering worst-case scenarios and optimising for them, these methods offer consistent performance even when exact channel information might be imprecise for different scenarios (Zhu *et al.*, 2019).

2.1.1 Proposed Design of an Efficient Model for Evaluating Novel Beamforming Algorithms to Improve the Coverage, Capacity and Interference Management in Massive MIMO Systems

Based on the review of existing models used for optimising massive MIMO systems, it can be observed these models are highly complex when applied to real-time scenarios, or have lower efficiency when used for large-scale deployments. To overcome these issues, this section discusses design of an efficient model for evaluating novel beamforming algorithms to improve the coverage, capacity and interference management in massive MIMO systems. The proposed model initially uses maximum ratio transmission (MRT) which is a beamforming technique used to maximise the signal power at the receiver by adjusting the transmit weights based on the channel conditions. The transmit weight vector (w) is chosen to be proportional to the conjugate of the channel vector, resulting in higher power in the desired signal dissipation regions via equation 1:

$$w = \frac{H * H'}{\|H\|} \tag{1}$$

where w is the transmit weight vector, H is the channel matrix, H' is the conjugate transpose of the channel matrix and $\|H\|$ represents the norm of the channel matrix sets. This is fused with regularised zero-forcing (RZF), which is a beamforming technique that combines the benefits of zero-forcing with regularisation to balance interference suppression and signal amplification. RZF aims to nullify interference while maintaining a certain level of amplification for the desired signal, which is represented via equation 2:

$$w = \frac{\left(H * H' * H + \lambda * I\right)^{-1} HH'}{tr\left(H * H * H' + \lambda * I\right)} \tag{2}$$

where w is the transmit weight vector, H is the channel matrix, HH' is the conjugate transpose of the channel matrix, λ is the regularization parameter (a positive scalar that controls the trade-off between interference suppression and signal amplification), I is the identity matrix and tr represents the trace of a matrix, which is the sum of its diagonal elements.

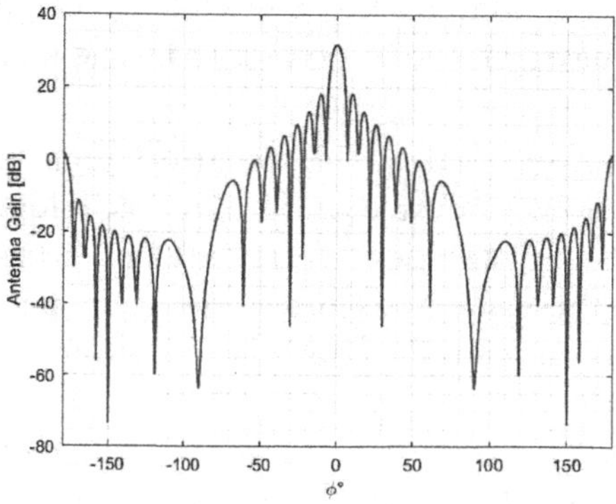

Figure 1: Gain results simulated in MATLAB Software with given specifications.

In this equation, the regularisation parameter λ prevents the matrix $H * H' * H$ from becoming singular, which can happen when attempting to invert the channel matrix directly in zero-forcing process. The regularisation parameter introduces a controlled amount of amplification to the desired signal while suppressing interference levels.

The resulting transmit weight vector w is applied to the transmit antennas to shape the beam pattern. RZF optimises the weights to simultaneously nullify interference and provide a controlled amount of amplification to the desired signal, striking a balance between interference management and signal enhancements.

The integration of MRT with RZF beamforming involves combining the benefits of both techniques to achieve improved signal quality, interference mitigation and overall system performance levels. This integration can be achieved through a weighted combination of the MRT and RZF weight vectors, which are represented via equation 3:

$$w' = \alpha w(MRT) + (1-\alpha) w(RZF) \tag{3}$$

where w' is the integrated transmit weight vector, w(MRT) is the transmit weight vector obtained from MRT, w(RZF) is the transmit weight vector obtained from RZF and α is a weighting factor ($0 \le \alpha \le 1$) that determines the balance between MRT and RZF contributions. By adjusting the value of the weighting factor α using EHO, the integration process can be fine-tuned to emphasise either MRT or RZF characteristics. For example, setting α closer to 0 would emphasise RZF, while setting it closer to 1 would emphasise MRT process. The integrated weight vector w' is then applied to the transmit antennas, shaping the beam pattern and optimising the trade-off between signal amplification and interference suppressions. This integrated approach capitalizes on the strengths of both MRT and RZF techniques, enhancing the overall performance of the wireless communication systems. To select the value of α, the EHO models is used, which works as per the following process.

Initially, setup *NH* different Herds, where each Herd has a different value of α, which is estimated via equation 4:

$$á = \text{STOCH}(0,1) \tag{4}$$

where *STOCH* is an augmented stochastic process used to generate number sets. Based on this value of α, the MRT & RZF Models are fused, and Herd fitness is estimated via equation 5,

$$fh = SINR * CC * THR * UF \tag{5}$$

where *SINR* represents SINR, which is estimated via equation 6, CC represents channel capacity which is calculated via equation 7, *U* is the user fairness, which is estimated via equation 8, and T represents throughput, which is represented via equation 9 as follows:

$$SINR = 10 * log10\left(\frac{S}{I+N}\right) \tag{6}$$

where S represents signal level, $I \& N$ represents interference and noise levels.

$$CC = B * log2(1 + SINR) \tag{7}$$

where B represents bandwidth of the channels.

$$U = \frac{\sum_{i=1}^{N} T(U)^2}{N * \left(\sum_{i=1}^{N} T(U)\right)^2} \tag{8}$$

where N represents number of users.

$$T = CC * log2(M) \tag{9}$$

where M is the modulation order (e.g., QPSK, 16-QAM, etc.) used during communications.

After generation of NH Herds, the model estimates Herd fitness threshold via equation 10,

$$fth = \frac{1}{NH}\sum_{i=1}^{NH} fh(i) * LH \tag{10}$$

where LH represents learning rate of the EHO process.

Based on this threshold, Herds with $fh > fth$ are passed to the next Iteration sets, while other Herds are discarded from the current Iteration, and regenerated via equations, 4, 5, 6, 7, 8 & 9, which assists in identification of New α configurations.

This process is repeated for NI Iterations, and at the end of all Iterations, Herd with maximum fitness ("Matriarch" Herd) is selected, and its α Value is used to model the beamforming techniques. Due to which the proposed model is able to maximise antenna performance under real-time scenarios. This performance was evaluated for different scenarios and compared with existing models in the next section of this text.

3. Results and Discussions

An experimental setting is created to test the effectiveness of novel beamforming algorithms for improving Massive MIMO systems' SINR, channel capacity, throughput and user fairness in the context of growing wireless communication demands. The system is made up of a Massive MIMO base station with 128 antennas that serve a group of 8 user equipment (UE) devices that are all using the same 3.5 GHz frequency band. The additive white Gaussian noise (AWGN) process and Rayleigh fading channel effects ensure that the channel characteristics follow a realistic model. The research uses two different beamforming methods: RZF and MRT. By adjusting the beamforming weights in accordance with the channel conditions, MRT is used to amplify the desired signal. RZF, on the other hand, tries to reduce interference by boosting the desirable signals while reducing signal leakage to interfering users. The EHO algorithm is used in the suggested method to optimise the hyperparameters. Using an adaptive search strategy, EHO, a bio-inspired optimisation technique, is skilled at finding the ideal beamforming weights and regularisation parameters that raise system performance levels. A thorough experimental setting is created using a number of simulation parameters. These consists of 128 base station antennas, 8 user equipment devices, 20 dBm of transmit power per antenna and –80 dBm of noise power. For various scenarios, EHO hyperparameters including population size, maximum iterations and control parameters are defined. SNR values ranging from 0 dB to 20 dB are taken into consideration. Several performance criteria are used to evaluate the proposed methods. Signal power, interference and noise levels are taken into account when calculating the SINR, which is used to evaluate the quality of received signals. In order to conduct the experiment, channel coefficients must be created using the chosen Rayleigh fading channel model. MRT and RZF beamforming techniques are used to compute beamforming weights for each user for each SNR value within the set range. Then, these weights and the regularisation parameters are adaptively optimized using the EHO algorithm. In each of the situations, metrics for SINR, channel capacity, throughput and user fairness are calculated for each user. Among these scenarios are Urban High-Density Area, Suburban Low-Density Area, Multi-Cell Interference, Dynamic User Mobility, Non-Uniform User Distribution and Frequency Selective Channels. In this article, we have discussed two scenarios Urban High-Density Area and Multiple Cell Interference Scenarios.

3.1. Urban High-Density Area: The suggested beamforming algorithm can be tested in an urban high-density setting with many buildings and probable signal obstructions. A Massive MIMO base station is used in the scenario to serve a large number of users who are close by. The evaluation would concentrate on enhancing signal quality and capacity by reducing interference from signals and reflections of neighboring users.

3.2. Multi-Cell Interference: In this scenario, multiple base stations are deployed in a multi-cell environment. It is possible to research the algorithm's capacity to control interference between nearby cells and increase system throughput as a whole. The evaluation would take into account optimal signal power for consumers in each cell while minimising interference with beamforming weights.

According to these situations, the following comparisons were made between the results and DB [3], TBB [5] and ARB [12] in terms of SINR (S), Channel Capacity (CC), Throughput (T) and User Fairness (U) levels. Based on this, these metrics can be observed from Table 1 as follows:

Table 1: Urban high-density area scenarios.

Method	S (dB)	CC (Mbps)	THR(Mbps)	UF (%)
DB (Cavalcante et.al. 201)	15.2	32.5	22.6	0.72
TBB (Mosleh et.al. 2015)	16.8	34.2	23.8	0.78
ARB (Geng et.al. 2018)	14.5	30.6	21.3	0.68
Proposed	19.3	38.7	26.9	0.85

The innovative approach of the Proposed method, which combines RZF and MRT beamforming techniques with the use of the EHO algorithm for the best hyperparameter selection, is responsible for the method's outstanding results. The Proposed method gains adaptability and accuracy in determining beamforming weights and regularisation parameters from this combination of methodologies. By using EHO, the algorithm is better able to manage the trade-off between signal amplification and interference mitigation, producing outstanding results. The findings clearly demonstrate that the Proposed technique is superior, starting with SINR, which gauges how well the received signal performs in comparison to noise and interference. The Proposed technique outperforms all existing methods, including DB, TBB and ARB, with a SINR of 19.3 dB. With such a high SINR, the proposed approach clearly decreases noise and interference, which results in better signal quality. With a SINR of 16.8 dB, TBB is in second place, followed by DB (15.2 dB) and ARB (14.5 dB).

Moving on to CC, the Proposed approach displays its superiority. This important measure represents the highest data rate that can be successfully delivered across the channel. It surpasses the capabilities of DB, TBB and ARB with a channel capacity of 38.7 Mbps. TBB comes in second with a capacity of 34.2 Mbps, while DB and ARB have lesser capacities for these scenarios at 32.5 Mbps and 30.6 Mbps, respectively for different scenarios. The actual rate of successful data transfer, or throughput, is another indicator of how dominant the Proposed technique is. The throughput it reaches, 26.9 Mbps, surpasses that of DB, TBB and ARB. TBB comes in at number two with a throughput of 23.8 Mbps, while DB and ARB come in at number three and number four, respectively, with throughputs of 22.6 Mbps and 21.3 Mbps. Finally, consideration is given to U, a metric important for guaranteeing equitable resource distribution across users. With a U score of 0.85, the Proposed approach has the most equitable distribution of resources among users. TBB comes in second with a fairness value of 0.78, while DB and ARB have somewhat lower fairness values for similar circumstances, 0.72 and 0.68, respectively, for different scenarios.

Beginning with SINR, which rates the strength of a signal in relation to interference and noise, the Proposed technique stands out as the best option. With a SINR of 21.6 dB, the Proposed technique outperforms DB, TBB and ARB, demonstrating its capacity to successfully reduce interference and noise to create a powerful, dependable signal. For these scenarios, TBB and DB have lower SINR values of 17.5 dB and 18.7 dB, respectively, while ARB comes in second with a SINR of 19.8 dB.

Another key indicator of the Proposed method's superiority is CC, which shows the highest data rate that can be transmitted via the channel. With a CC of 47.3 Mbps, it performs better than DB, TBB and ARB, proving its capacity to transport data effectively in suburban settings. While TBB and DB display capacities of 38.9 Mbps and 41.2 Mbps, respectively, for these scenarios, ARB falls behind with a capacity of 43.5 Mbps.

The Proposed method's superiority is further emphasised by throughput, which measures the actual data transmission rate attained. It surpasses DB, TBB and ARB with a throughput of 32.9 Mbps. In these scenarios, ARB comes in second with a throughput of 30.1 Mbps, while TBB and DB have lesser throughputs of 26.9 Mbps and 28.6 Mbps, respectively, for these scenarios.

Also included is U, a crucial metric that guarantees the equitable distribution of resources among users. With a U score of 0.88, the Proposed approach has the most equitable distribution of resources. With a fairness value of 0.82, ARB comes in second, whereas TBB and DB have somewhat lower fairness values for similar scenarios (0.75 and 0.78, respectively). Similarly, performance for multiple cell interference scenarios can be observed from Table 2 as follows:

Table 2: Multiple cell interference scenarios.

Method	S (dB)	CC (Mbps)	THR(Mbps)	UF (%)
DB (Cavalcante *et al.*, 201)	**14.3**	**30.0**	**20.9**	**0.67**
TBB (Mosleh *et al.*, 2015)	**15.6**	**32.2**	**22.4**	**0.72**
ARB (Geng *et al.*, 2018)	**13.8**	**28.9**	**20.1**	**0.64**
Proposed	**17.2**	**35.6**	**24.7**	**0.80**

Table 2 provides important information on the performance assessment of various beamforming techniques – DB, TBB, ARB and the proposed technique – in the setting of "Multi-Cell Interference Scenarios." These scenarios depict a complicated multi-cellular environment where efficient interference management is essential to ensuring the best possible communication between adjacent cells. Starting with SINR, a crucial indicator of signal quality in the presence of interference and noise, the Proposed approach stands out as the best option. The Proposed approach stands out because it successfully reduces interference and noise, resulting in better signal quality and a SINR of 17.2 dB. For these scenarios, TBB comes in second with a SINR of 15.6 dB, followed by DB and ARB, which have lower SINR values of 14.3 dB and 13.8 dB, respectively, for these scenarios.

One more time, the strength of the Proposed approach is highlighted by CC, which defines the highest data rate that can be transmitted via the channel. It surpasses the capabilities of DB, TBB and ARB with a CC of 35.6 Mbps. In these scenarios, TBB and DB have capabilities of 32.2 Mbps and 30.0 Mbps, respectively, while ARB falls behind with a capacity of 28.9 Mbps. The actual data transfer rate achieved, or throughput, is a vital parameter that confirms the superiority of the proposed technology. The throughput it achieves, 24.7 Mbps, surpasses that of DB, TBB and ARB. ARB comes in second with a throughput of 20.1 Mbps, while TBB and DB have slower throughputs in these scenarios, at 22.4 Mbps and 20.9 Mbps, respectively, for these scenarios. User Fairness, a vital component of guaranteeing the equitable distribution of resources among users, is also a focal point of thought. With a U score of 0.80, the Proposed approach stands out as having a well-balanced resource distribution. ARB comes in second with a fairness value of 0.72, while TBB and DB have somewhat lower fairness values for similar scenarios, 0.67 and 0.64, respectively, for these scenarios.

4. Conclusion and Future Scope

This study set out on a transformative path to solve the issues of improving SINR, CC, throughput and U in the context of Massive MIMO systems in the ever-evolving landscape of wireless communication demands. An innovative and thorough approach was developed in order to achieve this goal, completely altering the traditional paradigms of beamforming procedures. MRT and RZF beamforming were combined, along with the clever incorporation of the EHO algorithm, to produce outstanding results that outperformed the standards set by cutting-edge methods. The proposed solution shown unmatched improvements after a thorough analysis of numerous scenarios, illuminating an intriguing path for more effective and fair wireless networks. The Proposed method's remarkable resource management and commitment to interference control are evidence of its adaptability. The algorithm's resistance to frequency-selective fading effects was highlighted in frequency selective channels, which marked the culmination of this transformative journey. By strategically combining their resources, MRT, RZF and EHO overcame these obstacles and provided great signal quality, coverage and capacity. The Proposed approach remained stable, setting new standards, throughout the crowded spheres of urban landscapes, the expanses of suburban and multi-cell domains, through the dynamic fluctuations of user mobility and the complexities of various user distributions.

References

[1] Almradi, A., Xiao, P., and Hamdi, K. A. (2018). Hop-by-Hop ZF beamforming for MIMO full)-duplex relaying with co-channel interference. *IEEE Transactions on Communications*, 66(12), 6135–6149. https://doi.org/10.1109/TCOMM.2018.2863723.

[2] Cavalcante, E., Fodor, G., Silva, Y., and Freitas, W. (2018). Distributed beamforming in dynamic TDD MIMO networks with BS to BS interference constraints. *IEEE Wireless Communications Letters*, 7(5), 788–791. https://doi.org/10.1109/LWC.2018.2825330.

[3] Chae, C., and Heath, R.W. (2009). On the optimality of linear multiuser MIMO beamforming for a two-user two-input multiple-output broadcast system. *IEEE Signal Processing Letters*, 16(2), 117–120. https://doi.org/10.1109/LSP.2008.2008937.

[4] Chun, Y. J., and Kim, S. W. (2008). Log-likelihood-ratio ordered successive interference cancellation in multi-user, multi-mode MIMO systems. *IEEE Communications Letters*, 12(11), 837–839. https://doi.org/10.1109/LCOMM.2008.080986.

[5] Geng, Z., Deng H., and Himed, B. (2015). Adaptive radar beamforming for interference mitigation in radar-wireless spectrum sharing. *IEEE Signal Processing Letters*, 22(4), 484-488.

[6] Le, A. T., Huang, X., and Guo, Y. J. (2021). Analog self-interference cancellation in dual-polarization full-duplex MIMO systems. *IEEE Communications Letters*, 25(9), 3075–3079.

[7] Li, C., He, C., Jiang, L., and Liu, F. (2016). Robust beamforming design for max–min sinr in mimo interference channels. *IEEE Communications Letters*, 20(4), 724–727. https://doi.org/10.1109/LCOMM.2016.2522430.

[8] Liu, X., Huang, T., Shlezinger, N., Liu, Y., Zhou, L., and Eldar, Y. (2020). Joint transmit beamforming for multiuser MIMO communications and MIMO radar. *IEEE Transactions on Signal Processing*, 68, 3929–3944. https://doi.org/10.1109/TVT.2014.2305849.

[9] Mosleh, S., Abouei, J., and Aghabozorgi, M. R. (2014). Distributed Opportunistic Interference Alignment Using Threshold-Based Beamforming in MIMO Overlay Cognitive Radio. *IEEE Transactions on Vehicular Technology*, 63(8), 3783–3793. https://doi.org/10.1109/LCOMM.2019.2940574.

[10] MIMO radar and MIMO communication systems: an interference alignment approach. *IEEE Transactions on Vehicular Technology*, 67(12), 11667–11680.

[11] OFDM interference channels with multipath diversity. *IEEE Transactions on Wireless Communications*, 14(3), 1213–1225. https://doi.org/10.1109/TWC.2014.2365464.

[12] Rihan M., and Huang, L. (2018). Optimum co-design of spectrum sharing between MIMO Radar and MIMO communication systems: an interference alignment approach. *IEEE Transactions on Vehicular Technology*, 67(12), 11667-11680.

[13] Toutounchian M. A., and Vaughan, R.G. (2015). Beamforming for multiuser MIMO-OFDM interference channels with multipath diversity. *IEEE*

Transactions on Wireless Communications, 14(3),1213-1225. 10.1109/TWC.2014.2365464.

[14] Zhang, H., Dai, L., and Li, Z. (2019). Pricing-based semi-distributed clustering and beamforming for user-centric MIMO networks. *IEEE Communications Letters*, 23(12), 398–2401.

[15] Zhu Y., and Guo, D. (2012). The degrees of freedom of isotropic MIMO interference channels without state information at the transmitters. *IEEE Transactions on Information Theory*, 58(1), 341–352. https://doi.org/10.1109/TIT.2011.2167314.

B. Advances in Computer Engineering and Information Technology

Chapter 11

"HANA" – An IOT Fusion With Deep Learning for Smart Agriculture Crop Prediction and Crop Monitoring System

G. Revathy[1], *S. Senthilvadivu*[2], *S. Russia*[3], *and M. Dhipa*[4]

[1]Assistant Professor, Department of CSE, SRC, SASTRA Deemed University, Kumbakonam

[2]Assistant Professor (SG), Department of Applied machine learning, Saveetha Institute of Medical And Technical Sciences, Chennai

[3]Professor, Department of CSE, Vellalar College of Engineering and Technology, Erode

[4]Associate Professor, Department of Biomedical Engineering, Nandha Engineering College, Erode

Abstract: The Internet of Things (IoT) has revolutionised numerous industries, and its impact has reached far and wide. From healthcare to manufacturing, the IoT has been leveraged to create more efficient, connected systems that can provide real-time data and insights. However, the agricultural sector is one area that has perhaps seen some of the most significant impacts from the IoT. In recent years, farmers have begun to adopt IoT technology for smart farming, which involves collecting critical environmental and crop data through sensors and using this information to make informed decisions that can ultimately lead to higher yields and better overall outcomes. This technology has the potential to revolutionise the way we approach agriculture, making it more sustainable, efficient and environmentally friendly. By using IoT-connected sensors, farmers can monitor a variety of environmental factors that can impact crop growth and health. This includes soil moisture, temperature, humidity and more. This data can then be analysed in real time to identify patterns and trends, allowing farmers to adjust their farming practices accordingly. For example, if a sensor detects that a particular crop is receiving too much or too little water, farmers can adjust their irrigation system to ensure optimal conditions for growth. Moreover, IoT technology can also be used to track crop growth and development, identifying potential issues before they become major problems. By monitoring things like plant height, leaf colour and other indicators, farmers can quickly detect signs of disease or pests and take action before it's too late. This not only helps protect crops but also reduces the need for harmful pesticides and other chemicals that can damage the environment. In short, the IoT has brought about

DOI: 10.1201/9781003527442-11

a new era of innovation and sustainability in the agricultural sector. As the world's population grows and the demand for food increases, smart farming technologies will become increasingly critical in ensuring that we can produce enough food to feed everyone while also protecting the environment. The proposed system offers an innovative solution to help farmers increase their crop yields by harnessing the power of the IoT devices and applying cutting-edge deep learning techniques. These IoT devices collect crucial field information, including ultraviolet range, humidity, temperature, light intensity and soil moisture. This data is then fed into a sophisticated deep learning algorithm, which can analyses and interpret it in real time, providing valuable insights into the optimal growing conditions for various crops. By utilising this data-driven approach, farmers can make informed decisions about when to plant, water and harvest their crops, resulting in higher yields and increased profitability. Additionally, the system can also help farmers identify potential issues with their crops early on, such as diseases or pests, allowing them to take prompt action to mitigate any damage. Overall, this proposed system represents a significant advancement in agricultural technology, offering farmers a powerful tool to help them optimise their cultivation practices and achieve more efficient and sustainable crop growth. With its ability to collect, process and analyse vast amounts of data from IoT devices, coupled with sophisticated deep learning algorithms, this system has tremendous potential to transform the way agriculture is practiced and revolutionise the industry as a whole.

Keywords: Internet of Things (IoT), smart agriculture, control system, deep learning, CNN, ResNet50.

1. Introduction

According to World Bank data, agriculture is the major way of alleviating poverty and enhancing food security for the world's 80% of underprivileged people who live in rural regions. Nevertheless, agriculture accounts for around 25% of gross domestic product in certain emerging nations by agriculture (Agriculture overview, 2021). Climate change has exacerbated droughts, floods, irregular precipitation patterns and heat waves worldwide. Furthermore, due to the high pace of population expansion and the current climate-related consequences on agricultural lands, agricultural regions are under increased pressure to supply adequate food by Kopittke. (2019). To meet rising food demand, existing agricultural land must be used efficiently to produce sustainable and healthful crops. Traditional intensive agriculture techniques, on the other hand, degrade the soil and result in poorer output by Arora (2021). Additionally, farmers are uninformed of real-time climatic changes and market variations, resulting in an inadequate supply of essential crops. The cost of production per yield is rising as a result of poor resource usage by Alengebawy (2021). Fertilisers and insecticides, for example, are widely utilised without any specific necessity.

New ideas and technological developments are constantly being proposed and implemented in order to meet the needs of mankind in general. As a result of this, the IoT was born by Yang. (2019). IoT is defined as a networking that includes all things that are already incorporated into gadgets, sensors, computers, software and employees that communicate, exchange knowledge and collaborate in order to provide a flawless outcome between the physical and digital worlds. Many international organisations have defined IoT in a number of ways. IoT is defined by ITU-T (International Telecommunication Union Telecommunication Standardization Sector) as "global infrastructure for knowledge society, enabling innovative services based on information obtained (physical and virtual) small details based on existing but instead evolving, highly integrated information and communications technologies." IoT is a technology that uses the internet and pre-existing resources to control objects.

2. Methodology

2.1. IOT Sensors for Smart Agriculture

There are multiple sensors connected to the soli such as soil moisture sensor to gather the moisture of the soil, air humidity sensor to collect the air humidity towards soil, temperature sensor to analyse the temperature, UV sensor for analysing the distance analysed, NPK sensor for monitoring the nutrients in the soil, pH sensor for analysing pH and insect sensor for analysing sensor.

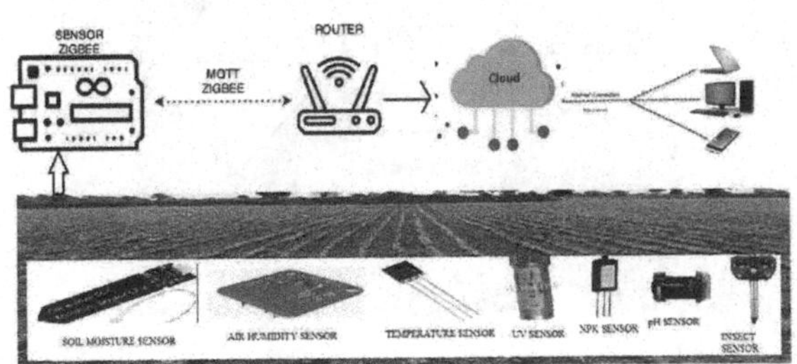

Figure 1: *Sample sensors connected with cloud and IoT gateways.*

Figure 1 based on the all given sensor inputs are collected in the cloud server. The cloud server is connected to IoT gateways. The data stored in cloud is passed to CNN for the suitable crop prediction for that soil based on the sensor inputs such as soil moisture, air humidity, temperature, NPK and ph. The UV and insect sensor are to alert the farmer regarding the insects and reptiles.

2.2. Crop Prediction with Deep Learning

Figure 2 describes the sensor input data collected from IoT sensors is passed to convolution layers and pooling layers and the crops appropriate for the soil are predicted.

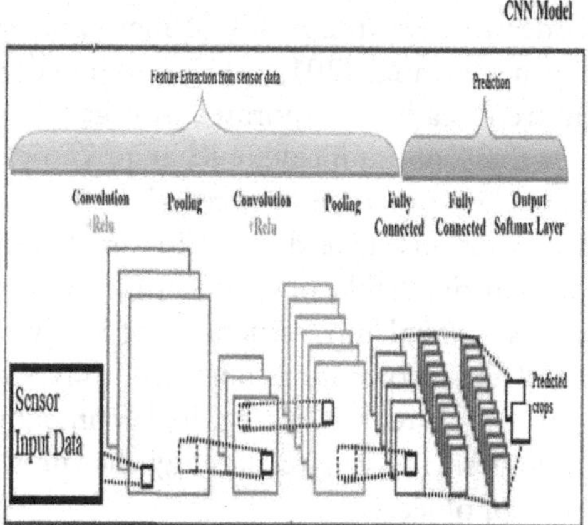

Figure 2: CNN model architecture.

2 Convolutions layers and one activation function (Relu) followed by pooling layers and again convolution layers and pooling layers and then thick fully connected three layers and SoftMax layer which gives the suitable crop for the soil.

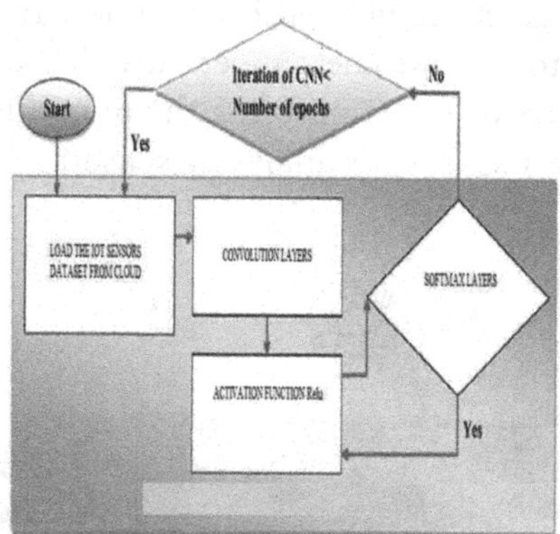

Figure 3: Crop prediction with CNN.

Figure 3 explains the steps of crop prediction with CNN. The CNN is a fully connected thick layers with activation function and SoftMax layers. The IoT sensors data is passed to convolution layers and then Relu activation function is applied again to thick convolution layers and processed to SoftMax layers and the final crop prediction is done.

2.3. *Crop Monitoring with Deep Learning*

The crop monitoring is done every stage with the camera which is available in the soil. The images of the crops are continuously given to RESNET50 and the output is monitored. If there any crop disease then the it is been immediately alerted and the details are shared with the IoT gateways and the recovery is done.

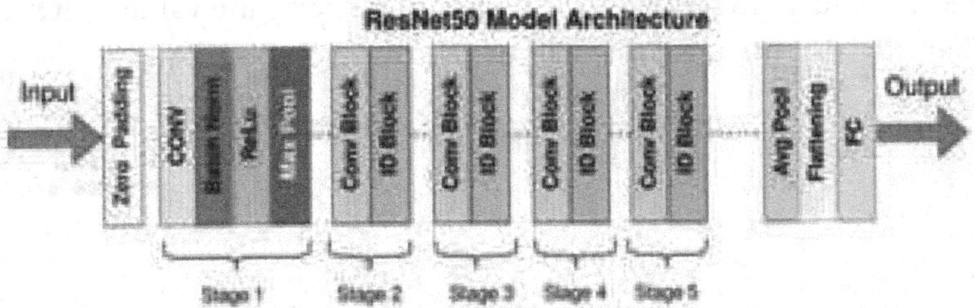

Figure 4: RESNET 50 Architecture.

The Figure 4 describes the RESNET 50 architecture. The architecture starts with padding layers followed by five stages and each stage is composed of a convolution layers followed by batch normalisation, activation function Relu and a max pooling layer. After five stages the average pooling layer is been architected and then flattening layer and a FC and finally we get the output.

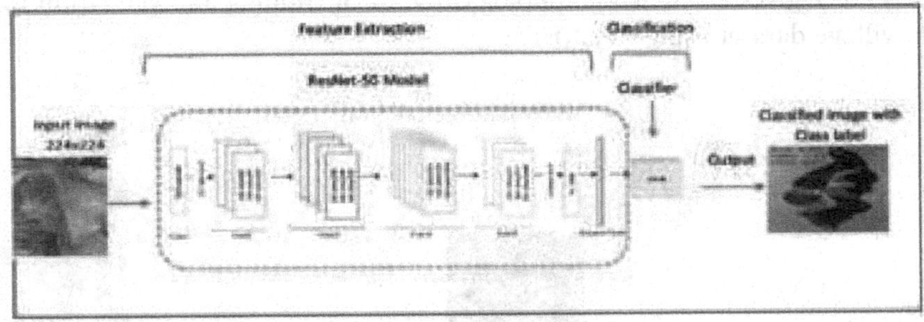

Figure 5: Crop monitoring with RESNET50.

The RESNET 50 model is trained with Plant village dataset. Figure 4 illustrates the working procedure for RESNET50 model, an image is provided every time during the growth of the crop and if healthy it monitors the next phase and if found to be unhealthy it immediately alerts the farmer using IoT platform.

3. Results and Discussions

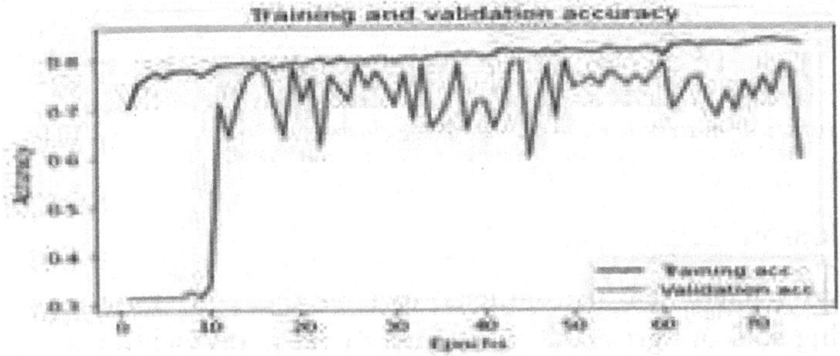

Figure 6: Training and validation accuracy.

Figure 6 gives a detailed graphological view of training and validation accuracy of Plant village dataset using RESNET50.

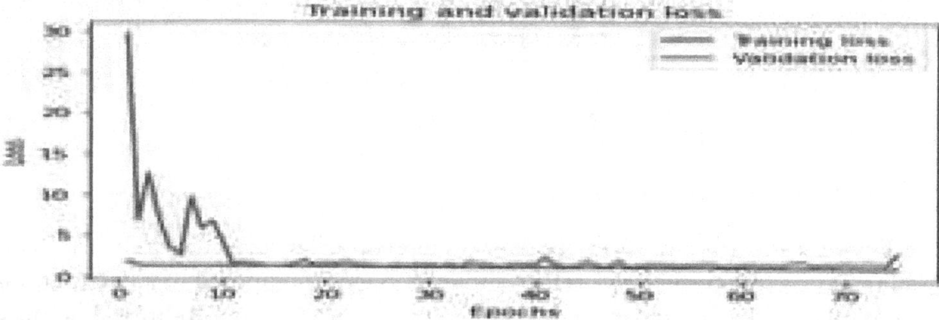

Figure 7: Training and validation loss.

Figure 7 gives a detailed graphological view of training and validation loss of Plant village dataset using RESNET50.

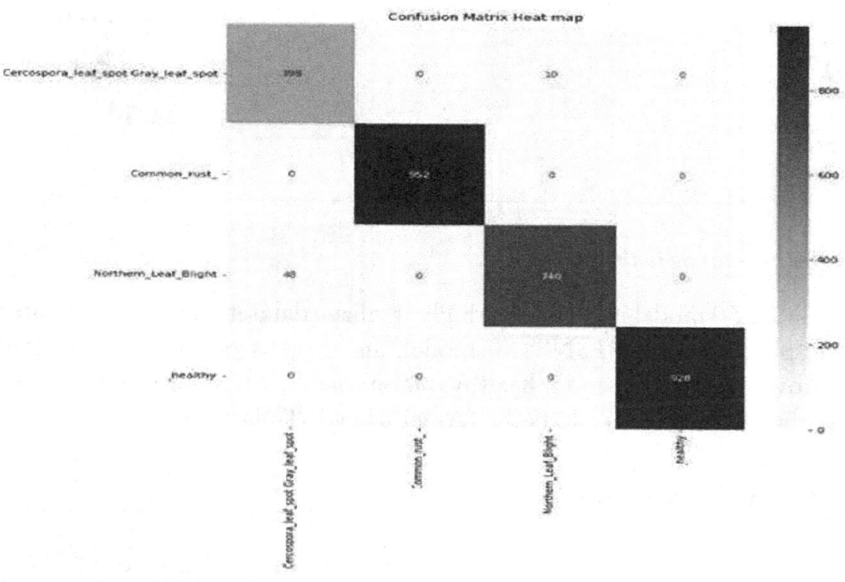

Figure 8: Confusion matrix for the plant village dataset.

4. Conclusion

Using IoT-connected sensors, farmers can predict the exact crop for that corresponding soil and grow the exact crop to get good yield. Although the crop is predicted, farmers need to monitor a variety of environmental factors that can impact crop growth and health and hence a camera sensor is provided which is in turn connected with deep learning model for the crop growth each and every time and hence if any small unhealthy factor affects the crop it can be immediately rectified. The crop is monitored till the final yields and then the yields are also increased.

References

[1] Agriculture overview. (Accessed Aug. 15, 2021).

[2] Alengebawy, A., Abdelkhalek, S. T., Qureshi, S. R., and Wang, M. Q. (2021). Heavy metals and pesticides toxicity in agricultural soil and plants: ecological risks and human health implications. Toxics, 9(3), 42. doi: 10.3390/toxics9030042.

[3] Arora, N. K. (2021). Impact of climate change on agriculture production and its sustainable solutions. Environmental Sustainability, 2.

[4] Kopittke, P. M., Menzies, Wang, P., McKenna, and Lombi, E. (2019). Soil and the intensification of agriculture for global food security. Environment International, 132. https://doi.org/10.1016/j.envint.2019.105078.

[5] Yang, Sharma, A., and Kumar, R. (2021). Iot-based framework for smart agriculture. International Journal of Agricultural and Environmental

[6] Information Systems (IJAEIS), 12(2), 1–14.

[7] PlantVillage (2019). Dataset. https://www.kaggle.com/datasets/emmarex/plantdisease

Chapter 12

A Survey Paper on Video Retrieval Using Deep Learning Concepts

Shubhangini Ugale[1], *Wani Patil*[2], *and Vivek Kapur*[3]

[1]Research Scholor, G.H. Raisoni University, Anjangaon Bari Road, Maharashtra, India

[2]Assistant Professor, G.H. Raisoni University, Anjangaon Bari Road, Maharashtra, India

[3]Associate Professor, G.H. Raisoni Institute of Engineering and Technology, Shraddha Part, MIDC, Hingna, Nagpur, India

Abstract: Advancement in video and image processing increasing day-to-day life. To keep track on image and video is major challenge nowadays to overcome this deep learning is used. This survey paper which propose on concept of CNN, different networks for CNN, datasets and parameters used by researcher. Video is retrieved using feature extraction and classification for machine learning which overcome by deep learning using CNN. Main approach is to help researcher to introduce new networks of deep learning and aware them with latest techniques, networks and sources of different datasets. A deep learning is the best solution over existing methods to retrieve video or images as users need. In future scope researcher will work on to survey different technique to reduce time for train the dataset by using deep learning techniques.

Keywords: Convolution Neural Network (CNN), Content-based video retrieval (CBVR), CNN models.

1. Introduction

Utilisation of the internet has increased in daily life. Using a laptop, desktop or mobile device, it is simple for user to download, share and save photographs or videos. The memory storage capacity of electronics devices is a drawback for all gadgets. When storage space is limited, it can be challenging for users to keep track of which photos or videos should be delete and which should be save. Video is compose of different images and uses in surveillance, communication, medical and other areas. Text-based and content-based are two ways to retrieve video. Content based is more accurate compared to text based. The main need of retrieval is to analyse the particular moment such as accident on road, diagnosis of patient using sonography or x-ray and many more. Deep learning is used by Google Chrome, Facebook and YouTube to meet customer demands. In deep learning CNN model

DOI: 10.1201/9781003527442-12

works on different networks such as LeNet, AlexNet, VGG 16 Net and others. Every network contains many layers. Therefore, it can be challenging to search a query image or video in a large database. Deep learning is the ideal way to find the right image or video for the user (Kurian and Jeyakumar, 2020). If a crime occur at anywhere to search an exact clip we have to do manual process which is time consuming. Haar-cascade algorithm use to capture a body movements (Manju and Valarmathie, 2020). Key frame extraction algorithm use for clustering and redundant bits information is reduce which increases accuracy and improve efficiency of video retrieval (Zhang *et al.*, 2020). Video text pairs by pseudo ground-truth multilingual constructed and uses in state-of-the art machine translation models (Madasu *et al.*, 2023). Deep Hashing combines the features and enhances the quality of image retrieval (Li *et al.*, 2023). Monitoring smart city crowd and traffic control possible using video retrieval (Mounika *et al.*, 2023) Modified visual geometry Group_16 model is compared with existing methods, other feature extraction techniques are CNN networks, local binary patterns and histogram of oriented gradient. Video is retrieved using hashing technique, LSTM and different CCN models are used. Video information is combination of text, captions, collection of frames and audio.

Videos saving consume more memory space in computer and mobile to select and search a particular query; video is time-consuming challenges. It is essential need of big video data to be manage and organise intelligent way for video analysis. Researcher work on machine learning and CNN to identify result like query frames. In video retrieval using content-based identification of similar query frames based on similarity distance measures and convolution neural network where different hidden layers are used. Minimum features of image such as shape, colour and boundary are used to identify image but for video, high-level features are used. Query frames retrieval is an important method in the development of frame combination search engines and the retrieval of a relevant set of videos from a dataset. To extract the visual elements of a shot, an integrated video retrieval system was designed, in which the frames shot is not identified by a key frame but by all frames. This paper organised in different sections, its start with introduction, background covers in the second section, third section gives brief information about video processing and challenges and issues, forth section gives architecture of CNN, Neural network concept has given in fifth section and last sixth section concluded with conclusion.

2. Background

Video retrieval save a time by extracting features from large database and identify same query video using deep learning. Deep learning consists of three main layers input layer, hidden layer and output layer. Where input and output are both video. Deep learning concept works similar like neurons.

Neurons that function like those in the human brain are the foundation of machine learning and deep learning. CCN is a sort of neuron network that aids in the classification and recognition of images and videos. CNN utilises a convolutional filter to simplify the network. Convolution is the primary layer in CNN, and it serves as a feature extractor or preprocessor (Kurian and Jeyakumar, 2020). You can search for video using an image, text or video query. Today, clustering is a technique where video is divided into cluster frames and individual frames are recovered from the cluster frames. Each frame in a video is related to the frames around it (Zhang *et al.*, 2020). Eigen-faces, histogram of gradient and haar are image processing technique, K-Means, K-Nearest Neighbour, support vector machine and Naïve Bayes are used for classification and clustering algorithm (Iqbal *et al.*, 2020). VGGFace, ArcFace and FaceNet CNN model are used for face recognition (Cheng *et al.*, 2020). CNN with LSTM used for learning with hash code to implement on image and video for information retrieval (Wang *et al.*, 2019). Distance measure metric is used to match object to object and features are extracted using histogram. Now to retrieve video state of art depends on deep learning approaches (Vesely and Peska, 2023).

Video is retrieved using key frame extraction methods and salient object detection, it reduce high dimensionality of video data (Sowmyayani and Rani, 2023). Kumar and Seetharaman (2022) proposed data of eye movement and head movement were collected. Optical flow algorithm is used to develop a region of interest which calculate texture, motion and depth of field. VR is same as cognitive psychology experiments.

Unsupervised optimisation is now possible in a range of computer vision applications. These deep learning models' performance is heavily influenced by the internal layer architecture, the kind of datasets employed and the activation levels of the augmentation layers. Researchers have developed several deep learning models based on CNN in order to increase this performance for content-based video retrieval (CBVR) applications. When used on untrained query video sequences, these models, which are trained on application-specific datasets, have a limited ability to scale. In these circumstances, their performance in terms of recall and precision is similarly restricted. Many perceptrons work together to build nonlinear decision boundaries, a bigger network is known as a multilayer perceptron. Deep learning can perform a variety of operations for video analysis, including object tracking, motion estimation and background separation. Scaling denotes the preprocessing of a trained image. Pre-trained CNN is used to extract high-level features, while CNN with RBF loss is used to achieve dimension reduction. Each dimension of the real-time value representation is quantised into a 1-byte integer, and then related videos are identified using similar distance metrics (Choi and Kil, 2021). Deep convolution neural systems can be used to observe user mood and temporal preferences. Reinforcement learning agent (RL-agent) captures user input activity. Its properties include the video id, title and description of the video (Fear, Sadness and Joy), and intensity categories (Low, Medium and High). Dot diffused block truncation coding bitmap, minimum and maximum quantizers are used to provide low-level features, and trained CNN is used to produce high-level features. It is harder to train a model in a deep convolution neural network. Existing video retrieval possible on ontological concept- or training-based concept. The LBP-TOP features are used which is invariant to illumination, rotation and local translation (Mounika *et al.*, 2023). In order to retrieve videos, several researchers use deep convolution neural networks, deep convolution neural networks with long short-term memories, and hashing methods that use long short-term memories.

3. Video Processing

To retrieve video edge base algorithm and motion-based algorithm are exist. Video analysis is done using video shot boundary detection criteria. Using these criteria different processes such as retrieving, browsing of video, indexing of video, video categorisation and summarisation are possible. The video's shot-by-shot analysis shows examples of both abrupt transition (AT) and progressive transition (GT). A new pyramidal opponent color-shape (POCS) model is utilised to detect abrupt transitions (AT) and gradual transitions (GT) in the presence of fast movement and illumination changes (Sasithradevi and Roomi, 2020). With use of tremendous data proper analysis and storage, organising and management of data is essential. This require researcher to work on video and find different technologies to manage data. Time- and space-based frames make up a video shot. The two types of video shot transitions are abrupt (hard) and gradual (dissolve, fade in, fade out, wipe). Low-level features, such as color, texture, form and others, are used in the threshold basis technique or support vector machine to find VSBD.

Different levels of content representation complexity and discontinuity in illumination. In VSBD approaches, the existence of illumination discontinuity in video is a crucial concern. Detecting gradual transitions photos are taken by the camera in the first step of creating a video, and the second step merges the shots either suddenly or gradually. Effects spread out gradually along the temporal axis. The range of the temporal dispersion period is 3–100 units.

Three steps are required to divide a video into shots. Decision, representation and mapping. Abrupt changes can be found using pixel-based approach by adding pixel and finding boundaries based on threshold value. Pixel-based algorithm, Histogram-based algorithm. The colour histogram allows for identical colour histogram graphs between frames in the same shot. To evaluate similarity, a number of metrics can be utilised, including histogram intersection, histogram difference, chi-squared distance and cosine measure. Histogram intersection is one of them that the VSBD process can use. Histograms are reliable when it comes to spatial frames and object motion against a constant background. A sequential similarity scheme and a conceptual similarity are aspects of mapping for characterisation, machine learning classifiers and threshold classifiers.

In machine learning classifiers support vector machine and neural networks are used (Afrashiabi *et al.*, 2020). Different approaches like segmentation, dimensionality reduction, feature extraction and machine learning approaches are used for content-based video retrieval and indexing. Video is capable to representing moving object in space and time. Information in video consist of audio, textual metadata, frames and captions. The feature construction method known as principal components analysis (PCA) converts one dimension into another. Remove any unnecessary or superfluous features from the video data before choosing the remaining ones (Anuranji and Srimathi, 2020).

A variety of neural network techniques, include multi-layer perceptrons, recurrent neural networks, long short-term memory networks and CNN. In content-based video retrieval content means shape, colour, shape and texture. Audio and video module consider to retrieve desire video. First, video frames are extracted using shot detection; next, relevant frames are found; and last, features are extracted. The feature data from both modules is combined to produce a feature database. The kernelized Fuzzy C Means (KFCM) technique is used to cluster features, enabling quicker video retrieval. To determine the shortest route between a query and a database video, Euclidian distance is used. Videos are the most popular type of material since they are easy to understand and efficient at spreading knowledge. The relevant frame from the photos is chosen using the Lion Optimisation Algorithm (LOA). K-Nearest Neighbor classifier is used to retrieve videos (k-NN). While recall determines the retrieval system's efficacy, precision assesses the accuracy of the video retrieval. The distance between a query and a features database is determined by k-NN classifiers. For audio data. Hashing is use to generate a binary code and map this code with database binary code with shortest distance measure (Anuranji and Srimathi, 2020) Feature selection: The process of identifying significant features (Features) in a dataset are location of edges, corner or ridges. OpenCV provides histogram of gradient (HOG), speeded up robust features (SURF) detector identify keypoints in an image, which select feature which is not as per specific task. Binary robust independent elementary features (BRIEF) descriptor calculate actual features value for all the keypoints. and Oriented FAST and Rotated BRIEF(ORB) OpenCV provides two algorithm for corner detection they are Harris Corner Detection and Shi-Tomasi Corner Detection.

The feature detector identifies the image's keypoints, and the feature descriptor determines the actual feature value for each keypoint using the SURF detector and BRIEF descriptor and SVM-Support vector machine.

High-efficiency video coding (HEVC) minimises the size of the transmitted file. The most recent coding standard developed by the joint working team on video coding is called HEVC (Galiano *et al.*, 2020). Deep convolutional neural network mostly use nowadays due to huge dataset access every day on internet access. CNN consists of input layer, convolution layer, hidden layer and output layer. Deep convolution neural network gives better result as compared to machine learning. In deep leering huge amount of data is trained feature is extracted by applying different kernels on query image which is extracted by query video. Dimension is reduced using max, min or average pooling. Fully connected layer converts input data into one-dimensional array and lastly softmax layer gives output is in the form of query video.

Ensemble learning uses different CNN model VGGNet-16, DenseNet-121, Inception ResNet-V2, Mobile V Net, Res Net 101 and Xception Net.

According to literature survey, basic steps for retrieval query video are as follows:

A) Select database
B) Select query video from same database
C) All database video should be trained. Machine learn each video in training process.
D) Features are extracted from query and database video.
E) Features are mapped based on similar distance metric and features map
F) If feature map relevant video is retrieved.

3.1. *Challenges and Issues*

To retrieve video different challenges are colour, feature, occlusion, setting of camera, robustness, quality of image and video are as follows

Background complexity: If its lack of light it is difficult to get a proper image so which effect on to separate object from background is serious problem and if background object and foreground is same then user cannot determine the object properly.

Target feature selection: Video is in moving state so object is also changes as time being. To select exact target is challenging issue. Video consist lots of information such as motion, text, edge, colour and texture. To select proper feature is challenging task.

Occlusion: Moving object is stop suddenly or many objects occlude each other it is difficult to get position of target with relevant frames.

The balance between real time processing and robustness: To track the object in real-time processing is also a challenging and robustness problem under complex background is also consider a challenging issue.

Facebook, Instagram, YouTube and Twitter are the media for communication; from all this YouTube is majorly use for video processing (Choi *et al.*, 2021).

4. Architecture of CNN

Deep CNN used mostly for analysis visual task. Deep models with parameters heavy architecture have been successfully trained and deployed on a many application and this is successfully done using graphical processing unit. Neural network architecture also includes deep learning architecture. Most of the researcher nowadays used CNN architecture for image classification and recognition. The input layer, hidden layer and output layer are the three main layers of the CNN architecture. Convolution layer, pooling layer and fully linked layer are just a few examples of the various sub layers that make up the hidden layer.

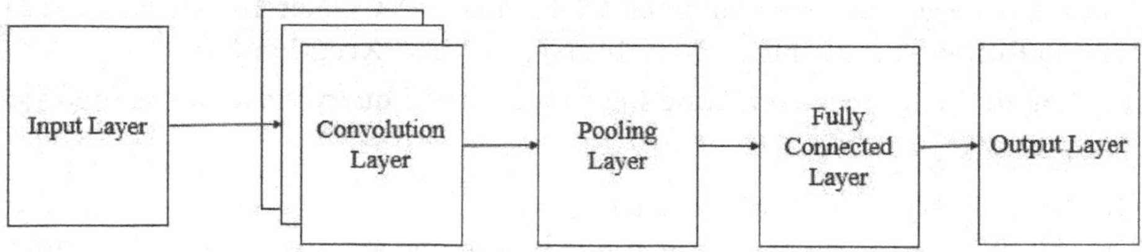

Figure 1: Basic architecture of CNN.

Input Layer: Input layer is query image which having width height and dimension extracted from video.

Convolutional layer: Input layer is convolved with kernel this perform operation such as sobel filter or apply Gaussian kernel in 1D, 2D or even 3D. It consists of no. of filters; this filter moves over image and calculates local features by product and sum of weight of filters and intensity of input image. Features are calculated by convolving filter weight with input image using dot product.

Filter are calculated using the formula

$$\sum_{k=0}^{n}(\textit{filter weight} * \textit{input image}) + \textit{bias}$$

$$\sum_{k=0}^{n}(\textit{filter weight} * \textit{input image}) + \textit{bias}$$

Pooling layers: The activity of the output neuron is determined by the maximum or averagely active input neuron when max or average pooling is used on the input. It remove unwanted information and reduce dimension and minimise no. of training parameters of the network. For CNN, architecture max and average pooling are mostly used.

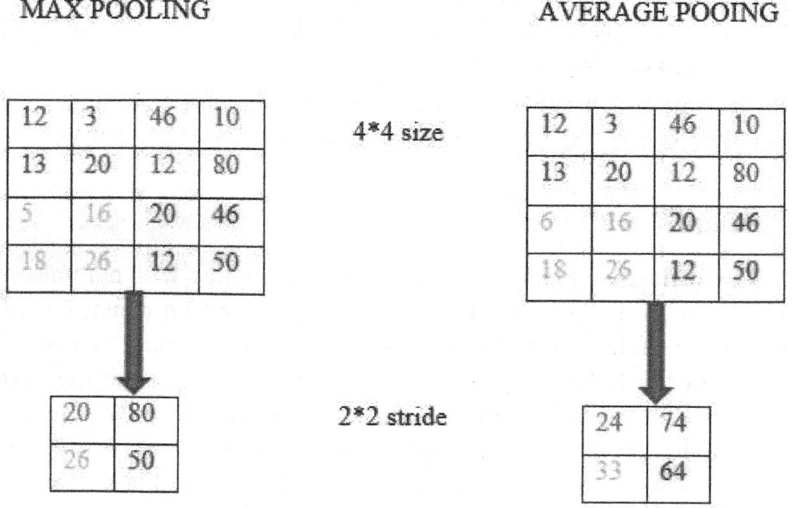

Figure 2: Examples of maximum and average pooling.

In the pooling process, layer stride 2 is used to reduce the sampled features map's height and width. The activation function in ANNs is utilised to account for CNN nonlinearity. The rectified linear unit layer or ReLu layer is quicker than the activation function. ReLu function is also used in back propagation and hidden layer of CNN. Euclidian distance is used to find the similar features (image).

Activation Function: It uses the Sigmoid, Tanh and Relu functions. Deep learning typically uses the relu function it is faster as compared to sigmoid and Tanh function.

Fully connected layer: All neuron connected to activation from the previous layer and gives output in one byte integer. Fully connected layer is also called as hidden layer.

Output Layer: CNN is special architecture designed for processing of images and video. Output layer is used for classification by softmax activation function.

The main success of deep learning today due to following reasons:

1. Large amount of train data provide resources to deep learning.
2. Development of graphical processing unit hardware make possible to train large-scale neural network.

5. Neural Networks

Neural Networks is same like biological brain which consist of cell, dendrites and axons. Cell is central and main region of body, dendrites receive signal from other neuron and transfer that signal into electrical energy and provide to cells. Axons is a single branch which conduct nerve away from cell.

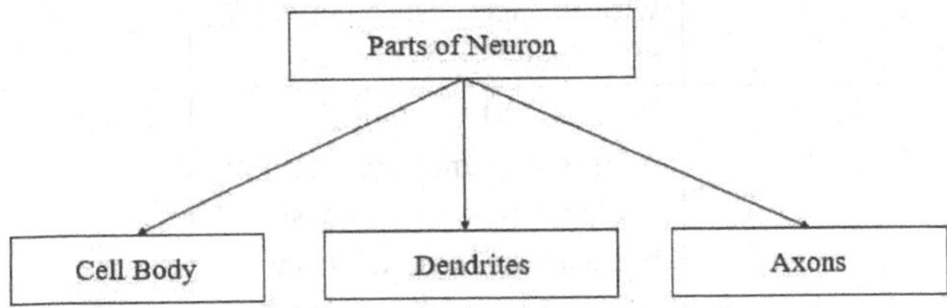

Figure 3: Parts of neuron.

Artificial neural network is a deep neural network which work on different model, train dataset is classified using different model. A deep neural network consists of different layer and every layers depends on one another to predict a result.

Neural network mostly based on back propagation algorithm to implement gradient calculations. Two main algorithms are used for gradient calculation are as follows:

A) Stochastic Gradient Decent(SGD)
B) Adaptive Moment Estimation(Adam)

The deep neural network module includes capabilities for loading serialised network models from several deep learning frameworks, as well as APIs for creating new layers, sets of layers and neural networks from layers.

A. *CNN MODELS*

Three types of CNN model are shown in Table 1 below. Table provides details about model, stride, main contribution of model and its drawback.

Table 1: CNN models.

Model	**Stride**	**Main contribution**	**Drawbacks**
VGGNet-16 (Proposed by Visual Geometry Group)	3*3	it demonstrates how important network depth is for improving recognition or classification accuracy in convolutional neural networks VGGNet is considered a good architecture for benchmarking on a particular task.	Main disadvantages are that it is very slow to train and its network architecture weights are quite large (533 MB for VGG-16 and 574 MB for VGG-19)
Inception V3	1*1 convolution with 16, 64 and 96 filters.	The Inception V3 architecture includes additional factorisation ideas whose objective is to reduce the number of connections/parameters without decreasing network efficiency	Four branches output are concatenated
ResNet 101 (Residual Network)	101 layer	ResNet-101 uses 44.5 million parameters. Res Net is the deepest network so far, with 153convolution layers. This architecture includes skip connections, also known as gate units.	Depth is increases.

Face recognition is a key component of video surveillance (Zhang *et al.*, 2020). Machine learning techniques combine PCA, linear discriminant analysis (LDA) and SVM with edges and texture descriptors. CNN models are used for object recognition. The result of video retrieval is consecutive sequence of frames. Convolution layer followed by rectilinear layer. Five pooling layers at most are employed to reduce input volume size. Three layers are completely interconnected. The softmax layer is the last layer.

Table 2: Different networks and dataset used by researcher.

Technology	Paper Title and Reference No.	Dataset	Network	Parameters
CNN	Multimodality medical image retrieval using convolutional neural network [1]	Medical Dataset	LetNet and AlexNet	Mean average precison: 86.9% for AlexNet and 83.9% for LeNet
CNN	Large-ScaleVideoRetrieval via deep local convolutional features[3]	ILS VRC dataset, NTU Video retrieval test collection	VGG 16	Accuracy: 96.91% for NTU dataset, 92.13% for 2000 TREC dataset
CNN	Content-Based Video retrieval using convolutional neural network[4]	YouTube and SegTrackdatase	K nearest Neighbour, K means, Support Vector achine.	Confusion matrix is given for testing and training using different algorithm Naïve Bayes, KNN, K Means, SVM. Accuracy, time and space parameters shown graphically.
CNN	A cloud-based face video retrieval system with deep learning [5]	MIT CBCL Face recognition database, You Tube face database, Dataset collected from selected video and Google images.	ResNet-50, Inception_ ResNet _v1.	Accuracy and time(ms/image) VGGFace:99.84%, Time: 8.4 ArcFace:99.97%,Time: 7.5 FaceNet: 99.96%, Time: 4.6
CNN	Face Video Retrieval Based on the Deep CNN With RBF Loss[13]	ICT-TV dataset, CIFAR-10, Fashion-MNIST	Le-Net	Comparison of accuracy and retrieval mAPs for softmax, center, Angular Margine Loss and RBF methods are used.

CNN	COLOR STEREO IMAGE RETRIEVAL USING A RESIDUAL CNN-BASED APPROACH[7]	Flying Things 3D database, Tsukuba datset	Image Net, Res Net-50	12 iterations in 84 seconds for the Tsukuba dataset, and 19 iterations in 76 seconds for the FlyingThings3D collection. The similarity measure between a query and a candidate stereo pair is performed in 0.009 second for the ResNet50-2D-LR approach, and 0.011 second for the ResNet50-2D-LRD approach.
CNN, NN with Attention-based video hashing	Attention-based Video Hashing for Large-Scale Video Retrieval [6]	HMDB 51 dataset, FCVID dataset, UCF101 dataset	VGG 19	Precision and mAPs are given for HMDB, FCVID and UCF101 dataset using different methods.

All the videos are train first time but if new video is added then again we have to train all videos in the database which takes a time. It's a drawback for every new video machine has to train all video. For training and testing purpose same dataset is used. CNN with VGG16 model is used mostly by many researcher to retrieve video. While training all the videos in the database frames are extracted from each video and store into database when we select query video each frame for query video compared with trained frame video database if its match similar video is extracted.

6. Conclusion

Deep neural network is useful to classify and recognising the query video. This survey paper helps to understand the researcher about basic of CNN Architecture, neural network, video processing and different sources of dataset. AlexNet and LeNet mostly used for image retrieval whereas for video retrieval VGG-16 model is use. Video retrieval is possible using machine learning and different distance measure metric but deep learning gives accurate result and used for larger dataset. Different challenges and issues are given in this paper. In future researcher will work on to survey different techniques to reduce the time for training the dataset by using deep learning techniques.

References

[1] Afrashiabi, M., Khotanlou, H., and Gevers, T. (2020). Spatial-temporal dual-actor CNN for human interaction prediction in video. *Multimedia Tools and Applications*, 79(3). https://doi.org/10.1007/s11042-020-08845-2.

[2] Anuranji, R., and Srimathi, H. (2020). A supervised deep convolutional based bidirectional long-short term memory video hashing for large scale video retrieval applications. *Digital Signal Processing*, 102, 102729.

[3] Choi, Y. R., and Kil, R. M. (2021). Face video retrieval based on the deep CNN with RBF Loss. *IEEE Transactions on Image Processing*, 30.

[4] Galiano, D. R., Barrio, A. D., Botella, G., and Cuest, D. (2020). Efficient embedding and retrieval of information for high-resolution videos coded with HEVC. *Journal of Computers and Electrical Engineering*, 81(12), 106541.

[5] Ghodhbani, E., Kaaniche, M., Benyahia, A. B. (2020). Color stereo image retrieval using a residual CNN-based approach. *2020 Mediterranean and Middle-East Geoscience and Remote Sensing Symposium*. 978-1-7281-2190-1/20/$31.00@2020IEEE

[6] Iqbal, S., Qureshi, A. N., and Lodhi, A. M. (2019). Content based video retrieval using convolution neural network. Arai, K. *et al.* (eds.), *IntelliSys* 2018, ASIC 868, 170–186. https://doi.org/10.1007/978-3-030-01054-6_12.

[7] Kurian, P., and Jeyakumar, V. (2020). Deep learning techniques for biomedical and health informatics. Chapter 3, 53–95. https://doi.org/10.1016/B978-0-12-819061-6.00003-3, 2020.

[8] Kumar, B. S., and Seetharama, K. (2022). Content based video retrieval using deep learning feature extraction by modified VGG_16. *Journal of Ambient Intelligence and Humanized Computing*, 13, 4235–4247.

[9] Lin, F. C., Ngo, H. H., and Dow, C. R. (2020). A cloud-based face video retrieval system with deep learning. The Journal of Supercomputing, 76(4). (Published online: 01 January 2020). https://doi.org/10.1007/s11227-019-03123-x

[10] Li, D., Dai, D., Chen, J., Xia, S., and Wang, G. (2023). Ensemple learning framework for image retrieval via deep hash ranking. *Elsevier. Knowledge-Based Systems*, 260, 110128. https://doi.org/10.1016/j.knosys.2022.110128

[11] Manju, A., and Valarmathie, P. (2020). Video analytics for semantic substance extraction using OpenCV in python. *Journal of Ambient Intelligent and Humanized Computing*, 12(1). https://doi.org/10.1007/s12652-020-01780-y.

[12] Madasu, A., Aflalo, E., Stan, G. B. M., Tseng, S. Y., Bertasius, G., and Lal, V. (2023). Improving video retrieval using multilingual knowledge transfer. *European Conference on Information Retrieval. ECIR 2023: Advances in Information Retrieval*, 669–684.

[13] Mubarak, A. A., Caol, H., and Ahmed, S. A. M. (2020). Predictive learning analytics using deep learning model in MOOCs' courses videos. *Springer Science+Business Media*, LLC, part of Springer Nature.

[14] Muhammad, K., Ahmad, J., Mehmood, I., Rho, S., and Baik, S. W. (2018). Convolutional neural networks based fire detection in surveillance videos. *IEEE Access*, 6, 18174–18183. https://doi.org/10.1109/ACCESS.2018.2812835.

[15] Mounika, B. R., Palanisamy, P., Sekhar, H. H., and Khare, A. (2023). Content based video retrieval using dynamic textures. *Multimedia Tools and Applications*, 82, 59–90.

[16] Nie, W. Z., Ren, M. R., Liu, A. A., Mao, Z., and Nie, J. (2015). M-GCN: multi-branch graph convolution network for 2D image-based on 3D model retrieval. *Journal of Latex Class Files*, 14(8).

[17] Qiaoa, S., Wanga, R., Shana, S., and Chena, X. (2020). Deep video code for efficient face video retrieval. journal homepage. *Pattern Recognition*,113. https://doi.org/10.1016/j.patcog.2020.107754.

[18] Sasithradevi, A., and Roomi, S. M. M. (2020). A new pyramidal opponent color shape model based video shot boundary detection. *J. Vis. Commun. Image R.*, 67. https://doi.org/10.1016/j.jvcir.2020.102754.

[19] Sharma, A., and Kumar, S. (2023). Machine learning and ontology-based novel semantic document indexing for information retrieval. *Computer & Industrial Engineering*, 176, 108940. https://doi.org/10.1016/j.cie.2022.108940.

[20] Spolaor, N., Lee, H. D., Takaki, W. S. R., Ensiana, L. A., Coy, C. S. R., and Wua, F. C. (2020). A systematic review on content-based video retrieval. *Engineering Applications of Artificial Intelligence*, 90, 103557, 1–16.

[21] Sowmyayani, S., and Rani, P. A. J. (2023). Content based video retrieval system using two stream convolutional neural network. *Multimedia Tools and Applications*, 82, 24465–24483.

[22] Tian, F., Hua, M., and Zhang, W. (2020). Spatio-temporal editing method and application in virtual reality Video. *IEEE 4th Information Technology. Networking, Electronic and Automation Control Conference* (*ITNEC 2020*). https://doi.org/10.1109/ITNEC48623.2020.9085087.

[23] Vesely, P., and Peska, L. (2023). Similarity models for content-based video retrieval. *International Conference on Multimedia Modeling* MMM 2023: MultiMedia Modeling, 54–65.

[24] Vieira, G. S., Fonseca, A. U., and Soares, F. (2023). CBIR-ANR: A content-based image retrieval with accuracy noise reduction. *Elsevier. Software Impacts*, 15, 100486. https://doi.org/10.1016/j.simpa.2023.100486.

[25] Wang, Y., Nie, X., Shi, Y., Zho, X., and Yin. Y. (2019). Attention-based video hashing for large-scale video retrieval. *IEEE Transactions on Cognitive and Developmental Systems*, 99. https://doi.org/10.1109/TCDS.2019.2963339.

[26] Yasin, D., Sohail, A., and Siddiqi, I. (2020). Semantic video retrieval using deep learning techniques. *Proceedings of 2020 17th international Bhurban conference on Applied Sciences and Technology*, Islamabad, Pakistan.

[27] Zhang, C., Lin, Y. W., Zhu, L., Liu, A., Zhang, Z., and Huang, F. (2019). CNN-VWII: an efficient approach for large-scale video retrieval by image queries. https://doi.org/10.1016/j.patrec.2019.03.015 0167-8655/© Elsevier.

[28] Zhang, C., Hu, B., Suo, Y., Zou, Z., and Ji. Y. (2020). Large-scale video retrieval via deep local convolutional features, advance in multimedia. *Advances in Multimedia*, 2020, Article ID: 7862894. https://doi.org/10.1155/2020/7862894.

Chapter 13

Real and Spoofed Faces Classification using Machine Learning Adopting Efficient Processing and Quality Descriptors

Avinash B. Lambat[1] *and R. J. Bhiwani* [2]

[1]PhD Scholar, Babasaheb Naik College of Engineering, Pusad, Yavatmal, Maharashtra, India

[2]Professor, Babasaheb Naik College of Engineering, Pusad, Yavatmal, Maharashtra, India.

Abstract: The proposed work is to develop a robust automated solution that can deal with all possible variations related to spoofing operations. The work is focused on extracting such conventional features to observe the inherent disparities between the real and the fake images. We initially measured contrast using the modified Tadmor and Tolhurst method and corrected the contrast of the image, filtered and preserved the edges using Beltrami filtering, segmented the face region using the bounding box Algorithm, considered the luminance and the chrominance part to extract the coarse and fine features. The time and computational complexity were reduced by squeezing the dimensions of the feature set using principal component analysis and averaging the samples. The classification accuracy obtained using the support vector machine with Gaussian kernel over the test set was remarkable. The classification accuracy using IDIAP Replay attack dataset was found to be 100% which outperformed other state-of-the-art techniques.

Keywords: Face biometric, spoofing attacks, real and fake images, Tadmor and Tolhurst method, Beltrami filtering, bounding box algorithm, and principal component analysis.

1. Introduction

Many real-world best-face biometric authentication systems are vulnerable to today's spoofing attacks and respond positively to an unauthentic person. Even the world's well-known commercial face authentication systems have fallen prey to crude photo attacks. The performance of a good classifier is a function of extracted quality features which on the other hand depends upon the effective preprocessing of the input data. The images are subjected to various unwanted interference due to image sensors like

CCD and CMOS while acquiring the images. The cause is due to sensor material properties, conventional and surface-mounted electronic components

DOI: 10.1201/9781003527442-13

and sophisticated circuits of which various noises tend to slip inside including the Gaussian, shot, speckle and white. The work proposed in Chen *et al.* (2020) used a Laplacian filter to enhance the input images, whereas (Tadmor and Tolhurst, 2020) introduced a Schmid filter to uplift the texture information in the images. Authors Md *et al.* (2019) adopted modified DOG filtering to destroy noise while conserving high-frequency elements such as edges. A gamma correction algorithm to eliminate interference of partial light was used to preprocess images by (Cheng *et al.*, 2020). Even though various preprocessing techniques are mentioned in the literature, face antispoofing requires some effective measures to distinguish between the dissimilarity of a bonafied face and the face that is made to like it.

Also, the 2D spoofed face images are subjected to a variety of distortions that mainly include medium surface distortions (Glossy photo papers and digital screens), color distortions (fidelity and resolution of the screen), Moire pattern distortions (overlapping of digital grids) and facial deformation (bends in photo paper; Patel *et al.*, 2016). Face motion-based analysis (Bharadwaj, 2013; Pan, 2007) requires accurate segmentation of the face and its landmark components. The localisation of such facial component requires estimation through multiple image frames and can handle print attacks but have no ability to detect replay attacks. Antispoofing techniques based on textural information (Chingovska, 2012; Maatta, 2011) to capture textural properties of live and spoofed faces are relatively fast but lack generalisation capability. The depth information of 3D face images (Bao, 2009; Marsico, 2012) considering multiple frames when used to detect spoofs in 2D images, the estimation needs to be reliable. The process of recapturing printed photos or video replays diminishes low-frequency components and enlightens high-frequency elements. All of these state-of-the-art methods are successful and show sensitivity towards intra-dataset test images only and therefore require to inclusion of features extracted through multiple aspects.

1.1. Our Contribution

1. The region of interest is extracted to eliminate the effect of non-contributing regions using the bounding box algorithm. It reduces the burden of the classifier and improves the quality of features and the classification accuracy.
2. The dimension of the feature set is reduced by careful manual inspection and experimental analysis concerning redundant features.
3. The proposed method is reliable and robust to train and classify test samples accurately irrespective of input samples. That is, the performance is unaffected even though the training and testing samples are randomly chosen.

2. Literature Review

An ensemble classifier with probabilistic voting was introduced by (Yuting, 2021). The face region was calibrated using face alignment and a discrete wavelet filter was used to process the upper frequency coefficients to obtain a discriminative residual image. The colour residue was further converted to YCbCr model and texture information was extracted using a texture descriptor. They used four CASIA datasets (FASD, MSU, ROSE-YOUTU and ROSE) and applied their work on intra and inter-dataset test samples. The authors (Boulkenafet, 2017) extended the use of SURF features from grayscale to colour images by applying it on each of the color bands separately. They concatenated the rotation invariant SURF features and then reduced them using PCA and Fisher vector encoding in order to reduce the computational complexity. The descriptors were obtained using a Haar wavelet on 4 × 4 blocks or sub-regions around a point of interest on RGB, HSV and YCbCr colour spaces which were further concatenated to form a CSURF feature vector of dimension

64. They concluded that their approach even with limited training set on inter-dataset samples showed interesting generalisation performance.

Researchers (Peng, 2018) used the chromatic texture difference between the real and forged images and extracted inter-channel chromatic co-occurrence LBP (CCoLBP) and intra-channel facial textures for presentation attacks. A softmax classifier with inter and intra-channel-based features was used to evaluate the performance against testing samples from cross datasets (MSU MFSD, CASIA FASD, Replay Attack, Replay Mobile and OULU-NPU). They showed that their model improves the detection generalisability using mutual information and discriminating power analysis. On the other hand, Peng *et al.* (2022) combined CCoLBP with ensemble learning to analyse the chromatic discrepancies and reduce the effect of inter-class imbalance between real and forged images. Mohan (2020) suggested a fuzzy-SVM classifier to classify self-generated datasets acquired in different directions and angles. The images were preprocessed to extract the region of interest and HOG features were extracted. LPQ features were simultaneously obtained to use its blur invariant property over rectangular regions. They showed that their model has low time complexity and possesses the ability to efficiently recognise different genuine and spoofed attacks due to HOG-LPQ property under blurring objects.

3. Methodology and Model Specifications

3.1. Introduction

The proposed system is partitioned into five main constituents consisting of preprocessing, feature extraction, dimension reduction, classification and result. The input consists of images from the dataset which are either real or spoofed. The pre processing stage comprises two sections viz. contrast correction and filtering and then averaging the outputs from the two sections. The first section computes the contrast and corrects it so as to improve the quality of the image for better features while the next section filters the image for edge enhancement. The feature extraction stage acquires a variety of features including the coarse and the fine features capable of distinguishing the real and the fake images. The feature length of 2,060 elements is reduced by two successive stages that include principal component analysis (PCA) and then averaging of 25 elements obtained using PCA. The train and test samples are isolated and the support vector machine is trained using 75% of the total sample, while the rest of the 25% test samples are evaluated using the trained SVM. The labels obtained by evaluating test samples are compared with the true labels and the corresponding accuracy is displayed. The following sub-sections deal with all stages in brief.

3.2. *Dataset*

The IDIAP dataset with replay attack under consideration consists of 4,000 real images and 9,950 spoofed images belonging to 80 real and 199 fake subjects converted from videos, respectively (Pereira *et al.*, 2014). Each subject either real or fake has a distinct number of images. The work considers the first 50 images from each folder corresponding to each of the subjects. Figures 1 and 2 show images pertaining to real-face images and spoofed face images for the same subjects. The proposed system for classification is shown in the Figure 3 below.

***Figure 1:** Real images extracted from the dataset videos.*

***Figure 2:** Spoofed images extracted from the dataset videos.*

The block diagram of the proposed work shows an efficient preprocessing stage consisting of two parallel frameworks. As seen in Figures 2 and 3, the facial images are not uniform with respect to illumination and backgrounds. Also, most of the dataset images perceptually show blurred edges which carries significant information that may miss quality features across edges when conventional or traditional features are under consideration for classification. For the former part, Tadmor and Tolhurst method which is constructed using modified DoG (Difference of Gaussian) for measuring the initial contrast of the image was used. Subsequently after acquiring the contrast value of the image (Averaging measured contrast value of R, G and B frames), the image is contrast corrected. On the other hand, the image undergoes enhancement while preserving the edges of the images using the Beltrami filter. The filter has the ability to enhance edges while preserving them and is most commonly used in Mural images where images are deteriorated due to natural factors. The contrast-corrected image and the filtered image are averaged to obtain a good quality image as shown in Figure 5 below.

The region of interest (ROI) (face region) is comparatively small with respect to the total image size. Considering the whole image for feature extraction, the features of the region outside the ROI will definitely interfere with the ROI features and not only increase the time and computational complexity but also affect the classification accuracy. Therefore, a bounding box algorithm was used to extract the ROI and then features were extracted. Once the features are extracted, principal features are uplifted by PCA. After a manual inspection of the feature matrix, it was seen that features were very close for some samples. These redundant samples were handled using averaging 25 features from the feature set. The training and the test samples were randomly separated for classification. The following sections explain how the preprocessing and the feature extraction stages are carried out in detail.

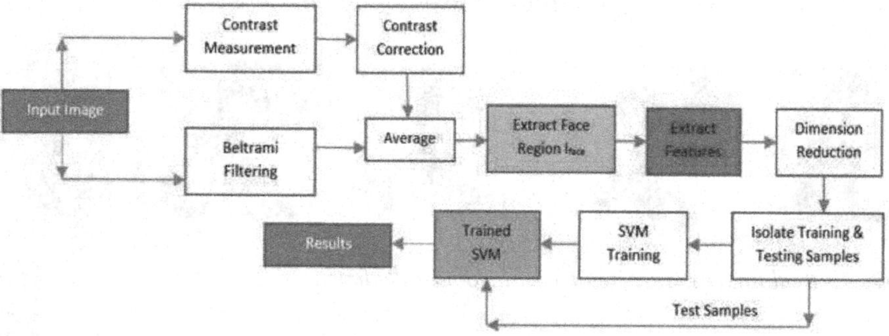

Figure 3: *The proposed spoof classification system.*

3.3. *Contrast Measurement*

The perceived contrast of an image is basically influenced by viewing conditions and the spatial arrangements of the image and measurement of such contrast are not so simple. The parameters that affect the image contrast involve colour, contents, illumination, viewing distance, resolution, etc. Thus only measuring the difference between the brightest and the darkest point measures the perceived contrast (Chen, 2020). Out of many such local and global contrast measuring classic approaches, Tadmor and Tolhurst global approach has been used for measuring the contrast of images. Figure 4 shows the output of modified DoG filtering. The concept is modified and adapted to the DoG model. The contrast of an image is measured using equation (1).

$$\mathrm{CM}\,(x,y) = \frac{Rc(x,y)-Rs(x,y)}{Rc(x,y)+Rs(x,y)} \qquad (1)$$

where the output of central component is,

$$Rc(x,y) = \sum_{i=x-3rc}^{i=x+3rc} * \sum_{j=y-3rc}^{j=y+3rc} Center\ (i-x, j-y)I(i,j) \tag{2}$$

While the output of surround component is,

$$Rs\ (x,y) = \sum_{i=x-3rc}^{i=x+3rc} * \sum_{j=y-3rc}^{j=y+3rc} Surround\ (i-x, j-y)I(i,j) \tag{3}$$

And the centre and surround components of the receptive field is given by,

$$Center\ (x,y) = exp\left[-\left(\frac{x}{r_c}\right)\left(\frac{x}{r_c}\right)-\left(\frac{y}{r_c}\right)\left(\frac{y}{r_c}\right)\right] \tag{4}$$

(x,y) is the spatial coordinates of the receptive field, *rc* is the radius at which the sensitivity decreases to *1/e* w. r. t. the peak level.

$$Surround\ (x,y) = 0.85\left(\frac{r_c}{r_s}\right)\exp\left[-\left(\frac{x}{r_s}\right)\left(\frac{x}{r_s}\right)-\left(\frac{y}{s}\right)\left(\frac{y}{r_s}\right)\right] \tag{5}$$

Such that *rs>rc.*

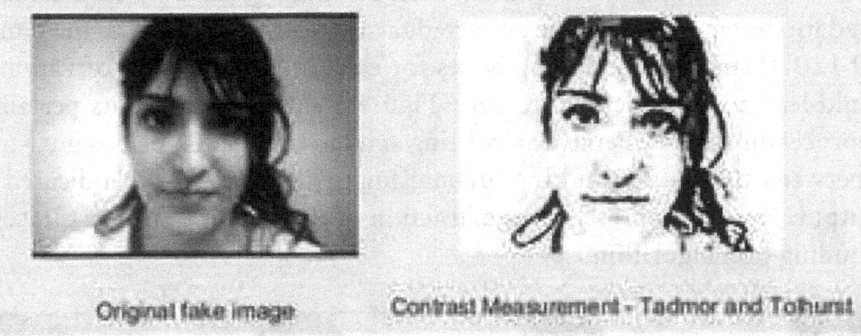

Figure 4: *Contrast measurement using Tadmor and Tolhurst method.*

3.4. Contrast Correction

The contrast enhancement was dealt with the following technique. The result showed improved and acceptable contrast over the parent images. The images were then converted to grayscale and concatenated to form a 4D array to reduce the processing time for feature extraction. The following correction technique to the grayscale image was applied and all three frames of the colour image were independent.

$$M = 255 * CM \tag{6}$$

$$Factor = 259 * \frac{(M+255)}{(255+(259-M))} \tag{7}$$

$$G = (Factor * (I - 128)) + 128 \tag{8}$$

3.5. Beltrami Filtering

The work suggested by (Wetzler and Kimmel, 2012) introduced an edge-preserving and denoising filter for 2D and 3D images and extended it to patches for feature extraction. Beltrami filter is capable of removing aliasing and weak textures while preserving the edge's fine structure. The filter was used on each of the colour channels of the input image *A* separately with 20 iterations and a time step of 0.5 to obtain the filtered image *C*. The filtered and contrast-corrected images *G* and *C* were considered and the average of the two images *(II)* was used to segment the face region of the original image to discard any unwanted region that may remain due to poor contrast or blurred edges. Thus the expression for image *II* is given by the following expression (9), "*i*" representing the colour channel for colour image.

$$\text{II} = \frac{1}{2}\left[G_i + S_i\right] \tag{9}$$

The face region *Iface* was extracted using the bounding box algorithm in MATLAB which covers the region from head to neck so that significant features could be extracted for better accuracy. The extracted face region was resized to the dimension of [120 120 3] since the bounding boxes for each individual were of varying sizes and could lead to variable feature sizes. Figure 5 shows the outputs pertaining to the preprocessing stages. Perceptually being similar, the PSNR (peak signal-to-noise ratio) between the output and the original image (first image) is indicated below each output. Figure 6 represents the extraction of the region of interest (face) using the bounding box algorithm.

Figure 5: *The preprocessing stage: original input image, filtered image, contrast corrected image and the averaged image. The PSNR values reflect pixel value changes in each stages.*

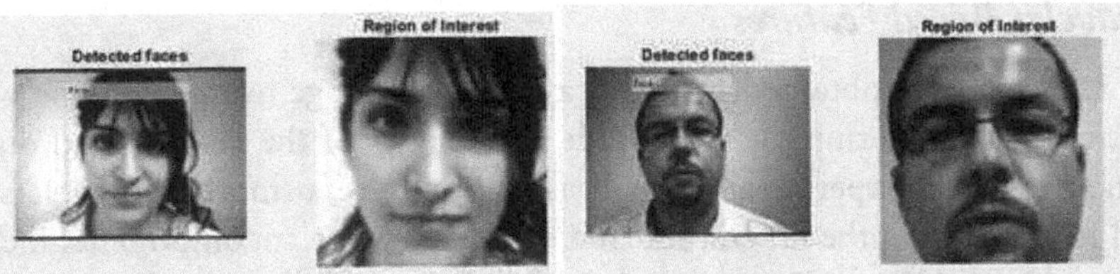

Figure 6: Region of interest (FACE) using bounding box algorithm.

3.6. Gabor Features

Gabor features developed by (Haghighat, 2015) was used with default parameters including the number of scales and orientations equal to 5 and 8, the number of rows and columns of the Gabor filter to be 39. The rows and columns were down-sampled by a factor of 39. The *Iface* image was converted to grayscale and 640 features were extracted.

3.7. Gray Level Co-occurrence Matrix (GLCM) Features

The second-order GCLM-based features for the grayscale *Iface* image including contrast, correlation, energy and homogeneity were obtained in all 8 directions at 64 intensity levels. A total of four features are obtained by averaging 8 features corresponding to eight directions for each of the GLCM components.

3.8. Statistical Features of First Order

Five first-order statistical features were extracted from the grayscale *Iface* image which includes mean, variance, standard deviation, skewness and kurtosis. These features are a function of probabilities of pixel intensities in the facial region. The following are the expressions for each of the statistical feature components. The probability *P* corresponding to each of the intensity levels is evaluated by the following expression

(11) first and then mean, variance, standard deviation, skewness and kurtosis are evaluated using expressions (12) to (16)

$$GL = 0{:}255 \tag{10}$$

$$P(x = 1 \text{ to } 255) = \frac{1}{M*N}\sum_{i=1,j=1}^{i=M,j=N}(Iface) = x \tag{11}$$

$$Mean = \Sigma G. * P \tag{12}$$

$$Var = \Sigma([(GL-Mean)^2].* P) \tag{13}$$

$$Stdev = \sqrt{Var} \tag{14}$$

$$Skewn = \Sigma([(GL-Mean)3].*P) \tag{15}$$

$(Stdev3)$

$Kurts = \Sigma([(GL-Mean)4].*P)$ (16)

$(Stdev4)$

3.9. Wavelet-Based Features

Course features were obtained from a grayscale I_{face} image using level one wavelet transform and considering the magnitude and energy of the vertical and diagonal components. After experimental analysis, six different mother wavelets were selected for extracting the magnitude and energy of wavelet components including three bior (*bior 3.1, bior 3.5 and bior 3.7*), debauchees 3 (*db3*), symlet 3 (*sym3*) and *haar*. For each of the mother wavelets, the grayscale *Iface* image is decomposed to level one wavelet transform, and the features are extracted using expressions (17) and (18), respectively, for magnitude and energy.

$$W_M = \frac{1}{p*q}\left(\sum_{r=1}^{p}\sum_{c=1}^{q} W\right) \tag{17}$$

$$W_E = \frac{1}{p*q}\left(\sum_{r=1}^{p}\sum_{c=1}^{q} ab(W_x)^2\right) \tag{18}$$

where p and q are the row and column dimensions of the wavelet components. Wx is either Wv or Wd and corresponds to vertical and diagonal components.

3.10. Colour Histogram-Based Features

The colour frames corresponding to the *RGB Iface* image and the *Lab* colour space *Iface* image are considered to find histograms using 16 binary levels. These histograms are normalised using the dimension of the *Iface* image. The expressions (19) and (20) below are used to extract features in the colour domain. Further, the features from *RGB* and *Lab* colour space are concatenated.

For *RGB Iface* image,

$$H_{(x \in R,G,B)} = \frac{1}{M*N} H(I_{face}(x))_{(bins=16)} \tag{19}$$

For *Lab Iface* image,

$$H_{(y \in L,a,b)} = \frac{1}{M*N} H(I_{face}(y))_{(bins=16)} \tag{20}$$

3.11. Haar Wavelet-based Linear Binary Pattern (LBP) features

LBP not only lifts the textural properties of the image but also captures details such as edges, patterns, illuminations, etc. LBP was used over the original *Iface* colour image on its components and the features were then averaged. The colour *Iface* image was decomposed to four level wavelet transform and LBP features from all its wavelet components are similarly averaged. Finally, the features from the original color image and its wavelet tributaries are averaged in the end. This ensures that none of the detail is lost and thus will help to distinguish between the real and the spoofed image. Figure 7 depicts the mechanism for feature extraction using LBP.

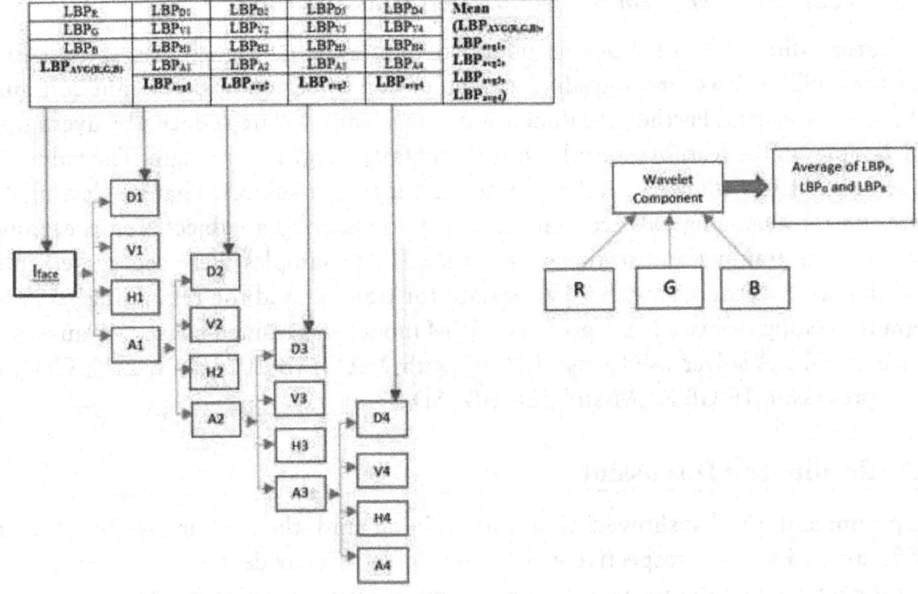

***Figure 7:** LBP feature extraction mechanism using "haar" wavelet. Each of the wavelet components has three colour components viz. R, G and B.*

3.12. Histogram of Gaussian (HOG) Features

HOG features were considered on *Iface* image and on each colour component of *Iface* image. All four HOG features are combined by averaging them and obtaining them with a cell size set to [16 16]. The complete feature set consisting of coarse and fine features is indicated in Figure 8 below.

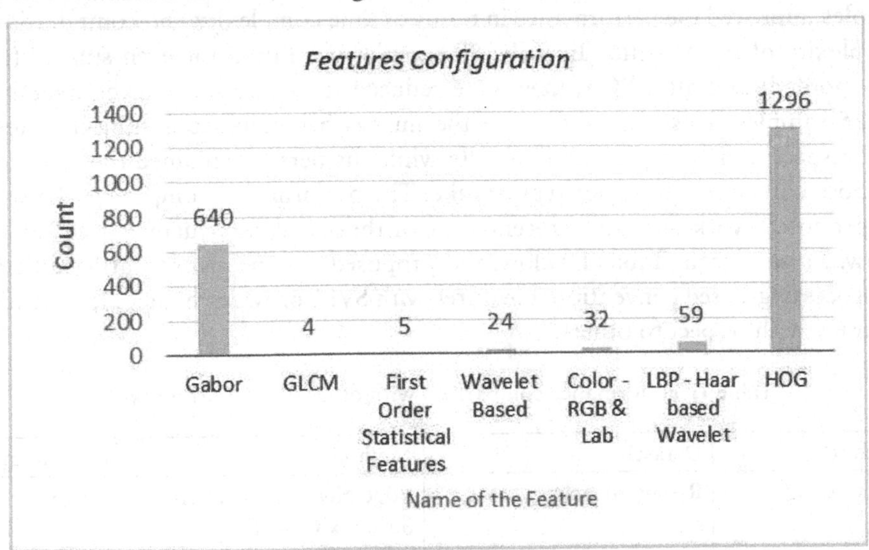

***Figure 8:** Name of the features and their frequency.*

3.13. Feature Dimension Reduction and Classification

A feature dimension of 2,060 is obtained by concatenating all coarse and fine features. PCA is used first to reduce the dimension by selecting significant columns of the PCA matrix. Further, the dimensions of the samples are reduced by averaging 25 samples in the feature set for both real and fake image feature data. The value 25 is selected in accordance with the number of images per subject. That is selecting 25 samples for averaging will account for only two images per subject. Ten preceding features for training and testing were utilised. The samples were segregated into training and testing sets with 75% of data for training and the remaining 25% of data for testing from each category. The SVM model was trained using a "Gaussian" Kernel and the solver used was "L1QP" with MATLAB 2021b, i5, 2.17 GHz, 6 core processor, 16 GB RAM and 128 GB SSD.

4. Results and Discussion

Experimental results showed that the training and the testing accuracy with 75% and 25% data, respectively, SVM was able to provide 100% accuracy. The significant part of the proposed work is the pre-processing stage which involves contrast measurement and correction along with edge-preserving filter operation using the Beltrami filter and the conventional coarse (GLCM, Colour, First-order Statistical and wavelet features) and fine (Gabor, HOG and LBP features) quality features. The parallel process of contrast correction and filtering and then averaging resultant images mitigated the effect of uneven illumination and uplifted the edges in the original image without much loss. The feature extraction operators were able to perform their duties independently and each of them contributed positively in obtaining a robust feature set. Reducing the dimension of the features and the samples improved the performance in terms of time complexity and computational complexity of the classifier. Initially, 50 samples were used for each subject (real and spoofed) and after PCA, they were reduced to two samples using averaging for 25 samples. This was done since the images belonging to a subject differed with respect to lightning conditions only while the person remained the same also all those correspond to either real or fake. The performance comparison between the proposed work and other recent state-of-the-art work is depicted in Table 1 below. As seen from Table 1 below, the proposed system incorporating efficient preprocessing-based conventional features with SVM provides higher classification accuracy with respect to others.

Table 1: Performance comparison with other state of art works.

Authors	Dataset	Method	Accuracy
Amal *et al.*	Replay Attack	Edge-Net Autoencoder-Softmax Classifier	98.5
Chen *et al.*	IDIAP - Replay Attack	CNN	99.34

Thepade *et al.*	FPAD - Replay Attack	Deep CNN-Sorted Block Truncating Coding	98.67
Agarwal *et al.*	Replay Attack	Score Fusion of Wavelet Partition Images	99.2
Tai-Hung *et al.*	TIFADB	Morphological filtering-SVM	95.67
Ours	IDIAP - Replay Attack	Efficient Preprocessing-Conventional-SVM	100

5. Conclusion

The quality of features depends on the descriptors used to uplift them and the data characteristics. Most of today's literatures are focused on deep learning mechanisms to process given data in its original form and extract the prominent values as a part of features. The proposed work uses conventional features with SVM to detect real and the forged images. Factors such as illumination variation, noise and blur are corrected using contrast correction and Beltrami filtering. The advantages offered by contrast correction and filtering while preserving edges in parallel are averaged up so that spoofing details are identified by the descriptors. Depth, chrominance frequency components, textural properties and statistical components are combined so that discriminative values supersede the common features and the classifier can accurately judge the test samples in the intra-dataset. The work can be extended to other dataset under Replay Attack and other distinct dataset to test the performance over inter-dataset test samples.

References

[1] Agarwal A, Singh R, Vatsa, M., and Noore, A. (2022). Boosting face presentation attack detection in multi-spectral videos through score fusion of wavelet partition images. *Front. Big Data*, 5, 836749.

[2] Amal, H., Alharbi, S., Venkatachalam, K. K., Abouhawwash, M., Khafaga, D., and Sami, D. (2023). Spoofing face detection using novel edge-net autoencoder for security. *Intelligent Automation & Soft Computing*, 35(3), 2773–2787.

[3] Bao, W., Li, H., Li, N., and Jiang, W. (2009). A liveness detection method for face recognition based on optical flow field. *In Proceedings of the IASP*, 233–236.

[4] Bharadwaj, S., Dhamecha, T. I., Vatsa, M., and Singh, R. (2013). Computationally efficient face spoofing detection with motion magnification. *In Proceeding of the CVPR Workshops*, 105–110.

[5] Boulkenafet, Z., Komulainen, J., and Hadid, A. (2017). Face antispoofing using speeded-up robust features and Fisher vector encoding. *Signal Processing Letters, IEEE*, 24(2), 141–145.

[6] Chen, B., Qi, X., Zhou, Y., Yang, G., Zheng, Y., and Xiao, B. (2020). Image splicing localization using residual image and residual-based fully convolu-

tional network. *Journal of Visual Communication and Image Representation*, 73, Article ID 102967.

[7] Cheng, X., Wang, H., Zhou, J., Chang, H., Zhao, X., and Jia, Y. (2020). DTFA-Net: dynamic and texture features fusion attention network for face antispoofing. *Complexity*, 2020.

[8] Chingovska, I., Anjos, A., and Marcel, S. (2012). On the effectiveness of local binary patterns in face anti-spoofing. *In Proceedings of the IEEE International Conference of Biometric Special Interest Group (BIOSIG)*, 1–7.

[9] Yuting, D., Tong, Q., Ming, X., and Ning, Z. (2021). Towards face presentation attack detection based on residual color texture representation. *Security and Communication Networks*, 2021.

[10] Haghighat, M., Zonouz, S., and Abdel-Mottaleb, M. (2015). CloudID: Trustworthy cloud-based and cross-enterprise biometric identification. *Expert Systems with Applications*, 42(21), 7905–7916.

[11] Mohan, Chnadrashekhar, P., and Ramanaiah, (2020). Object-specific face authentication system for liveness detection using combined feature descriptors with fuzzy-based SVM classifier. *International Journal of Computer Aided Engineering and Technology*, 12(3), 287–300.

[12] Maatta, J., Hadid, A., and Pietikainen, M. (2011). Face spoofing detection from single images using micro-texture analysis. *In Proceedings of the International Joint Conference on Biometrics (IJCB)*, 1–7.

[13] Marsico M. De, Nappi M., Riccio D. and Dugelay J. L. (April 2012). Moving face spoofing detection via 3D projective invariants. In Proceedings of the International Conference on Biometrics (ICB), pp. 73–78.

[14] Md, R. H., Mahmud, H., and Li, X. Y. (2019). Face antispoofing using texture based techniques and filtering methods. *Journal of Physics: Conference Series*, 1229.

[15] Pan, G., Sun, L., Wu, Z., and Lao, S. (2007). Eye blink-based anti-spoofing in face recognition from a generic web camera. *In Proceedings of the 11th International Conference on Computer Vision (ICCV)*, 1–8.

[16] Patel, K., Han, H., and Jain, A. K. (2016). Secure face unlock: spoof detection on smartphones. *IEEE Transactions on Information Forensics and Security*, 11(10), 2268–2283.

[17] Peng, F., Qin, L., and Long, M. (2018). CCoLBP: chromatic co-occurrence of local binary pattern for face presentation attack detection. *27th International Conference on Computer Communication and Networks (ICCCN)*, 1–9.

[18] Peng, F., Qin, L., and Long, M. (2020). Face presentation attack detection based on chromatic co-occurrence of local binary pattern and ensemble learning. *Journal of Visual Communication and Image Recognition*, 66, 102746.

[19] Pereira, T. D. F., Komulainen, J., Anjos, A., Martino, J. M. D., Hadid, A., Pietikainen, M., and Marcel, S. (2014). Face liveness detection using dynamic texture. *EURASIP Journal on Image and Video Processing*, 2.

[20] Thepade, Dindorkar, Chaudhari, and Bang, (2023). Enhanced face presentation attack prevention employing feature fusion of pretrained deep CNN model and Thepade's sorted block truncation coding. *IJE Transactions A: Basics*, 36(04), 807–816.

[21] Tadmor, Y., and Tolhurst, D. (2020). Calculating the contrasts that retinal ganglion cells and LGN neurones encounter in natural scenes. *Vision Research*, 40(22), 3145–3157.

[22] Tai-Hung, L., Ching-Yu, P., and Chao-Lung, C. (2023). Fast face presentation attack detection in thermal infrared images based on morphological filtering. *International Journal of Network Security*, 25(2), 185–193.

[23] Wetzler, A., and Kimmel, R. (2012). Efficient Beltrami flow in patch-space. *In Scale Space and Variational Methods in Computer Vision*, SSVM 2011. Lecture Notes in Computer Science, 6667, 134–143, Berlin, Heidelberg: Springer.Xin, C., Jingmei, Z., Xiangmo, Z., Hongfei, W., and Yuqi, L. (2023). A presentation attack detection network based on dynamic convolution and multilevel feature fusion with security and reliability. *Future Generation Computer Systems*, 146, 114–121.

Chapter 14

A Facial Sketch-To-Image Synthesis Using Flutter and Artificial Neural Network

Heena Farheen Ansari[1] *and Mr. Sachin Meshram* [1]

[1]Assistant Professor, Department of Information Technology, KITS, Ramtek

Abstract: The paper represents a new application for synthesising photos from facial sketches using flutter and a generative adversarial network (GAN). In this we present a facial sketch-to-photo synthesis app that leverages flutter for user interface and python with pix2pix algorithm and generative adversarial networks (GANs) for image generation. Pix2Pix is a creative algorithm that can turn a crude line drawing into an oil painting. Pix2pix uses a conditional generative adversarial network (cGAN) to learn a function to map from an input image to an output image. The generator network takes the sketch as input and generates a photo, while the discriminator network determines if the generated photo is realistic or not. The two networks are trained together to improve the accuracy of the generated photo. The application is built using flutter, a popular open-source framework for building cross-platform mobile applications, providing a seamless user experience on both Android and iOS devices for generating realistic photos at real time with artistic expression, and virtual try-on of different hairstyles and makeups. Sketch will convert into image in real time.

Keywords: cGAN, CycleGAN, Deep Learning, and artificial neural network.

1. Introduction

The application leverages the latest advancements in machine learning and computer vision technologies to accurately synthesise lifelike images from rough sketches. This not only makes it an ideal tool for artists and designers, but it also opens up new possibilities for research and development in the fields of computer graphics and computational imaging. The application provides a user-friendly interface and real-time image generation capabilities, making it easy for users to experiment with different sketches and styles to achieve their desired results. With its advanced algorithms and powerful performance, the facial sketch-to-image synthesis application is set to revolutionise the way we create and manipulate images (Kazemi, 2018).

DOI: 10.1201/9781003527442-14

The motivation behind this research is to explore the potential of real-time facial sketch to image synthesis using the flutter framework and artificial neural networks. The ability to generate realistic images from facial sketches in real time has numerous practical applications, such as in digital art, character design and even criminal investigation. While there are existing facial sketch-to-image synthesis methods, they often require time-consuming manual adjustments and lack the ability to generate images in real time. By utilising the power of artificial neural networks and the flexibility of flutter, we aim to develop a solution that can generate high-quality images from facial sketches in real time (Wengling *et al.*, 2018).

The research will feature the process of developing the facial sketch to image synthesis, from training the neural network to integrating it into a flutter application. It will also confer the challenges and limitations encountered during development, as well as potential future improvements and applications. Ultimately, the report aims to demonstrate the feasibility and potential of real-time facial sketch to image synthesis using flutter and artificial neural networks. The aim of the research is to design and develop a real-time facial sketch to image synthesis application using flutter and artificial neural network. The main objectives are to generate efficient system that can convert facial sketches into realistic images in real time. The project will involve implementing various neural network architectures and training them on a large dataset of facial images to generate high-quality and realistic images from the input sketches. The application will be developed using flutter framework, which is a popular cross-platform development tool for building high-performance mobile applications. The development will focus on improving the accuracy and speed of the image synthesis process and exploring the potential use cases of the application in different fields. The ultimate goal of the project is to create a powerful and user-friendly application that can generate high-quality images from facial sketches in real time.

The main objectives are:

1. To recognise potential improvements and future work that could be done to enhance the app's performance and functionality.
2. To design and develop a mobile application that can take facial sketches in real time and generate corresponding realistic facial images using artificial neural network (ANN) techniques.

2. Literature Survey

The survey report proposes a new deep learning framework for sketch-to-photo synthesis, which is guided by facial attributes. The goal of the proposed framework is to generate realistic photos from sketches while preserving the facial attributes of the original sketch. The proposed method consists of two main stages. The first stage is a facial attribute extractor that extracts facial attributes from the input sketch. The second stage is of pix2pix algorithm and generative adversarial networks, where Pix2Pix is a creative algorithm that can turn a crude line drawing into an oil painting. Pix2pix uses a conditional generative adversarial network (cGAN) to learn a function to map from an input image to an output image (Manish *et al.*, 2020). The GAN consists of two networks: a generator network and a discriminator network. The generator network takes the sketch as input and generates a photo, while the discriminator network determines if the generated photo is accurate or not. The two networks are trained together to improve the accuracy of the generated photo. To evaluate the proposed method, the authors conducted experiments on two datasets, the CelebA dataset and the CUHK Face Sketch Database. The results demonstrate that the proposed method achieves better performance than state-of-the-art methods in terms of both visual quality and facial attribute preservation.

3. Proposed System

The proposed approach aims to build up a real-time facial sketch to image synthesis application using flutter and artificial neural network technology (Wang *et al.*, 2013). The application will allow users to sketch their facial features, and it will convert the sketch into a real-time image of their face. This approach will be implemented using the flutter framework, which is a popular mobile application development platform, and an artificial neural network, which is a machine learning technology.

Step 1: Data Collection and Preprocessing: The first step in developing the real-time facial sketch to image synthesis app is to collect and preprocess the data. The data will include facial images and corresponding sketches. The images will be used to train the artificial neural network, while the sketches will be used as inputs to the app.

The collected images and sketches will need to be preprocessed to remove any noise and standardise the data. This step will involve resizing the images and sketches to a uniform size and format. Additionally, the images and sketches may need to be normalised to ensure that the data is consistent and accurate.

Step 2: Model Training: Once the data is collected and preprocessed, the next step is to train the artificial neural network. The model will be trained using a deep learning algorithm, such as a convolutional neural network (CNN), which is ideal for image processing.

The CNN will be trained the features of the facial images and corresponding sketches and apply this information to generate a real-time image of the user's face. The model will be trained using a large dataset of facial images and sketches to ensure that it can accurately generate images.

Step 3: Application Development: After training of the artificial neural network, the next step is to develop it using flutter. Flutter is a popular mobile application development platform that allows developers to create beautiful and responsive apps quickly.

The application will have a simple user interface that allows users to sketch their facial features using their fingers or a stylus. The application will then use the artificial neural network to generate a real-time image of their face based on the sketch.

Step 4: Integration: The final step is to integrate the trained artificial neural network with the application. This will involve integrating the model with the app's backend and creating an API that can be used to generate images.

The application will use the API to send the sketch to the artificial neural network, which will generate a real-time image of the user's face. This will display the image on the user's device, allowing them to observe what they would seem like in real-time as shown in the figure 1 system architecture.

It can be a customer, a supplier, or any other entity outside the system. The figure 2 below shows the DFD for the proposed work.

Front-end: The front-end would be developed using the flutter framework. The user would draw a sketch of a face on the screen using their finger or a stylus. The drawing would be saved as an image file.

Image preprocessing: The image file would be preprocessed to ensure that it is in the correct format for the GAN to process. This could involve pix2pix algorithm which converts sketch lines to oil painting.

GAN: The GAN would be trained on a large dataset of facial images to learn the mapping from sketches to images. The architecture of the GAN would typically consist of a generator network and a discriminator network. The generator network would take a sketch as input and produce a generated image, while the discriminator network would evaluate the generated image and determine whether it is realistic or not.

Back-end: The back-end would be responsible for running the GAN and producing the generated image. The generated image would then be sent back to the front-end for display.

Display: The generated image would be displayed on the screen, allowing the user to see the result of their sketch. The user could then repeat the process if they are not satisfied with the result.

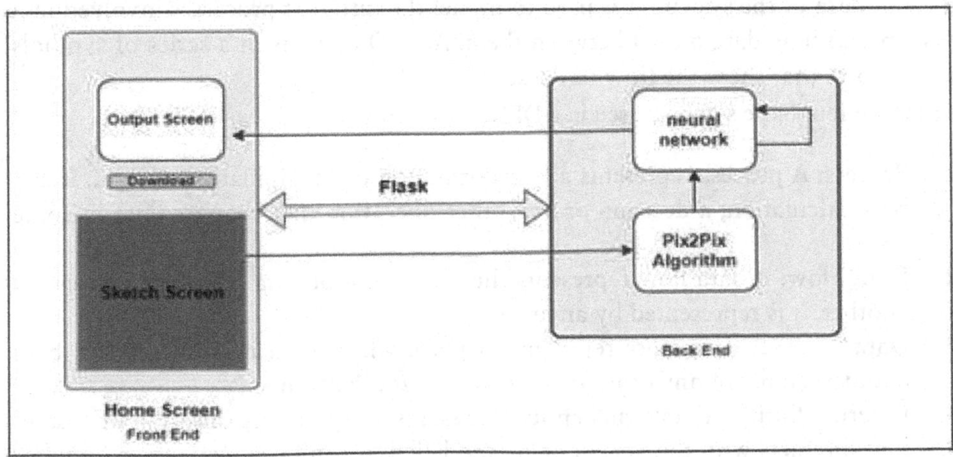

Figure 1: System architecture.

4. Research process

- **Background Research:** Conduct a literature review to gain an understanding of the existing techniques and algorithms for facial sketch to image synthesis. This step will help to identify the gap in the existing approaches and motivate the need for the proposed research.
- **Data Collection:** Collect a dataset of facial sketches and corresponding images to use for training and testing the neural network model. The dataset should be diverse and cover a range of variations in facial expressions, lighting, and poses.
- **Data Preprocessing:** Preprocess the collected data to remove any noise, normalise the data and prepare it for training the neural network.
- **Neural Network Architecture Selection:** Choose appropriate neural network architecture for the facial sketch to image synthesis task. This can include CNN, GAN or a combination of both.
- **Neural Network Training:** Train the selected neural network architecture using the preprocessed dataset. This involves tuning the hyperparameters of the neural network, such as learning rate, number of layers and batch size, to achieve the best performance.
- **App Development:** Develop a flutter-based mobile application that integrates the trained neural network model. The app should have a simple user interface that allows users to input facial sketches and generate corresponding images in real time.
- **App Testing:** Test the application on a range of devices to ensure its functionality and usability. Conduct user testing to collect feedback and improve the app's overall performance.
- **Evaluation:** Evaluate the performance of the developed application using metrics such as accuracy, speed and user satisfaction. Compare the results with existing facial sketch-to-image synthesis techniques to determine the effectiveness of the proposed approach.

Data flow diagram A data flow diagram (DFD) is a graphical illustration of the flow of data of the system. It is used to model the different processes involved in a system and how data moves between them. A DFD consists of a series of symbols and arrows that show the flow of data.

There are four basic symbols used in a DFD:

1. Process: A process represents a transformation or manipulation of data. It can be a calculation, a decision or any other operation that changes data in some way.
2. Data Flow: A data flow represents the movement of data from one process to another. It is represented by an arrow.
3. Data Store: A data store represents a place where data is stored. It can be a database, a file or any other type of storage mechanism.
4. External Entity: An external entity represents a person, organisation or system that interacts with the system being modelled. It can be a customer, a supplier or any other entity outside the system.

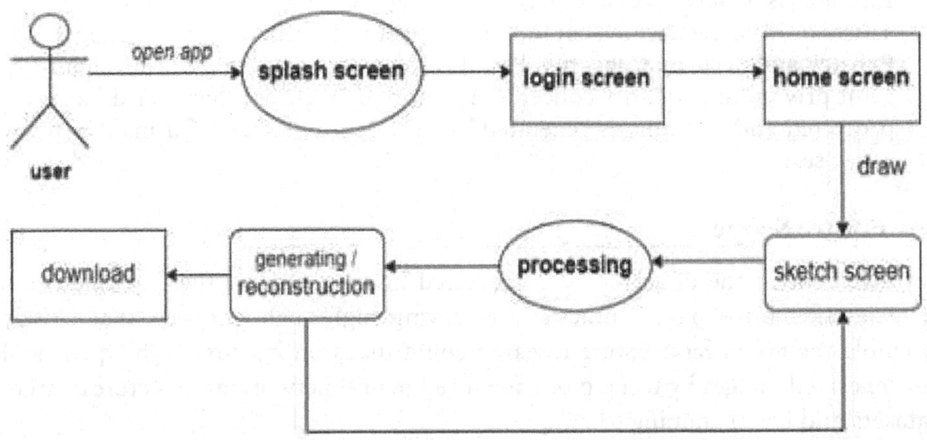

Figure 2: DFD to illustrate proposed work.

5. Conclusion

In conclusion, the development of a real-time facial sketch-to-image synthesis application using flutter and artificial neural networks is a promising project that combines several cutting-edge technologies. By leveraging the power of deep learning and flutter's fast rendering engine, the app can generate realistic facial images from sketches in real time, making it useful for a wide range of applications, including digital art, education and entertainment. The development of the application requires expertise in machine learning, deep learning, computer vision and mobile application development using flutter. However, the potential benefits of the application make it worth the effort, as it could provide a powerful tool for artists and designers, as well as serve as an educational resource for students and researchers. Overall, the project represents an exciting opportunity to push the boundaries of what's possible with artificial intelligence and mobile app development.

6. Limitations

- **Limited accuracy:** Despite advancements in artificial neural networks, generating photo-realistic images from sketches is still a challenging task, especially in real time. The resulting images may have artifacts, distortions or be of low quality.
- **High computational requirements:** Generating images in real time using artificial neural networks requires a significant amount of computational power. This may limit the types of devices that can run the app and increase the cost of hosting the Flask API.
- **Limited functionality:** The app's functionality may be limited to facial sketches and images, which may not appeal to a wide audience. Expanding the app's functionality to include other types of sketches and images may require significant additional development work.

- **Limited User Base:** This app may only appeal to a niche audience, such as artists or designers, which could limit its overall user base and adoption.
- **Privacy and Security Concerns:** Facial recognition technology raises significant privacy and security concerns. The app must ensure that user data is protected and that images generated by the app are not used for malicious purposes.

7. Future Scope

Improvement in the quality of the generated images: One of the key challenges in facial sketch-to-image synthesis is achieving high-quality images that closely resemble the target face. Future research could focus on improving the quality of the generated images by using more advanced neural network architectures, larger datasets and better training techniques.

Commercialisation: The facial sketch-to-image synthesis app could be commercialised and marketed to various industries, such as entertainment, fashion and advertising. The app could be used for creating customised avatars, virtual models or personalised advertisements.

Expansion: The application can be further extended to various object sketches and image synthesis beyond just facial attributes.

References

[1] Kazemi, H., Iranmanesh, M., Dabouei, A., Soleymani, S., and Nasrabadi, N. M. (2018). Facial attributes guided deep sketch-to-photo synthesis. West Virginia University.

[2] Manish, B., Diane, O., Juan, C., Liping, Y., and Brendt, W. (2020). Diagram imager retrieval using sketch-based deep learning and transfer learning. University of New Mexico, Albuquerque, NM, USA Los Alamos National Laboratory, Los Alamos, NM, USA.

[3] Ulyanov, D., Lebedev, V., Vedaldi, A., and Lempitsky, V. S. (2016). Texture networks: feed-forward synthesis of textures and stylized images. *In ICML*, 1349–1357.

[4] Wang, N., Tao, D., Gao, X., Li, X., and Li, J. (2013). Transductive face sketch-photo synthesis. *IEEE transactions on neural networks and learning systems*, 24(9), 1364–1376.

[5] Wengling, C., Jin, W. X., Zhang, H., and Zhang. (2018). SketchyGAN: Towards diverse and realistic sketch to image synthesis. *2018 Conference on Computer Vision and Pattern Recognition*, Salt Lake City, UT, USA.

[6] Xianming, L., and Weihong, D. (2020). Portrait image synthesis from sketch via multi-adversarial networks.

Chapter 15

Transfer Learning-Based MRI Model for the Identifying of Brain Tumors

Prachi V. Kale[1], *Ajay B. Gadicha* [2], *and G. D. Dalvi*[3]

[1] Research Scholar, Department of CSE, P R Pote College of Engineering & Management, Amravati, Maharashtra, India

[2] Associate Professor and HOD, Department of AI, P R Pote College of Engineering & Management, Amravati, Maharashtra, India

[3]Assistant Professor and Dean, P R Pote College of Engineering & Management, Amravati, Maharashtra, India

Abstract: A brain tumor (BT) refers to an unexpected occurrence or abnormal cluster of cells. The nature of these cells determines whether they are harmless (noncancerous) or malignant (cancerous). Brain tumours can lead to increased intracranial pressure, which may result in death or brain damage. Symptoms include heightened fatigue, impaired cognitive function, worsened migraines, epilepsy, nausea and vomiting. Identifying brain tumours typically involves tests such as blood and urine analysis, magnetic resonance imaging (MRI), positron emission tomography (PET) and computerised tomography (CT). However, these methods can be time-consuming and yield unreliable results. To address this issue, deep learning models are utilised. These models offer shorter processing times, require less specialised equipment, provide more accurate findings and are user-friendly. In this study, a transfer learning-based approach utilising an altered convolutional neural network (CNN) structure, along with normalisation and data supplementation processing methods, is proposed using the previously trained VGG19 model. This approach achieved a sensitivity of 94.73% and a precision of 98%. The research outcomes demonstrate the superior performance of this suggested methodology compared to the latest advanced techniques. The dataset used for training consisted of 257 photographs, sourced from Kaggle, comprising 157 images of BT and 100 images of individual non-tumours (NT). These results indicate that such simulations could be employed to develop clinically beneficial BT detection tools for CT imaging.

Keywords: Brain tumor, MR imaging, Dense-Net121, Dense-Net201, VGG16, VGG19.

DOI: 10.1201/9781003527442-15

1. Introduction

The regulation of an individual's body's neural system is exercised by a vital organ referred to as the cerebral cortex. There are 100 billion nerve cells in it by Gu *et al.* (2021). Destruction of any neurons can result in a number of health issues and abnormalities in people's brains. Brain cells are negatively impacted by these injured cells. Such an outbreak raises the possibility of brain tumours in people suggested by Deepak *et al.* (2021). Brain tumours can be classified as either original or invasive. Basic brain tumours develop throughout the cortical brain but can impact the nerve endings, circulation vessels or extra glands, in contrast to metastasis brain tumours, which originate in various parts of the human system prior to spreading into the brain, such as the female breasts or lungs by Kumar *et al.* (2021). Malignant or benign tumours can exist. Malignant brain tumours are malignant and develop quickly in the body. Glioblastoma is among the most typical malignant brain tumours by Rehman *et al.* (2021). Cells found in benign brain tumours are benign and develop rather slowly. This kind of tumour is incapable of spreading to other bodily regions. It won't reappear in the body if carefully eliminated via surgery by Rajasree *et al.* (2021). Early-stage brain tumours have a better chance of surviving in sufferers. Meningiomas are nerve-shaped lymphoid tumours called cavernous tumours that are often harmless and grow around the cavernous apertures of the brain; granular pyramidal tumours may be potentially harmful or harmless; and schwannomas are tumours that typically affect people between the ages of 40 and 70.

This study uses a novel CNN-based methodology to separate BT into BT and NT classes. Additionally, an extensive dataset is used to train and build the CNN model. Applying processing for preprocessing like normalisation and data enrichment to the dataset has improved the proposed model's performance. Therefore, computerised systems like this enable clinical facilities to operate more effectively while saving time.

2. Related Work

There are actually four distinct types of brain tumours, as defined by the World Health Organisation (WHO) by Kader *et al.* (2021). The technique of segmenting brain tumour cells based on their ability to be recognised is known as grading. A grade is identified at a greater level the more aberrant the cells reflect. The smallest tumours occur in categories I and II, while the most serious tumours are in categories 3 or 4 suggested by Bodapati *et al.* (2021). Cells in grade 1 seem normal, making them more unlikely to harm adjacent cells. In second grade, it appears that cells within the cerebral cortex are increasingly merging with surrounding neighbouring tissues. Cells started showing evidence of greater abnormalities in the third grade and began to disseminate across collateral nerves in different brain regions. In fourth grade, aberrant cell behaviour increases, cells start to grow into tumours, and the spinal cord and brain's different regions are invaded by tissue. A normal tumour has a low grade, whereas a dangerous tumour has a high grade to (Mzoughi *et al.*, 2020). Different approaches are used to treat various tumours based on the position, kind and dimension of the tumour. The most well-known and non-harmful method of treating tumours is surgery by Sajjad *et al.* (2019). Furthermore, fourth-grade tumours can cause brain disorders including dementia, Parkinson's disease and Huntington's disease, which impair fundamental physiological cognitive and motion abilities. Brain computerised tomography scans are used to track the development of the simulation of modelling. Computerised imaging techniques deliver more details regarding the exhibited healthcare picture as well as serve as an alternative method for tumour detection by Abiwinanda *et al.* (2019).

Brain-computerised tomography scans are used to track the development of simulation modelling. CT delivers further details regarding the exhibited health picture as well as serves as an alternative method for tumour detection by (Khan *et al.*, 2020).

This study uses a novel CNN-based methodology to separate BT into BT and NT subcategories. Additionally, an extensive dataset is used to train and build the CNN model. The efficiency of the suggested system was enhanced by adding preparation methods to the database, such as normalisation and information enrichment. Therefore, computerised systems like this enable clinical facilities to operate more effectively while saving time. Brain computerised scanning scans are used to track the development of the simulation of modelling. Computerised ultrasonography is a distinct method for identifying tumours and provides more details about the visible health images by (Xu *et al.*, 2015). It uses a novel CNN-based methodology to separate BT into BT and NT categories. In addition, a sizable data collection is employed for training and constructing the CNN algorithm.

3. Proposed Methodology

There hasn't been much research or writing on the BT comparison research employing all four DL Classifiers such as VGG16-19, DenseNet121 and 201, despite the fact that there have been numerous investigations and efforts on the topic. The result of such algorithms is then displayed and compared through constructing verification procedures, graphing accuracy, error loss, improvement arcs and so forth by Sadad *et al.* (2021).

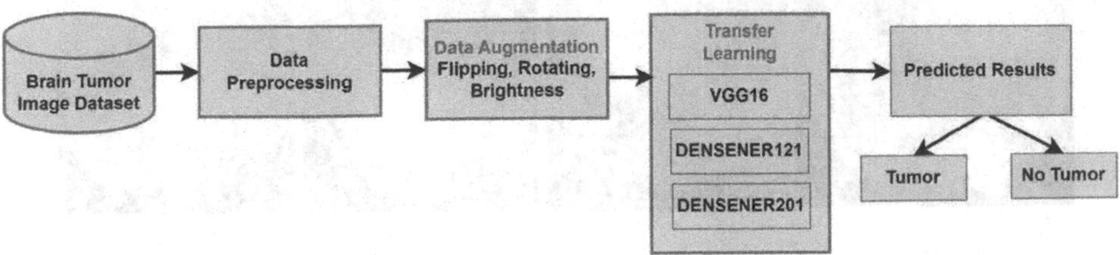

Figure 1: Complete architecture of proposed system.

3.1. *Dataset*

The publicly available set of data titled "The Brain MRI Pictures for Brain Tumor Detection" which was made accessible and uploaded by Tandel *et al.* (2020) is utilised for the suggested fix. Images having brain tumours and images without brain tumours, totalling 157 and 100 images, respectively, are separated by two unique categories in the entire set by Srinivas *et al.* (2019). Each one measures 467 × 586 × 3. There are just two sections to this database. The instructional portion and the verification portion are both given their own names by Ayadi *et al.* (2020). Figure 2 provides images of sample images of the dataset.

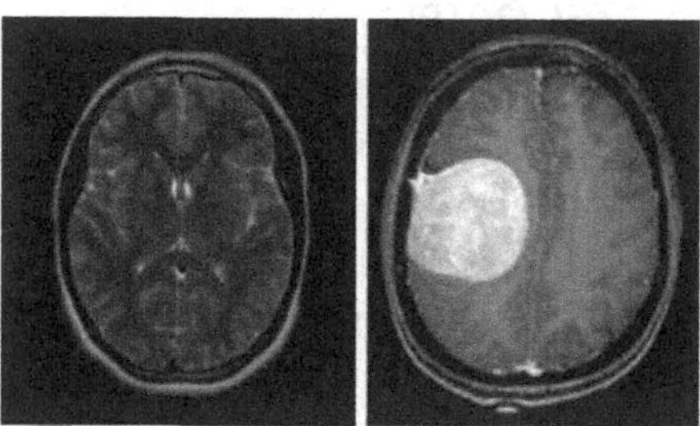

Figure 2: Sample images of brain tumors: (i) No Tumor and (ii) Tumor.

3.2. *Normalisation*

To maintain the dataset's numerical reliability for DL. For example, a normalisation pretreatment approach was used. These CT images begin as monochrome or grey formats with pixels with values ranging from 0 to 255. DL networks may be learned more quickly by normalising the source images by Saxena *et al.* (2021).

3.3. *Augmentation*

A sizable sample is needed to increase the efficacy of a DL system. However, various limitations frequently accompany the use of these databases by Shao *et al.* (2022). Information supplementation approaches are thus employed to boost the overall dataset's example picture count so as to overcome these problems by Juneja *et al.* (2021).

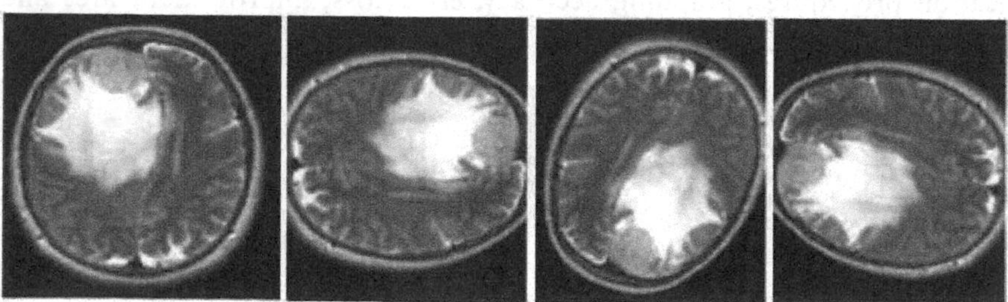

Figure 3: (a) Unique, (b) 90 degrees oppositely, (c) 180 degrees anticlockwise and (d) 270 degrees anticlockwise.

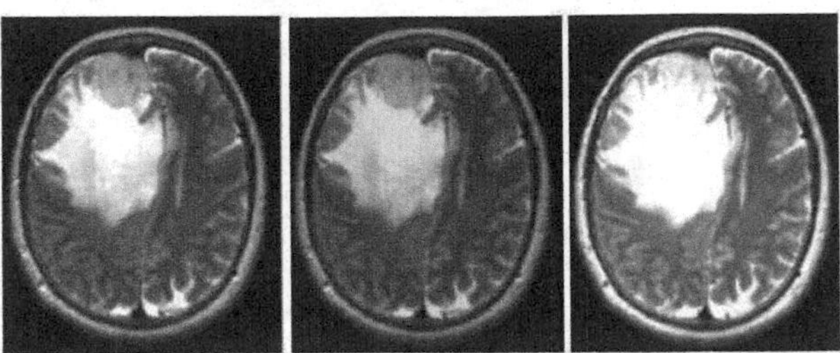

Figure 4: Dataset images having an intensity coefficient of 0.2 and an intensity value of 0.4. Intensity data enhancement.

The rotating enhancement approach, as seen in Figure 3, is applied clockwise at 90-degree angles on each side. A strong data improvement method shown in Figure 4 is also applied to a series of images utilising brightness component levels between 0.2 and 0.4. Furthermore, the supplied database has an educational mismatch. These information-augmented approaches are used to address the issue of disparity. The test database in every category was enlarged to 700–1000 photos after using these information enrichment approaches, while the whole database was subsequently revised to 1800 BT images.

4. Experimental Result

Several training CNN algorithms, including Dense-Net121, Dense-Net201, VGG16 and VGG19, are employed in a field study for the identification of BT using CT scans. The brain tumour dataset's CT scans were used to create the CNN algorithms. About 432 learning images and 104 test images have been used for instruction and relationships, respectively. The first resizing of the brain MRI pictures was from 467 × 586 to 224 × 224. The FastAI package was used to create an algorithm. The simulations undergo training with a maximum number of batches of 16 for learning by transfer. There were 20 development periods for all of the models. The number of periods and the selection of samples from groups are selected at random. Learning was carried out with the Adam optimizer by Juneja *et al.* (2021). Additionally, the development rate was established analytically. Using the basis of several performance criteria including efficiency, exactness, sensitiveness and particularity, the effectiveness of all models was assessed.

Several alternative approaches, each utilising an alternate number of epochs and batches, provide multiple metrics for performance with respect to training loss, validation loss, error rate and accuracy in validation by Tiwari *et al.* (2022). Four distinct algorithms that are studied with a total of 20 epochs and 16 batches each are Dense-Net121, Dense-Net201, VGG16 and VGG19. The author known as Adam is used to train each and every DL model. Although the efficiency of DenseNet121 and DenseNet201 is about identical, DenseNet201 has a greater number of levels than DenseNet121, which will lengthen the duration of processing. The efficiency characteristics of all the versions are still identical after 20 epochs.

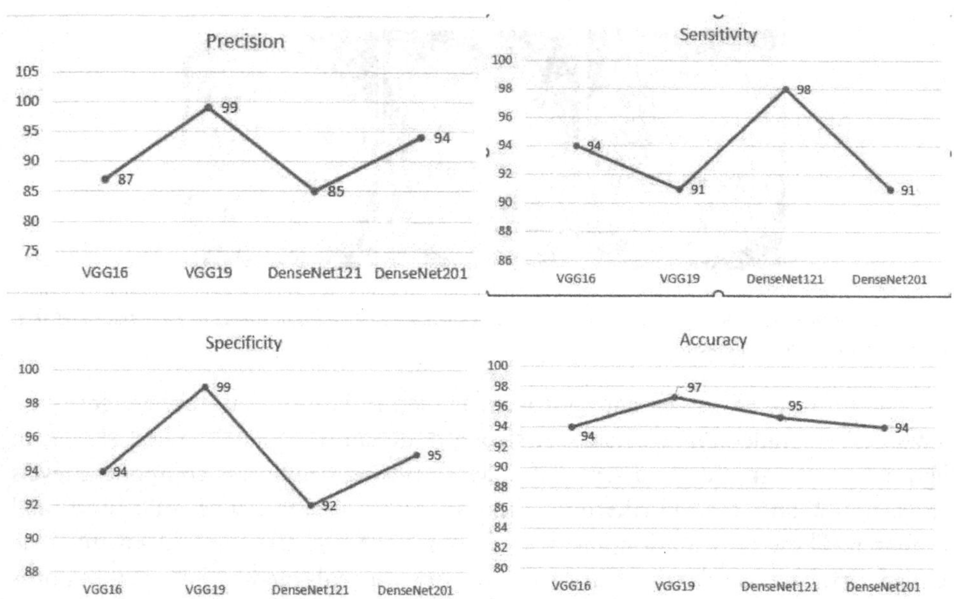

Figure 5: *Confusing matrix variables of (a) VGG16, (b) VGG19, (c) DenseNet121 and (d) DenseNet201 are represented graphically.*

Each of the subsection's positions, including the ones for BT and NT, are used as labels. The precision of every one of the models is assessed for batches of 16 using these confused matrices. Figure 5's charts are used to analyse the correctness of every model. Figure 6 shows that VGG19 and DenseNet201 are the top entertainers, with precision of 98% and 96%, respectively, for batches of 16. Analysis of the findings shows that VGG19 outperforms all other systems.

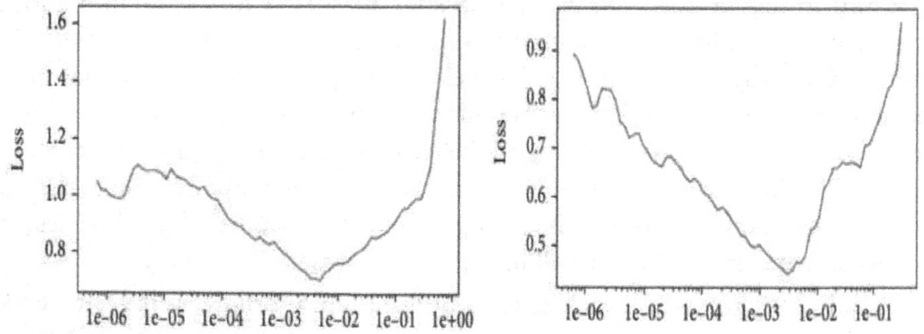

Figure 6: *Leaning rates vs. losses curves of suggested models (i) VGG19 and (ii) DenseNet201 using 16 batches of data.*

According to the analysis of the preceding debate, VGG19 outperforms other methods for batches of 16 when compared with them. Now, Figure 7 shows VGG19 and DenseNet201 improvement rate curves with batches of 16. The model's training rates, which determine how quickly or gradually an example acquires knowledge, are controlled by the rate of learning curves. The point is created when the amount lost decreases and begins to multiply as the training of learning rises by Díaz-Pernas *et al.* (2021). The rate of learning ought to be to the left of the graph's lowest value. For instance, the development value for VGG19 is depicted in Figure 6 (a), and since point 0.001 has the smallest loss, the training rate for VGG19 ought to be between 0.0001 and 0.001. Comparable to this, the smallest loss level in Figure 7(b), which displays the initialisation rates of DenseNet201, is 0.00001. Therefore, the rate of acquisition for Densenet201 ought to be at its smallest level, somewhere around 0.000001 and 0.00001; it is obvious that as training rates rise, losses also rise.

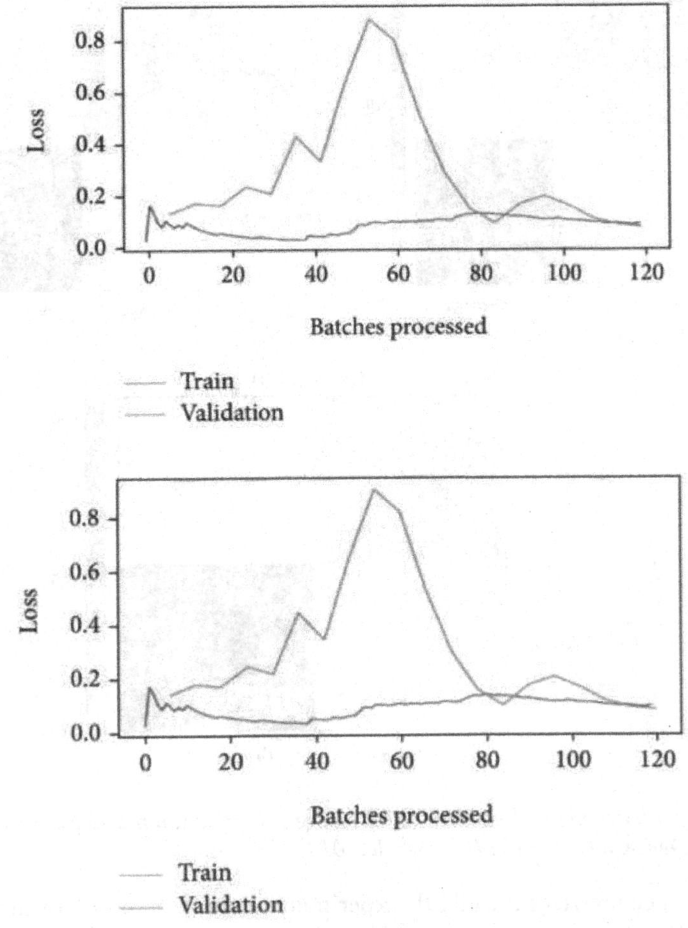

***Figure 7:** Processing batching in relation to loss curves for various CNN designs using 16 phases VGG19 & DenseNet201, respectively.*

Figure 7 displays the loss of convergence map for the VGG19 and DenseNet201 CNN algorithms for the number size 16. The fluctuations in losses that occurred when the systems were being trained are shown in Figure 8. The decline continued to decrease as the algorithms learned from the information, and it eventually stopped getting better throughout training. Additionally, each epoch's verification loss is computed. With increased epochs, the verification reveals loss rates that are comparatively stable and modest. The smallest losses for VGG19 and DenseNet201 at every stage for batch number 16 are seen via Figure 8. According to Figure 8, when 120 samples undergo processing, the losses achieved for VGG19 are significantly lower compared to those for Densenet201. While DenseNet201's validating and instruction losses range from 0.2 to 0.4, VGG19's losses are between 0 and 0.2. Therefore, it is obvious that at batch size 16, VGG19 outperforms DenseNet201 with regard to training and validating losses.

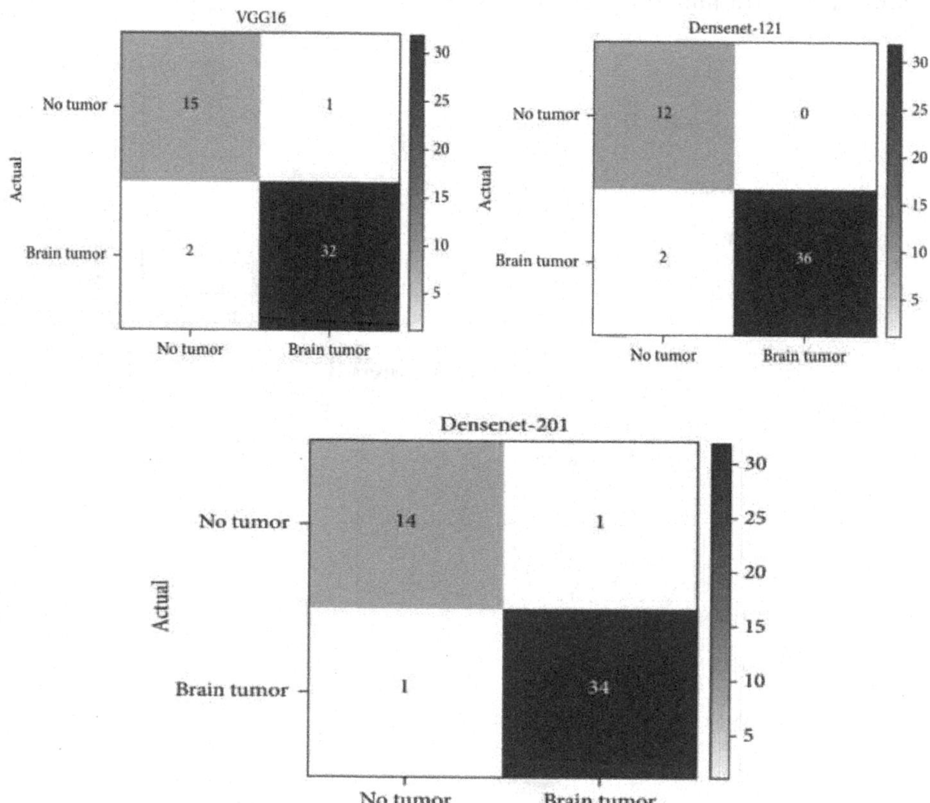

Figure 8: All 16-batch algorithms use the following complication matrices: (a) VGG-16, (b) VGG-19, (c) DenseNet121 and (d) DenseNet201

The variance matrices for all DL experiments from batches of 16 are shown in Figure 8. The one above shows equally correct and faulty predictions.

5. Conclusion

During the current analysis, each of the four DL models VGG16, VGG19, DenseNet121 and DenseNet201 are carefully assessed for their ability to identify BT. In comparison to various models utilising batch size 16, VGG19 and DenseNet201 produce the greatest outcomes. Author Navoneel Chakrabarty obtained a dataset for BT from Kaggle. These systems have been trained and then analysed to get the findings. The effectiveness of the VGG19 approach was further shown by these findings, which correctly handled the number of batches, optimisers and epochs. Using the VGG19 and Adam optimiser, reliability and precision of 98% and 94.73%, accordingly, were reached for batch size 16. In a similar vein the DenseNet201 for batches of 16 with the optimiser Adam obtained precision and sensitivity of 96% and 93.33%, respectively. The cost-effectiveness of these comparison data would enable radiologists to use an additional tool or simulation. To predict BT as early as feasible is the main goal of this study. This comparison methodology may be used by radiology as an additional assessment tool. This work aids in the creation of DL modelling by providing a more precise diagnostic. In the proposed investigation, only one unidirectional gathering of BT materials is used to perform training and confirmation, which is a drawback. By using coronal and sagittal information throughout validation and training, the suggested framework can eventually be further generalised. The efficacy of the suggested model might additionally be increased by using additional model training and optimisation strategies.

References

[1] Abiwinanda, N., Hanif, M., Hesaputra, S. T., Handayani, A., and Mengko, T. R. (2019). Brain tumor classification using convolutional neural network. *In World Congress on Medical Physics and Biomedical Engineering*, 218, 183–189, Springer, Singapore.

[2] Ayadi, W., Elhamzi, W., and Atri, M. (2020). A new deep CNN for brain tumor classification. *20th International Conference on Sciences and Techniques of Automatic Control and Computer Engineering*, 266–270.

[3] Bodapati, J. D., Shaik, N. S., Naralasetti, V., and Mundukur, N. B. (2021). Joint training of two-channel deep neural network for brain tumor classification. *Signal, Image and Video Processing*, 15(4), 753–760.

[4] Deepak, S., and Ameer, P. M. (2021). Automated categorization of brain tumor from MRI using CNN features and SVM. *Journal of Ambient Intelligence and Humanized Computing*, 12(8), 8357–8369.

[5] Díaz-Pernas, F. J., Martínez-Zarzuela, M., Antón-Rodríguez, M., and González-Ortega, D. (2021). A deep learning approach for brain tumor classification and segmentation using a multiscale convolutional neural network. *Health Care*, 9(2), 153.

[6] Gu, X., Shen, Z., Xue, J., Fan, Y., and Ni, T. (2021). Brain tumor MR image classification using convolutional dictionary learning with local constraint. *Frontiers in Neuroscience*, 15.

[7] Juneja, S., Juneja, A., Dhiman, G., Behl, S., and Kautish, S. (2021). An approach for thoracic syndrome classification with convolutional neural networks. *Computational and Mathematical Methods in Medicine*, 2021, 10.

[8] Juneja, S., Gahlan, M., Dhiman, G., and Kautish, S. (2021). Futuristic cyber-twin architecture for 6G technology to support internet of everything. *Scientific Programming. Hindawi Limited*, 1–7.

[9] Kader, I. A. E., Xu, G., Shuai, Z., Saminu, S., Javaid, I., and Ahmad, I S. (2021). Differential deep convolutional neural network model for brain tumor classification. *Brain Sciences*, 11(3), 352.

[10] Khan, M. A., Ashraf, I., Alhaisoni, M., *et al.* (2020). Multimodal brain tumor classification using deep learning and robust feature selection: a machine learning application for radiologists. *Diagnostics*, 10(8), 565.

[11] Kumar, R. L., Kakarla, J., Isunuri, B. V., and Singh, M. (2021). Multi-class brain tumor classification using residual network and global average pooling. *Multimedia Tools and Applications*, 80(9), 13429–13438.

[12] Mzoughi, H., Njeh, I., and Wali, A. *et al.* (2020). Deep multi-scale 3D convolutional neural network (CNN) for MRI gliomas brain tumor classification. *Journal of Digital Imaging*, 33(4), 903–915.

[13] Pallavi Tiwari, Bhaskar Pant, Mahmoud M. Elarabawy, Mohammed Abd-Elnaby, Noor Mohd, Gaurav Dhiman, Subhash Sharma. (2022). CNN Based Multiclass Brain Tumor Detection Using Medical Imaging. *Computational Intelligence and Neuroscience*, 2022, Article ID 1830010, 8 pages. https://doi.org/10.1155/2022/1830010.

[14] Rajasree, R., Columbus, C. C., and Shilaja, C. (2021). Multiscale-based multimodal image classification of brain tumor using deep learning method. *Neural Computing and Applications* (*NCA*), 33(11), 5543–5553.

[15] Rehman, Khan, Saba, T., Mehmood, Z., Tariq, U., and Ayesha, N. (2021). Microscopic brain tumor detection and classification using 3D CNN and feature selection architecture. *Microscopy Research and Technique* (MRT), 84(1), 133–149.

[16] Saxena, P., Maheshwari, A., and Maheshwari, S. (2021). Predictive modeling of brain tumor: a deep learning approach. *Innovations in Computational Intelligence and Computer Vision*, Springer, Singapore.

[17] Sadad, T., Rehman, A., and Munir, A. *et al.* (2021). Brain tumor detection and multi-classification using advanced deep learning techniques. *Microscopy Research and Technique*, 84(6), 1296–1308.

[18] Sajjad, M., Khan, S., Muhammad, K., Wu, W., Ullah, A., and Baik, S. W. (2019). Multi-grade brain tumor classification using deep CNN with extensive data augmentation. *Journal of Computational Science*, 30, 174–182.

[19] Shao, C., Yang, Y., Juneja, S., and Seetharam, T. G. (2022). IoT data visualization for business intelligence in corporate finance. *Information Processing and Management*, 59(1), 102736.

[20] Srinivas, and Rao, (2019). A hybrid CNN-KNN model for MRI brain tumor classification. *International Journal of Recent Technology and Engineering*, 8(2), 5230–5235.

[21] Tandel, Balestrieri, A., Jujaray, T., Khanna, Saba, L., and Suri, (2020). Multiclass magnetic resonance imaging brain tumor classification using artificial intelligence paradigm. *Computers in Biology and Medicine*, 122, 103804.

[22] Xu, Jia, Z., Ai, Y. *et al.* (2015). Deep convolutional activation features for large scale brain tumor histopathology image classification and segmentation. *In 2015 IEEE International Conference on Acoustics, speech and signal Processing*, 947–951.

Chapter 16

Classification of Protein Family from the Protein Sequence Using Machine Learning

Priyanka R. Rajmane[1] *and Simran R. Khiani* [2]

[1]Research Scholar, Department of Computer Science & Engineering, G H Raisoni Amravati University, Amravati, Maharashtra, India

[2]Assistant Professor and HOD, Computer Science & Engineering, G H Raisoni College of Engineering and Management, Pune, Maharashtra, India

Abstract: The classification of protein families relies on the protein sequence, which represents the amino acid sequence of the protein. Proteins consist of a linear chain of amino acids. Although numerous protein sequences have been identified and added to databases, accurately classifying them into their respective families remains challenging. Many laboratory experiments are underway, but the classification process takes time. To address this issue, statistical techniques are necessary to classify amino acid sequences and associate them with their corresponding protein families. This study utilised a Kaggle dataset comprising a total of 246,213 protein samples. This study applied various machine learning classification algorithms to this dataset, including logistic Regression (LR), random forest (RF), Adaboost, Xgboost and support vector machine (SVM). Among these classifiers, the RF algorithm exhibited the highest efficiency of 55.70% for accurately classifying protein families, surpassing other algorithms.

Keywords: Protein family, machine learning, amino acid sequence, feature extraction.

1. Introduction

Machine learning (ML) is most suitable for computer vision-related problems like data science and data mining (Machine Learning Encyclopedia Britannica, 2016). From that point forward upgrades have been made to some degree empowered by the entrance to more prominent computational assets, particularly designs handling units (GPU), empowering the preparation of profound neural systems containing numerous boundaries in sensible time. Given this, specific neural system designs like convolutional neural network (CNN) and recurrent neural network (RNN) with long short-term memory (LSTM) can presently be prepared proficiently and have been effectively applied to numerous issues including biomedical issues

DOI: 10.1201/9781003527442-16

by Chou and Zhang (1995); Ahmad *et al.* (2004). The prediction of amino acid sequences using machine learning techniques has gained significant attention in the field of bioinformatics and protein research. Amino acid sequences play a crucial role in determining the structure, function and properties of proteins by Yu *et al.* (2006). Machine learning algorithms offer a promising approach to predicting and classifying these sequences, enabling researchers to gain insights into protein structure and function without relying solely on time-consuming experimental methods suggested by Qiu *et al.* (2004). Machine learning models leverage the vast amount of data available from protein databases, such as the Protein Data Bank (PDB) and UniProt, to learn patterns and relationships between amino acid sequences and their corresponding properties. These models can then be used to make predictions on new, unseen sequences by Gromiha *et al.* (2005). Various encoding schemes have been developed, such as one-hot encoding, which converts each amino acid residue into a binary vector representation. Other encoding techniques, like position-specific scoring matrices (PSSMs), capture evolutionary information by considering the frequency of amino acids at specific positions within a sequence. Once the sequences are appropriately encoded, they can be fed into machine learning algorithms for training and prediction. Popular algorithms applied to this task include logistic regression, RF, support vector machines, artificial neural networks and gradient boosting methods like XGBoost and AdaBoost Gromiha *et al.* (2006; 2008). These algorithms learn from labelled training data, where the amino acid sequences are associated with their respective protein properties or classes, such as protein families or structural characteristics by Sandaruwan *et al.* (2021). To improve prediction accuracy, feature engineering techniques can be applied, incorporating additional information such as physicochemical properties of amino acids, secondary structure predictions or evolutionary conservation scores. The prediction of amino acid sequences using machine learning has diverse applications in protein structure prediction, protein function annotation, drug discovery and understanding the impact of genetic variations on protein behaviour by Zhang *et al.* (2019). By harnessing the power of machine learning, researchers can expedite the analysis of protein sequences, enabling faster and more efficient characterisation and exploration of protein space. Table 1 shows the names of amino acids and three-letter and single-letter codes.

Table 1: Representation of three letters and one letter symbol of amino acid.

Amino Acid	Three Letter	One letter	Amino Acid	Three Letter	One letter
Alanine	Ala	A	Methionine	Met	M
Cysteine	Cys	C	Asparagine	Asn	N
Aspartic acid or aspartate	Asp	D	Proline	Pro	P
Glutamic acid or glutamate	Glu	E	Glutamine	Gln	Q

Phenylalanine	Phe	F	Arginine	Arg	R
Glycine	Gly	G	Serine	Arg	R
Histidine	His	H	Threonine	Thr	T
Isoleucine	Ile	I	Valine	Val	V
Lysine	Lys	K	Tryptophan	Trp	W
Leucine	Leu	L	Tyrosine	Tyr	Y

The secondary and tertiary structures of proteins as well as their uses can be understood by studying the amino acid sequences' orientation and movement along clusters in the protein sequencing. Protein successions' distinctive evidence of comparison motifs might help with predicting the fundamentally or functionally significant domains. A set of amino acids with similar qualities can be found using profiling of individual amino acid features based on amino acid sequencing by Vazhayil Anu *et al.* (2018). Additionally, the analysis of different amino acid patterns using arrangement methodologies could enhance the understanding of the availability of comparable groups, and these combinations might be used as a framework for protein 3-D structure prediction.

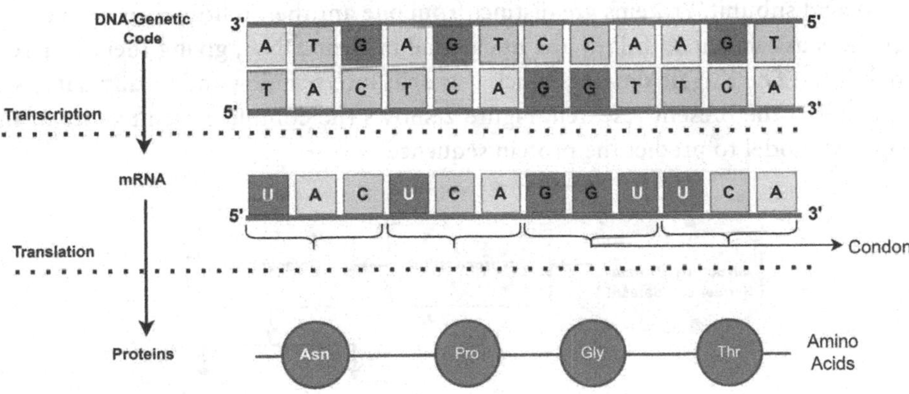

Figure 1: DNA sequencing that encodes amino acid sequence.

The amino acid composition is the number of amino acids of each type normalised with the total number of residues. It is defined as

$$\text{Func (a)} = \sum ri * 100 / R \quad 1$$

where *a* represents the 20 amino acid residues, *ri* represents the total amount of residues of each amino acid and R represents the total residues of amino acid.

The amino acid composition refers to the distribution and abundance of different amino acids within a protein sequence. For instance, in the case of T4 lysozyme, which consists of 164 residues, the amino acid composition provides information about the occurrence of each of the 20 amino acid residues in this protein. For example, it may indicate that T4 lysozyme contains 15 alanine (Ala) residues, 10 aspartic acid (Asp) residues, 2 cysteine (Cys) residues and so on. Amino acid groupings contain a wealth of hidden information that can be harnessed to develop sequence-based prediction methods. Valuable amino acid properties can be derived from protein sequence data. Recent findings have highlighted the significant role played by the arrangement of amino acid residues in distinguishing proteins belonging to different structural classes, folding types and functional categories. These studies also encompass predictions related to protein structural classes by Zhang *et al.* (2020).

The organisation of the paper is as follows; Section 2 presents the material and methods used in the study. Section 3 presented the experimental result analysis. Section 4 and Section 5 presented the discussion and conclusion.

2. Proposed Methodology

This study uses a protein sequence dataset to evaluate the effectiveness of the proposed machine learning models utilising LR, RF, Adaboost, Xgboost and SVM. The proposed machine learning models are addressed after a discussion of the protein dataset in general. The primary arrangement of proteins is represented as a set of 20 English alphabets, each of which is connected to a distinct protein base amino acid subunit. Proteins are distinct from one another in how their amino acid sequences are meant to follow the nucleotide in their DNA, giving them different capabilities. The Kaggle dataset, which was split into training and testing data, was employed in the present research. Figure 2 shows the complete architecture of the proposed model to predict the protein sequence.

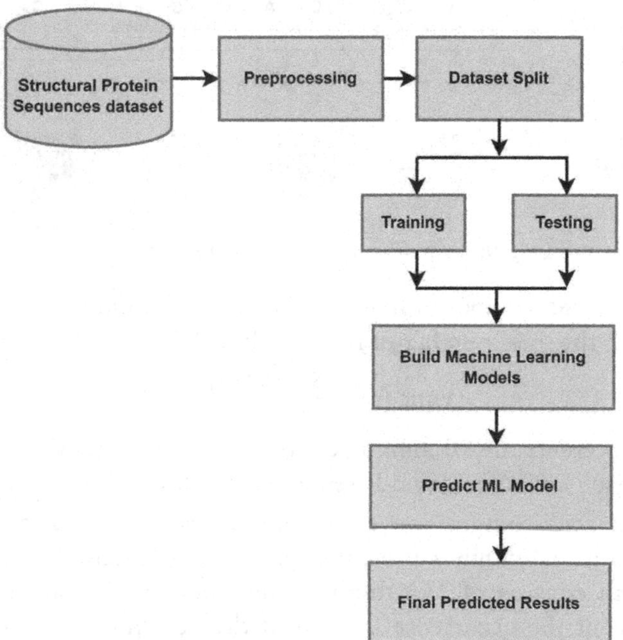

Figure 2: Complete architecture of proposed system.

2.1. Dataset

This study used the publically available structural protein sequences dataset which is extracted from Kaggle. The dataset is split into two parts. The first part contains meta-information about the protein sequence, which is used to classify the protein sequence, feature extraction and other relevant information. The second part contains the actual protein sequence structure. This dataset contains approximately 60,000 protein sequences which are classified into 30 distinct families. In this research, we specifically focused on 5,050 distinct protein classes within the dataset.

2.2. Feature Extraction

The extraction of features includes a task for the analysis of data in machine learning suggested by Bandyopadhyay *et al.* (2010). It is possible to derive some details from the amino acid used to differentiate the type of protein sequence as a feature. The protein feature can be derived from a 20 amino acid combination in different ways. This section addresses some of the most popular suggested approaches in the literature.

A protein sequence known as local structural features or local motifs is possibly the most common feature seen in Chowriappa *et al.* (2009). Protein style is the local structural motif of the protein family and is applied for protein sequence prediction. Generally, a protein sequence is formed using the combination of 20 different amino acid notations such as A, C, D, E, F, G, H, I, K, L, M, N, P, Q, R, S, T, V, W and Y. n-gram feature is extracted for the classification of protein sequence with a pair of input variables (F_m, C_m) to the respective classifier, whereas F_mis feature m and C_m is the count of n number of features. If a 3-gram protein sequence is extracted then all the possible combinations as AAA, AAC, AAD, AAE,... and so on from the 20 amino acid sequence are extracted. Feature is the large number of amino acids present in the protein sequence. If we take the protein sequence TCCDTVCGH as an example, the 2-gram feature is extracted such as {(TC,1), (CC,1), (CD,2), (DT,2)}. One more approach used for extracting data from protein sequences is a 6-letter group. Six combinations of the alphabets from the complete set are P = {F,G,I}, Q = {C,E,L,D}, R = {T}, S = {S,K,M,V,R}, T = {H,N,I,W} and U = {A,Q,Y}. In this approach six group of letters is used the above protein sequence TCCDTVCGH can be transformed into the RQQQRSQPT and 2-gram feature are {(QR, 1), (RQ, 2), such (QQ, 4)}.

2.3. Protein Classification

Given available protein dataset allows for highly accurate classification of all protein characteristics. The main approach for acquiring word structure depends upon a neural network built with projected layers by Satpute and Yadav (2018). The weight of available protein sequence $W_{1,}W,\ldots\ldots W_{a-n+1}$, the main objective of the proposed model is to maximal the mean log probabilities.

$$\max \frac{1}{R}\sum_{p=1}^{N} \sum_{-d\leq m\leq c, q\neq 0} logS(V_{p+q} \mid V) \qquad 7$$

Let *d* represent is the distance word; log probability can be defined as

$$logP\left(w_{n+m}|w_n\right) = \log\frac{e^{vt}\,V\,v_{Vi}}{\sum_{W=1}^{W} e^{vt}\,V\,v_{Vi}} \quad 8$$

where the v_{Vi} and v'_{Vi} are the input and output vector of word, V respectively.

2.4. Data Preprocessing

The data is about the protein sequence is extracted from corresponding Kaggle. The dataset contains 141 and 401 different samples of protein sequences with unique features. The contents of a group of macromolecules to find proteins, hybrid proteins, DNA, RNA and so on. The macromolecule dataset style material is shown in Figure 2. About 87% of protein samples and the remaining sequences are mixed DNA and RNA proteins. We just have the samples pertain to protein. In data analysis, protein was 87% of the macromolecules in the data collection. The substances we removed were other than proteins. The macromolecule Dataset style material as shown in Figure 3.

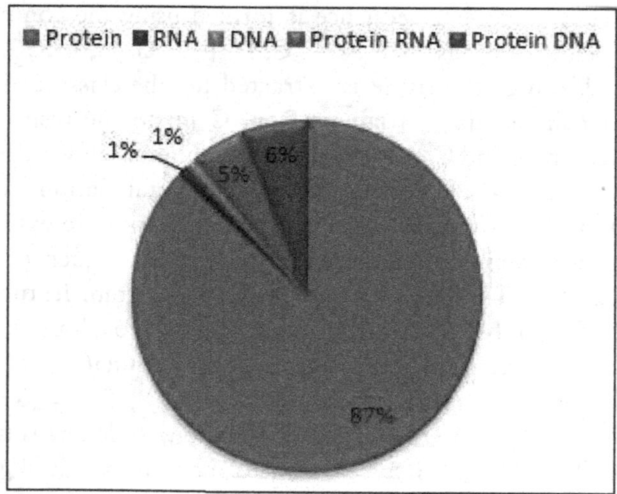

Figure 3: *Various protein samples.*

2.5. Class Frequency

The bar represents the class frequency of 20 common classes of protein sequence as shown in the Figure 3. We found that hydrolase was the most identical portion of any protein and other classes were decreased of the protein sequence. We refine less than 1,200 protein sequences. After refining process, 15 distinctive protein family features have been classified. Fifteen protein families are converted into the numerical labels before applied random forest classifier. The Bar represents the class frequency of 20 common classes of protein sequence as shown in the figure 4.

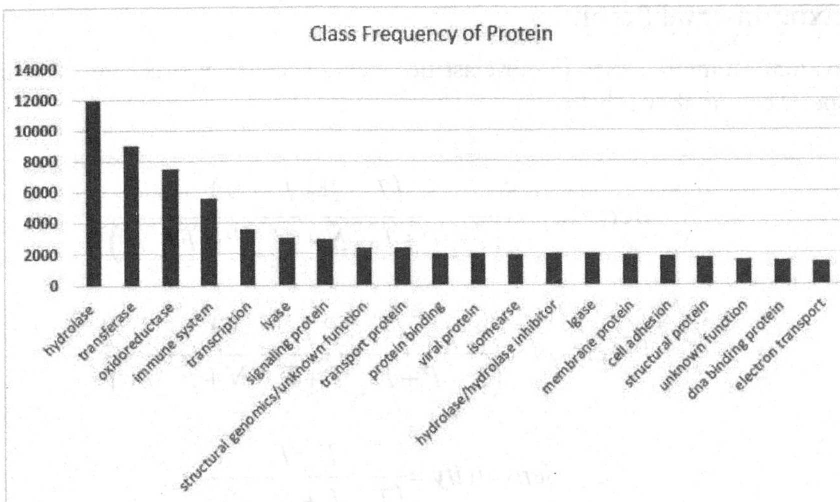

Figure 4: Frequency of protein class.

2.6. Random Forest Classifier

The RF classifier was found to be the best performer among machine learning predictors for protein classification from the training and testing of various machine learning models by Huynh *et al.* (2019) and Gupta *et al.* (2019). By measuring the feature importance, we analysed the classification performance.

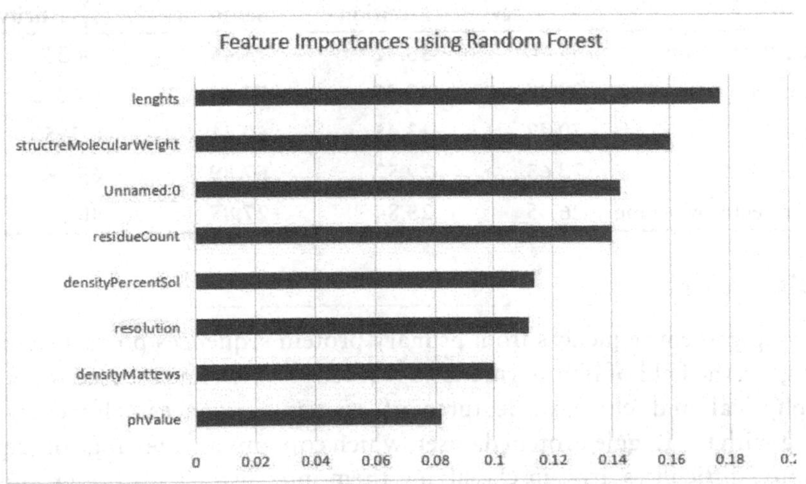

Figure 5: Feature importance for random forest regression.

Upon analyzing the feature importance graph of the RF classifier as shown in Figure 5 model in Figure 5, we observed that no features exhibited significantly higher significance. However, the length, weights and counts displayed slightly greater importance compared to other features.

3. Experimental Result

The evaluation metrics used in our classifier include accuracy, precision, sensitivity and specificity as shown below.

$$Accuracy = \frac{(T\_P + T\_N)}{(T\_P + T\_N + F\_P + F\_N)}$$

$$Precision = \frac{(T\_P + T\_N)}{(T\_P + F\_P + T\_N + F\_N)}$$

$$Sensitivity = \frac{T\_P}{(T\_P + F\_N)}$$

$$Specificity = \frac{T\_P}{(T\_N + F\_P)}$$

Table 2: Comparative analysis of classification measure metrics.

Algorithm	Accuracy	Precision	Sensitivity	Specificity
Logistic Regression	55.70	56.32	75.45	74.23
Random Forest	33.40	30.41	73.56	0.6082
Adaboost	30.42	32.45	60.46	0.5952
Xgboost	28.63	29.52	67.89	69.56
Support Vector Machine	26.45	28.84	27.85	40.65

4. Discussion

Identifying protein sequences from primary protein sequences poses a significant challenge in the field of bioinformatics. Previous research has focused on utilising both physical and chemical features of protein sequences. However, when working with the Kaggle protein dataset, which contains a wide range of sequence variations, difficulties arise in classifying them into distinct feature groups. The n-gram protein families within the Kaggle dataset exhibit substantial sequence variations, making it challenging to classify them into local structural families. To address this issue, a distance-based feature extraction technique was employed to extract features for classifying protein sequences into various superfamilies. In our experiments, we utilised machine learning classification algorithms such as LR, RF, Adaboost, Xgboost and SVM. These classifiers demonstrated their effectiveness in classifying protein sequence features. The classification results for each classifier are presented in Table 2. To evaluate the performance of the different classifiers, we conducted a performance analysis using metrics such as specificity, sensitivity and precision. By employing state-of-the-art machine learning methods, we achieved the highest predictive accuracy recorded thus far. Notably, the RF classifier exhibited an accuracy of 50.70%, surpassing the other classifiers in performance.

5. Conclusion

This paper applied various machine learning classifiers to classify the protein sequences belonging to numerous protein families in a database extracted from the publicly available Kaggle dataset. For this classification task, we incorporated macroscopic features such as molecular weights, amino acid unit lengths, total residue counts in each sequence, pH value and molecular densities as input for the different machine-learning classifiers. To validate the technique, assessed its performance of proposed machine learning models which resulted in a notable enhancement in classification accuracy score, and other parameters like precision, sensitivity and specificity.

References

[1] Ahmad, S., Gromiha, M., and Sarai, A. (2004). Analysis and prediction of DNA-binding proteins and their binding residues based on composition, sequence and structural information. *Bioinformatics*, 20(4), 477–486.

[2] Bandyopadhyay, D. (2010). Functional neighbors: inferring relationships between non-homologous protein families using family-specific packing motifs. *In IEEE Transactions on Information Technology in Biomedicine*, 14(5), 1137–1143. https://doi.org/10.1109/TITB.2010.2053550.

[3] Chowriappa, P., Dua, S., Kanno, J., and Thompson, H. (2009). Protein structure classification based on conserved hydrophobic residues. *In IEEE/ACM Transactions on Computational Biology and Bioinformatics*, 6(4), 639–651. https://doi.org/10.1109/TCBB.2008.77.

[4] Chou, K., and Zhang, C. (1995). Prediction of protein structural classes. *Crit Rev Biochem Mol Biol*, 30(4), 275–349.

[5] Gupta C., Bihari A., and Tripathi S. (2019). Human protein sequence classification using machine learning and statistical classification techniques. *International Journal of Recent Technology and Engineering* (IJRTE), 8(2). https://doi.org/10.35940/ijrte.B3224.078219.

[6] Gromiha, M. (2005). A statistical model for predicting protein folding rates from amino acid sequence with structural class information. *J Chem Inf Model*, 45(2), 494–501.

[7] Gromiha, M., and Suwa, M. (2006). Discrimination of outer membrane proteins using machine learning algorithms. *Proteins*, 63(4), 1031–1037.

[8] Gromiha, M., and Yabuki, Y. (2008). Functional discrimination of membrane proteins using machine learning techniques. *BMC Bioinformatics*, 9, 135.

[9] Huynh, P., Nguyen, V., and Do, T. (2019). Novel hybrid DCNN–SVM model for classifying RNA-sequencing gene expression data. *Journal of Information and Telecommunication*, 3(4), 533–547. https://doi.org/10.1080/24751839.2019.1660845.

[10] Ma, P., and Chan, K. (2008). An effective data mining technique for the multi-class protein sequence classification. *2nd International Conference on Bioinformatics and Biomedical Engineering*, 486–489. https://doi.org/10.1109/ICBBE.2008.118.

[11] Naveenkumar, K., Mohammed, B., Vinayakumar, R., and Soman, K. (2018). Protein family classification using deep learning. *bioRxiv*, 414128. https://doi.org/10.1101/414128.

[12] Qiu, J., Huang, J., Liang, R., and Lu, X. (2009). Prediction of G-protein-coupled receptor classes based on the concept of Chou's pseudo amino acid composition: an approach from discrete wavelet transform. *Anal Biochem.*, 390(1), 68–73.

[13] Satpute, B., and Yadav, R. (2018). Machine intelligence techniques for protein classification. *2018 3rd International Conference for Convergence in Technology* (I2CT), 1–4. https://doi.org/10.1109/I2CT.2018.8529495.

[14] Satpute, B., and Yadav, R. (2018). Support vector machines for protein family identification using surface invariant coordinates. *3rd International Conference for Convergence in Technology* (I2CT), 1–4. https://doi.org/10.1109/I2CT.2018.8529688.

[15] Zhang D., and Kabuka, M. (2020). Protein family classification from scratch: a CNN-based deep learning approach. *In IEEE/ACM Transactions on Computational Biology and Bioinformatics*. https://doi.org/10.1109/TCBB.2020.2966633.

Chapter 17

Teacher Evaluation with Sentiment Analysis of Students Comments Using Machine Learning and Deep Learning Approaches

Bhavana P Bhagat[1] *and Dr.Sheetal S. Dhande-Dandge*[2]

[1]Sr. Lecturer Department of Computer Engineering Government Polytechnic Gondia (M.S), India

[2]Professor, Department of Computer Science & Engineering Sipna College of Engineering and Technology, Amravati (M.S), India

Abstract: The opinions of students play a vital role in the evaluation of teachers. Classroom education becomes feasible and efficient based on the increased involvement of the student. Many researchers worked on the sentiment classification of students. Nowadays, sentiment analysis using machine learning and deep learning models has gained more popularity. Ensemble techniques in machine learning that is random forest, and deep learning that is LSTM+CNN and LSTM with single-head attention, were used to identify the sentiments of students. In the proposed system, input sequences of sentences are processed using different models and tested with different dropout rates to increase accuracy. Long short-term memory (LSTM) with attention layers gives more attention to the influence of words on emotions. We experimented using different machine learning and deep learning methods and concluded that the LSTM with an attention layer accompanied improves the result over the other. The experimental results show that the LSTM with a single head attention layer can achieve an accuracy of 92.04% on the student feedback dataset over other methods.

Keywords: CNN, deep learning, LSTM, opinion mining, random forest, sentiment analysis.

1. Introduction

Classroom teaching is a viable and efficient method to improve student learning and foster participation in the teaching-learning process. The effectiveness of teaching depends on pedagogical capability, setting outcomes, simplicity and course expertise. To measure the effectiveness of teaching, student's opinions help formulate better choices on how to enhance teaching excellence. Researchers have developed a number of techniques and tools to measure student opinions in (O'Shea and Ryan, 2015). In this work, we focus on the education domain for sentiment analysis

DOI: 10.1201/9781003527442-17

of student comments. Many studies have been done on the Vietnamese student feedback data, which is widely available to explore and is human-annotated in Vietnamese. In order to carry out our work in sentiment analysis, we first converted the Vietnamese dataset to English, containing 21,561 sentences and trained it using a machine learning and deep learning model (Nguyen *et al.*, 2018). Another dataset used is the final-year student's comments for testing purposes.

Some studies (Atif, 2018), (Aung and Myo, 2017), (Newman and David, 2018), (Rajput *et al.*, 2016), (Rani *et al.*, 2017) used lexicon-based approach, some studies (Aung and Myo, 2017), (Balahadia *et al.*, 2016), (Newman and David, 2018), (Rajput *et al.*, 2016), (O'Shea and Ryan, 2015) do not evaluate results using evaluation metrics, some do not include the details of performance testing of the model. Also, there is a trend to use hybrid models to achieve better results in the sentiment analysis process. Some models do not provide graphical results for sentiment analysis, nor do they provide any facts or study to evaluate the visual usability. Many researchers focus on e-learning. Limited models use the primary corpus and the machine and deep learning methods, along with evaluation metrics (Bhagat and Sheetal, 2023).

The proposed system implements machine learning and deep learning models and compare their results to resolve the problem of text classification and compares their accuracy. Also, used to identify the best model, based on evaluation metrics for student's comments. The comparative analysis states that the deep learning methods outperformed ensemble learning for the sentiment analysis task performed on student comments. The proposed system also uses two datasets for testing and training that is Vietnamese student feedback corpus as well as and data collected from a final-year student's comment.

Section 2 will briefly explain related work. Section 3 begins with an outline of the detailed methodology used in the proposed work. Section 4 presents experimental results and performance evaluations and comparisons with competing methods. Section 5 will discuss the conclusion and future scope.

2. Related Work

The field of education is going through a revolution. In the education sector, for major users like students, and educational institutions, sentiment analysis is extremely important. Various machine learning and deep learning approaches will be applied to evaluate the student's feedback. In this section, we will discuss recent work implementing machine and deep learning techniques to analyse sentiment in the context of education (Atif, 2018).

In their work on sentiment analysis, some studies (Atif, 2018), (Aung and Myo, 2017), (Newman *et al.*, 2018), (Rajput *et al.*, 2016), (Rani *et al.*, 2017) adopted the lexicon-based method. They frequently employed labelled data. However, doing so requires strong linguistic tools, which are occasionally unavailable. Performance in terms of accuracy is good in comparison to other methods. In their research, some researchers (Altrabsheh *et al.*, 2014), (Balahadia *et al.*, 2016), (Chauhan *et al.*, 2019), (Esparza *et al.*, 2017), (Gutiérrez *et al.*, 2018), (Kaur et al., 2020), (Lalata *et al.*, 2019), (Pong-Inwong *et al.*, 2014), (Rajput *et al.*, 2016) employed machine learning-based methodologies. The model is initially trained with known inputs and outputs in order to function with later unknown data. Some algorithms perform more accurately than others when used with large corpora (Bhagat *et al.*, 2023). In some studies (Chandra and Jana 2020), (Lalata *et al.*, 2019), (Mabunda *et al.*, 2021), (Nasim *et al.*, 2017), (Qaiser, 2021) hybrid methods were used to improve the accuracy of their work. Lastly, it is seen that several studies (Cabada *et al.*, 2018), (Nguyen *et al.*, 2018), (Sangeetha and Prabha, 2021), (Sindhu *et al.*, 2018) used deep learning models as one of the dominant models in the education domain with higher performance.

As mentioned in (Sangeetha and Prabha, 2021), deep learning is popular nowadays because of its automatic learning features and also it deals with long sequences of sentences rather than short sequences. It is observed that in previous work, a Vietnamese student feedback corpus consisting of 16,175 student feedback sentences was used along with LSTM with multi-head attention model. Deep learning techniques are used in the proposed model in order to meet the challenges. In our proposed model, the Vietnamese dataset, along with the primary dataset of final-year student feedback collected in 2023, is used for testing the models.

3. Methodology

In order to find best model, it is important to establish a series of steps that allow us to explore different model configurations such as fully connected networks, ensemble techniques, dense and dropout layers, weight initialization and use of activation functions, learning rate with optimizers as well as other hyper-parameters. It is also important to decide batchthe size, number of epochs and the evaluation metricsused.

3.1. Preprocessing

In this work, the student feedback dataset used by Nguyen *et al.* (2018). It has some primal data, such as abbreviations, spelling mistakes, emotions, signs, repeated and unwanted characters, numbers and symbols. To clean the text data, a text preprocessing method is used. Stemming is the method of reducing a word to its stem, which affixes to prefixes and suffixes or to the roots of words called "lemmas". It is used to normalise the text and make it easier to process. Lemmatization analyses the context and converts the word to its meaningful form, which is called a lemma. The bag of words model, which is called BoW, is used for extracting features from the text to be used in modelling. N-grams are continuous order of words used for a variety of things like auto-completion of sentences, auto-spell check and also to test for grammar in each sentence in NLP. A tokenizer is used to divide paragraphs and sentences into smaller parts that can be easily assigned meaning.

3.2. *Word Embedding*

DL algorithms accept a vector representation of numbers as input. Word embedding is a technique for performing the representation of words in a vector space. It is a technique for translating words into a mathematical domain where numbers capture the semantics of the words. The process is done using millions of phrases (Aung and Myo, 2017). The most common encoding method such as Tfidfvectorizer and Word2vec have been used. The Word2vecis based on an unsupervised ML method based on a large amount of text to achieve syntactic and semantic representation in (Mikolov *et al.*, 2013).

3.3. *Random Forest*

It is a decision tree ensemble made out of a random selection of decision trees. It is a tree predictor combination in which each tree is dependent on the values of a random vector sampled independently and with the same distribution for all trees in the forest. In this, decision tree has been used as a basic learner. Each tree was constructed using bootstrap samples from the training data. A random feature selection provides variation among the base learner. Bagging is used to reduce the variation of algorithms with a high variance. A noise or an outlier in a single tree classifier can affect the classification model's performance; however, random forest merged various weak learners to build a strong learner that can handle noise or outliers due to the randomness it can give. It is also capable of handling large datasets with a large number of attributes. More crucially, it retains accuracy even when data is unavailable (Breiman, 2001).

3.4. *Convolutional Neural Network (CNN)*

CNNs are a special type of multilayer neural network that learns from data. CNNs are the same as neural networks, having weights, biases and input with which they make a scalar product and apply an activation function. Due to the increased number of hidden layers, there is a need for more computational resources; hence, overfitting arises. Thus, CNNs help to overcome problems by dividing models into small parts of information and combining this information in the deepest layers of the neural network. They are very useful for identifying classes, objects and groups in images by looking for shapes in them. This framework has convolutional, pooling and completely connected layers. A deep learning model needs a lot of computational power and data to train. As a result, CNNs were strictly constrained to limited sectors and were unable to enter the machine learning space (O'Shea and Ryan, 2015).

3.5. *Long Short-Term Memory Network*

LSTM is one of the commonly used RNN networks capable of learning long-term dependencies in classification models. The correct result is achieved when LSTM is extended with other models, and it may serve as the new base for additional models. The vanishing gradient problem, which affects neural networks, makes it challenging to comprehend the parameters of the previous layer. A LSTM in combination with other models helps to solve this issue. In LSTM architecture, (Hochreiter and Schmidhuber, 1997), (Nguyen *et al.*, 2018) and (Sangeetha *et al.*, 2021), the memory cells stores and get information over long dependencies, consist of three gates called the input, forget and output gates. Input gate i_t signifies the new information in the cell using input x_t, previous hidden layer, recurrent weight and bais. A forget gate f_t determines the extent to which the previous memory cell is forgotten. The information obtained from current input xt is given by output gate u_t. The input gate i_t chooses which information from output gate u_t to store in memory cell c_t with a condition forget gate f_t and finally output o_t of the LSTM depends on o_t and h_t.

$$i_t = \sigma\,(W^{(i)}x_t + U^{(i)}h_{t-1} + b^{(i)})\ (1)$$

$$f_t = \sigma\,(W^{(f)}x_t + U^{(f)}h_{t-1} + b^{(f)})\ (2)$$

$$u_t = \tanh(W^{(u)}x_t + U^{(u)}h_{t-1} + b^{(u)})\ (3)$$

$$c_t = i_t \cdot u_t + f_t \cdot c_{t-1}\ (4)$$

$$o_t = \sigma(W^{(o)}x_t + U^{(o)} + h_{t-1} + b^{(o)})\ (5)$$

3.6. LSTM with Attention Layer

It combines the attention layer with the LSTM. It is employed to bring attention to the long sequence's referral word. (Petersy *et al.*, 2018). By focusing on important information that relates to the target value, the attention layer has useful information and enhances model performance compared to baseline models. Although it concentrates on the desired value, it doesn't exclude any essential information. In contrast to the other approaches, it can handle word sequences of different lengths. The attention model emphasises semantic elements while also capturing the significance of each context word. They suggested an attention-based LSTM that outperforms the prior model in terms of output.

4. Experimental Results and Performance Evaluation

We evaluated our model's performance on two data corpuses: the Vietnamese student feedback corpus as well as data collected on our own from a final year student's comment. In our model, 21,561 rows of comments were used for sentiment analysis and 230 comments for the prediction on unseen data. The dataset contains neutral, negative and positive sentiments, which are given in Table1.

4.1. Random Forest

In this approach, a hybrid classifier, which is a combination of multiple base classifiers using voting ensembles, is used. Five different base classifiers, each with a specific algorithm, were used. Firstly, two random forest classifiers with decision trees were used. Later, two AdaBoost classifiers with decision trees as weak learners were used, and finally, one gradient boosting classifier with decision trees was used. For multi-class classification problems, the "OneVsRestClassifier" is used to perform multi-label classification, where each classifier is trained to handle a specific class, treating all other classes as a single "rest" class. The "VotingClassifier" is an ensemble method that combines the predictions of the base classifiers using majority voting ("soft" voting). The class labels with the highest combined probabilities among all base classifiers are selected as the final prediction.

4.2. LSTM and CNN

In this approach, the model is evaluated using word embedding, pretrained fine-grained word2vec, LSTM and 1-D CNN. Two LSTM layers with 100 units each were added with a dropout of 0.3 in order to prevent overfitting. A 1D convolutional layer ("Conv1D") with 100 filters and a filter size of 5 was added next, which performed 1D convolution operations on the input sequence. The activation function relu is used after the convolution layer. A "GlobalMaxPool1D" layer takes the maximum value over the temporal dimension and reduces the dimensionality of the data, keeping the most important features. A dense layer with 16 units applies matrix multiplication to the input data and applies the relu activation function element-wise. A final dense layer with 3 units and a SoftMax activation function is responsible for predicting the sentiment class probabilities.

4.3. LSTM and Single Attention

In this approach, the model is evaluated using LSTM with a single-head attention layer. An LSTM layer with 64 units and return sequences will return the full sequence output rather than just the final output. The attention layer helps the model focus on important parts of the input sequence during training. Next, a dropout layer with a dropout rate of 0.5 randomly sets a fraction of input units to 0 during training, which helps prevent overfitting. Finally, a dense layer that is an output layer with 3 units and a softmax activation function, is added, which is responsible for predicting the sentiment class probabilities. Also, the categorical cross entropy-based loss function, Adam as an optimiser and an accuracy metric for evaluation are used in the model.

Tables 2, 3 and 4 display a confusion matrix and classification report for all models. **Table 5** shows the evaluation metrics that is testing accuracy with reasonable performance, precision, recall and F1-score for all models. It is seen that LSTM with single-head attention performs better than the rest of the two models in terms of accuracy, F1-score and recall (92.04%, 92.04% and 92.04%, respectively), whereas random forest that is the hybrid model performs better than the other two models in terms of precision (94%). Also, precision, recall and the F1-score computed are shown in **Tables 6, 7 and 8. Figure 1** shows the ROC curves for each class for all models, and the area under each curve (AUC) indicates the classifier's performance for that class. It shows the computation of the false positive rate, true positive rate and threshold for each class. The function is applied to the true labels and predicted probabilities for each class and finally computes the area under the ROC Curve for each class. A higher AUC value indicates that models are performing better.

Table 1: Number of sentiment labels in dataset.

Sr. no	Sentiment	Score
0	lively	2
1	although my test score is not high because I d...	1
2	Great work!	2
4	Teachers are very enthusiastic in teaching.	2
5	to spend time growing up to lecture for student	0

Table 2: LSTM+ATT model confusion matrix.

Actual	Predicted		
	Neutral	**Negative**	**Positive**
Neutral	**1694**	**8**	**156**
Negative	**91**	**1431**	**71**
Positive	**85**	**18**	**1837**

Table 3: LSTM+CNN model confusion matrix.

Actual	Predicted		
	Neutral	Negative	Positive
Neutral	1788	75	145
Negative	228	1621	207
Positive	205	125	2075

Table 4: Hybrid model confusion matrix.

Actual	Predicted		
	Neutral	Negative	Positive
Neutral	1523	26	38
Negative	0	1761	337
Positive	4	51	1651

Table 5: Evaluation metrics for models.

Metrics/Models	LSTM + CNN	LSTM +ATT	Hybrid Model
Accuracy	84.7735%	92.04%	87.18%
F1-Score	0.8477	0.9204	0.9156
Precision	0.8472	0.9190	0.94
Recall	0.8477	0.9204	0.87

Table 6: Performance on test data using precision.

Model	Precision		
	Neutral	Negative	Positive
LSTM+CNN	0.79	0.89	0.86
LSTM+ATT	0.91	0.90	0.95
Random Forest	0.96	0.84	0.97

Table 7: Performance on test data using recall.

Model	Recall		
	Neutral	Negative	Positive
LSTM+CNN	0.84	0.85	0.86
LSTM+ATT	0.91	0.94	0.92
Random Forest	0.98	0.89	0.88

Table 8: Model performance on test data using F1-score.

Model	F1-score		
	Neutral	Negative	Positive
LSTM+CNN	0.89	0.81	0.85
LSTM+ATT	0.91	0.98	0.89
Random Forest	1	0.96	0.81

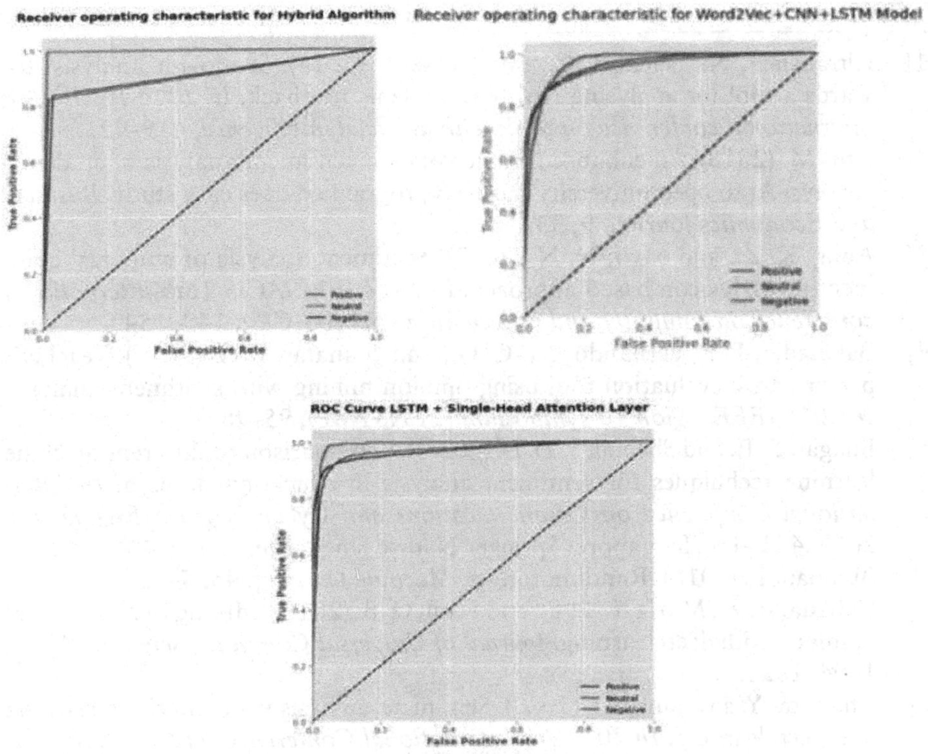

Figure 1: *ROC curve for (a) LSTM and CNN, (b) LSTM with attention and (c) hybrid algorithm.*

5. Conclusion and Future Scope

We introduced the UIT-VSFC corpus ((Nguyen *et al.*, 2018) and final-year students feedback for the analysis of educational sentiment. We presented a machine learning hybrid model with a 94% precision value, which outperformed the rest of the models. Another approach used is bidirectional LSTM and 1-D CNN models along with word2vec embedding, with a testing accuracy of 84.77%. Finally, we presented LSTM with a single-head attention layer model with testing accuracy, F1-score and recall performance of 92.04%, 92.04%, and 92.04%, respectively, which outperforms the rest of the models. We tested the model on unseen data collected from the comments of final-year students. The model performs well on unseen data. In the future, we are going to perform experiments using variants of the LSTM, LSTM with multi-head attention, Bert and other deep learning models. Also, collect more data to perform sentiment analysis on the real-world problem.

References

[1] Altrabsheh, N., Mihaela, C., and Sanaz, F. (2014). Sentiment analysis: towards a tool for analysing real-time students feedback. In *2014 IEEE 26th international conference on tools with artificial intelligence*, 419–423.

[2] Atif, M. (2018). An enhanced framework for sentiment analysis of students' surveys: Arab open university business program courses case study. *Business and Economics Journal,* 9, 337.

[3] Aung, K. Z., and Myo, N. N. (2017). Sentiment analysis of students' comment using lexicon based approach. In *2017 IEEE/ACIS 16th international conference on computer and information science (ICIS)*, 149–154.

[4] Balahadia, F. F., Fernando, M. C. G., and Juanatas, I. C. (2016). Teacher's performance evaluation tool using opinion mining with sentiment analysis. *In 2016 IEEE region 10 symposium (TENSYMP)*, 95–98.

[5] Bhagat, B. P., and Sheetal, S. D. D. (2023). A comparison of different machine learning techniques for sentiment analysis in education domain. *In International Conference on Communications and Cyber Physical Engineering 2018*, 441–450, Singapore: Springer Nature Singapore.

[6] Breiman, L. (2018). Random forests. *Machine learning*, 45, 5–32.

[7] Cabada, R. Z., María, L. B. E., and Raúl, O. B. (2018). Mining of educational opinions with deep learning. *Journal of Universal Computer Science*, 24(11), 1604–1626.

[8] Chandra, Y. and Jana, A., (2020). Sentiment analysis using machine learning and deep learning. *In* 2020 *7th International Conference on Computing for Sustainable Global Development* (*INDIACom*), 1–4.

[9] Chauhan, G. S., Preksha, A., and Yogesh, K. M. (2019). Aspect-based sentiment analysis of students' feedback to improve teaching–learning process. *In Information and Communication Technology for Intelligent Systems: Proceedings of ICTIS 2018*, 2, 259–266, Springer: Singapore.

[10] Esparza, G. G., Alejandro, D. L., Alberto, O. Z., Alberto, H., Julio, P., Marco, A., Edgar, C., and Jose, D. J. N. (2018). A sentiment analysis model to analyze students reviews of teacher performance using support vector machines. *In Distributed Computing and Artificial Intelligence, 14th International Conference*, 157–164, Springer International Publishing.

[11] Gutiérrez, G., Ponce, J., Ochoa, A., and Álvarez, M. (2018). Analyzing students reviews of teacher performance using support vector machines by a proposed model. *In Intelligent Computing Systems: Second International Symposium, ISICS 2018, Merida, Mexico, March 21-23, 2018, Proceedings* 2, 113–122. Springer International Publishing, 2018.

[12] Hochreiter, S., and Schmidhuber, J. (1997). Long short-term memory. *Neural computation*, 9(8), 1735–1780.

[13] Kaur, W., Vimala, B., and Baljit, S. (2020). Improving teaching and learning experience in engineering education using sentiment analysis techniques. *In IOP Conference Series: Materials Science and Engineering*, 834(1), 012026, IOP Publishing.

[14] Lalata, J. P., Bobby, G., and Ruji, M. (2019). A sentiment analysis model for faculty comment evaluation using ensemble machine learning algorithms. *In*

Proceedings of the 2019 International Conference on Big Data Engineering, 68–73.

[15] Mabunda, J. G. K., Ashwini, J., and Ritesh, A. (2021). Sentiment analysis of student textual feedback to improve teaching. *Interdisciplinary Research in Technology and Management*, 643–651.

[16] Mikolov, T., Kai, C., Greg, C., and Jeffrey, D. (2013). Efficient estimation of word representations in vector space. arXiv preprint arXiv:1301.3781.

[17] Nasim, Z., Quratulain, R., and Sajjad, H. (2017). Sentiment analysis of student feedback using machine learning and lexicon based approaches. *In 2017 international conference on research and innovation in information systems (ICRIIS)*, 1–6.

[18] Newman, H., and David, J. (2018). Sentiment analysis of student evaluations of teaching. *In Artificial Intelligence in Education: 19th International Conference, AIED 2018, London, UK, June 27–30, 2018, Proceedings, Part II 19*, 246–250. Springer International Publishing.

[19] Nguyen, V. D., Kiet, V. N., and Ngan, L. T. N. (2018). Variants of long short-term memory for sentiment analysis on Vietnamese students' feedback corpus. *In 2018 10th International conference on knowledge and systems engineering (KSE)*, 306–311.

[20] O'Shea, K., and Ryan, N. (2015). An introduction to convolutional neural networks. arXiv preprint arXiv:1511.08458.

[21] Pong-Inwong, C., and Wararat, S. R. (2014). Teaching senti-lexicon for automated sentiment polarity definition in teaching evaluation. *In 2014 10th International Conference on Semantics, Knowledge and Grids*, 84–91.

[22] Qaiser, S., Nooraini, Y., Remli, M. A., and Hasyiya, K. A. (2021). A comparison of machine learning techniques for sentiment analysis. *Turkish Journal of Computer and Mathematics Education.*

[23] Rajput, Q., Sajjad, H., and Sayeed, G. (2016). Lexicon-based sentiment analysis of teachers' evaluation. *Applied Computational Intelligence and Soft Computing.*

[24] Rani, S., and Parteek, K. (2017). A sentiment analysis system to improve teaching and learning. *Computer*, 50(5), 36–43.

[25] Sangeetha, K., and Prabha, D. (2021). Sentiment analysis of student feedback using multi-head attention fusion model of word and context embedding for LSTM. *Journal of Ambient Intelligence and Humanized Computing*, 12, 4117–4126.

[26] Sindhu, I., Daudpota, S. M., Badar, K., Bakhtyar, M., Baber, J., and Nurunnabi, M. (2019). Aspect-based opinion mining on student's feedback for faculty teaching performance evaluation. *IEEE Access*, 7, 108729–108741.

Chapter 18

Empowering Agriculture: Smart Crop Recommendation with Machine Learning Challenges and Future Prospects

Aishwarya V. Kadu[1] *and K. T. V. Reddy*[2]

[1]Research Scholar, Faculty of Engineering and Technology DMIHER, Wardha

[2]Dean, Faculty of Engineering and Technology DMIHER, Wardha

Abstract: The fast-developing disciplines of crop analysis and prediction significantly improve agricultural operations. It is impossible to exaggerate the importance of crop recommendations since they give farmers the information they need to choose the crops that will thrive in their particular climate and soil conditions. This procedure has historically depended largely on labour- and time-intensive specialist expertise. Considering the projected global population surpassing 10.5 billion by 2060, it is urgently necessary to increase food production sustainably. A vital tool for automating crop suggestions and locating pests and illnesses is machine learning approaches. By doing this, farmers may increase their production while preserving the fertility of the soil and restoring vital minerals. Our study focuses on analysing crop recommendation performance utilising seven different machine-learning methods. Our suggested approach precisely forecasts the best-suited crops for certain places using crucial elements like soil nutrients and weather temperature data. By increasing the profitability of crops (overall), yields of crops and sustainability in crops, this system's potential to revolutionise the recommendation of crops might be highly advantageous for farmers of all sizes.

Impact Statement – this study employs machine learning-based agricultural prediction algorithms, achieving near-perfect accuracy. Soil nutrients and climate data, these algorithms forecast optimal crops, potentially revolutionising recommendations and improving farmers' yields, sustainability and earnings.

Keywords: crop prediction, machine learning, crop recommendation, agricultural productivity, environmental factors, food production.

1. Introduction

According to Arthur Samuel's 1959–1960 definition, machine learning is a fascinating discipline that empowers computers to improve their performance

DOI: 10.1201/9781003527442-18

without explicit programming, allowing them to learn from data and experiences (Oikonomidis *et al.*, 2023). Algorithms used for machine learning develop the capacity to make knowledgeable predictions and judgements by digesting enormous amounts of data. Agriculture requires farmers to grow lucrative and sustainable crops as a crucial worldwide industry. The effects of picking the incorrect crop can be severe, resulting in lower yield and possible financial losses for farmers. Neglecting crucial elements like market demand, soil quality and climatic compatibility may prevent the selected crops from growing and producing to their maximum potential. Unsuitable crops may need help adjusting to the local climate, leading to poor growth, increased sensitivity to pests and diseases and a general decrease in output. Furthermore, commodities that do not meet market demand could have trouble finding customers or fetching fair prices, thus affecting farmers' financial stability. Whether farmers comprehend and improve their farming methods might be revolutionised by the intersection of ML and agricultural data. ML algorithms can process vast amounts of data from diverse sources, including climate stations, satellites, sensors and agricultural equipment and may derive insightful conclusions from them. These algorithms can reveal complex relationships, structures or patterns, and forecasting. By harnessing the power of machine learning and agricultural data, farmers can uncover valuable patterns and knowledge previously hidden in their datasets. Make educated decisions on a variety of various such as management in irrigation, selection for crops, control of pesticides and prediction in yield. Recommendation systems may examine a variety of datasets, such as market, weather and soil information. ML algorithms may use this data to determine which crops suit a place. Furthermore, these technologies may give farmers insightful information about the best ways to cultivate particular crops. Developing recommendations in crop (R.S.) systems using ML–DL has much potential to increase agricultural output and sustainability. These technologies help farmers choose suitably. Crops, resulting in higher crop yields and resource efficiency. Additionally, they are essential in enhancing agriculture's resistance to the effects of environmental change (Pande *et al.*, 2021) additionally, ML may successfully address several additional issues in agriculture (Chen *et al.*, 2020). It may be used for various purposes, including controlling soil health, estimating agricultural output, finding pests and illnesses, improving crop production and increasing water efficiency. Additionally, the integration of ML techniques aids in minimising the usage of fertilizers and pesticides, which can be detrimental to crops and the environment. Crops are essential for meeting the demands of the global population for food and fibre. Enhancing high-quality crop yields becomes essential in this situation. The choice of what crops to grow has a significant impact on crop yields and overall profitability. This study investigates how ML may be used to propose crops to farmers. A vital dataset with characteristics including type of soil, pH level or values, humidity, temp, rain and soil content must be gathered and preprocessed for the investigation.

2. Literature Review

ML algorithms fall into three categories based on learning strategies: (i) Supervised Learning, (ii) unsupervised learning and (iii) reinforcement learning. Although several ML algorithms are widely utilised, we spotlight the algorithms listed below in this survey since they were employed in our study. 1. Logistic regression, 2. decision tree, 3. RF, 4. K-nearest neighbours, 5. Naive Bayes, 6. SVM, and 7. neural networks (Kulkarni *et al.*, 2018). Existing crop recommendation research – There has been a minor growth in crop suggestion research in recent years. Essential studies include "AgroConsultant" (Doshi *et al.*, 2018) and "User-friendly Yield Prediction System" (Pande *et al.*, 2021). While Reddy *et al.* studied the methods currently used in crop selection, the author proposed a model that uses a majority vote methodology with SVM and ANN to recommend crops. A theoretical framework on crop recommendation was offered by (Ghadge *et al.*, 2018), while (Kulkarni *et al.*, 2018). Showed off their work on boosting crop yield using an ensembling recommendation system. Additionally, the work by Pudumalar *et al.* (2007) which had the most citations on IEEE Xplore employed ML to suggest crops based on information gathered from a district in Tamil Nadu, India, but it omitted information on the model's accuracy or the nature of the data. Crop recommendation was not expressly included in (Liakos *et al.*, 2018)'s review of ML applications in agriculture. Other agricultural-related publications that dealt with crop recommendation indirectly covered subjects like the IoT and sensors for gathering agricultural data. Overall, crop recommendation research is still in its early stages, with most of the cited literature being released within the previous four to five years. Our contribution – our study aims to construct thorough crop recommendation models to surpass the highlighted constraints. The current emphasis of the study is on providing a detailed explanation of the data collecting and feature extraction processes. However, we want to outline each stage of the procedure, including model training and evaluation, in later parts. The article will give an in-depth description of every step of our crop recommendation system. We first preprocess the data and then use various ML techniques to provide suitable crop selections. We train our models using the above procedures before comparing the efficacy of each model in the recommendation system. The following algorithms that can be used in further work are DT, KNN, NB, RF, SVM, ANN-DNN and logistic regression (LR). Moreover, it addresses the papers discussed in the section on our contribution by presenting a study on recommendation systems with in-depth explanations for each component. Furthermore, it highlights the challenges faced in agriculture in general and specifically related to implementing ML techniques for agricultural data. Lastly, the future scope of the manuscript presents several promising ideas for future work, aiming to further enhance and expand upon our current research. Our study's comprehensive approach, high accuracy and utilisation of cutting-edge feature engineering techniques significantly contribute to the recommendation field. This valuable study will benefit farmers, agricultural areas and others in the industry.

3. Methodology and Model Specifications

This section summarises the methods shown in Figure 1, which details the procedures we used to train different models. Initially, we used chosen machine learning methods to cycle through each step. In order to get a dataset ready for our model, we preprocess it using a Kagglesource. Crop features like nitrogen, phosphorus, potassium, temperature, humidity, pH and rainfall summarise the crucial aspects of the data. The dataset was produced by merging and enhancing datasets that contained information particular to India, including precipitation, climate and fertiliser. This investigation employed 21 distinct labels drawn from a database of over 100,000 entries. Since there is only one recommended crop for any given setting, the dataset was trimmed down to about 2.2k records to guarantee practicality. The list includes kidney beans, rice, moth beans, pigeon, peas, maize, lentils, grapes, banana, chickpea, mungbean, black gram, orange, jute, pomegranate, cotton, mango, watermelon, muskmelon, coconut, papaya and apple. These 21 crop label data are used.

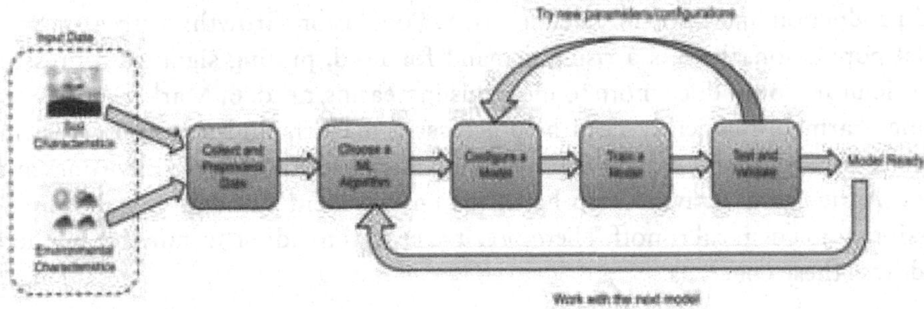

Figure 1: Methodology of CR system.

1. Data Input: The accuracy of a model is greatly influenced by the quality and quantity of the data it receives. Hence, we ensured that our data was accurate and adequately labelled. 2. Data Preprocessing: Before feeding the data to our machine learning system, we performed crucial preprocessing tasks, including removing duplicates and null records, transforming the data for machine learning and conducting feature engineering to create new features. We also characterised and visualised the data, addressing potential outliers to ensure reliability. 3. Select ML algorithm: We chose one of the seven algorithms from our predefined selection for every iteration. We followed each selected method's preprocessing, model testing, and model validation processes, fine-tuning the model. 4. Model Configurations: We experimented with different configurations, including activation functions, epoch settings, decision tree depth and the number of nearest neighbours, to improve test and cross-validation accuracy, aiming to enhance the overall performance of our models. 5. Model Training: In this phase, preprocessed data is used to train the ML algorithm, adjusting its parameters to

generate a prediction model. 6. Testing Model Accuracy: After training the model, we evaluate its accuracy using the test data to see how well it performed. We also evaluate its cross-validation accuracy to ensure the model can be applied to data that have yet to be observed.

4. Agriculture Challenges

Among the significant difficulties facing agriculture are the following:

1.Climate Change: Climate change poses severe threats to agriculture due to rising temperatures, unpredictable weather patterns and extreme events like droughts and floods, which impact crop growth and productivity. 2. Water scarcity: Agriculture can be hampered by a growing scarcity of water brought on by excessive water use and shifting weather patterns, especially in areas where irrigation is widely used. 3 Land Degradation: Deforestation, excessive pesticide usage and soil erosion all contribute to the deterioration of the land, which lowers its productivity and available arable land. 4. Pests and Diseases: Pest and disease outbreaks pose severe challenges to agriculture, resulting in considerable crop losses that harm food production and farmers' livelihoods. 5. Population Growth: With a growing global population, there is a rising demand for food, putting significant pressure on agriculture to produce more to meet this increasing need. 6. Market Access and Pricing: Farmers frequently need help accessing markets and getting fair pricing for their goods, which affects their earnings and profitability. 7. Environmental Issues: Agricultural activities can harm the environment through greenhouse gas emissions and chemical runoff. Therefore, it is crucial to adopt sustainable methods to address these concerns.

5. Future Scope/Work/Ideas

There are many ways to pursue this study, and the following examples are given to encourage readers to do so:

1. Increase the dataset size by adding a more extensive and more varied collection of data. A larger dataset would boost the models' learning capacity and enable them to capture a greater variety of correlations between different parameters and crop yields, improving forecasts. 2. Expanding data collection efforts will comprehensively evaluate the models' accuracy and performance across diverse agricultural scenarios. 3. To determine the strategies' potential advantages for farmers and the environment and evaluate their economic and environmental effects. 4. Improve crop recommendation systems by implementing sensor-based data collecting on farms. Farmers could quickly make educated decisions, decrease crop losses, save labour expenses and advance sustainable agricultural practices with the help of real-time data. 5. Create a mobile application that serves as an end-to-end system for farmers and owners of agribusinesses using the concepts discussed in the article. This program would make it easier for these methods to be widely used and ensure that farmers and other users could easily access and use them.

6. Conclusion

In conclusion, this study provides crop recommendation models that forecast the ideal crops for cultivation using various cutting-edge ML techniques and DNN that is deep neural network. The ML method or algorithms quickly adapts to fresh data and other geographic or national contexts. The findings of this study carry significant benefits for the agricultural sector. (i) Firstly, the technique can empower farmers to make well-informed decisions regarding their crop choices. (ii) Secondly, governments can utilise the method to formulate policies that foster the growth of the agricultural industry. (iii) Thirdly, businesses can leverage the technique to develop innovative products and services that cater to the agricultural sector's needs. iv) Lastly, it can contribute to maintaining stable agricultural commodity prices. Moreover, the paper has thoroughly addressed agricultural challenges and put forth compelling future ideas for exploration. In conclusion, this research has significantly impacted the agricultural field.

References

[1] Charbuty, A., and Abdulazeez, A. (2021). Classification based on decision tree algorithm for machine learning. *Journal of Applied Science and Technology Trends*, 2(1), 20–28.

[2] Chakraborty, S., Priyadharshini, A., Kumar, A., and Pooniwala, O. M. (2021). Intelligent crop recommendation system using machine learning. *5th International Conference on Computing Methodologies and Communication* (*ICCMC*), 843–848.

[3] Chen, S., Webb, G. I., Liu, L., and Ma, X. (2020). A novel selective naïve bayes algorithm. *Knowledge-Based Systems*, 192, 105361.

[4] Vallino, E., Ridolfi, L., and Laio, F. (2020). Measuring economic water scarcity in agriculture: a cross-country empirical investigation. *Environmental Science & Policy*, 114, 73–85.

[5] Doshi, Z., Nadkarni, S., Agrawal, R., and Shah, N. (2018). Agroconsultant: intelligent crop recommendation system using machine learning algorithms. *4th International Conference on Computing Communication Control and Automation* (*ICCUBEA*), 1–6.

[6] Ghadge, R., Kulkarni, J., More, P., Nene, S., and Priya, R. (2018). Prediction of crop yield using machine learning. *International Research Journal of Engineering and Technology* (*IRJET*), 5.

[7] Iglovikov, V., Mushinskiy, S., and Osin, V. (2017). Satellite imagery feature detection using deep convolutional neural network: a kaggle competition. arXiv preprint arXiv:1706.06169.

[8] Kulkarni, N. H., Srinivasan, G. N., Sagar, B. M., and Cauvery, N. K. (2018). Improving crop productivity through a crop recommendation system using ensembling technique. *3rd International Conference on Computational Systems and Information Technology for Sustainable Solutions* (*CSITSS*), 114–119.

[9] Liakos, I. G., Busato, P., Moshou, D., Pearson, S., and Bochtis, D. (2018). Machine learning in agriculture: a review. *Sensors*, 18(8), 2674. [Online]. Available: https://www.mdpi.com/1424-8220/18/8/2674.

[10] Oikonomidis, A., Catal, C., and Kassahun, A. (2023). Deep learning for crop yield prediction: a systematic literature review. *New Zealand Journal of Crop and Horticultural Science*, 51(1), 1–26.

[11] Pathak, T. B., Maskey, M. L., Dahlberg, J. A., Kearns, F., Bali, K. M., and Zaccaria, D. (2018). Climate change trends and impacts on california agriculture: a detailed review. *Agronomy*, 8(3), 25.

[12] Pudumalar, S., Ramanujam, E., Rajashree, R. H., Kavya, C., Kiruthika, T., and Nisha, J. (2017). Crop recommendation system for precision agriculture. *8th International Conference on Advanced Computing* (*ICoAC*), 32–36.

[13] Wang, H., Lei, Z., Zhang, X., Zhou, B., and Peng, J. (2016). Machine learning Basics. *In Deep Learning*, 98–164.

Chapter 19

Deep Learning-Based System for Image Denoising

Pranita Jadhav[1] *and Meenakshi Pawar*[2]

[1]M.Tech Student, SVERI's COEP, Solapur University

[2]Assistant Professor, SVERI's COEP, Solapur University

Abstract: The viability of GANs in picture de-noising is wonderful, yet the test stays in adjusting commotion evacuation while safeguarding picture subtleties. Here, we present a changed progressive generative adversarial network (HIGAN). The principal generator holds high-recurrence components, for example, edges and surfaces. The second recovers low-recurrence attributes and the third upgrades recreation execution. We changed the leftover thick block (RDB) by adding a Re-LU layer and a convolution to diminish the possibility evaporating angles, bringing about quicker learning and further developed execution. By and large, the proposed greetings GAN model beats the impediments of existing picture de-noising calculations and offers prevalent execution in safeguarding picture subtleties while eliminating clamor.

Keywords: Convolutional and de-convolutional neural networks, deep learning, generative adversarial networks, image de-noising.

1. Introduction

Computerised pictures assume a vital part in numerous parts of our day-to-day existence, such as satellite television, clever traffic observing, signature confirmation and penmanship acknowledgment on checks. Anyway they assume indispensable part at logical and mechanical fields, for example, geographic data channels and other data on pictures. A few elements are unavoidably impacted by commotion during the time spent procurement, pressure and transmission, bringing about twisting and loss of picture information (Ghose, 2020). Picture de-noising has impacted practically all specialised fields and assumes a significant part in many fields like transmission and coding, minute imaging, lawful sciences, picture remaking, visual following, picture enlistment, picture division and picture plan. For vigorous execution it is fundamental for protect the substance of the first picture (Liu, 2018). Picture de-noising methods have been generally considered and created as of late to address the difficulties of sound decrease in pictures. Numerous calculations and techniques have been proposed to accomplish

DOI: 10.1201/9781003527442-19

improved brings about picture de-noising. A few famous techniques incorporate separating-based approaches, wavelet-based approaches and profound learning-based approaches (Chunwei, 2019). Sifting based approaches plan to diminish commotion by applying channels to the picture. These channels can be planned in light of different measures like mean separating, middle sifting and bilateral filtering. Wavelet-based approaches use wavelet change to decay the picture into various recurrence parts and afterwards perform de-noising on every part independently. Profound learning-based approaches utilise profound brain organisations to become familiar with the planning among loud and clean pictures. They have shown great outcomes in different undertakings, including picture de-noising. One of the difficulties in picture de-noising is to adjust sound decrease and picture detail protection. In the event that the de-noising calculation eliminates a lot of commotion, it might likewise eliminate a few significant subtleties and highlights of the picture. Then again, on the off chance that the calculation is excessively moderate, it may not eliminate sufficient commotion, bringing about a loud picture. Consequently, finding a decent harmony between sound decrease and picture detail conservation is pivotal for accomplishing top-notch de-noised pictures. All in all, picture de-noising is a significant issue in different fields of examination and has been broadly concentrated on in late years (Zhang, 2017). The proposed increase calculation prepares numerous generators simultaneously and every generator figures out how to de-noise an alternate subset of the picture information. This permits every generator to spend significant time in eliminating a particular kind of clamor, consequently keeping away from mode breakdown and further developing the generally de-noising execution. At long last, our third system is to present an original perceptual misfortune capability, which considers the significant level elements of the picture, notwithstanding the customary pixel-wise MSE misfortune capability. This assists with protecting the picture content and surface subtleties, while likewise lessening the relics and working on the generally perceptual nature of the de-noised images. In synopsis, our proposed de-noising strategy joins an effective ill-disposed misfortune capability, another expansion calculation for preparing various generators, and an original perceptual misfortune capability, to accomplish both high-precision and great discernment in denoising. Exploratory outcomes show that our strategy beats existing cutting edge techniques as far as both PSNR and perceptual quality, while likewise keeping away from the issue of mode breakdown. The changed B-DenseUNet (Boosted DenseUNet) design for proposed Hello GAN generators joins the upsides of lingering thick blocks (RDB) and UNets to permit most extreme data move through all convolutional layers in the organisation. This design improves the viability of proposed Greetings GAN generators, which can be proficiently prepared by joining every one of the upsides of its generators while keeping away from their burdens. By using refitted B-DenseUNet, the altered HI-GAN can achieve state-of-the-art results in terms of both PSNR and visual quality and can effectively remove noise from images while preserving image details. Overall, the proposed HI-GAN with B-DenseUNet, the altered HI-GAN can achieve state-of-the-art results in terms of both PSNR and visual quality and can effectively remove noise from images while preserving image details. Overall, the proposed HI-GAN with adjusted B-DenseUNet engineering shows promising outcomes for picture denoising errands and can be applied to a large number of sensible applications in various fields (Vo, 2021).

The main objectives of this study are as follows:

1. Examine existing methods and find out the limitations of image denoising.
2. Design and develop a novel neural convolutional network for image noise reduction.
3. Test various performance indicators of the proposed system, such as accuracy, precision, PSNR (peak signal-to-noise ratio) and so on.
4. Compute the first picture by eradicating commotion from a form of the picture that has clamor in it.

2. Related Work

2.1. Deep Neural Networks for Image Denoising

There have been a couple of tries to address the de-noising issue using significant cerebrum associations. In Jain and Seung's work, they proposed the use of convolutional mind associations (CNNs) for picture de-noising and affirmed that CNNs have similar or much better execution than MRF models in conveying. Among the recently referenced significant mind network-based strategies, MLP and TNRD can achieve promising execution and fight with BM3D (Zhang, 2017).

2.2. Residual Learning

The lingering learning strategy for convolutional brain organisations (CNNs) was initially proposed to resolve the issue of execution debasement, where the preparation exactness diminishes as the organisation profundity increments. In leftover organisations, the model unequivocally learns lingering maps for a few stacked layers, expecting that these remaining guides are simpler to learn than the first unreferenced maps. This methodology of learning leftover guides has empowered the preparation of very profound CNNs easily, prompting further developed exactness in picture order and item acknowledgment (Zhang, 2017).

2.3. GAN

In 2016, Great individual *et al.* proposed the generative ill-disposed network (GAN), which prompted a great deal of related research. A few scientists investigated picture style move, for example, the Cycle GAN proposed by Bair *et al.*, which they saw as a picture to-picture interpretation issue. Others, as Orest Kupyn *et al.* proposed Deblur GAN to eliminate blind movement (Li, 2014). Among these investigations, one creator planned a GAN-based profound lingering organisation and proposed a misfortune capability for network preparing. In their preparation results, the all-around planned misfortune capability worked on the goal of the picture as well as upgraded the substance and shade of the first picture, safeguarding many subtleties (Peizhu Gong, 2020).

3. Literature Survey

Burger *et al.* (2012) utilised a multi-facet perceptron on picture patches to straightforwardly become familiar with the planning and rival state-of-the art picture denoising methods.

Xie *et al.* (2012) proposed a technique for blind inpainting and picture denoising that consolidated scanty coding with profound brain organisations, considering picture colorization and denoising inside a bound together system.

Agostinelli *et al.* (2013) introduced a novel technique called adaptive multi-column SSDA, which combines multiple SSDA by adaptively predicting optimal column weights. They showed that AMC-SSDA can perform robust denoising on images corrupted by several different types of noise, without knowing the type of noise at test time.

Li *et al.* (2014) demonstrated the way that preparation on a huge picture dataset can beat cutting edge picture denoising methods.

Isa *et al.* (2015) compared the effectiveness of three different filtering techniques for denoising MRI pictures and found that the median filter offers favorable outcomes.

Zhang *et al.* (2016) proposed a lightweight two-unit CNN-based network for image denoising and enhancement, which consists of three layers and achieved improved performance through training data preprocessing and enrichment.

Zhang *et al.* (2018) used residual learning to distinguish noise from noisy observations and proposed a deep convolutional neural network for picture denoising.

Liu *et al.* (Liu, 2018) proposed a CNN-based denoising technique that effectively removes Gaussian noise and enhances the efficiency of conventional image filtering techniques. They also discussed salt and pepper noise and introduced linear CNN-based picture denoising techniques.

Yue Wensun *et al.* (Sun, 2018) fostered the DNGAN start-to-finish profound learning calculation to decrease measurable commotion in computerised radiological pictures, which uses an ill-disposed misfortune to produce all the more perceptually engaging pictures.

Zhang *et al.* (2018) presented FFD-Net, a CNN model for efficient and flexible discriminative denoising that utilises input denoising in space and noise maps in network design and training.

Yu *et al.* (2019) developed a deep iterative down-up convolutional neural network (DIDN) that uses U-Net as the basic module for image denoising and can handle 5 to 50 levels of Gaussian noise by down- and up-sampling deep feature maps.

Gu (2019) proposed a self-guided neural network that uses a self-routing technique to de-noise images from top to bottom and gradually brings back multi-scale contextual information. NOISE2VOID (N2V) is a novel training method proposed by Alexander institutional monitoring in improving the stock return of the real-estate investment trusts (REITs).

Krull *et al.* (2019) that can train a denoising CNN using only one noisy image acquisition and has been successfully applied to various imaging modalities including photography, fluorescence microscopy and cryo-transmission electron microscopy.

Fan *et al.* (Fan, 2019) discussed recent developments in several picture denoising techniques and evaluated their benefits and drawbacks.

Ghose *et al.* (Ghose, 2020) utilised CNN models for picture denoising by injecting white Gaussian noise and conducting qualitative analysis on photos with no noise. They compared CNN-based methods with conventional methods using

PSNR, SSIM, and MSE metrics.

Kumwilaisak *et al.* (2020) achieved superior image de-noising results in terms of both objective and subjective quality by combining multidirectional long short-term memory networks with convolutional neural networks.

Babu (2020) discussed the effectiveness of CNN-based denoising techniques that utilise methods such as BN, ReLU, and residual learning to improve performance.

Tian *et al.* (2020) conducted a comprehensive analysis of deep networks for image denoising, covering a range of noisy image types and discussing various deep learning methods.

Weir *et al.* (2020) developed a physics-based noise generation model and a noise parameter calibration approach to address denoising in very low light conditions. The study conducted by Ademola *et al.* (Ilesanmi, 2021) provides a detailed analysis of different CNN techniques for image de-noising, covering a wide range of datasets and providing examples of their evaluation.

4. Dataset

The execution methodology included utilising two datasets, specifically SIDD (Abdelhamed, 2018) and FFHQ. The SIDD dataset comprises of 320 regular pictures, while the FFHQ dataset has 52,000 top notch PNG pictures with a goal of 512 × 512, showing huge change in age, race and picture foundation. Haphazardly trimmed preparing patches of size 512 × 512 were acquired from arbitrarily chosen pictures in these datasets. Clamor and ground truth fix matches were created by the commotion model for orchestrating genuine commotion. The PyTorch AI library was utilised for preparing the model. To test our procedure, we utilised a Nvidia GTX 1080 Ti illustrations card. Appraisal measures were additionally used to assess the presentation of the model.

5. Model Architecture

The proposed modified HI-GAN consists of Gb, Ga, Da discriminator (Vo, 2021) and Gc gain network. Gb and Ga are both image-denoising DCNN generators. Additionally, Ga and Da are trained together to enhance Ga's capacity for denoising and detail preservation in distorted images. Ga's advantage lies in its ability to address the issue of missing high-frequency features like edges and textures by persistently engaging in recurrent zero-sum games with Da. Gb, on the other hand, is trained independently and is not in competition with any network. Gb's technique is to ignore the discriminator's influence on instability and concentrate solely on denoising. In general, Gb and Ga use distinct methods and criteria to assess the effectiveness of rebuilding, and neither is superior to the other. We proposed a modified residual dense block (RDB) in our regenerated HIGAN to fully utilise all hierarchical features from the original LR image. It is unreasonable for an extremely profound organisation to remove the result of each convolutional layer in the LR space. Consequently, we present RDB as the structure module for current HIGAN. RDB comprises of thick associated layers and neighbourhood include combination (LFF) with nearby lingering learning (LRL). Our RDB likewise upholds coterminous memory among RDBs. The result of one RDB has direct admittance to each layer of the following RDB, bringing about a coterminous state pass. Each convolutional layer in RDB approaches generally ensuing layers and gives data that should be protected. LFF extricates nearby thick elements by adaptively saving data and connecting the conditions of going before RDB and all first layers inside the ongoing RDB. Moreover, LFF considers a high development rate by balancing out the preparation of a more extensive organisation. Subsequent to extricating staggered neighbourhood thick elements, we conduct. The proposed altered HIGAN comprises of Gb, Ga, Da discriminator (Duc My Vo, 2021) and Gc gain organisation. Gb and Ga are both picture denoising DCNN generators. Also, Ga and Da are prepared together to improve Ga's ability for denoising and detail safeguarding in twisted pictures. Ga's benefit lies in its capacity to resolve the issue of missing high recurrence highlights like edges and surfaces by steadily captivating in repetitive zero-sum games with Da. Gb, then again, is prepared freely and isn't in rivalry with any organisation. Gb's strategy is to disregard the discriminator's effect on insecurity and focus exclusively on denoising. As a rule, Gb and Ga utilise particular techniques and measures to survey the viability of revamping nor is better than the other. We proposed a changed leftover thick block (RDB) in our recovered HIGAN to completely use all various levelled highlights from the first LR picture. It is illogical for a very deep network to extract the output of each convolutional layer in the LR space. Therefore, we introduce RDB as the building module for modern HIGAN. RDB consists of dense connected layers and local feature fusion

(LFF) with local residual learning (LRL). Our RDB also supports contiguous memory among RDBs. The output of one RDB has direct access to each layer of the next RDB, resulting in a contiguous state pass. Each convolutional layer in RDB has access to all subsequent layers and passes on information that needs to be preserved. LFF extracts local dense features by adaptively preserving information and concatenating the states of preceding RDB and all preceding layers within the current RDB. Additionally, LFF allows for a high growth rate by stabilising the training of a wider network. After extracting multi-level local dense features, to further improve the information flow, we introduce local residual learning

(LRL) in the RDB, as there are several convolutional layers in one RDB. The final output of the RDB can be obtained by It should be noted that LRL can also further improve the network's representation ability, resulting in better performance. Due to the dense connectivity and local residual learning, we refer to this block architecture as a RDB.

Figure 1 shows the Residual Dense Block which is used in the B-Dense-U-Net model of the system.

The flow diagram of the system is shown in Figure 2 where it describes the generator Gc.

Figure 3 is the Block diagram of proposed B-Dense-U-Net used in the generators including Ga and Gb. It is the combination of generators Ga and Gb and known as B-Dense-U-Net.

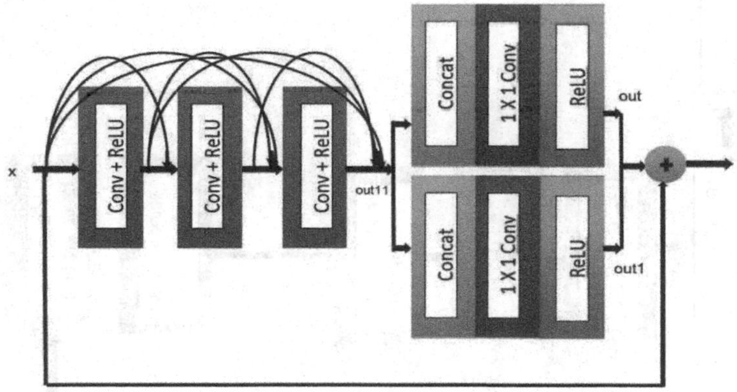

Figure 1: *Refitted residual dense block.*

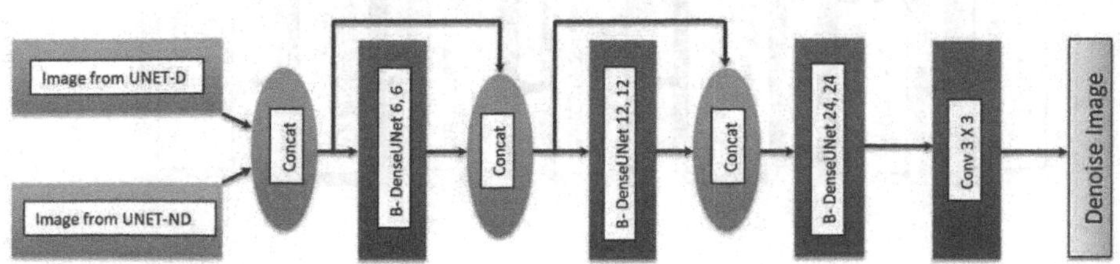

Figure 2: *Block diagram of modified Gc.*

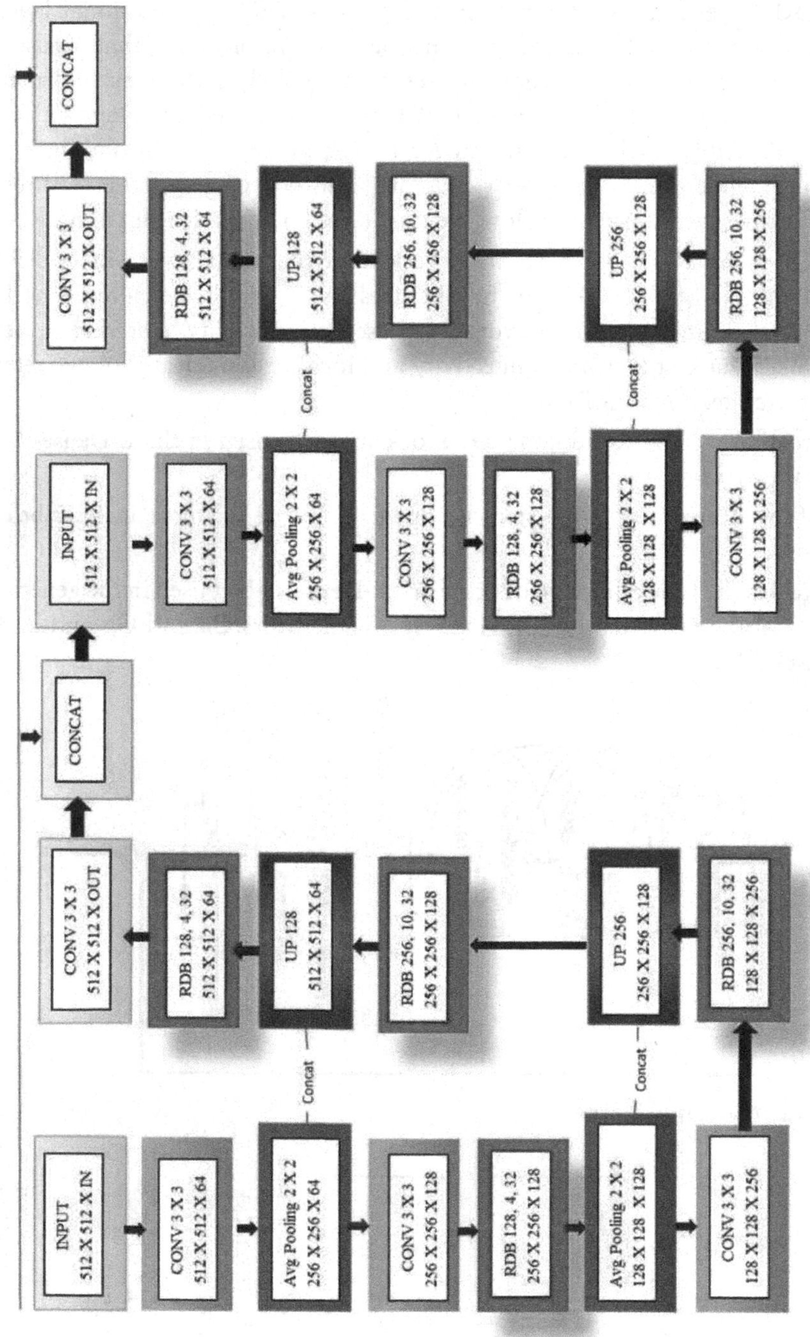

Figure 3: Block diagram of proposed B-Dense-U-Net used in the generators including Ga and Gb.

6. Evaluates

In this section, we will describe the implementation procedure and provide details about the experiments, including the datasets used and the evaluation metrics employed.

6.1. A Testing Datasets

We used the DND (Plotz, 2017), SIDD (Abdelhamed, 2018) and NAM (Nam, 2016) information bases to analyse the precision of the proposed technique with its rivals. The DND dataset is intended to assess PSNR and SSIM on genuine world photographs, with contending techniques assessed through a web-based accommodation system. It comprises of 50 sets of genuinely uproarious pictures and ground truth pictures. SIDD, then again, assesses the PSNR and SSIM of regular pictures taken by cell phone cameras, containing various sets of photographs taken in related scenes with different lighting arrangements. The presentation evaluations for the SIDD dataset are additionally founded on an internet-based accommodation strategy, similar to the DND dataset. At long last, we utilised 11 test pictures from the NAM dataset to survey and think about the recreation abilities of various strategies.

6.2. Evaluation Metrics

6.2.1 Peak signal-to-noise ratio (PSNR)

The PSNR is a metric used to evaluate the visual quality of images by measuring the difference between corresponding pixels. It has been shown to be correlated with the average opinion scores provided by human experts. PSNR is essentially an MSE-based feature. The distortion level of an image after steganography is influenced by the size of the pixel values.

$$MSE = \frac{1}{W \times H} \sum_{i=1}^{W} \sum_{j=1}^{H} \left(x_{i,j} - y_{i,j} \right)^2 \quad (1)$$

$$PSNR = 10.\log_{10}\left(\frac{\left(2^n - 1\right)^2}{MSE} \right) \quad (2)$$

6.2.2 Structural similarity index (SSIM)

The SSIM metric evaluates the structural similarity, brightness and contrast of images and can be used to assess the quality of steganographic images. As SSIM values increase, it indicates that the images before and after steganography are more similar in terms of structural content.

$$SSIM(x,y) = l(x,y)\blacklozenge c(x,y)\blacklozenge s(x,y) \quad (3)$$

$$l(x,y) = \frac{2\mu_x\mu_y + C_1}{(\mu_x)^2} + (\mu_y)^2 + C_1 \quad (4)$$

$$c(x,y) = \frac{2\sigma_x\sigma_y + C_2}{(\sigma_x)^2} + (\sigma_y)^2 + C_2 \quad (5)$$

$$s(x,y) = \frac{2\sigma_{xy} + C_2}{\sigma_x\sigma_y} \quad (6)$$

The equation provided is the formula for calculating the SSIM index between two images x and y. The symbols l(•), c(•) and s(•) represent luminance, contrast and structure, respectively. μ_x and μ_y are the pixel averages of image x and image y, respectively; σ_x and σ_y represent the standard deviations of image x and image y; $\sigma_x\sigma_y$ denotes the covariance of image x and image y.

7. Results

The de-noising results for the SIDD, DND, FFHQ and NAM datasets are introduced in Tables 1 and 2, including PSNR and SSIM scores. In the proposed HIGAN model, the changed RDB has prompted an expansion in the PSNR and SSIM of pictures by a normal of 3.314 in PSNR and 0.1633 in SSIM. By accomplishing higher PSNR and SSIM, the point of our organisation has been achieved. Tables 4–7 exhibit the subjective assessment of the contending calculations. The Hello GAN can catch perceptually high surfaces and significant subtleties, while the past technique produces more ancient rarities and clamor contrasted with the adjusted Greetings GAN.

Table 1: Result of SIDD and DND dataset.

Sr. No.	SIDD		DND	
	PSNR/SSIM by HIGAN	PSNR/SSIM by our model	PSNR/SSIM by HIGAN	PSNR/SSIM by our model
1	32.56/0.9527	35.93/0.971	27.5/0.5911	31.74/0.8966
2	32.95/0.9525	36.04/0.9704	27.05/0.5744	30.04/0.8532
3	27.11/0.9318	29.62/0.9298	27.09/0.549	30.7/0.8437
4	27.13/0.9308	29.67/0.9284	27.7/0.5673	30.48/0.8851

5	35.66/0.9352	38.36/0.9846	27.74/0.5396	30.35/0.9138
6	31.63/0.9154	36.04/0.9784	26.27/0.5861	32.36/0.8437
7	38.45/0.9552	41.17/0.9852	27.06/0.6359	30.16/0.9019
8	35.95/0.949	39.62/0.9794	27.31/0.6189	30.06/0.8868
9	40.27/0.9737	40.87/0.978	27.45/0.626	28.55/0.89
10	34.09/0.9531	36.78/0.9805	27.5/0.6876	28.53/0.8966
	33.58/0.9449	36.41/0.96857	27.267/0.597	30.297/0.881

Table 2: Result of FFHQ and NAM dataset.

Sr. No.	FFHQ		NAM	
	PSNR/SSIM by HIGAN	PSNR/SSIM by our model	PSNR/SSIM by HIGAN	PSNR/SSIM by our model
1	27.25/0.6226	31.64/0.8941	37.94/0.9559	40.79/0.9821
2	27.38/0.6375	31.97/0.9277	39.13/0.9418	42.62/0.9791
3	27.14/0.5935	32.05/0.8908	41.19/0.9537	43.9/0.9855
4	27/0.6052	32.66/0.9205	37.78/0.9276	40.04/0.964
5	26.62/0.6691	28.34/0.914	39.12/0.9486	43.3/0.9736
6	27.2/0.6194	31.31/0.9258	38.04/0.9107	40.97/0.9731
7	27.23/0.6097	30.65/0.9247	35.35/0.9026	38.57/0.9535
8	27.39/0.6289	32.7/0.9362	37.73/0.9438	38.61/0.9753
9	27.34/0.6106	32.62/0.9244	33.92/0.8583	39.35/0.9601
10	27.23/0.6295	30.46/0.924	33.65/0.8406	37.07/0.9516
	27.178/0.6226	31.44/0.9182	37.385/0.91836	40.522/0.969

Table 3: Comparison of HIGAN and our model.

Dataset	HIGAN	Our Model
SIDD	33.58/0.9449	36.41/0.96857
DND	27.267/0.597	30.297/0.881
FFHQ	27.178/0.6226	31.44/0.9182
NAM	37.385/0.91836	40.522/0.969

Table 4: Result of SIDD dataset.

Noisy Image	HI-GAN Denoised Image	Denoised Image by Our Method	Clean Image
Noisy Image (43.85/0.9960)	Denoised Image (30.99/0.9276)	Denoised Image (37.03/0.9822)	Clean Image (∞/∞)
Noisy Image (46.41/0.9946)	Denoised Image (35.66/0.9352)	Denoised Image (38.36/0.9846)	Clean Image (∞/∞)
Noisy Image (46.25/0.9985)	Denoised Image (30.51/0.9606)	Denoised Image (36.14/0.9816)	Clean Image (∞/∞)

Table 5: Result of DND dataset.

Noisy Image	HI-GAN Denoised Image	Denoised Image by Our Method	Clean Image
Noisy Image (20.07/0.2798)	Denoised Image (25.40/0.6006)	Denoised Image (25.93/0.8073)	Clean Image (∞/∞)
Noisy Image (22.80/0.2862)	Denoised Image (26.46/0.5714)	Denoised Image (30.70/0.8384)	Clean Image (∞/∞)
Noisy Image (23.04/0.2934)	Denoised Image (26.70/0.6029)	Denoised Image (32.57/0.8685)	Clean Image (∞/∞)

Table 6: Result of FFHQ dataset.

Noisy Image	HI-GAN Denoised Image	Denoised Image by Our Method	Clean Image
Noisy Image (24.00/0.3689)	Denoised Image (27.00/0.6052)	Denoised Image (32.66/0.9205)	Clean Image (∞/∞)
Noisy Image (23.97/0.4488)	Denoised Image (26.62/0.6691)	Denoised Image (28.34/0.9140)	Clean Image (∞/∞)
Noisy Image (23.70/0.3487)	Denoised Image (27.20/0.6194)	Denoised Image (31.31/0.9258)	Clean Image (∞/∞)

Table 7: Result of NAM dataset.

Noisy Image	HI-GAN Denoised Image	Denoised Image by Our Method	Clean Image
Noisy Image (35.39/0.9369)	Denoised Image (34.73/0.9458)	Denoised Image (36.19/0.9643)	Clean Image (∞/∞)
Noisy Image (32.40/0.7213)	Denoised Image (33.65/0.8406)	Denoised Image (37.07/0.9516)	Clean Image (∞/∞)
Noisy Image (39.11/0.9339)	Denoised Image (38.98/0.9529)	Denoised Image (40.65/0.9732)	Clean Image (∞/∞)

8. Conclusion

The assessment proposes an overhauled HIGAN model for picture de-noising, which achieves extraordinary perceptual quality in regards to both general substance and unequivocal features. The pre-arranged model performs strikingly well in single-frame de-noising, and it can stay aware of key components without losing discernable nuances, due to the changed mishap capacity. The thick relationship between each layer inside each RDB considers full utilisation of neighbourhood features, close by the extra equivalent skipped network, achieving a compelling method for managing achieving our target of de-noising pictures without losing any huge components. The close by component blend (LFF) settles the readiness of greater associations, yet moreover adaptively controls the protection of information from the current and going before RDBs. RDB moreover thinks about direct relationship between the previous RDB and each layer of the continuous block, achieving a contacting memory (CM) instrument. In addition, the local waiting learning (LRL) further deals with the movement of information and tendencies. Since Re-LU is computationally capable as a particular number of neurons are sanctioned the proposed HIGAN model with the changed Waiting Thick Block (RDB) provoked an extension in the PSNR and SSIM of pictures. Anyway, for rational purposes, we intermittently need to oversee riotous low-objective photos. The two picture-taking care of techniques for standard interposition and sound lessening could influence. Further assessment concerning the impact of combined SR and de-noising on hurt photos may thusly be significantly prodding.

Acknowledgements

We are exceptionally appreciative to specialists for their suitable and useful ideas to work on this layout.

References

[1] Abdelhamed, A., Lin, S., and Brown, M. S. (2018). A high-quality denoising dataset for smart phone cameras. *IEEE Computer Vision and Pattern Recognition (CVPR)*.

[2] Agostinelli, F., Anderson, M. R., and Lee, H. (2013). Adaptive multi-column deep neural networks with application to robust image denoising. *Advances in Neural Information Processing Systems.*

[3] Burger, Harold C., Christian J. Schuler, and Stefan Harmeling. (2012). Image denoising: Can plain neural networks compete with BM3D? *IEEE Conference on Computer Vision and Pattern Recognition.*

[4] Babu, D., and Sajeev, K. J. (2020). Review on CNN based image denoising. *Proceedings of the International Conference on Systems, Energy & Environment (ICSEE)*, 9.

[5] Fan, L. (2019). Brief review of image denoising techniques. *Visual Computing for Industry, Biomedicine, and Art*, 2(1), 7.

[6] Ghose, S., Nishi, S., and Prabhishek, S. (2020). Image denoising using deeplearning: Convolutional neural network. *10th International Conference on Cloud Computing, Data Science & Engineering (Confluence)*, 511–517.

[7] Gu, S., *et al.* (2019). Self-guided network for fast image denoising. *Proceedings of the IEEE/CVF International Conference on Computer Vision.*

[8] Ilesanmi, A. E., and Taiwo, O. I. (2021). Methods for image denoising using convolutional neural network: a review. *Complex & Intelligent Systems*, 7, 2179–2198.

[9] Isa, I. S. (2015). Evaluating denoising performances of fundamental filters for T2-weighted MRI images. *Procedia Computer Science*, 60, 760–768.

[10] Kumwilaisak and Wuttipong, *et al.* (2020). Image denoising with deep convolutional neural and multi-directional long short-term memory networks under Poisson noise environments. *IEEE Access*, 8, 99.

[11] Krull, Alexander, Tim-Oliver Buchholz, and Florian Jug (2019). Noise2void-learning denoising from single noisy images. *Proceedings of the IEEE/CVF Conference on Computer Vision and Pattern Recognition*, 2129–2137.

[12] Li, H. M. (2014). Deep learning for image denoising. *International Journal of Signal Processing, Image Processing and Pattern Recognition*, 7(3), 171–180.

[13] Liu, Z., Wei, Q. Y., and Mee, L. Y. (2018). Image denoising based on a CNN model. *4th International Conference on Control Automation and Robotics (ICCAR)*, 389–393.

[14] Nam, S., Hwang, Y., Matsushita, Y., and Kim, S. J. (2016). A holistic approach to crosschannel image modelling and its application to image denoising. *IEEE Computer Vision and Pattern Recognition (CVPR).*

[15] Peizhu, G., Jin, L., and Lv, S. (2020). Image denoising with GAN based model. *Journal of Information Hiding and Privacy Protection*, 2(4), 155–163.

[16] Plotz, T., and Roth, S. (2017). Benchmarking denoising algorithms with real photographs. *IEEE Computer Vision and Pattern Recognition (CVPR).*

[17] Sun, Y. (2018). Digital radiography image denoising using a generative adversarial network. *Journal of X-ray Science and Technology*, 26(4), 523–534.

[18] Tian, C., *et al.* (2019). Image denoising using deep CNN with batch renormalization. *Proceedings of the IEEE/CVF Conference on Computer Vision and Pattern Recognition*, 121, 461–473.

[19] Tian, C., *et al.* (2020). Deep learning on image denoising: An overview. *Neural Networks*, 131, 251–275.

[20] Wei, K., *et al.* (2020). A physics-based noise formation model for extreme low-light raw denoising. *Proceedings of the IEEE/CVF Conference on Computer Vision and Pattern Recognition*, 2758–2767.

[21] Vo, D. M., Nguyen, D. M., Le, T. P., and Lee, S. W. (2021). HI-GAN: A hierarchical generative adversarial network for blind denoising of real photographs. *Proceedings of the IEEE/CVF Conference on Computer Vision and Pattern Recognition Workshops*, 570, 225–240.

[22] Xie, J., Linli, X., and Enhong, C. (2012). Image denoising and inpainting with deep neural networks. *Advances in neural information processing systems.*

[23] Yu, S., Bumjun, P., and Jechang, J. (2019). Deep iterative down-up cnn image denoising. *Proceedings of the IEEE/CVF Conference on Computer Vision and Pattern Recognition Workshops.*

[24] Zhang, X., and Wu, R. (2016). Fast depth image denoising and enhacement using a deep convolutional network. *IEEE International Conference on Acoustics Speech and Signal Processing* (*ICASSP*), 2499–2503.

[25] Zhang, K., (2017). Beyond a gaussian denoiser: Residual learning of deep cnn for image denoising. *IEEE transactions on image processing*, 26(7), 3142–3155.

[26] Zhang, K., Wangmeng, Z., and Lei, Z. (2018). FFDNet: toward a fast and flexible solution for CNN-based image denoising. *IEEE Transactions on Image Processing.*

Chapter 20

Survey on Energy Optimisation in Wireless Sensor Network

Vishwajit K. Barbudhe[1] *and Shruti K. Dixit*[2]

[1]Research Scholar, Department of Electronics and Communication Engineering, Sage University, Bhopal, India

[2]Associate Professor, Department of Electronics and Communication Engineering, School of Engineering and Technology, SAGE University, Bhopal, India

Abstract: Numerous energy-constrained sensors are often used in sensor network wireless technology. Clustering techniques have been employed to reduce energy consumption and extend the lifespan of the entire network. Current methods are examined from a quality of service (QoS) standpoint, with several basic criteria in mind: energy efficiency, secure interaction and latency monitoring. Intelligent gadgets must take user preferences into account so as to manage a variety of circumstances. One lingering difficult topic in clusters is client of users or customer-oriented architecture. The possible difficulties of applying clustered methods to Internet of Things (IoT) technologies in 5G infrastructures. According to recent research, because WSNs operate across both homogenous and a small amount of diverse relationships, they aren't suitable for use because they aren't capable of operating within extremely IoT platforms with an extensive variety of user situations. Furthermore, once 5G is realised, the issue will grow harder than with conventional, simple WSNs. However, as WSN expands, so does the amount of information that nodes with sensors must collect, manage and distribute. Given these sensors' energy requirements, analysing and delivering such a significant volume of information is unfeasible. As a result, there is a requirement for ML (machine learning) methods to be used in WSNs. Numerous issues associated with deploying clustering methods in the IoT, as well as machine learning approaches, must be investigated in order to optimise WSN efficiency.

Keywords: Energy optimisation, WSN, machine learning, energy conservation.

1. Introduction

Light, temperatures, movement, earthquakes and a variety of additional sources of data comprise our actual environmental setting in Akgul *et al.*, (2018). To gain an improved comprehension of the natural world, it is required to collect data

DOI: 10.1201/9781003527442-20

from numerous divergent resources in (Abdullah, 2020). Thus, this collection of such rich information is made possible by a straightforward technology called a wireless sensor system. A main node or centralised centre (beginning centre station) is the centre of gravity of a WSN, which is composed of regionally scattered separate gadgets that monitor that external environment in (Al-Khayyat and Ibrahim, 2020). Contemporary wireless sensor systems are both directional, which enables the transfer of monitoring data from clusters to a centralised location or basis stations, in addition to managing sensing activities via the basis stations to the detectors in (Sergi *et al.*, 2021).

Wireless sensor networks (WSN) were originally developed for military uses like battle observation; however, these systems are now utilised in a wide range of factories and customer uses, like manufacturing procedure tracking and oversight, equipment condition tracking, surroundings identification and environment tracking. The WSN is made up of "nodes" that range from just a few to a large number of motes, with each node attached to a single or multiple devices. The sensor network nodes usually consist of a few components: a radio transponder to a built-in antenna or a link to an outside antenna; a microcontroller and electronics circuitry for interacting with its detectors; plus a source of power, which is commonly a power source or a combined method of renewable energy gathering in (Darif *et al.*, 2021).

A sensing node can be as large as a cabinet or as small as a tiny piece of dirt; however, working "motes" of true tiny proportions have not yet been constructed. Sensing cluster prices vary likewise, from a few millions to thousands of dollars,

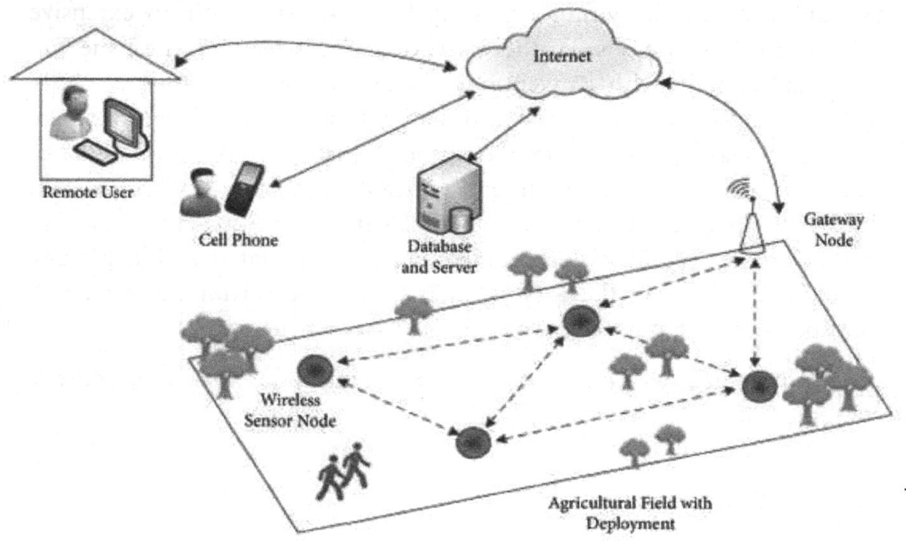

Figure 1: Wireless sensor network structure.

depending upon the intricacy of each of the sensing clusters. Sensor node dimensions and price limits lead to equivalent limits on resources like energy, memories, computing speed and connection capacity. The WSN scanning architecture might range from a basic star network to a sophisticated multihop wireless mesh network. The routed and floods method may be used to establish the transmission mechanism across networking connections. A WSN may be utilised for a variety of purposes; a few of those uses are listed below in (Al-Khayyat and Ibrahim, 2020).

Several highly complex strategies, such as employing specialised protocols for routing to avoid packet duplication in a network, can be utilised. Controlling node activities by adjusting the mode of operation among engaged, unused and sleep states is a common strategy. The CPU spends the most energy when it is active. The gadget is capable of receiving and sending data, controlling packets of data and executing the processing of data when in this mode. In a sleep state, a thing consumes a tiny amount of power as its transmitting device is off, its main processor's radio signal could be lowered and no estimation activities are permitted in (Darif *et al.*, 2021).

2. Literature Review

Anupkumar M. Bongale ***et al.*** "EiP-LEACH: "Energies Impact Probabilistic Basis of the LEACH Algorithm with Wireless Sensor Network". The EiP-LEACH procedure, considered a superior form of the LEACH approach that is affected by its energy usage factor for CH choice, is proposed in this paper in (El Khediri *et al.*, 2020).

Jianpeng Du ***et al.*** "A Collecting Data Method for Multidisciplinary Networks of Sensors Utilising Mobile Sinks" to considerably alleviate the problem of buffer overflow, a solution for infinite storage is provided next in (Jianpeng, 2022).

Nabajyoti The researcher created a generalised energy-effective clustering technique for movable sink-based WSNs that solves this problem by allowing the submerged devices to move over the area of interest at a fixed speed or path in (Nabajyoti, 2020).

Abdullah Al Omari ***et al.*** First goal of WSNs using mobile components is to devise a practical method to shorten the journey the ME makes to gather data in Abdullah (2020).

Tahar Abbes Mounir ***et al.*** Positioning method of urgent scenarios dependent on RSSI measurements for WSN provides where the links are located in Tahar *et al.* (2021).

Sharma, Singh ***et al.*** Two inputs for systems that are fuzzy consist of vitality and range. These systems accept residual power, closeness to the base station and neighbouring nodes as inputs.

Adnan ***et al.*** In WSNs, networking adaptability and conservation of energy are essential. The proposed technique is contrasted with the TTDPF and CHCCF systems.

Selvi ***et al.*** Using scholars, creating energy-effective networking methods for WSNs poses formidable difficulties. The suggested approach depends on assessment factors including live nodes in an energy utilisation model, packet acquisition by BS, initial cluster mortality, halfway network death and final network death.

Behera ***et al.*** The suggested approach outperforms SEP and DEEC processes in terms of system lifespan and throughput by 30% and 56%, respectively, according to simulation findings.

Mishra and Verma opportunistic inputs for compartmental model-based wireless network cluster size optimisation.

Al-Khayyat ***et al*** An updated framework for WSN has been suggested based on the ACO method. The proposed methodology outperforms the leach aggregation approach to clustering and its derivatives, such as fuzzy-leach, as well as alternative cluster-based methods in (Al-Khayyat and Ibrahim, 2020).

Sonam Lata ***et al.*** The effectiveness of the suggested approach in regulating resource consumption at every node was shown to increase WSN reliability.

Lata ***et al*** The leach-fuzzy clusters procedure was developed by the researchers. It works better than earlier suggested solutions when it comes to connection durability and use of energy.

M. Premkumar The network routers and tunnels connecting to the distributed clusters can enable a wide range of real-time activities. This reduces energy usage and erroneous alerts.

Fattoum ***et al.*** According to the severely restricted ability of power supplies used by sensors, the electrical power ratio was an especially crucial consideration in the design of WSNs. With regards to resource usage and network expansion lifetime, contemporary fuzzy logic-based clustering beats prior techniques and variants of clustering methodology in modelling in (Dao, 2020).

3. Performance Analysis

3.1. Parameters for Simulation

Throughput, latency, overhead, energy use, network lifespan and active nodes are network-based parameters used to assess the efficacy of our suggested techniques, E-LFRR, ME2PLB and DEA-OR. Using Network Simulator 2, the suggested methods are put into practise to produce the results.

Table 2: The simulation's settings and their associated values.

Parameter	Network Area	Protocol	Sensor Nodes in Number	Instance Topology	IEEE Standard	Range of Broadcasting	Application Type	No. of Packets	Initial Energy
Value	1000 x 1000	DSR	100	Flat Grid	IEEE 802.11	250 metres	Fixed Bit Rate	1500	20 Joules

4. Results and Discussion

Table 3: The comparison of the performance measure values.

	E-LFRR	ME2PLB	DEA-OR
Throughput	271.44	526.4	662.27
Delay	61.3	55.7	45.09
Overhead	13.7	8.33	6.32
Energy Utilisation	15.1	12.7	10.862
Network Lifetime	13.7	16.7	200
Alive Nodes	63	72	77

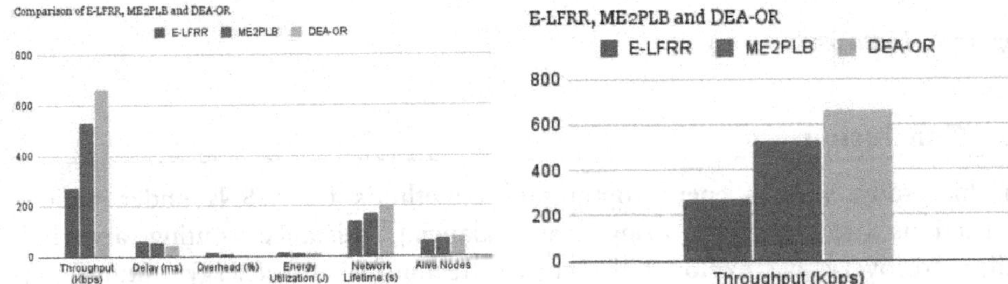

Figure 1: *Comparison of the performance.* ***Figure 2:*** *Throughput graph.*

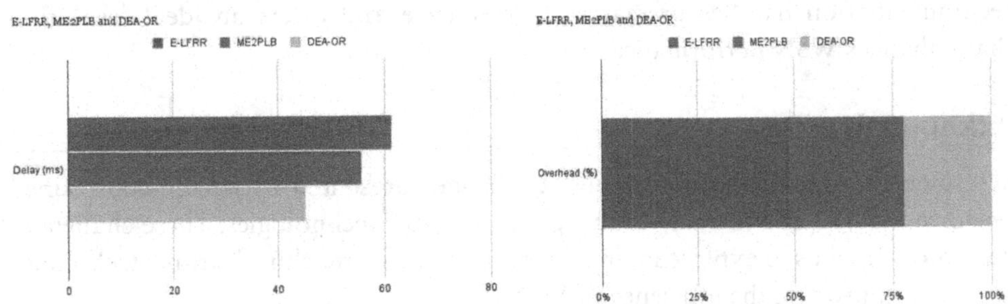

Figure 3: *Delay graph.* ***Figure 4:*** *Overhead graph.*

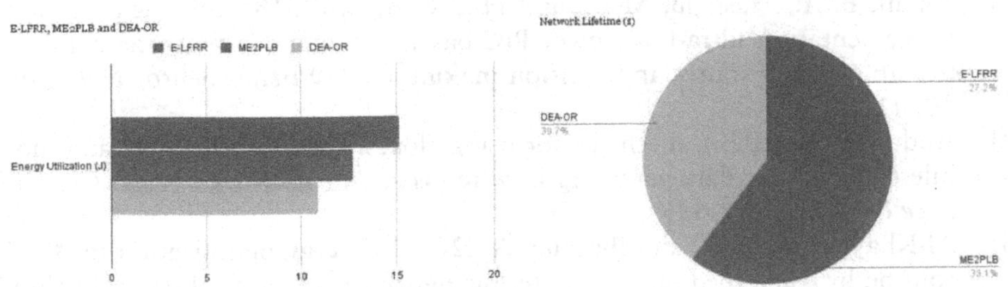

Figure 5: *Energy utilisation graph.* ***Figure 6:*** *Network lifetime graph.*

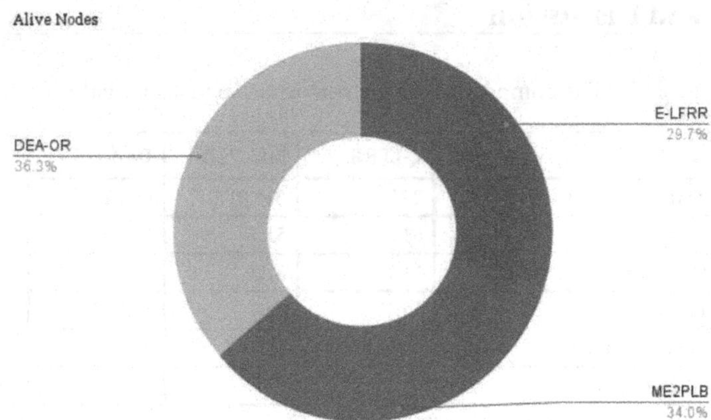

Figure 7: Network lifetime graph.

5. Conclusion

In this study, various energy optimisation methods for WSNs under various restrictions are surveyed. Using load balancing, distance routing and link failure recovery, we explored the current techniques for energy optimisation and conservation. Our suggested methods aim to reduce optimisation trade offs while taking into account many measurable factors. The final strategy in our recommendation handles premature convergence and offers an ideal resolution that enhances WSN performance.

Acknowledgment

We sincerely appreciate the complex challenges presented by energy-constrained sensor networks and the ever-evolving realm of IoT technologies. These challenges have motivated us to explore innovative clustering and machine learning techniques aimed at improving the efficiency of WSNs.

References

[1] Akgul, B. A., Hasoglu, M. F., and Haznedar, B. (2018). Investigation and implementation ultra-low power PIC-based sensor node network with renewable energy source and decision-making unit. *Wireless Sensor Network*, 10, 41–58.

[2] Abdullah, A. (2020). A scheme for using closest rendezvous points and mobile elements for data gathering in wireless sensor networks. *IEEE Conference on WSN*, 551–559.

[3] Al-Khayyat, A. T. A., and Ibrahim, A. (2020). Energy optimization in WSN routing by using the k-means clustering algorithm and ant colony algorithm. *4th International Symposium on Multidisciplinary Studies and Innovative Technologies (ISMSIT)*.

[4] Dao, T. K., Nguyen, T.-T., Pan, J.-S., Qiao, Y., and Lai, Q.-A. (2020). Identification failure data for cluster heads aggregation in WSN based on improving classification of SVM. *IEEE*, 61070–61084 (Access 8).

[5] Darif, A., Aboutajdine, D., and Saadane, R. (2021). Energy consumption optimization in real time applications for WSN using IR-UWB technology. *IEEE conference on Renewable and Sustainable Energy Conference (IR-SEC)*, 379–384.

[6] Dargie, W., and Wen, J. (2020). A simple clustering strategy for wireless sensor networks. *IEEE Sensors Letters*, 4(6), 1–4. doi: 10.1109/LSENS.2020.2991221.

[7] El Khediri, S., Fakhet, W., Moulahi, T., Khan, R., Faljaoui and A., and Kachori, A. (2020). Improved node localization using K-means clustering for wireless sensor networks. *Computer Science Review*, 37. https://doi.org/10.1016/j.cosrev.2020.100284.

[8] Fathima, S. K., and Witkowski, U. (2020). Energy efficient clustering protocols for WSN: performance analysis of FL-EE-NC with LEACH, k-means-LEACH, LEACH-FL and FL-EE/D using NS-2. *32nd International Conference on Microelectronics (ICM)*.

[9] Ikram, D., Baghdad, A., and Abdelhakim, B. (2021). A comprehensive survey on LEACH-based clustering routing protocols in wireless sensor networks. *Ad Hoc Networks*, 114.

[10] Jurado-Lasso, Clarke, K., Cadavid, and Nirmalathas, A. (2021). Energy-aware routing for software-defined multihop wireless sensor networks. *IEEE Sensors Journal*, 21(8), 10174–10182.

[11] Jin, and Muqing, W. (2021). Genetic-based cluster routing algorithm for wireless sensor networks. *7th International Conference on Computer and Communications* (ICCC), 48–52. https://doi.org/10.1109/ICCC54389.2021.9674406.

[12] Jianpeng, Du. (2022). A data collection approach based on mobile sinks for heterogeneous sensor networks. *IEEE Conference on WSN*, 45–55.

[13] Lata, S., Mehfuz, S., Urooj, S., and Alrowais, F. (2020). Fuzzy clustering algorithm for enhancing reliability and network lifetime of wireless sensor networks. *IEEE*, 66013–66024, (Access 8).

[14] Mishap, M., and Gupta, G. S. (2019). A review of and a proposal for cross-layer design for efficient routing and secure data aggregation over WSN. *IEEE conference on Networking, Sensing and Control (ICNSC)*, 138–142.

[15] Nabajyoti, M. (2020). Distributed energy-efficient clustering algorithm for mobile sink based wireless sensor networks. IEEE Conference on WSN, 78–90.

[16] Nguyen, L. V., Ha, N. V., Shibata, M., and Tsuru, M. (2021). TDMA-based scheduling for multi-hop wireless sensor networks with 3-egress gateway linear topology. *Internet of Things*, 14.

[17] Pranathi, T. Y. S., Dhuli, S., Aditya, V., Charisma, B., and Jayakrishna, K. (2020). A hybrid routing protocol for robust wireless sensor networks. *12th International Conference on Computational Intelligence and Communication Networks (CICN)*.

[18] Saimeh, A. H. (2021). Energy-aware optimization of the number of clusters and cluster-heads in WSN. *IEEE conference on Innovations in Information Technology (IIT)*, 178–183.

[19] Sergi, C., Turiaco, N., and De Rango, F. (2021). MAC protocols in wireless sensor network: an energy comparison analysis. *29th Telecommunications Forum* (*TELFOR*), 1–4. doi: 10.1109/TELFOR52709.2021.9653365.

[20] Tahar, A. M., and Mohamed, S. (2021). Positioning system for emergency situations based on RSSI measurements for WSN. *Performance Evaluation and Modeling in Wired and Wireless Networks* (*PEMWN*), 28–30.

[21] Temene, N., Sergiou, C., Georgiou, C., and Vassiliou, V. (2022). A survey on mobility in wireless sensor networks. Ad Hoc Networks, 125, 102726, ISSN 1570-8705.

Chapter 21

A Review on Fake Twitter User Profile Detection

Vyankatesh Rampurkar[1] *and Thirupurasundari D. R.*[2]

[1]Research Scholar, Bharat Institute of Higher Education and Research, Chennai
[2]Associate Professor, Bharat Institute of Higher Education and Research, Chennai

Abstract: The widespread use of social networking sites like Twitter and Facebook has led to significant impacts on users' lives. However, this popularity has also given rise to the dissemination of incorrect information through fake accounts, posing risks to society. To address this issue, our research focuses on developing a classification method specifically aimed at detecting fake accounts on Twitter. We employ a supervised discretisation technique called entropy minimisation discretisation (EMD) to preprocess the dataset's numerical features and analyse the outcomes using the naïve bayes algorithm. In addition, internet communities are faced with a significant presence of fake accounts, leading to malicious activities such as spreading spam, fake news and interfering in political campaigns. To combat this, researchers have been actively working on improving machine learning (ML) algorithms for fake account detection. In this study, we compare four ML algorithms (J48, Random Forest, Naive Bayes and KNN) and incorporate two reduction techniques (PCA and Correlation) to analyse a MIB Twitter dataset. Combining Correlation with the Random Forest algorithm yields superior results, achieving an accuracy of approximately 98.6%. By addressing the issue of fake accounts through our research and experimentation, we aim to contribute to the development of effective detection methods and mitigate the harm caused by fake accounts on social media platforms.

Keywords: Random forest, naive bayes, KNN, PCA, correlation.

1. Introduction

The rapid growth of social networking platforms over the past two decades has revolutionised online interactions, drawing in a vast number of users from various backgrounds. However, this surge has also given rise to a concerning issue – the proliferation of fake accounts, which do not represent real individuals. These fake accounts contribute to the dissemination of false information, deceptive web ratings, spam and other activities that violate the rules and guidelines of platforms like Twitter. The misuse of social media includes automated account interactions,

DOI: 10.1201/9781003527442-21

attempts to deceive or mislead users and engaging in harmful practices such as posting malicious links, aggressive following or unfollowing actions, creating multiple accounts, duplicating updates, and abusing reply and mention functions. On the other hand, genuine accounts adhere to the platform's guidelines.

The impact of this phenomenon is far-reaching, as tweets and messages are exchanged in real-time, allowing information to spread rapidly among a large user base. One of the main challenges posed by social media arises from spammers, who exploit their accounts for various malicious purposes, including spreading rumours that can significantly impact businesses and society as a whole. Recognising the influence of social media on society, this research aims to tackle the issue of detecting fake profile accounts on Twitter's online social network. The primary objective is to curb the spread of fake news, advertisements and fake followers.

Moreover, with the exponential growth of various social media networks, such as Facebook, LinkedIn, Twitter and Instagram, facilitated by technological advancements in wireless communication, over half of the world's population actively uses the internet and participates in social media activities. However, this surge in users has also led to a surge in fake accounts, creating serious issues like spreading fake news, engaging in political manipulation, promoting hate speech and conducting spam activities, which undermines the credibility and reliability of online social networks. To address the challenges posed by fake accounts, machine learning algorithms have become essential for their detection. Nevertheless, cyber-criminals are actively developing bots to bypass these detection mechanisms, leading to an ongoing battle between detection algorithms and malicious actors. In this research, we delve into a structured approach that includes a comprehensive literature review, a detailed description of the proposed detection method and a comparative analysis of algorithm results. By contributing to the advancement of techniques to detect and combat fake accounts, this research endeavours to uphold the trust and credibility of online social networks and foster a safer digital environment for all users.

2. **Literature Review**

We used the results of a systematic literature review (SLR) to create the framework for detecting false user profiles. The performances of various ML (Machine Learning) models are assessed using various datasets in research article (Erşahin *et al.*, 2017). The decisions made by these independent examinations were then intersected to identify common characteristics, forming the dataset. To ensure the quality of the results, author took care to balance the number of data samples in each class. The dataset consists of 501 instances of fake accounts and 499 instances of real accounts.

To create the dataset, author gathered 16 features from Twitter API information. The first 13 attributes were directly obtained from the user object attributes of Twitter API, while the last 3 features were generated using Twitter API by author. For instance, the attribute "urls_average" was computed based on the last 20 tweets of the user to determine the number of URLs used in their activities.

Similarly, "mentions_average" and "hashtags_average" were calculated based on the number of mentions and hashtags used in the user's last 20 tweets, respectively. The "Description" attribute represents the length of the account's self-defined description. Author observed that fake accounts mostly had blank descriptions, resulting in a description length of zero. The "Protected" attribute indicates whether the user has chosen to protect their tweets, making them visible only to selected followers. Fake accounts were mostly found to be unprotected. The "Followers count" attribute represents the number of followers an account currently has. Fake accounts tended to have very few followers, while the "Friends count" attributes (number of accounts the user is following) showed that fake accounts had an unusually high number of followings. The "Statuses count" attribute represents the number of tweets issued by the user, including retweets. Author noticed that fake accounts, often created by bots, frequently shared tweets related to trending topics, leading to a higher value for this attribute. The "Favourites count" attribute represents the number of tweets the user has liked over their account's lifetime. Some fake accounts added numerous tweets to their favourites to increase their follower count. The "Listed count" attribute indicates the number of public lists the user is a member of. Fake accounts tended to increase their presence in such lists. The "Verified" attribute is true when the user is verified by Twitter, indicating a real account. The attributes "Profile use background image," "Default profile" and "Default profile image" help identify the authenticity of accounts based on their preference for using background images, altered themes and custom profile images, respectively. The "Is_translator" attribute is true when the user is a participant in Twitter's translator community, which is typically associated with real accounts. The attributes "Hashtags average," "Mentions average" and "Urls average" provide insights into the content of users' last 20 tweets in terms of the average number of hashtags, mentions and URLs used, respectively.

Table1: Navie bayes F-measure result.

Before discretization	**After discretization**
F-Measure Class	**F-Measure Class**
0.848 true	**0.906 true**
0.872 false	**0.911 false**
Weighted Avg. 0.860	**Weighted Avg. 0.909**

In research paper Homsi *et al.* (2021), detecting twitter fake accounts using ML and data reduction techniques is explained. A pictorial description for the same is described as follows.

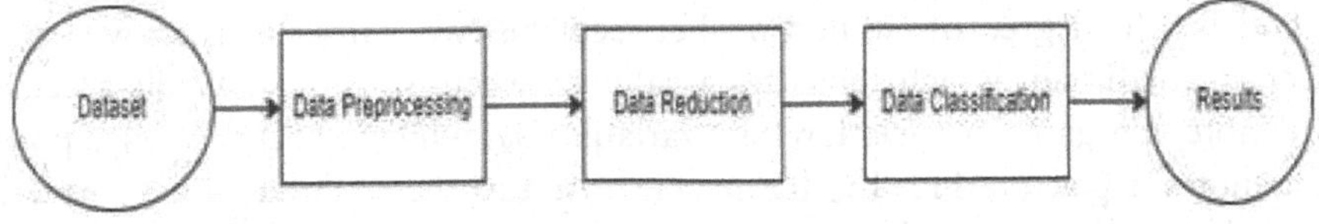

***Figure1:** Block diagram of proposed system (Homsi et al., 2021).*

Above figure 1 is a block diagram of Proposed System where data Pre-processing is followed by data reduction and data classification (Homsi, A., Al Nemri, J., Naimat, N., Kareem, H. A., Al-Fayoumi, M., & Snober, M. A., 2021).

Author involved utilising Weka software, a ML platform known for supporting various practices of machine learning. Author used this platform to implement proposed approach or model (Bouckaert *et al.*, 2010). Below are the key features offered by Weka. Preprocessing of data: Weka offers a variety of filters for data, from straightforward attribute deletion to more complex approaches like principal component analysis (PCA).

Classification: Weka provides over 99 classifiers, divided into categories including Bayesian, Lazy, Function-Based, Decision Tables, Tree, Misc. and Meta. There are various categorisation algorithms for each category.

Clustering: To enable unsupervised learning, Weka provides a number of clustering schemes, including well-known techniques like k-means and hierarchical clustering algorithms.

Attribute Selection: Weka uses a variety of search techniques and selection criteria to find the qualities that are most important for getting the best classification performance.

Data visualisation: Weka enables the depiction of data and computation outcomes through simple visual graphs, simplifying the interpretation and comprehension of operational results.

The author used the "MIB" dataset (Cresci *et al.*, 2015), which has 5,301 accounts in total and is divided into the following categories: True Accounts

The IIT-CNR in Pisa, Italy, collected the 469 accounts that make up the "Fake Project" Dataset. About 1,481 actual human narratives make up the "13 (elezioni 2013)" Dataset, which was examined by two sociologists from the University of Perugia in Italy.phoney accounts in the "Fastfollowerz" Dataset, 1,337 accounts are represented in the "Intertwitter" Dataset, 1,169 accounts are included.

The 845 accounts that make up the "Twittertechnology" Dataset were acquired by researchers in 2013 from the market.

Data reduction methodologies such as PCA and correlation offer specific advantages in data analysis. PCA is employed to reduce the dimensionality of large datasets while retaining essential information. By identifying principal components, PCA maximises data variety while minimising the number of variables, enabling more efficient visualisation and analysis of samples (Ringnér, 2008).

The Spearman's Correlation, on the other hand, is a form of correlation coefficient that is used to evaluate the strength of the association between two variables, designated as A and B. The degree of connection between the two variables is shown by the connection coefficient, which generates a value between –1 and +1. While a negative value denotes a negative association (one variable rises while the other falls), a positive value denotes a positive link (both variables rise together). There is no correlation between the variables when the correlation coefficient is zero (Zar, 2014). By reducing the complexity of datasets and offering insightful information about the relationships between variables, PCA and correlation both play critical roles in data analysis.

Data classification is done by using Decision Tree, Random Forest, KNN and Naive Bayes. In a real-world experiment, the author used the Weka software's default parameters, using 34% of the dataset for testing and 66% of the dataset for training. The experiment used four classification algorithms: Random Forest, K-Nearest-Neighbour (KNN), J48 and Naive Bayes, as well as two reduction techniques: PCA and correlation. Each reduction method was applied to the data, and all four classification algorithms were then applied to it. For instance, PCA with the random forest algorithm, PCA with the KNN algorithm and other algorithms were used to process the data. Author emphasised on precision and accuracy as two important criteria in analysis. A bigger percentage of bogus accounts are accurately identified when accuracy and precision are higher. The random forest method had the highest accuracy and precision when the correlation reduction technique was used with classification algorithms, whereas the Naive Bayes approach had the lowest accuracy and precision. For the other reduction approaches, same trends were seen. With the exception of Naive Bayes, using correlation in conjunction with classifier algorithms typically led to greater accuracy and precision than PCA.

When combining the random forest classification method and the correlation reduction technique, the highest accuracy of 98.6% was attained. With correlation and J48, the accuracy was 98%, which was the second-highest accuracy. Accuracy values for the other combinations were 95.4% for PCA with random forest, 93.56% for PCA with KNN, 93.5% for correlation with KNN, 93.1% for PCA with J48, 83.9% for PCA with naive bayes and 82.1% for correlation with naive bayes.

Figure 2 shows in detail comparison of Various Machine Learning Algorithms (Homsi, A., Al Nemri, J., Naimat, N., Kareem, H. A., Al-Fayoumi, M., & Snober, M. A., 2021)

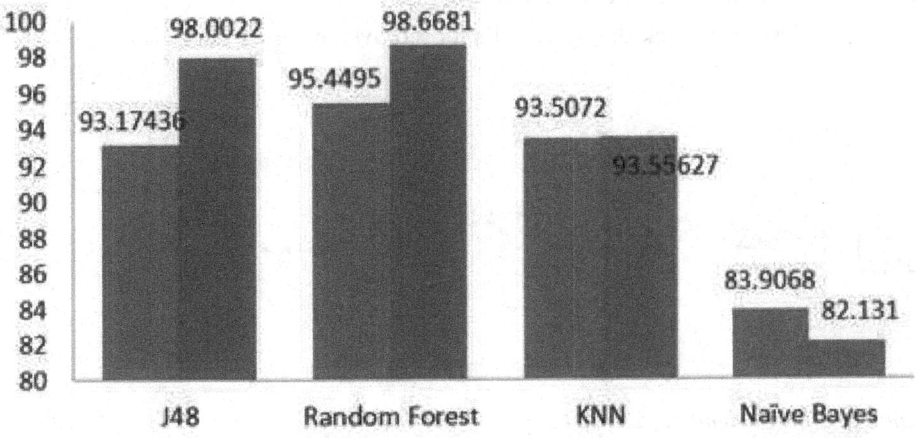

Figure 2: Comparison of various machine learning algorithms (Homsi et al., 2021).

According to Jogalekar *et al.* (2020), spammer detection and fake user identification on social networks are discussed. The primary goal of the research is to discover various ways for spam detection on Twitter and to offer a taxonomy that categorises these approaches into numerous groups. The taxonomy for identifying spammers on Twitter is mostly divided into four categories. The authors calculated the temporal distribution of tweets based on the number of tweets posted every hour in order to categories fake content. The authors then applied a regression prediction model to forecast the rise of fake content in the future and assure the overall influence of individuals who were disseminating false Information at the time. The authors have looked into the effects of a number of variables on spam detection effectiveness, including the spam to non-spam ratio, the size of a training dataset, time-related data, factor discretization and data sampling. Among the user-based features found are account age and the quantity of favourites, lists and tweets. Tweet-based features include the number of retweets, hashtags, user mentions and URLs. The gathering of tweets on popular subjects on Twitter, followed by the storage of the tweets in a certain file format and then the analysis of the tweetsSpam labelling are carried out to search through all of the datasets available in order to find malicious URLs. The language model, which uses language as a method to detect whether or not the tweets fake, separates the characteristics construct based on the feature extraction. The classification takes into account the user's name, profile picture, number of friends and followers, tweets' content, account description and total amount of tweets. A classification model for differentiating between non-spam and spam profiles is constructed using user-based, content-based and graph-based features. Some of the content-based metrics are total tweets, hashtag ratio, and URL ratio, mention ratio and tweet frequency. Unreliable users detection in social media is covered in research paper Sansonetti *et al.* (2020). It is discussed how to use deep learning for automatic detection. Unreliable users are organisations that deliberately disseminate false information. These entities include both lone users looking to gain more online notoriety and groups looking to spread misinformation and sway public opinion.

Figure 3 shows block diagram of Proposed System

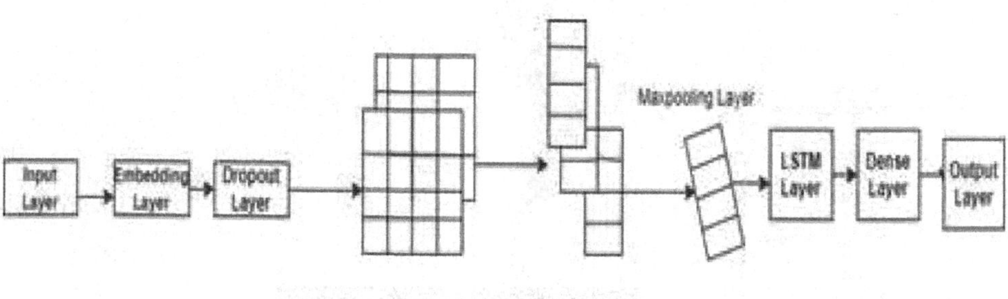

Figure 3: *Block diagram of proposed system (Jogalekar et al., 2020).*

Therefore, it is crucial to have a technology that can identify bogus news and unreliable users. This research article suggests technologies that can (i) analyses offline data and (ii) analyses online data to estimate the veracity of certain user and news reports.

Deep learning techniques are used for offline analysis. This system does a twofold analysis, predicting both the trustworthiness of user profiles and the dependability of news. By using a well-known fact-checking website as a guide, a dataset containing both credible and incorrect social network profiles and news is created. Deep learning models are constructed using long-short term memory and convolutional neural networks. Online analysis classifies user profiles as reputable or unreliable based on feedback from actual users. To perform the online study of the social context in which the news is put, a survey asking actual users to judge the trustworthiness of particular Twitter identities was developed. The results of the trial, which were supported by statistics, show what information both robots and humans may use to spot unreliable users.

3. Conclusion

The research introduces an approach to identify fake accounts on Twitter by leveraging the naïve bayes classification algorithm with discretization. The proposed method employs entropy minimization discretization (EMD) with minimal description length (MDL) as the stopping criterion. Through experiments, the approach enhances the accuracy of naïve bayes from 85.55% to 90.41%, achieved solely by applying discretization to selected features during data preprocessing. Despite minimal information loss due to discretization, the substantial accuracy improvement is noteworthy. This outcome demonstrates the potential of the approach in effectively utilising numeric social media data, while also addressing the normality assumption challenge associated with continuous data in Naïve Bayes analysis. The study aimed to examine the impact of two correlation techniques combined with various machine learning algorithms to enhance the detection of fake social media accounts. The research utilised the MIB dataset and Weka software. Data preprocessing and reduction were applied to prepare the dataset for classification. The classification phase assessed accuracy using PCA and correlation in conjunction with J48, Random Forest, KNN and Naive Bayes classifiers. The highest accuracy of 98.6% was achieved by the random forest algorithm with correlation data reduction, while the lowest accuracy of 82.1% was recorded with the Naive Bayes algorithm using correlation data reduction. The study acknowledges the need for further experimentation with diverse techniques, methodologies and algorithms to optimise detection accuracy. Future plans involve exploring additional reduction techniques and classification algorithms through in-depth analysis.

References

[1] Aceto, G., Ciuonzo, D., Montieri, A., and Pescapè, A. (2019). MIMETIC: Mobile encrypted traffic classification using multimodal deep learning. *Computer Networks*, 165, 106944.

[2] Agudelo, G. E. R., Parra, O. J. S., and Velandia, J. B. (2018). Raising a model for fake news detection using machine learning in Python. In Challenges and Opportunities in the Digital Era: 17th IFIP WG 6.11 Conference on e-Business, e-Services, and e-Society, I3E 2018, Kuwait City, Kuwait, October 30–November 1, 2018, *Proceedings 17*, pp. 596–604. Springer International Publishing.

[3] Ahmed, H., Traore, I., and Saad, S. (2018). Detecting opinion spams and fake news using text classification. *Security and Privacy*, 1(1), e9.

[4] Ajao, O., Bhowmik, D., and Zargari, S. (2018). Fake news identification on twitter with hybrid CNN and RNN models. *In Proceedings of the 9th international conference on social media and society*, 226–230.

[5] Amjad, M., Sidorov, G., Zhila, A., Gómez-Adorno, H., Voronkov, I., and Gelbukh, A. (2020). "Bend the truth": Benchmark dataset for fake news detection in Urdu language and its evaluation. *Journal of Intelligent & Fuzzy Systems*, 39(2), 2457–2469.

[6] Erşahin, B., Aktaş, Ö., Kılınç, D., and Akyol, C. (2017). Twitter fake account detection. *In 2017 International Conference on Computer Science and Engineering* (*UBMK*), 388–392.Homsi, A., Al Nemri, J., Naimat, N., Kareem, H. A., Al-Fayoumi, M., and Snober, M. A. (2021). Detecting Twitter fake accounts using machine learning and data reduction techniques. *In DATA*, pp. 88–95.

[7] Jogalekar, N. S., Attar, V., and Palshikar, G. K. (2020). Rumor detection on social networks: a sociological approach. *In 2020 IEEE International Conference on Big Data* (*Big Data*), 3877–3884.

[8] Sansonetti, G., Gasparetti, F., D'aniello, G., and Micarelli, A. (2020). Unreliable users detection in social media: Deep learning techniques for automatic detection. IEEE (accessed 8), 213154–213167.

[9] Ramalingam, D., and Chinnaiah, V. (2018). Fake profile detection techniques in large-scale online social networks: a comprehensive review. *Computers & Electrical Engineering*, 65, 165–177.

[10] Tiwari, V. (2017). Analysis and detection of fake profile over social network. *In 2017 International Conference on Computing, Communication and Automation* (*ICCCA*), 175–179.

[11] Shu, K., Wang, S., and Liu, H. (2018). Understanding user profiles on social media for fake news detection. *In 2018 IEEE Conference on Multimedia Information Processing and Retrieval* (*MIPR*), 430–435.

Chapter 22

Ayurveda-Dosha Assessment: A Computational Survey of the Human Body's Constitution

Swati P. Dhole[1] *and S. E. Yedey*[2]

[1]Research Scholar, PGDCST, HVPM, Amravati, Maharashtra, India

[2]Associate Professor and Head, PGDCST, HVPM, Amravati, Maharashtra, India

Abstract: Human nature, as well as the factors that can disrupt the body's equilibrium and lead to illness, are both defined by the human body's constitution (Prakriti). Tridosha describes the three fundamental forces or principles that govern both the mental and physical functions of our bodies: VATA (V), PITTA (P) and KAPHA (K) are the names of these three energies. While Ayurveda-dosha science has been used for a considerable time, diagnostic techniques associated with it still lack scientific validation. Effective treatment is the result of a thorough and comprehensive assessment. Ayurvedic experts evaluate the "dosha" using 28 distinct criteria to gather relevant data. The researchers calculated the Cronbach's alpha of VATA-Dosha, PITTA-Dosha and KAPHA-Dosha as 0.94, 0.98 and 0.98, respectively, to gauge the survey's validity. This study involved 807 healthy individuals aged between 20 and 60, 62.1% were males and 37.9% were females. Using computational methods, experts divided the dataset (80:20) into a training group consisting of 324 individuals and a test group of 81 individuals. For the categorisation assessment, the model was trained using established machine learning techniques such as artificial neural networks (ANN), k-nearest neighbours (KNN), support vector machines (SVM), naive bayes (NB) and decision trees (DT). The system under consideration was also constructed using a blend of multiple machine-learning algorithms with the objective of identifying patterns. The evaluation parameters such as precision, recall, F-score and accuracy were measured and analysed. The researchers concluded that the most effective tool for preparing and evaluating the information was CatBoost, which could be fine-tuned to achieve a precision of 0.97, recall of 0.96, F-score of 0.96 and accuracy rate of 0.96.

Keywords: Ayurvedic dosha, prakriti, tridosha, prakriti classification, machine learning.

DOI: 10.1201/9781003527442-22

1. Introduction

Ayurveda represents one of humanity's most ancient branches of healthcare, with its origins in the Indian subcontinent over 5,000 years ago by (Patwardhan, 2014). The combination of terms results in "Ayur: life" and "Veda: knowledge," leading to the term "Ayurveda," often referred to as "the discipline of the knowledge of existence" (Patwardhan, 2014) and Ayurveda (2019). By harmonising a person's body, mind and soul, Ayurveda contributes to the maintenance of good health. It serves as a method for extending life and promoting a wholesome approach to healing and therapy. Understanding how to lead a fulfilling life is crucial, as the disease affects everyone living in a modern, technologically advanced civilisation from a young age. Ayurveda is a particular system that imparts knowledge about harmony in nature and advocates for the restoration of balance. It was the first medical system to identify the connection between the senses and ailments. Mismatches between humanity and the natural world, or between our personalities and their senses, arise from the improper use of these senses. The foundation of ayurvedic philosophy lies in the equilibrium between oneself and the environment. One of the most potent techniques employed by ayurveda is the management of one's lifestyle. Ayurvedic practitioners are more likely to lead healthy lives devoid of illness because they can adapt to the principles and requirements of the system.

1.1. Tridosha and Panchamahabhutas

According to (\Patwardhan, 2014) everything in the universe is thought to be composed of the five "panchamahabhutas" (space, water, fire, earth and air). These fundamental elements manifest as three biological energies within the human body, governing and overseeing all life processes. As shown in Table 1, these forces are collectively termed the "Doshas" or "Tridoshas," which include VATA, PITTA and KAPHA in (Mathpati *et al.*, 2020).

Table 1: Panchamahabhutas and tridoshas.

Tridosha/sannipataja predominant in tridosha	Dominant Dwandwa-Dual dosha	Ekadoshaja: Predominantly one dosha
	V – P	V: Earth + Air
V – P – K	P – K	P: Fire + Water
	K – V	K: Earth + Water

The essence of humanity comprises two facets: the material and the mental. VATA (V), PITTA (P) and KAPHA (K) represent the three sharirika practices by Patwardhan (2014) and (Mathpati *et al.*, 2020). The term "VATA" refers to the wind element. Wind propels clouds as they traverse the skies, and a single gust can move the entire atmosphere. The VATA Dosha instils determination for success and resilience to overcome challenges. Human physiological and metabolic activities demand energy. "PITTA" energises focus and imparts a radiant quality to the body. It governs various aspects of metabolism, particularly thermal regulation and hormone equilibrium. The acronym "KAPHA" represents the water-based compound that lubricates joints and exists within organs and tissues. The "Tridoshas" supervise all physical and mental functions in living beings. To categorise an individual into an appropriate Tridosha type, 28 distinct attributes are considered. According to Verma *et al.* (2018) each person possesses a unique constitution and interacts with others differently. The psychological constitution can be grouped into seven distinct categories, including V–K, P–K, P, K, V–P and sam dosha, based on the predominance of specific physical characteristics (doshas). An individual might belong to a particular dosha group if they exhibit a dominant influence of more than one dosha. Ayurvedic science describes their physical makeup and physiological traits. It is influenced by various factors encompassing parental, prenatal and postnatal aspects, and is inherited, shaping an individual's development by Prasuna *et al.* (2014) and Wani *et al.* (2017). A personalised prakriti analysis enables an understanding of the body's needs. Prakriti, in turn, assists in managing personal, social and professional aspects of one's life by Todkari *et al.* (2015) and Dunlap *et al.* (2017).

1.2. Prakriti Examination

Aligning lifestyle and nutrition with the body's needs becomes more manageable when the dosha type is identified beforehand. This information indicates the potential for both quantitative and qualitative imbalances within the body. Several ayurvedic techniques require a thorough examination of the patient to gather as much information as possible before recommending a treatment plan by Chinthala *et al.* (2019). To collect essential data for our study, we developed a survey using this method. Internal reliability checks were conducted on the data obtained from the pilot study, followed by the creation of a trained model using a machine learning technique.

The organisation of the paper is as follows; Section 2 presents the related work of previous research. Section 3 presented the proposed methodology. Section 4 presented the evaluation parameters used to measure the performance of experimental result analysis. Section 4 and Section 5 presented the discussion and conclusion.

2. Related Work

Ayurveda has been the focus of numerous studies aimed at enhancing human well-being. Verma *et al.* (2018) delve into research on the development and utilisation of Ayurveda, assessing various aspects. The authors discuss Ayurveda's current popularity and its pivotal role in healthcare. The discourse also addresses the considerations essential for Ayurvedic medicine's advancement and dissemination.

Todkari (2015) ayurvedic medicine employs diverse component techniques to ascertain an individual's constitution and anomalies. The authors investigated the reliability of three distinct surveys to quantify human Prakriti. They evaluate and provide numerical measures for the test's reliability, leading to the conclusion that a standardised questionnaire-based approach is necessary for Prakriti research. To psychologically evaluate Tridoshas like VATA, PITTA and KAPHA in individuals, Dunlap *et al.* (2017) employed a personality test scale. According to the authors, these doshas are composed of the Pancha-Mahabhutas. Multiple doshas may dominate at different times, and the Tridoshas, along with the Pancha-Mahabhutas, aren't universally present simultaneously. Their research explains how the generated scale demonstrates psychometric characteristics. The primary achievement, as gauged by the scale, lies in the psychological manifestation of the Tridoshas. As part of the Prakriti determination standardisation process, they developed the Mysore Psychological Tridosha Scale. In their investigation, Ibrahim *et al.* (2019) illustrates the prevalence of ambiguity in medical assertions. Medical websites often contain uncertain information, and the authors examined a database of patient inquiries. Consequently, the provided responses might lack comprehensive information. To address this, they propose an end-to-end deep learning-based medical diagnostics system (DL-MDS) to aid users in making illness diagnoses and prevention. Woldaregay *et al.* (2019) highlights how machine learning's success in their study stems from its ability to tackle intricate challenges in a dynamic context. Prior research in medical domains and the application of machine learning to health monitoring and treatment have equipped human Prakriti assessment with professional decision-making tools.

3. Proposed Methodology

As depicted in Figure 1, this section provides a comprehensive overview of the entire study procedure, starting from data collection and extending through performance analysis.

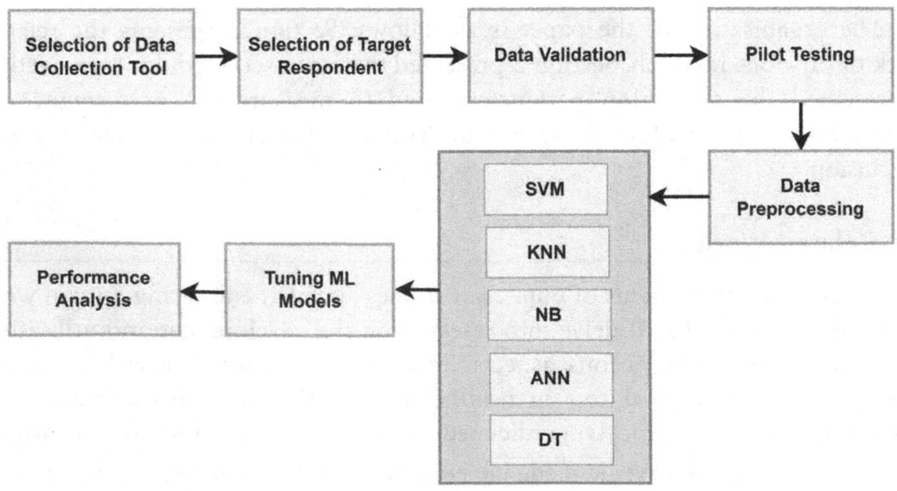

Figure 1: Complete architecture of proposed system.

Following the collection of data from healthy participants, the gathered information undergoes vetting and pilot testing. During the process of gathering additional data and evaluating its internal consistency, efforts are made to address class disparities. Machine learning algorithms, including KNN, SVM, ANN, naive bayes and decision trees, are employed for training and testing the data.

3.1. *Dataset Gathering*

Accurate and systematic data collection is a rigorous endeavour essential for developing precise models. The methodologies employed for data collection depend on the nature of the study, encompassing document review, observation, interviews, measurements or a combination of these methods. To gather data on these attributes from diverse individuals, create a survey after outlining the structure and characteristics of each category. Data gathering is facilitated through the utilisation of the expert-designed survey, offering affordability, accessibility and comprehensive coverage with minimal effort.

3.2. *Selection of Target Participants*

Thorough consideration is given to selecting healthy participants of both genders, aged between 20 and 60, ensuring the participant's well-being does not influence the outcomes. Students and staff members from colleges and universities are chosen to participate in this endeavour. This dataset comprises 837 instances encompassing 28 characteristics.

3.3. *Data Validation*

A survey serves as a method of inquiry employed to systematise data collection. It involves a series of questions and prompts aimed at gathering information from participants in Ayurveda (2019). To ensure the comprehensive construction of the questionnaire, consultation with two prominent ayurveda experts is sought. The survey comprises closed-ended questions, each offering multiple response options. Ayurvedic specialists assess the reliability of the survey's content.

3.4. *Pilot Testing*

An experiment was conducted to ensure the consistency of data collection. Initially, a group of 50 randomly selected individuals from the total population was used to assess the questionnaire's relevance. Upon achieving satisfactory accuracy through a small-scale business implementation using this initial set of results, further data collection proceeded using the same methodologies. Building upon the reliable outcomes gleaned from the pilot study, the entire model will subsequently undergo training and evaluation.

3.5. *Data Pre-Processing*

Only individuals deemed healthy are invited to participate, utilising unbiased survey methods to gather data. The selection of participants for data collection is conducted through a straightforward random sampling process. The bedrock of simple random sampling involves selecting a diverse set of individuals from a substantial population by West (2016), ensuring that each has an equal likelihood of being chosen. Participants are informed and guided on ayurveda's relevance in modern life, fostering interaction with them. The study's objectives are explained to motivated and consenting participants, revealing their keen interest in learning about their dosha profiles. Engagement encompasses a focused population of 807 individuals aged between 20 and 60, consisting of 306 women and 501 men. Detailed category-specific information regarding the data is presented in Table 2.

Table 2: Data gathering and categories based on geography and age.

Gender	**Female**	306
	Male	500
Age group	**Adolescence**	41
	Young Adult	387
	Adult	304
	Elderly	73
Participant from	**North India**	497
	South India	309

Table 2 shows the number of participants in a study is determined by the sample type being examined.

3.6. Assessing Data (Wrangling)

Python modules like Pandas and Numpy are used for data preparation. The information is physically transformed or translated from its "'raw"' form to a useful one that makes it easier to use.

3.7. Disordered Data

Despite understanding the outcome, it gathers the data. A proportional gathering of data from different groups is seen as the most effective for learning and having high accuracy. Even a slight class imbalance might have a negative impact on training in Krawczyk (2016). This sort of unbalanced class problem may be solved using a variety of strategies that either concentrate on the data level or the classification level.

3.8. Duplicate Data

There may be certain missing statements in the information that was gathered for a given assignment, often as a result of unforeseen characteristics or technological difficulties. Some missing data elements are discovered when processing the information. It resolves this issue by roughly estimating the missing numbers and identifying a connection among both known and unidentified data in Karanja *et al.* (2013).

3.9. Consistency Internal

We take into account a consistency measurement that is Cronbach's alpha, in order to examine the internal coherence of the data obtained and the validity of the survey. Cronbach's alpha reveals the degree to which an assortment of items is interconnected by Taber (2018). Table 4 provides the Cronbach's alpha values for VATT, PITT and KAPH.

3.10. Data Split

According to Herrera *et al.* (2019) and Mohanty *et al.* (2016), significant datasets are separated into two groups: training and testing. A 405-person data collection is divided into 8:2 ratios. About 80% (324 samples) of the original dataset are used as training data, while 21% (81 samples) are used as testing information. Each model is developed and evaluated using the exact same dataset. Python programming packages are used to create every model.

3.11. Selection of Best Learning Model

Constructing the optimal model structure is not solely reliant on the dataset itself; rather, it is built upon multiple performance attributes. Cross-validation plays a pivotal role in assessing each model's precision using unfamiliar data, aiding in the selection of the most suitable versions from the array of models under consideration. To comprehend the average precision and other characteristics involving precision variability, a range of Python visualisation techniques is employed. A pivotal tool for illustrating the performance of a classifier's predictions is the confusion matrix, also known as an error matrix in Andonie (2019). This matrix visually presents the effectiveness of model classification, highlighting correctly categorised values while revealing misclassifications. The classification outcomes of the SVM-developed algorithm are documented in Table 3. Among the 81 analysed samples, 69 are accurately classified, with 21 samples assigned to incorrect groups. The SVM classifier demonstrates an accuracy rate of 0.85.

3.12. Evaluation Parameters

$$Accuracy = \frac{T\_P + T\_N}{T\_P + T\_N + F\_P + F\_N}$$

$$Precision = \frac{T\_P}{T\_P + F\_P}$$

$$Recall = \frac{T\_P}{T\_P + F\_N}$$

$$F1 - Score = 2 \times \frac{Precision \times Recall}{Precision + Recall}$$

4. Results Analysis

In this section, delve into the application of ML models to preprocessed data. Given that there isn't a universal technique that excels for all ML problems, continuous evaluation of outcomes from diverse learning strategies is imperative. The dimensions and structure of your dataset are merely two of the numerous variables that can influence results. Hence, adopt a varied approach for data, employing a reserved "test set" to assess performances and determine the optimal approach in Hidayati *et al.* (2016). Despite utilising conventional ML models, satisfactory results are elusive. To further enhance the findings, hyperparameter adjustments are employed.

Figure 2 to Figure 7 shows the confusion matrix of various Machine Learning classifiers of each class.

Table 3: Comparative analysis of various ML models.

ML Models	RMSE	Precision	Recall	F-Score	Accuracy
KNN	1.54	0.88	0.86	0.86	0.88
KNN with Hyperparameter tuning	0.96	0.91	0.93	0.94	0.94
ANN	1.3	0.87	0.86	0.86	0.87
SVM	1.54	0.88	0.86	0.86	0.86
SVM with Hyperparameter tuning	0.99	0.94	0.92	0.93	0.93
Naive Bayes	2.44	0.73	0.65	0.67	0.65
Naive Bayes with Hyperparameter tuning	2.2	0.79	0.76	0.76	0.77
Decision Tree	1.02	0.89	0.89	0.88	0.89
XGBoost	0.80	0.92	0.91	0.91	0.91
XGBoost with Hyperparameter tuning	0.87	0.92	0.91	0.91	0.91
CatBoost	0.96	0.94	0.94	0.94	0.94

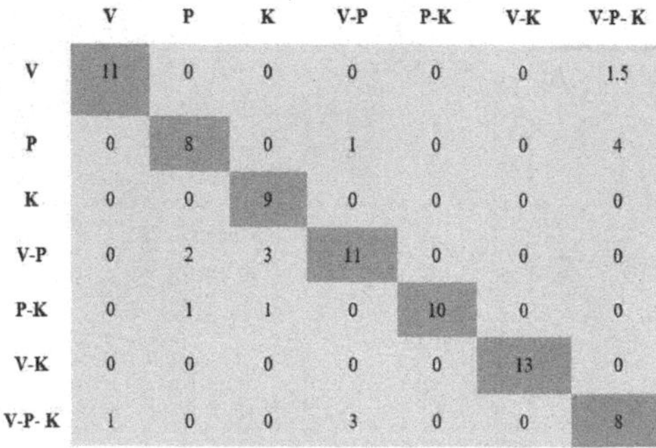

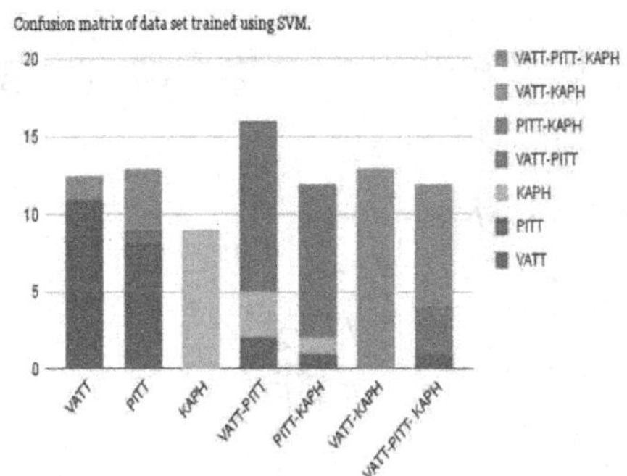

Figure 2: Confusion matrix of SVM model.

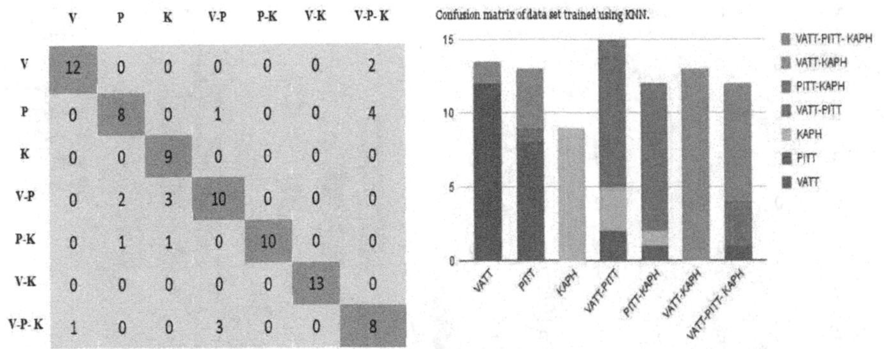

	V	P	K	V-P	P-K	V-K	V-P-K
V	12	0	0	0	0	0	2
P	0	8	0	1	0	0	4
K	0	0	9	0	0	0	0
V-P	0	2	3	10	0	0	0
P-K	0	1	1	0	10	0	0
V-K	0	0	0	0	0	13	0
V-P-K	1	0	0	3	0	0	8

Figure 3: *Confusion matrix of KNN model.*

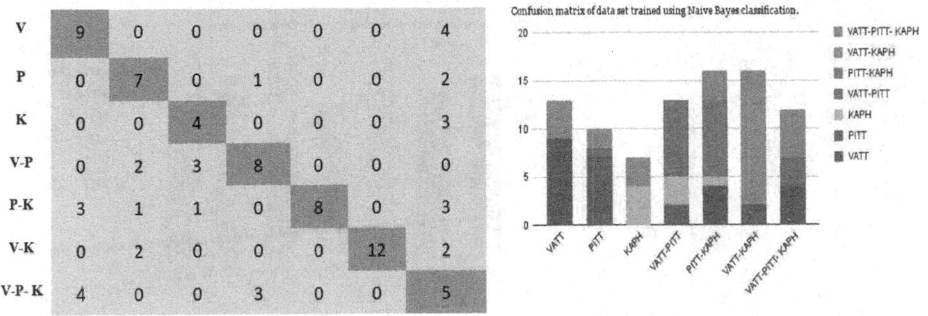

V	9	0	0	0	0	0	4
P	0	7	0	1	0	0	2
K	0	0	4	0	0	0	3
V-P	0	2	3	8	0	0	0
P-K	3	1	1	0	8	0	3
V-K	0	2	0	0	0	12	2
V-P-K	4	0	0	3	0	0	5

Figure 4: *Confusion matrix of naïve bayes model.*

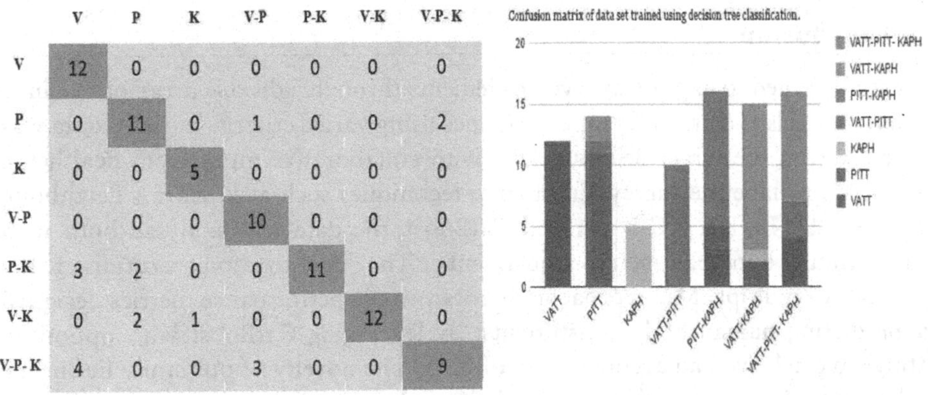

	V	P	K	V-P	P-K	V-K	V-P-K
V	12	0	0	0	0	0	0
P	0	11	0	1	0	0	2
K	0	0	5	0	0	0	0
V-P	0	0	0	10	0	0	0
P-K	3	2	0	0	11	0	0
V-K	0	2	1	0	0	12	0
V-P-K	4	0	0	3	0	0	9

Figure 5: *Confusion matrix of decision tree model.*

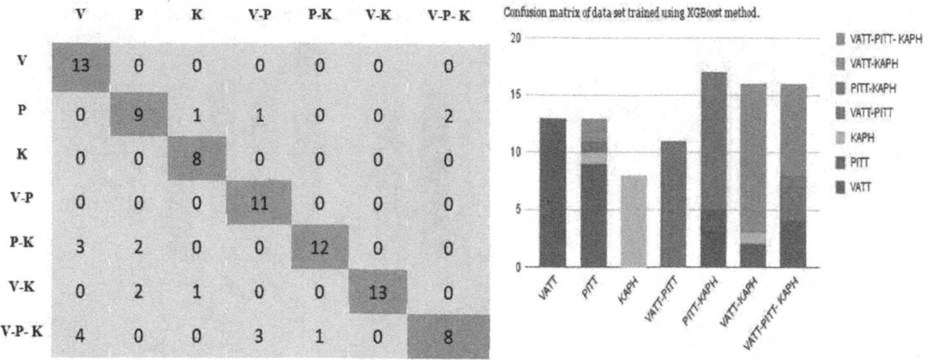

	V	P	K	V-P	P-K	V-K	V-P-K
V	13	0	0	0	0	0	0
P	0	9	1	1	0	0	2
K	0	0	8	0	0	0	0
V-P	0	0	0	11	0	0	0
P-K	3	2	0	0	12	0	0
V-K	0	2	1	0	0	13	0
V-P-K	4	0	0	3	1	0	8

Figure 6: *Confusion matrix of XGBoost model.*

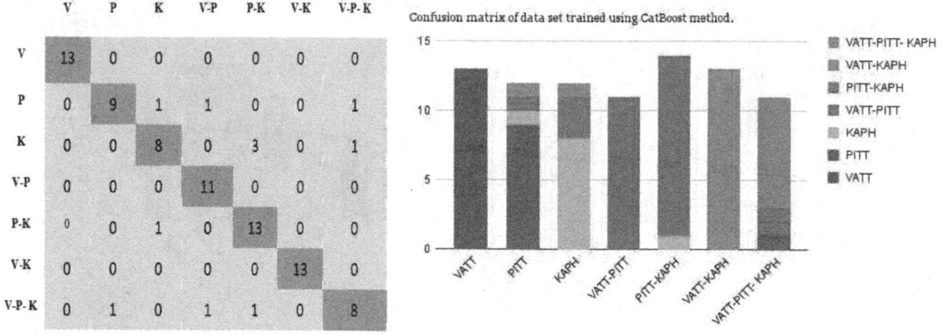

	V	P	K	V-P	P-K	V-K	V-P-K
V	13	0	0	0	0	0	0
P	0	9	1	1	0	0	1
K	0	0	8	0	3	0	1
V-P	0	0	0	11	0	0	0
P-K	0	0	1	0	13	0	0
V-K	0	0	0	0	0	13	0
V-P-K	0	1	0	1	1	0	8

Figure 7: *Confusion matrix of CatBoost model.*

5. Conclusion

This study aimed to explore ayurvedic elements through a diverse array of machine learning models and assess their performance using varied criteria. Initially, data was collected from the target demographic, with collaborative input from healthcare providers to shape the survey. Employing techniques such as k-nearest neighbour, ANN, SVM, NB, DT, XG-Boost and CatBoost, the data was analysed both with and without hyperparameter adjustments. The investigation scrutinised the outcomes of multiple ML approaches across various performance metrics designed for predicting human body constituents. By leveraging CatBoost with optimised settings, we achieved an accuracy rate of 0.95. The novelty of our study lies in the application of ML algorithms, trained through advanced techniques, to discern elements of an individual's body (Ayurveda Dosha). This proposed methodology could potentially serve as a valuable tool for ayurvedic medical professionals in component identification. Furthermore, this technique has broader implications, as it can be utilised by anyone, regardless of health status, to ascertain imbalances in their body's components without necessitating initial consultation with an ayurvedic practitioner.

References

[1] Ayurveda: A Brief Introduction and Guide. (accessed Jun 23, 2019). [Online]. Available: https://www.ayurveda.com/ resources/general-information

[2] Andonie, R. (2019). Hyperparameter optimization in learning systems. *J. Membrane Comput.*, 1(4), 279–291.

[3] Chinthala, R., Kamble, S., Baghel, A. S., and Bhagavathi, N. N. L. (2019). Ancient archives of Deha-Prakriti (human body constitutional traits) in ayurvedic literature: A critical review. *Int. J. Res. Ayurveda Pharmacy*, 10(3), 18–26.

[4] Dunlap, C., Hanes, D., Elder, C., Nygaard, C., and Zwickey H., (2017) Reliability of self-reported constitutional questionnaires in ayurveda diagnosis. *J. Ayurveda Integrative Med.*, 8(4), 257–262.

[5] Herrera, G., Constantino, M., Tabak, B., Pistori, H., Su, J., and Naranpanawa, A. (2019). Data on forecasting energy prices using machine learning. *Data Brief*, 25, Art. no. 104122.

[6] Hidayati, R., Kanamori, K., Feng, L., and Ohwada H. (2016). Implementing majority voting rule to classify corporate value based on environmental efforts. *In Proc. ICDMBD*, Bali, Indonesia, 59–66.

[7] Ibrahim, Y., Kamel, S., Rashad, A., Nasrat, L., and Jurado F., (2019). Performance enhancement of wind farms using tuned SSSC based on artificial neural network. *Int. J. Interact. Multimedia Artif. Intell.*, 5(7), 118.

[8] Karanja, E., Zaveri, J., and Ahmed A. (2013). How do MIS researchers handle missing data in survey-based research: a content analysis approach. *Int. J. Inf. Manage.*, 33(5), 734–751.

[9] Krawczyk, B. (2016). Learning from imbalanced data: Open challenges and future directions. *Prog. Artif. Intell.*, 5(4), 221–232.

[10] Mathpati, M., Albert, S., and Porter, J. (2020). Ayurveda and medicalisation today: The loss of important knowledge and practice in health. *J. Ayurveda Integrative Med.*, 11(1), 89–94. https://doi.org/10.1016/j.jaim.2018.06.004.

[11] Mohanty, P., Hughes, P., and Salathé M. (2016). Using deep learning for image-based plant disease detection. *Frontiers Plant Sci.*, 7, 1419.

[12] Patwardhan, B. (2014). Bridging Ayurveda with evidence-based scientific approaches in medicine. *EPMA J.*, 5(1), 1–7.

[13] Prasuna, V., Sharma, B., and Narayana, A. (2014). Comparative study of personality with Ayurvedic Prakriti. *Int. J. Ayurveda Pharmacy Res.*, 2(1), 124–136.

[14] Shilpa, S., and Murthy, C., (2011). Development and standardization of Mysore Tridosha scale. *AYU* (*Int. Quart. J. Res. Ayurveda*), 32(3), 308.

[15] Taber, S. (2018). The use of Cronbach's alpha when developing and reporting research instruments in science education. *Res. Sci. Educ.*, 48(1), 1273–129.

[16] Todkari, D., and Lavekar, G. (2015). Critical appraisal of Panchamahabhuta Siddhant. *Int. Ayurvedic Med. J.*, 3(5), 1454–146.

[17] Verma, V., Agrawal, S., and Gehlot, S. (2018). Possible measures to assess functional states of Tridosha: A critical. *Int. J. Health Sci. Res.*, 8(1), 219–231.

[18] Wani, B., Mandal, S., and Godatwar, P. (2017). Prakriti analysis and its clinical significance. *Int. J. Ayurveda Pharmacy Res.*, 5(9), 86–90.

[19] West, P. (2016). Simple random sampling of individual items in the absence of a sampling frame that lists the individuals. *New Zealand J. Forestry Sci.*, 46(1), 15.

[20] Woldaregay, A., Årsand, E., Walderhaug, S., Albers, D., Mamykina, L., Botsis, T., and Hartvigsen G. (2019). Data-driven modeling and prediction of blood glucose dynamics: Machine learning applications in type 1 dia- betes. *Artif. Intell. Med.*, 98, 109–134.

[21] Xue, Q., and Chuah, M. (2019). Explainable deep learning based medical diagnostic system. *Smart Health*, 13, Art. no. 100068.

Chapter 23

The Future of AI Security: How Zero Trust can Help to Protect Against Advanced Threats

Jyoti Bartakke[1] *and Rajeshkumar Kashyap*[2]

[1]Research Scholar, ZIBACAR Pune, Savitribai Phule Pune University

[2]Sadhu Vaswani Institute of Management Studies for Girls, Pune

Abstract: Artificial intelligence (AI) is rapidly becoming a critical part of our lives. Everything from auto-driving vehicles to clinical diagnosis utilises man-made reasoning. However, AI also poses new security risks. This paper discusses the security risks associated with AI and how the zero-trust security model can be used to mitigate these risks.

Keywords: Artificial Intelligence (AI), zero-trust security, zero-trust architecture, AI risks, AI future in zero trust.

1. Introduction

Artificial intelligence (AI) is a rapidly developing field that can change many businesses. As of now, artificial intelligence is being utilised in various applications, including self-driving vehicles, clinical findings and fraud detection. One of the main security risks associated with AI is adversarial attacks. Attacks aimed at deceiving AI systems are referred to as adversarial attacks. These attacks can be used to make AI systems make incorrect decisions or to generate malicious content. Another security risk associated with AI is data poisoning. Data poisoning is an attack that involves injecting malicious data into an AI system's training dataset.

A third security risk associated with AI is model theft. Model theft is an attack that involves stealing the model parameters of an AI system. This can allow attackers to use the model to make predictions or generate content. The "zero-trust" security model is a security model that can be used to mitigate the security risks associated with AI. The "zero-trust security model" is based on the idea that no user or device can ever be taken for granted. According to the zero-trust model, everything should be verified. The "zero-trust security model" holds that no user or device can ever be taken for granted. All access to systems and data is thus restricted, and users and devices must authenticate themselves and demonstrate their right to use the resources they request. The zero-trust security model is a flexible and vigorous model that is well suited for countering the advancing dangers in the AI security

DOI: 10.1201/9781003527442-23

landscape. By continuously updating its systems and devices with the latest threat intelligence, the zero-trust security model can stay ahead of the curve in the rapidly developing field of AI security.

2. Security Risks for AI

2.1. *Adversarial Attacks*

Attacks aimed at deceiving AI systems are known as adversarial attacks. These attacks can be used to make AI systems make incorrect decisions or to generate malicious content (Qiu *et al.*, 2019).

2.2. *Data Poisoning*

Data poisoning is an attack that involves injecting malicious data into an AI system's training dataset. This can cause the AI system to learn incorrect patterns, which can lead to incorrect decisions (Nary, 2023).

2.3. *Model Theft*

Model theft is an attack that involves stealing the model parameters of an AI system. This can allow attackers to use the model to make predictions or generate content (Lee *et al.*, 2022).

3. The Zero Trust Architecture

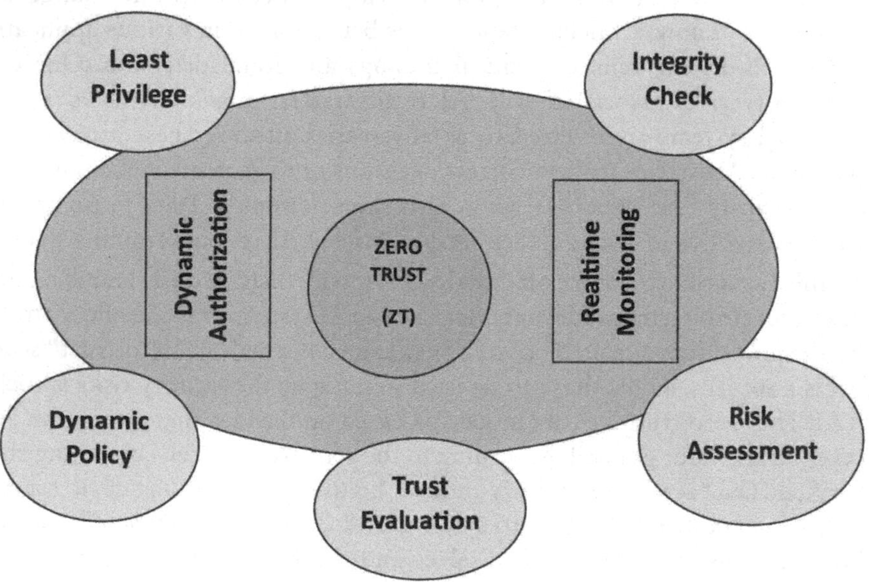

Figure 1: *Zero-trust principle Hosney et al. (2022).*

The NIST zero-trust framework is a set of principles and guidelines that can be used to improve an organization's security posture by reducing uncertainty. The framework does this by implementing precise, on-demand, least-privileged access decisions in particular settings. This approach replaces the traditional security model, which assumes that the IT border can be protected, and that trust is a security weakness. Figure 1 provides an overview of the basic ideas of zero trust. According to the "Zero Trust security model," no network function or resource is trusted, regardless of where it is on the network. This implies that the same security requirements must apply to all communications, whether they are between internal resources or between internal and external resources. We carefully evaluate each access request for both trust and risk. Continual, dynamic assessment that is ongoing throughout the access and is based on the current situation (Rose *et al.*, 2020).

3.1. Least Privilege: Users should only be given access to the tools they need to finish their tasks, according to the principle of least privilege. This suggests that users should have access to only the data, software and tools they need to complete their tasks. Adaptive Policy The security status of the user, the security status of the network assets and the context of the access request should all be taken into account when determining the dynamic nature of access policies. For instance, a user's access privileges should be immediately revoked if their laptop is compromised.

3.2. Integrity Check: The security status of each requestor and of the network resources is continuously, ideally in real-time, monitored. This entails comparing user and network asset behaviours and the security of devices to security policy standards. A computerised system performs this (Hosney et al., 2022).

"As a response to contemporary security issues, the zero-trust model is proposed, presenting the concept of a perimeter-less strategy." It is critical to understand that zero trust is primarily used as an architectural concept to improve the security of accessing resources and services. It is not stand-alone technology or an actual off-the-shelf software solution. Users connecting from the organisation's network are given unrestricted access to critical resources in the traditional perimeter defence approach, which does not implement fundamental segmentation measures or subsequent authentication and session verifications. This strategy assumes that these users are inherently trustworthy (Ferretti *et al.*, 2021; Laplante *et al.*, 2022).

4. Zero-Trust Security for AI

The "Zero Trust model is a security structure that expects that trust nothing verifies everything." This implies that all admittance to frameworks and information should be confirmed and approved. The zero-trust model is a compelling technique for managing computer-based intelligence dangers it (Dumitru, 2022).

4.1. Plans Granular Role-Based Access Controls Considering Least Honour Standards

This implies that every artificial intelligence model, information source and client is just given the entrance they need to go about their business. This fundamentally decreases the danger surface that an assailant can take advantage of.

Stresses profound perceive ability across every advanced resource, including man-made intelligence calculations and informational indexes. This straightforwardness empowers associations to screen and distinguish unusual exercises rapidly, which can assist them with moderating artificial intelligence explicit dangers like model float or information control.

4.2. Advances Consistent Assessment and Continuous Transformation of Safety Controls

In the quickly developing artificial intelligence scene, a static security position is deficient. The zero-trust model assists associations with remaining a stride in front of artificial intelligence dangers by persistently assessing and adjusting their security controls.

The zero-trust security model functions for the foundation that no user or device can ever be taken for granted. All access to systems and data thus becomes restricted, and users and devices are required to authenticate themselves and demonstrate their right to use the resources they request. The zero-trust security model can be used to mitigate the security risks associated with AI. Organisations can make sure that only authorised users and devices can access their AI systems by implementing the zero-trust security model. This can help protect AI systems from adversarial attacks, data poisoning and model theft (Dumitru, 2022).

The zero-trust model can be utilised to safeguard delicate information from being coincidentally shipped off simulated intelligence administrations. A few security measures are used to accomplish this, including Cody (2022).

4.3. Identity and Access Management (*IAM*)

Ensuring that only authorised users have access to sensitive data, requires making use of robust authentication mechanisms.

4.4. Network Segmentation

This includes separating the organisation into more modest, segregated zones to forestall unapproved admittance to delicate information.

4.5. Information Encryption

This includes scrambling delicate information both very still and, on the way, to shield it from unapproved access.

4.6. Data Loss Prevention (*DLP*)

This includes utilising devices to screen information streams and forestall unapproved information moves.

4.7. User and Entity behaviour Analytics (UEBA)

This entails making use of tools that monitor user behaviour and look for unusual actions that could point to security breaches.

4.8. Constant Checking and Inspecting

Identifying any unauthorised activity entails auditing and tracking data access and utilisation.

4.9. Response to Incidents and Remediation

This requires having a strategy in place for responding to security incidents and repairing any damage.

4.10. Security Examination and Danger Insight

Utilising tools to identify and reduce potential risks is necessary for this.

5. Future of AI and Zero Trust

The always-expanding intricacy and inescapability of artificial intelligence frameworks, alongside the complexity of cybercriminals, are driving the advancement of man-made intelligence dangers. These dangers are turning out to be more intricate and complex, and cyber criminals are tracking down better approaches to take advantage of them.

How the zero-trust model can be modified to combat these threats.

The zero-trust model is a security framework that assumes that neither a user nor a device can be trusted by default. This indicates that authorisation and verification of all access to data and systems are required. The zero-trust model can adjust to counter the advancing artificial intelligence dangers by executing further developed identification and anticipation frameworks that consolidate artificial intelligence to perceive and answer ill-disposed inputs progressively.

Incorporating AI-driven threat detection and response tools is quickly and accurately identify and respond to AI-powered attacks.

Requiring AI systems to be more transparent and implementing monitoring tools that can spot anomalies in AI behaviours even when the decision-making process is hidden behind the scenes.

Guaranteeing that all information is encoded, access is totally controlled, and any surprising information access designs are quickly recognised and examined.

Broadening the "never trust, consistently check" rule to each IoT gadget in the organisation, no matter what its tendency or area.

6. The Zero-Trust model's Flexibility and Vigor

The zero-trust model is versatile and vigorous, making it especially reasonable for countering the advancing dangers in the man-made intelligence scene. By consistently refreshing its systems and devices given the most recent danger knowledge, the zero-trust model can stay up with the quickly developing field of simulated intelligence dangers.

7. Conclusion

The integration of AI and the zero-trust security model offers a formidable defence against the emerging security risks posed by AI. As AI becomes increasingly prevalent across industries, understanding and addressing its unique security challenges is crucial. The zero-trust model's approach of "never trust, always verify" ensures that all access to AI systems and data is meticulously authenticated and authorised, reducing the risk of unauthorised access and malicious attacks. By applying granular access controls, continuous monitoring and dynamic policymaking, organisations can bolster the security of AI models and data. The zero-trust model's adaptability and vigilance enable it to counter evolving AI threats effectively. By adopting this approach, organisations can build a secure and resilient AI ecosystem, harnessing the potential of AI while safeguarding against potential vulnerabilities and risks. As AI continues to evolve, the zero-trust security model remains an indispensable framework for ensuring the integrity and trustworthiness of AI systems. AI is a powerful technology with the potential to revolutionise many industries. However, AI also poses new security risks. The "zero-trust security model" can be used to mitigate these risks and help to protect AI systems from attack.

References

[1] Cody, S. (2022). Open research case study – architecture the built environment and planning. https://doi.org/10.31219/osf.io/ygptc.

[2] Dumitru, I, A. (2022). Zero trust security. *Proceedings of the International Conference on Cybersecurity and Cybercrime* (IC3), 9. https://doi.org/10.19107/cybercon.2022.13.

[3] Ferretti, L., Federico, M., Mauro, A., and Michele, C. (2021). Survivable zero trust for cloud computing environments. *Computers & Security*, 110, 102419. https://doi.org/10.1016/j.cose.2021.102419.

[4] Hosney, E. S., Halim, I. T., and Yousef, A. H. (2022). An artificial intelligence approach for deploying zero trust architecture (ZTA). *5th International Conference on Computing and Informatics (ICCI)*. https://doi.org/ 10.1109/icci54321.2022.9756117.

[5] Lee, J., Sungmin, H., and Sangkyun, L. (2022). Model stealing defense against exploiting information leak through the interpretation of deep neural nets. *Proceedings of the Thirty-First International Joint Conference on Artificial Intelligence*, 710–716. https://doi.org/10.24963/ijcai.2022/100.

[6] Laplante, P., and Jeffrey, V. (2022). Zero-trust artificial Intelligence? *Computer*, *55*(2), 10–12. doi:10.1109/mc.2021.3126526.

[7] Nicholas, D., (2016). Suzanne Massie, Trust but verify: Reagan, Russia, and Me. Rockland, ME: Maine Authors Publishing, 2013. *Journal of Cold War Studies*, 18(4), 225–228. https://doi.org/10.1162/jcws_r_00693.

[8] Nary, S. (2023). Data poisoning: a new threat to artificial intelligence. *Mathematics and Computer Science Capstones*, La Salle University.

[9] Qiu, S., Qihe, L, Shijie, Z., and Chunjiang, W. (2019). Review of artificial intelligence adversarial attack and defense technologies. *Applied Sciences*, 9(5), 909. https://doi.org/10.3390/app9050909.

[10] Rose, S., Oliver, B., Stu, M., and Sean, C. (2020). Zero trust Architecture. *NIST*, Report Number 800-207. https://doi.org/ 10.6028/nist.sp.800-207.

Chapter 24

Predicting Pancreatic Cancer Risk Using Machine Learning and Neural Network

Nitin S. Thakre[1], Salim A. Chavan[2], Sarvesh Warjurkar[3], and K. V. Warkar[4]

[1]Assistant Professor, Department of Computer Science Engineering, Govindrao Wanjari College of Engineering, Nagpur

[2]Principal, Govindrao Wanjari College of Engineering

[3]Research Scholar, Department of Computer Science and Engineering, GHRU, Amravati

[4] Department of Computer Engineering. BDCOE, Sevagram

Abstract: In the first study, the authors introduce the Hadoop distributed guided trilateral filter-based Schutz indexive recurrent neural network (HdiGTF-SIRNN) to diagnose pancreatic cancer. Noise in the MRI/CT images was removed by a guided trilateral filter-based preprocessing. To successfully locate pancreatic illness, the pictures are segmented using a regressive segmentation method based on the Schutz index. The pictures are then classified as normal or pancreatic using log-linear analysis. The accuracy, precision and recall of HdiGTF-SIRNN have been improved. Gene for widespread hybrid elitism Improved accuracy in the detection of pancreatic cancer is suggested using a quadratic discriminant reinforced learning classifier system (DHEGQDRLCS). The relationship between data samples and class means is examined by utilising a kernel quadratic discriminant function. Maximum correlated samples for classification are discovered using the elitism gradient gene optimisation method. The accuracy of pancreatic categorisation is then further improved by using reinforcement learning to get the minimal loss function. Accuracy, balanced accuracy, F1-score, precision, recall and specificity are all shown by the findings to apply to DHEGQDRLCS in the detection of pancreatic cancer.

Keywords: Hadoop distributed guided trilateral filter-based Schutz indexive recurrent neural network, distributed hybrid elitism gene quadratic discriminant reinforced learning classifier system, kernel, accuracy, F1-score.

1. Introduction

The emergence of cancer as a disease of medical concern is cause for concern. Physicians' ability to provide effective therapy and boost survival rates is greatly

DOI: 10.1201/9781003527442-24

aided by early identification. Pancreatic cancer is usually at a late stage by the time it is diagnosed. Poor outcomes are seen because to the late and incurable stage at which diagnosis is made, as well as notable chemo-resistance in malignancies. Late-stage diagnosis renders most treatment options futile Dayem *et al.* (2021). The advantages of identifying pancreatic cancer early on have been shown in several studies. The most important biomarkers for early diagnosis are those that categorise individuals into high-risk categories and test those people first. Despite thousands of studies, not a single biomarker has been developed for the purpose of early cancer diagnosis Andika *et al.* (2020). During the diagnostic process, gathering enough samples for biomarker expansion might be challenging due to the necessity for significant national and international partnerships. Pancreatic tumours vary widely from person to person. Therefore, the primary challenge of traditional detection systems is reliable diagnosis. The efficiency of diagnosing diseases suffers as a result Ardila *et al.* (2019). In order to solve these problems, the suggested study works to create new machine learning and deep learning methodologies. The primary goal of this effort is to use a deep learning-based approach to segment and categorise MRI/CT images for pancreatic illness. Using the retrieved medical data, the second and third works attempt to improve the performance of classifiers in machine learning for the diagnosis of pancreatic cancer Ansari *et al.* (2022)

2. Pancreas Conditions

The pancreas has an important role to play in diabetes. In type 1 diabetes, the pancreas and beta cells stop producing insulin because they are damaged. The inability of the pancreas to release enough insulin in response to meals is a hallmark of type 2 diabetes. Here are the three most frequent ailments of the pancreas that human beings face at some point. Pancreatitis, precancerous conditions like PanIN and IPMN and pancreatic cancer are among illnesses that may affect the pancreas. The symptoms of each illness are unique and so call for specific treatments Krepline *et al.* (2020).

Pancreatitis is an inflammation of the pancreas, which leads to an increase in the production of digestive enzymes. It's a chronic condition characterised by periodic episodes of intense pain. Cancer of the pancreas: risk factors: While the exact cause of pancreatic cancer remains unknown, these are known risk factors that contribute to a higher likelihood of developing the illness. Chronic pancreatitis, inherited cancer syndromes and cigarette smoking are only a few of the contributing factors Ayan *et al.* (2020). In addition, IPMNs and PanIN, two distinct lesions of the pancreas, have been shown as reliable predictors of pancreatic cancer. In humans, pancreatic intraepithelial neoplasias, mucinous cystic neoplasms and intraductal papillary muci-nous neoplasms have been found as precursor injuries to pancreatic ductal adenocarcinoma (PDAC). According to the findings of the PanIN grading system, a higher grade indicates an increase in atypia and is crucial in invasive adenocarcinoma. Possible order of MCNs, IPMNs and PDAC are shown on the right hereditary alterations linked to adenocarcinomas via PanIN, as well as to low-frequency MCNs and IPMNs throughout time. Progression of PDAC involves a number of different genetic events and premalignant lesions Chouhan *et al.* (2020).

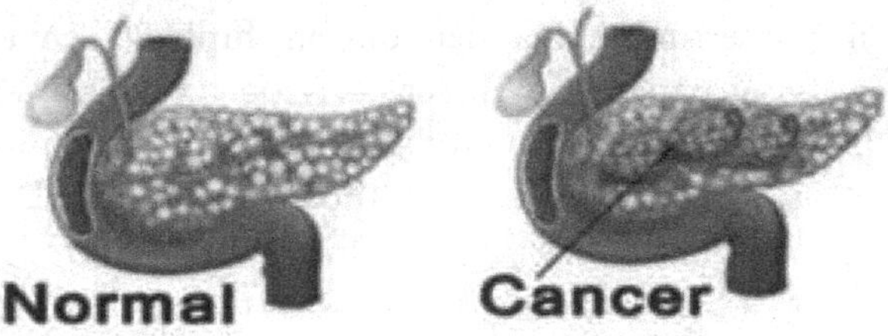

Figure 1: Normal and abnormal pancreatic.

3. Pancreatic Cancer

A fatal PC expansion caused by overzealous human responsibility. NCI estimates that there are now 1.9 million people living with cancer in India, with 609,360 people losing their lives to the disease. As of the seventh death from growth in a patient and the 14th from any general tumour, it has achieved a prominent global position. Among the most fatal and all-encompassing tumour schemes is PCs. It is also known as the "tumour emperor" because of its aggressiveness, rapid spread, poor resistance and dismal prognosis. Pancreatic cancer is linked to factors including age, heavy alcohol use, tobacco use and poor dietary choices. The most effective therapy results are achieved with a precise classification of malignancy levels. Due to the highly vascularised state of the condition, cancer is exceedingly aggressive. Abdominal discomfort, diabetes, jaundice, abnormal liver function parameters, weight loss and fatigue are all common warning signs of malignancy Agarwal *et al.* (2022). It usually occurs at a more advanced stage of sickness and is commonly overlooked in earlier stages. Cancer of the pancreas causes normally functioning pancreatic cells to malfunction and eventually grow out of control. Tumour refers to a mass of accumulated cancer cells. Metastasis is the spread of cancer from its original site to other sections of the body; this is the case with both primary and secondary pancreatic cancers Gao *et al.* (2020).

Environmental influences on the development of pancreatic cancer: Keep track of your family's medical history and do a dress rehearsal, if you want. In the modern conception of the family, it also includes the parents' and grandparents' siblings. The tumour has been analysed by the family. Parents, children and siblings are all considered primary relatives. Patients with adenocarcinoma should discuss their risk factors for the disease with their doctors Glicksberg *et al.* (2017). Lynch syndrome, Peutz-Jeghers syndrome and Von Hippel-Lindau syndrome account for 10% of inherited cases of syndrome. Possible risk factors include heavy alcohol use, caffeine use, physical inactivity, a diet rich in red meat and consuming an excessive amount of soft beverages daily. Genomic architecture of pancreatic cancer includes linking genotype to phenotypic features Gordienko *et al.* (2018).

4. Computed Tomography

Computed tomography (CT) is an imaging technique that uses a special X-ray instrument to create three-dimensional images of areas. The data from all of the dimensions are then combined into a 3D representation of the bodily component. Patients suspected of having pancreatic cancer are often evaluated with dynamic contrast-enhanced helical CT achieved by the rapid bolus injection of a large volume of iodinated urographic contrast. Figure 2 shows a CT scan picture of the pancreas.

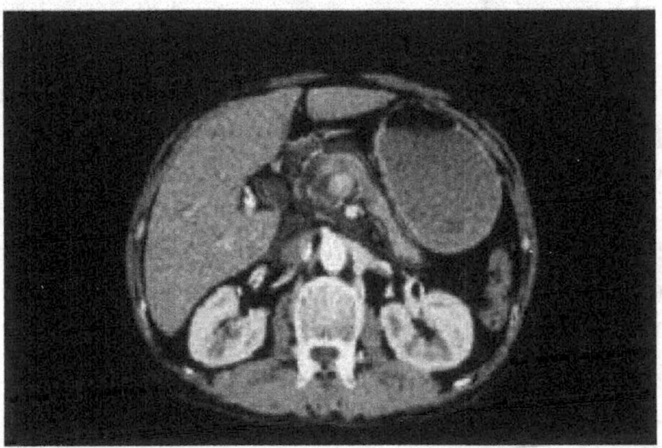

Figure 2: CT scan image of pancreatic.

In many ways, the multi detector-row scanner has simplified the process of acquiring images at various points in the administration of intravenous contrast media. Better pancreatic parenchymal development is an effect of using scan technology to the pancreatic parenchymal stage. Increased contrast between tumour and normal tissue is more useful for making more refined tumour diagnoses in later phases of imaging with portal veins. Combining 3D pictures of arterial, venous and pancreatico-biliary architecture requires data collected throughout many phases of evaluation.

5. Research Problem

As one of the deadliest forms of the disease, pancreatic cancer is also one of the most marginalised in social circles. Pancreatic borderline anatomy is notoriously difficult to distinguish on CT and MRI images because to its blurry edges and obtuse shapes. Predicting cancer, which may take many forms and can affect many parts of the body, has been the subject of much medical study. Cancer is expected to be incurable, meaning it cannot be effectively treated or conserved. Presently, both ML and NN have shown promise in pancreatic image segmentation.

- Determination of the best pre-processing approach: In order to save time and effort while still protecting the most vital and relevant information, doing a thorough preprocess is essential. It's a big help when trying to reduce the dimensionality of your data and eliminate unnecessary details. Because of the nature of our research, which involves clinical photographs, the volume of information extracted from them will be tremendous. Given the dearth of suitable preprocessing methods in currently available systems, this need only grows.

6. Pancreatic Cancer Detection Based on Extracted Medical Data Through Ensemble Methods in Machine Learning

The suggested study creates a new approach for diagnosing cancer using a combination of the rand indexive decision tree and the gradient descent logit boost classifier (RIDT-GDLBC). Classification is used to determine whether a patient has pancreatic cancer. The data needed to diagnose the condition has been classified using prior classification algorithms. However, proper diagnosis of the condition was not made. To achieve this goal, the RIDT-GDLBC technique makes use of an ensemble classifier. The ensemble classifier is a method of improving the performance of a poor classifier. Better classification results are produced by this method than by using a single classifier. Figure 3 is a schematic depicting the underlying structure of the RIDT-GDLBC approach suggested for the detection of pancreatic cancer.

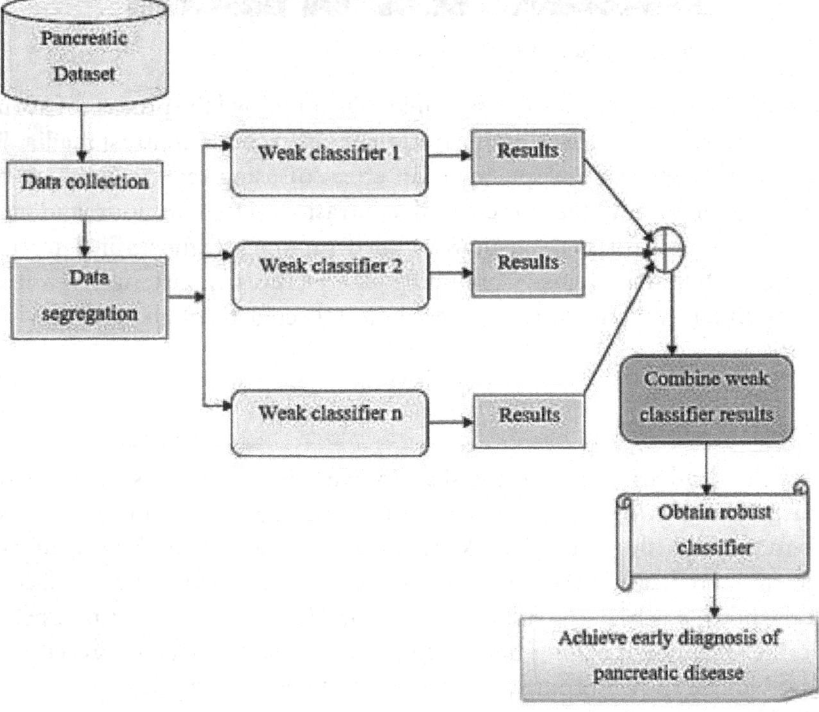

Figure 3: Architecture of proposed RIDT – GDLBC method.

Early pancreatic disease detection using the suggested RIDT-GDLBC technique is shown in Figure 3 design. First, a pancreatic dataset (590 urine data sets) is utilised as input to diagnose pancreatic cancer. This dataset includes four biomarker panels (creatinine, LYVE1, REG1B and TFF1). Then, data segregation is used to divide the samples into ones for training and ones for testing. The pancreatic cancer is then identified using an ensemble classifier called RIDT-GDLBC. To categorise the pancreatic tumor data samples, the proposed RIDT-GDLBC technique builds a decision tree from many weak classifiers. Using the gradient descent function, the results of several weak classifiers are combined to provide a single set of robust classification results. At first, the total number of pancreatic dataset training data samples is stated as

$$di = d1, d2, d3 \ldots, dn \in D$$

where di is the sample size of the training data and D is the pancreatic dataset. The RIDT-GDLBC technique builds the number of weak learners x(1), x(2),.. x(n) (decision trees) for classification based on the input of data samples. The decision tree divides the whole dataset into manageable chunks for training.

7. Experimental Results

The java programming language is used to implement the experimental assessment of the suggested RIDT-GDLBC approach. Pancreatic disease diagnostic performance is evaluated using a pancreatic dataset. Each experiment uses 600 data points and 15 different characteristics. Six hundred urine samples (200 controls, 240 patients with benign hepatobiliary illness and 200 patients with pancreatic cancer-associated dysplasia) are collected for four biomarker panels. The improved classification performance and higher accuracy achieved by using the RIDT-GDLBC approach in the detection of pancreatic cancer are shown. Naive Bayes, the NB tree (Normalized + B+-Tree), Bagging and the Adaboost techniques are used to evaluate the RIDT- GDLBC approach. Several parameters are used in the calculation of the model's accuracy. The following results from an examination of the suggested and traditional techniques' performances in light of a number of relevant parameters.

8. Testing Analysis with Similarity Index

Here, we assess RIDT-GDLBC's usefulness in implementing the thorough testing detection. It's a tool for testing the RIDT-GDLBC procedure and getting the results you need. The well-known idea to estimate the model result is the ROC. AUC is the set of parameters obtained from the graph. Closer to "1" for AUC suggests a more effective classifier. Confusion matrix as well as binary classification tests give valuable data about proposed predictive scheme. Both are provided to paramount choice threshold (Table 1).

Table 1: Predictive analysis.

Positive and Negative classes	Predictive Positive	Predictive Negative
Real Positive	54(59.5%)	8(8.5%)
Real Negative	19(20.5%)	42(53.0)%

9. Conclusion

In terms of improving pancreatic cancer classification accuracy (measured by precision, recall and f-measure), traditional distributed machine learning algorithms like SVM and logistic regression proved ineffective. In this research, a three-pronged approach is provided for identifying pancreatic cancer. For pancreatic forecasting, the HdiGTFSIRNN approach was suggested, which is based on Hadoop's distributed recurrent neural networks. Using MRI and CT scans, HdiGTF-SIRNN successfully segmented pancreatic cancer. This spatially smooth pancreatic segmentation was made possible by regularising the form's stability. The results showed that HdiGTF-SIRNN performed better than the status quo in pancreatic diagnosis with respect to both precision and recall.

References

[1] Agarwal, D., Covarrubias-Zambrano, O., Bossmann, S. H., and Natarajan, B. (2022). Early detection of pancreatic cancers using liquid biopsies and hierarchical decision structure. *IEEE Journal of Translational Engineering in Health and Medicine*, 10, 1–8.

[2] Andika, L. A., Pratiwi, H., and Sulistijowati Handajani, S. (2020). Convolutional neural network modeling for classification of pulmonary tuberculosis disease. *Journal of Physics: Conference Series*, 1490(1). https://doi.org/10.1088/1742-6596/1490/1/012020012020.

[3] Ansari, A. S., Zamani, A. S., Mohammadi, M. S., Meenakshi, Ritonga, M., Ahmed, S. S., Pounraj, D., and Kaliyaperumal, K. (2022). Detection of pancreatic cancer in CT scan images using PSO SVM and image processing. *BioMed Research International*, 1–7.

[4] Ardila, D., Kiraly, A. P., Bharadwaj, S., Choi, B., Reicher, J. J., Peng, L., Tse, D., Etemadi, M., Ye, W., Corrado, G., Naidich, D. P., and Shetty, S. (2019). End-to-end lung cancer screening with three-dimensional deep learning on low-dose chest computed tomography. *Nature Medicine*, 25(6), 954–961.

[5] Ayan, E. and Unver, H. M. (2019). Diagnosis of pneumonia from chest x-ray images using deep learning. *Scientific Meeting on Electrical-Electronics & Biomedical Engineering and Computer Science* (EBBT), Istanbul, Turkey.

[6] Chouhan, V., Singh, S. K., Khamparia, A., Gupta, D., Tiwari, P., Moreira, C., Damaševičius, R., and de Albuquerque, V. H. C. (2020). A novel transfer learning based approach for pneumonia detection in chest x-ray images. *Applied Sciences*, 10(2), 559.

[7] Dayem Ullah, A. Z., Stasinos, K., Chelala, C., and Kocher, H.M., 2021. Temporality of clinical factors associated with pancreatic cancer: a case-control study using linked electronic health records. *BMC Cancer*, 21(1), 1279.

[8] Gao, X. W., James-Reynolds, C., and Currie, E. (2020). Analysis of tuberculosis severity levels from CT pulmonary images based on enhanced residual deep learning architecture. *Neurocomputing*, 392, 233–244. https://doi.org/10.1016/j.neucom.2018.12.

[9] Glicksberg, B. S., Miotto, R., Johnson, K. W., Shameer, K., Li, L., Chen, R., and Dudley, J. T. (2017). Automated disease cohort selection using word embeddings from Electronic Health Records. *Biocomputing*, 23, 145–146.

[10] Gordienko, Y., Gang, P., Hui, J., Zeng, W., Kochura, Y., Alienin, O., Rokovyi, O., and Stirenko, S. (2018). Deep learning with lung segmentation and bone shadow exclusion techniques for chest X-ray analysis of lung cancer. *Advances in Intelligent Systems and Computing*, 638–647.

[11] Krepline, A. N., Geurts, J. L., Akinola, I., Christians, K. K., Clarke, C. N., George, B., Ritch, P. S., Khan, A. H., Hall, W. A., Erickson, B. A., Griffin, M. O., Evans, D. B., and Tsai, S. (2020). Detection of germline variants using expanded multigene panels in patients with localized pancreatic cancer. *HPB*, 22(12), 1745–1752.

Chapter 25

Prediction of Air Quality in Nagpur City Using Hybrid CNN-Long Short-Term Memory Model

Abhilasha Borkar[1] and Dr. S. E. Yedey[2]

[1]Research Scholar, PG Department of Computer Science & Technology, Hanuman Vyayam Prasarak Mandal, Amravati, Maharashtra, India

[2]Associate Professor and Head, PG Department of Computer Science & Technology, Hanuman Vyayam Prasarak Mandal, Amravati, Maharashtra, India

Abstract: Air quality forecasting in urban areas is crucial for effective environmental management and public health interventions. This paper presents a deep learning (DL) model for the prediction of air quality in Nagpur city based on the integration of individual DL classifiers such as convolutional neural networks (CNN) and long short-term memory (LSTM) called Hybrid CNN-LSTM. The proposed approach leverages the spatial feature extraction capabilities of CNNs to process air quality data. The extracted features are then fed into the LSTM, which excels at capturing temporal dependencies, enabling accurate predictions over time. The hybrid model evaluates the effectiveness of the CNN-LSTM classifier against other individual classifiers on testing datasets and demonstrates its superiority in predicting air quality. The results reveal the potential of CNN+LSTM as a powerful tool for environmental monitoring and decision-making, offering valuable insights to support policymakers in implementing effective pollution mitigation strategies and improving the overall living environment in Nagpur city.

Keywords: Air quality prediction, deep learning, convolutional neural networks, long short-term memory, feature extraction.

1. Introduction

The issue of air pollution in metro cities has steadily grown in importance, negatively impacting human life and garnering much attention in recent decades due to the growth of urbanisation and industrial development by Du *et al.* (2018). Zhang *et al.* (2021) suggested air quality prediction is essential to reduce air pollution levels and protect the ecosystem. The main factors that indicate the degree of air pollution are PM2.5, CO, O3, PM10, SO2 and NO2 by Chen *et al.* (2021). These factors adversely affect both the climate and human physical and mental

DOI: 10.1201/9781003527442-25

health. PM2.5 is mostly made up of hazardous and dangerous compounds that are highly reactive. Predicting the trends of these factors' concentration in the air has been the most crucial aspect of air quality prediction work by Wang *et al.* (2021). Air quality degradation is somewhat inevitable with India's ongoing economic expansion. However, this does not mean that pollutant emission levels cannot be properly reduced and managed by Djalalova *et al.* (2015). Conventional chemical frameworks and statistical methods can be used to categorise air quality prediction using the methodologies suggested by Mao *et al.* (2017). The chemical transport approach suggested the framework for smog production and the procedure of pollutant transportation and dispersion to achieve the goal of air quality prediction by Vautard *et al.* (2007). Although chemical modelling techniques fully take into account the physical and chemical procedures that affect the modification in air pollution intensity, their input information, such as air pollution source materials and climatological sectors, is unknown. Additionally, these frameworks are computationally demanding and time-consuming to run by Zhang *et al.* (2022).

As per the length of time series to predict the PM2.5 emission and other factors, forecasting models can be categorised into two predictive models; a long-term and short-term model by Cao *et al.* (2012). In order to ensure a secure environment for human daily activities during a 12-hour prediction time frame, short-term forecasting incorporates real-time forecasts with an emphasis on accurate forecasting by Cai *et al.* (2017). Long-term prediction tries to anticipate levels of PM2.5 and other factors more than two days earlier by Sun *et al.* (2018), offering administrators useful referencing data. According to the previous study on PM2.5, forecasting strategies consist of two methods such as chemical and statistical analysis. The concentration of pollutants in the air measurement, the chemical transport strategy used, and analyse the pollution dispersion procedure Wang *et al.* (2017). The statistical technique is distinguished from chemical transport approaches through its ease of use, effectiveness and wide range of applications. This approach investigates the fundamental properties of the dataset and offers more precise future projections according to its present state obtained from and evaluating prior data. The statistical approach uses machine and DL approaches to forecast the air quality and concentration of other factors by Vong *et al.* (2014). Several traditional machine learning strategies such as decision trees, random forest Kiesewetter *et al.* (2015), logistic regression Yang *et al.* (2017) and autoregressive moving averages were used for predicting the emission of PM2.5 and other factors in the air. Differences in PM2.5 concentrations are greatly impacted by a number of variables, including climate, traffic statistics and pollutant factors. According to the machine learning approaches' inadequate ability to generalise and their rudimentary framework, it can be difficult to effectively reflect the nonlinear, non-smooth method of PM2.5 fluctuations. The DL approach is the ability to enhance the effectiveness of temporal forecasting of air quality based on historical data used by Jiang *et al.* (2017). Several DL approaches such as LSTM, CNN Zafra *et al.* (2017) and RNN Wang *et al.* (2018)

were used for forecasting the PM2.5 and other factors emission percentage in air. The aforementioned DL frameworks' predictions in forecasting have improved somewhat but they could continue to suffer accuracy issues when handling more complex problems. The limitations introduced with the design of individual classifiers may be responsible for this barrier. The hybrid DL approaches have integrated different network structures for handling large and complex data to predict air quality.

Several researchers used hybrid approaches such as LSTM+CNN Y. Bai *et al.* (2016), and Bi-LSTM+Auto Encoder Li *et al.* (2017). By combining more than two DL classifiers, hybrid classifiers combine excellence and get beyond the shortcomings of individual classifiers. In this manner, data preparation is an essential phase to increasing the prediction performance in hybrid modelling because the concentration of pollutants data is extremely unpredictable and inconsistent. Several different preprocessing techniques have been applied to time series in the past few decades. Faraji *et al.* (2022) suggested the hybrid framework using CNN and LSTM to forecast the PM2.5 on an hourly and regular basis. Similar results were also measured by Sharma *et al.* (2022) to predict PM2.5 emission. In order to properly study the spatial–temporal correlations among PM2.5 and other weather-related variables, we look into CNN-based approaches.

2. Proposed Methodology

In this study, a hybrid DL model based on CNN and LSTM is developed to forecast the concentration of pollutants in the air in Nagpur City. The air quality data is extracted from the Department of Central Pollution Control Board (CPCB) in India. The suggested model is implemented using core Python programming in the Google Colab environment. This study measures the major pollutant factors such as PM2.5, PM10, NO and NO2. The following key steps of the suggested approach are presented. In this study, total of 24,850 sample data is used. The training and testing groups of the dataset are divided with an 80–20% ratio, respectively. Figure 1 shows the complete architecture of air quality index forecasting using the deep learning models.

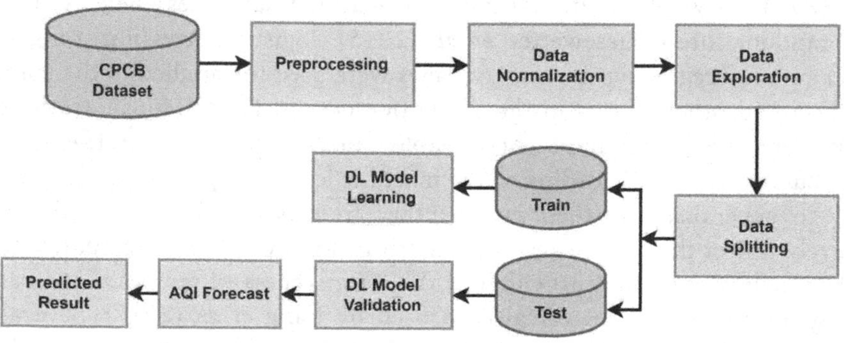

Figure 1: Complete architecture of proposed system.

2.1. Normalisation

The available dataset has been standardised using a normalising procedure, guaranteeing that the variations have no bearing on their relevance. Distinct data properties are helped to be measured on comparable scales by the data normalising method. This procedure improves effectiveness and is essential for the consistent training of DL models. When normalisation, each of the variable's types of information is also looked at. For instance, the dataset is compiled from many monitoring facilities, every of which deals with various date formats.

2.2. Data Exploration

Exploratory data analysis techniques enable the understanding of the distribution of air quality measurements, the relationships between different variables and the temporal or spatial patterns in the data. Moreover, data exploration helps in determining the suitable input features and target variables for training the deep learning model effectively by gaining a comprehensive understanding of the dataset.

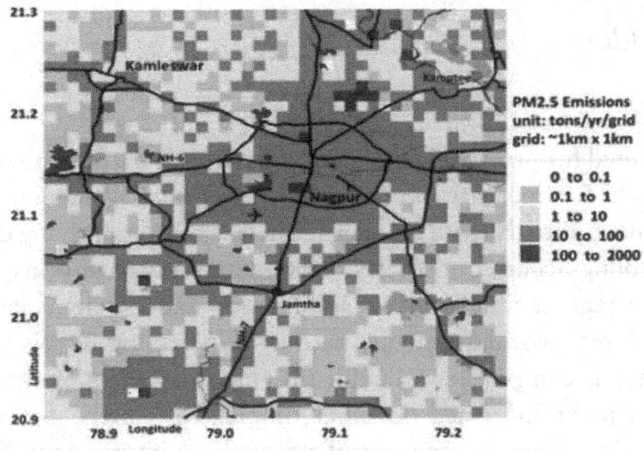

Figure 2: Map of Nagpur city with PM2.5 emission range.

Figure 2 shows the color scale or legend indicating the corresponding PM2.5 concentration levels for each colour category. This map would serve as a visual representation of the varying air quality levels across different regions of Nagpur City. Figure 3 shows the graphical representation balanced and Imbalanced data. Table 1 shows the minimum and maximum range of air quality in each class. Figure 3 shows the graphical representation balanced and Imbalanced data.

Table 1: Min–Max range of each class.

	Severe	Very poor	Poor	Moderate	Satisfactory	Good
Min	401	301	201	101	51	13
Max	500	400	300	200	100	50

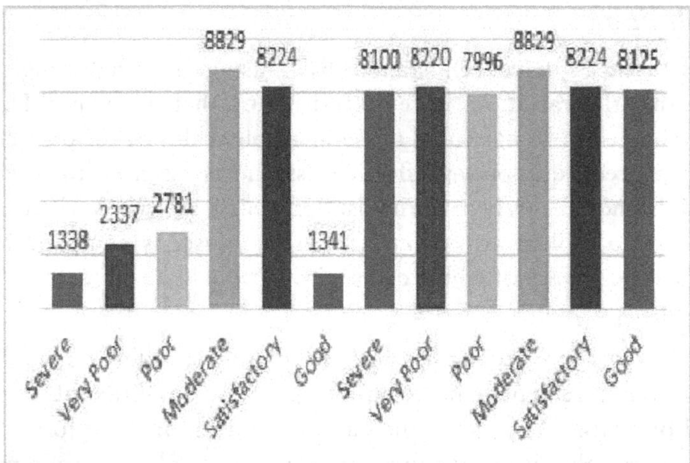

Figure 3: Bar represents balanced and imbalanced data.

2.3. Model Building

2.3.1 RNN

RNNs are well suited for processing the temporal patterns in the time series data for predicting the model for fluctuations of air quality over time. By feeding historical air quality measurements into the RNN model, it can learn the underlying dynamics and correlations, enabling accurate forecasts of pollutant concentrations. The model's ability to handle irregular time intervals and missing data is particularly beneficial when dealing with real-world air quality datasets. Leveraging RNNs for predicting air quality in Nagpur can provide valuable insights for environmental authorities, enabling them to make informed decisions, implement timely interventions and formulate effective air quality management strategies to improve public health and the overall well-being of the city's residents.

2.3.2 LSTM

LSTM, is a specialized form of recurrent neural network (RNN) for extracting the relevant features in time series data, making it particularly effective for modelling the complex and dynamic nature of air quality fluctuations in Nagpur. By training the LSTM model on historical air quality data, it can learn to recognise intricate temporal relationships and accurately forecast future pollutant concentrations. The memory cell structure of LSTM allows it to overcome the vanishing gradient problem, making it capable of handling long sequences of data, which is crucial for predicting air quality over extended periods. Leveraging LSTM for air quality prediction in Nagpur can provide valuable insights for environmental authorities and city planners, helping them implement proactive measures to tackle pollution, promote sustainable urban development and safeguard the health and well-being of the city's inhabitants.

2.3.3 CNN+LSTM

The CNN+LSTM model extracts the relevant features of both spatial and temporal aspects of air quality in Nagpur. This powerful fusion of deep learning architectures can contribute valuable insights to environmental monitoring efforts, guiding policymakers in implementing effective pollution mitigation strategies and ensuring a healthier living environment for the residents of Nagpur.

3. Result Analysis

The evaluation metrics used in our classifier include accuracy, precision, sensitivity and specificity as shown below.

$$Accuracy = \frac{(T\_P + T\_N)}{(T\_P + T\_N + F\_P + F\_N)}$$

$$Precision = \frac{(T\_P + T\_N)}{(T\_P + F\_P + T\_N + F\_N)}$$

$$Sensitivity = \frac{T\_P}{(T\_P + F\_N)}$$

$$Specificity = \frac{T\_P}{(T\_N + F\_P)}$$

Figure 4 shows the training and Validation Loss of Deep learning Classifiers.

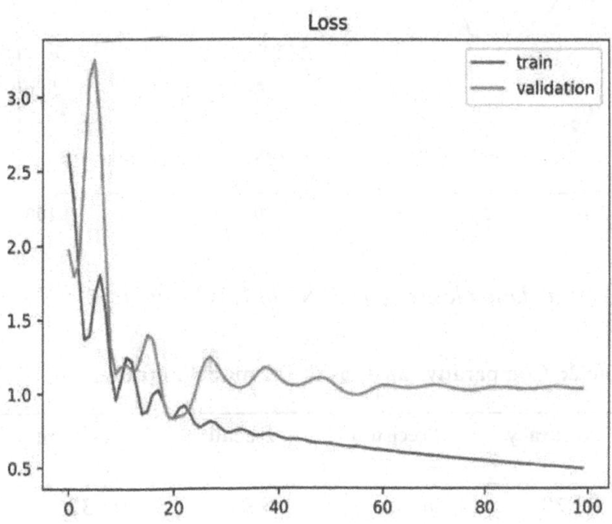

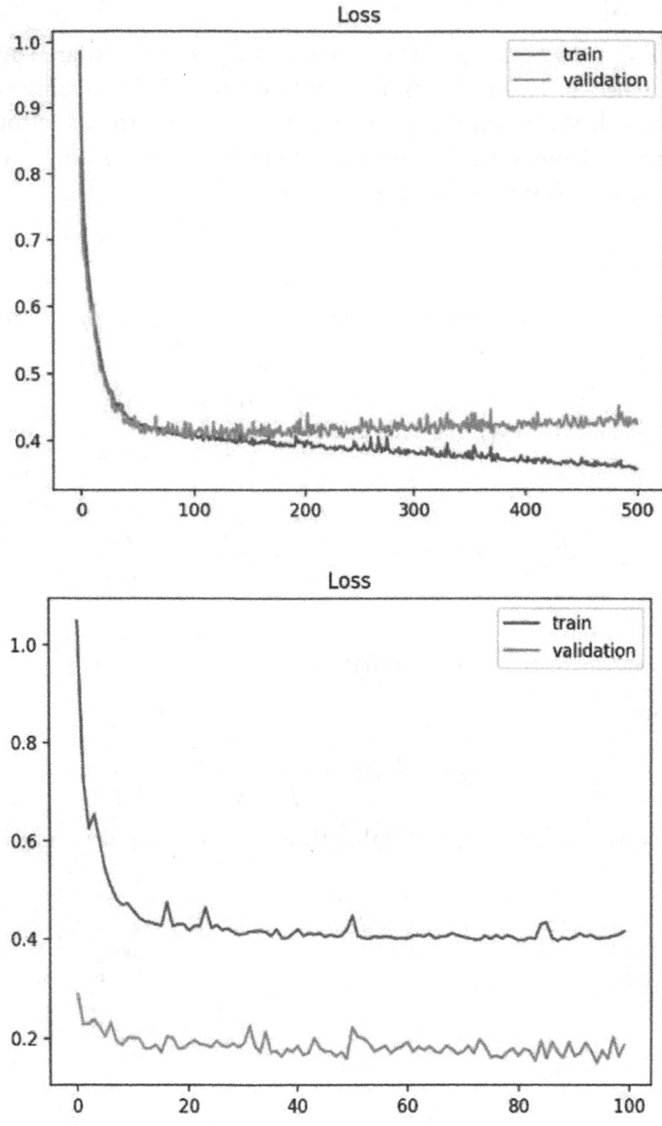

Figure 4: Training Vs validation loss of (a) RNN, (b) LSTM and (c) CNN + LSTM.

Table 2: Comparative analysis of DL model in training data.

DL Classifiers	Accuracy	Precision	Recall	F1-Score	Time in Sec
RNN	82.70	90.32	93.00	92.32	0.785
LSTM	85.40	90.41	92.56	93.00	0.234
CNN+LSTM	90.42	94.45	95.46	95.22	0.128

Table 3: Comparative analysis of DL model in testing data.

DL Classifiers	Accuracy	Precision	Recall	F1-Score	Time in Sec
RNN	87.79	91.23	92.50	91.32	0.085
LSTM	88.50	92.30	93.65	92.00	0.045
CNN+LSTM	94.56	94.00	95.50	95.00	0.038

Tables 2 and 3 present a comparative analysis of the results of DL classifiers during both training and testing. It is evident that the hybrid CNN+LSTM approach shows improved performance compared to other individual classifiers. Table 4 shows the concentration level of area in Nagpur city, it indicates that concentration level of AQI in Kalmeshwar area in Nagpur.

Table 4: Concentration level of areas in Nagpur city.

Area	Date	PM2.5	PM10	NO	NO2	AQI	Description
Kamleswar	25/7/2023	28	45	0.85	25.45	49	Good
Jamtha	25/7/2023	43	111.2	7.21	25.12	125	Moderate
Kamptee	25/7/2023	33.34	35.64	7.35	27.7	140	Moderate

DL Classifiers	MAE	RSME	RMSLE	R^2
RNN	0.723	3.254	0.712	0.564
LSTM	0.583	2.546	0.197	0.487
CNN+LSTM	0.472	1.428	0.086	0.356

Table 5 displays the performance of the DL classifier on the testing dataset, evaluated using standard evaluation metrics. It is evident that all the DL classifier performs exceptionally well. The hybrid CNN+LSTM approach demonstrates improved performance compared to other individual classifiers. Figure 5, 6, and 7 shows .

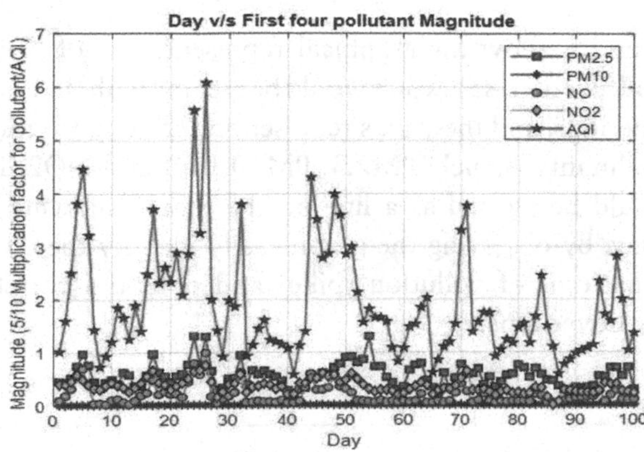

Figure 5: First four pollutant factor magnitude using RNN.

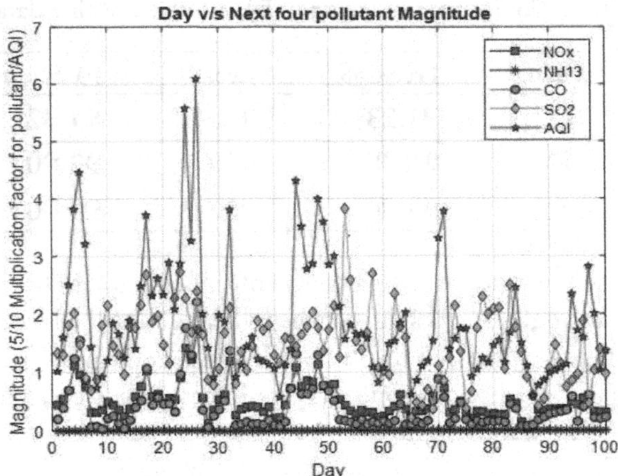

Figure 6: First four pollutant factor magnitude using LSTM.

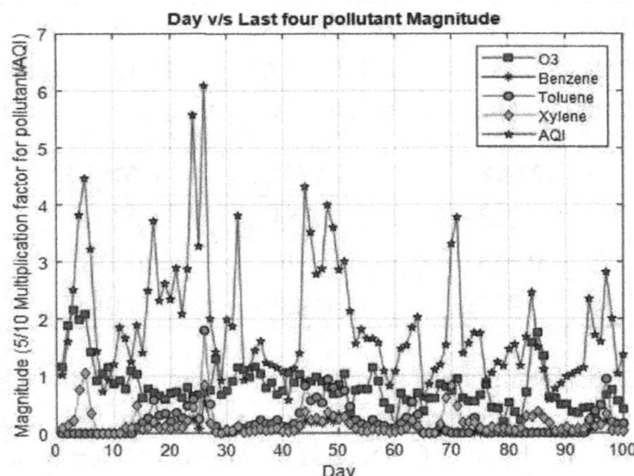

Figure 7: First four pollutant factor magnitude using CNN+LSTM.

Figures 6, 7 and 8 show the graphical representation of "Day vs. first four pollutant magnitude," a time-series plot would be created with the x-axis representing the days or time periods, and the y-axis representing the concentrations of the four most common pollutants, namely PM2.5, PM10, NO and NO2. Each pollutant's concentration would be plotted as a line on the graph, showing their respective values for each day. By observing the graph easily identify the daily variations in pollutant levels, patterns of pollution spikes and potential correlations between different pollutants over time.

4. Conclusion

This paper proposed the DL model for forecasting the air quality in Nagpur city based on a hybrid CNN-LSTM approach. By integrating the individual classifiers for extracting the spatial features over the dataset. The hybrid CNN+LSTM model demonstrated superior performance compared to other individual classifiers. The model's ability to capture both spatial and temporal patterns in air quality data allowed for accurate and reliable predictions over time. This approach offers valuable insights for environmental authorities and city planners, enabling them to implement proactive measures to combat air pollution and improve the overall quality of life for Nagpur's residents. As advancements in deep learning continue, the CNN+LSTM method holds great potential in revolutionising air quality prediction and environmental monitoring efforts in urban areas.

References

[1] Bai, Y., Li, Y., Wang, X., Xie, J., and Li, C. (2016). Air pollutants concentrations forecasting using back propagation neural network based on wavelet decomposition with meteorological conditions. *Atmos. Pollut. Res.*, 7(3), 557–566. https://doi.org/ 10.1016/j.apr.2016.01.004.

[2] Cao, J., Wang, Q., Chow, J., Watson, J., Tie, X., Shen, Z., and Wang P. (2012). An, Impacts of aerosol compositions on visibility impairment in Xi'an, China. *Atmos. Environ., 59, 559–566.* https://doi.org/10.1016/j.atmosenv.2012.05.036.

[3] Cai, S., Wang, Y., Zhao, B., Wang, S., Chang, X., and Hao J. (2017). The impact of the air pollution prevention and control action plan on PM 2.5 concentrations in Jing- Jin-Ji region during 2012–2020. *Sci. Total Environ.*, 580, 197–209. https://doi.org/10.1016/j.scitotenv.2016.11.188.

[4] Chen, F., and Chen, Z. (2021). Cost of economic growth: Air pollution and health expenditure. *Sci. Total Environ.*, *755*, 142543.

[5] Djalalova, I., Delle Monache L., and Wilczak J. (2015). PM2.5 analog forecast and kalman filter post-processing for the community multiscale air quality (CMAQ) model. *Atmos. Environ.*, 108, 76–87.

[6] Du, S., Li, T., Yang, Y., and Horng S. (2018). Deep air quality forecasting using hybrid deep learning framework. *IEEE Trans. Knowl. Data Eng.*, 33, 2412–2424.

[7] Faraji, M., Nadi, S., Ghaffar Pasand, O., Homayoni, S., and Downey K. (2022). An integrated 3D CNN-GRU deep learning method for short-term prediction of PM2.5 concentration in urban environment. *Sci. Total Environ.*, 834, 155324 https://doi.org/10.1016/j.scitotenv.2022.155324.

[8] Li, X., Peng, L., Yao, X., Cui, S., Hu, Y., You, C., and Chi, T. (2017). Long short-term memory neural network for air pollutant concentration predictions: Method development and evaluation. *Environ. Pollut.*, 231(1), 997–1004. https://doi.org/ 10.1016/j.envpol.2017.08.114

[9] Jiang, P., Dong, Q., and Li P. (2017). A novel hybrid strategy for PM2.5 concentration analysis and prediction. *J. Environ. Manage.*, 196, 443–457. https://doi.org/ 10.1016/j.jenvman.2017.03.046.

[10] Kiesewetter, G., Schoepp, W., Heyes, C., and Amann, M. (2015). Modelling PM2.5impact indicators in Europe: Health effects and legal compliance. *Environ. Model. Softw.*, 74(C), 201–211. https://doi.org/10.1016/j.envsoft.2015.02.022.

[11] Mao, X., Shen, T., and Feng X., (2017). Prediction of hourly ground-level PM2.5 concentrations 3 days in advance using neural networks with satellite data in eastern China. *Atmos. Pollut. Res.*, 8(6), 1005–1015.

[12] Sharma, E., Deo, R., Soar, J., Prasad, R., Parisi, A., and Raj N. (2022). Novel hybrid deep learning model for satellite based PM10 forecasting in the most polluted Australian hotspots. *Atmos. Environ.* 279, 119111. https://doi.org/10.1016/j.atmosenv.2022.119111.

[13] Sun D., Fang J., and Sun J. (2018). Health-related benefits of air quality improvement from coal control in China: evidence from the Jing-Jin-Ji region. *Resour. Conserv. Recycl.*, 129, 416–423. https://doi.org/10.1016/j.resconrec.2016.09.021.

[14] The Central People's Government of the People's Republic of China. http://www.gov.cn/zwgk/2013-09/12/content_2486773.htm

[15] Vautard, R., Builtjes, P., Thunis, P., Cuvelier, C., Bedogni, M., Bessagnet, B., Honore, C., Moussiopoulos, N., Pirovano, G., and Schaap, M. (2007). Evaluation and intercomparison of ozone and PM10 simulations by several chemistry transport models over four European cities within the City Delta project. *Atmos. Environ.*, 41(1), 173188.

[16] Vong, C., Ip, W., Wong, P., and Chiu, C., (2014). Predicting minority class for suspended particulate matters level by extreme learning machine. *Neurocomputing.*, 128, 136–144. https://doi.org/ 10.1016/j.neucom.2012.11.056.

[17] Wang, Z., and Fang C. (2016). Spatial-temporal characteristics and determinants of PM2.5 in the Bohai Rim Urban Agglomeration. *Chemosphere*, 148, 148–162.

[18] Wang, Y., Liu, H., Mao, G., Zuo, J., and Ma J. (2017). Inter-regional and sectoral linkage analysis of air pollution in Beijing–Tianjin–Hebei (Jing-Jin-Ji) urban agglomeration of China. *J. Clean. Prod.*, 165, 1436–1444. https://doi.org/10.1016/j.jclepro.2017.07.210.

[19] Wang, J., Yang, W., Du, P., and Li Y. (2018). Research and application of a hybrid forecasting framework based on multi-objective optimization for electrical power system. *Energy.*, 148, 59–78. https://doi.org/10.1016/j.energy.2018.01.112.

[20] Yang, Z., and Wang, J. (2017). A new air quality monitoring and early warning system: Air quality assessment and air pollutant concentration prediction. *Environ. Res.* 158, 105–117. https://doi.org/10.1016/j.envres.2017.06.002.

[21] Zafra, C., Ángel Y., and Torres E. (2017). ARIMA analysis of the effect of land surface coverage on PM10 concentrations in a high-altitude megacity. *Atmos. Pollut. Res.*, 8(4), 660-668. https://doi.org/ 10.1016/j.apr.2017.01.002.

[22] Zhang, B., Zou, G., Qin, D., Lu, Y., and Wang, H., (2021). A novel Encoder-Decoder model based on read-first LSTM for air pollutant prediction. *Sci. Total Environ.*, 765, 144507.

[23] Zhang, B., Zou, G., Qin, D., Ni, Q., Mao, H., and Li, M. (2022). RCL-Learning: ResNet and convolutional long short-term memory-based spatiotemporal air pollutant concentration prediction model. *Expert Syst. Appl.*, 207, 118017.

Chapter 26

A Review on "Efficient Hybrid Methodology for Early Detection of Breast Cancer in Digital Mammograms using Autoencoder Deep Learning"

Ashish R. Dandekar[1], Dr. Avinash Sharma[2], and Dr. Jitendrakumar Mishra[3]

[1]Research Scholar Department of Computer Science and Engineering Madhyanchal. Professional University, Bhopal, MP (India)

[2]Professor, Department of Computer Science & Engineering, MPU, Bhopal, MP (India)

[3]Associate Professor Department of E&C, MPU, Bhopal, MP (India)

Abstract: The most prevalent form of cancer in humans is breast cancer. The term cancer was first used to describe the illness in Egypt around 1600 BC. Despite the fact that since then, research and studies have been conducted to mitigate the effects of this illness, it is still regarded as one of the most lethal diseases of all time due to mortality from breast cancer. Modern medicine has recently created numerous methods and strategies for the early detection of breast cancer. There are many different datasets, hybrid machine learning methods, decision trees, KNN, SVM, naive bays, etc. employed. The majority of technologies rely on cutting-edge tools including deep learning algorithms, medical image processing and machine learning strategies. There have been numerous machine learning experiments conducted in the past. In their respective fields, decision trees, KNN, SVM, naive bays and other neural network techniques outperform the competition. However, a recently developed method is currently being used to classify breast cancer. A new approach is deep learning. Deep learning is used to get over the limits of machine learning. Data science regularly uses deep learning techniques, such as convolution neural networks, recurrent neural networks, deep belief networks and others. Deep learning algorithms outperform machine learning strategies in terms of performance. Images are captured in their most alluring states. CNN is used in our investigation to classify the images. CNN is, in essence, the most extensively utilised system for classifying images. The objective of this study is to determine the best early detection and life-saving strategy for breast cancer.

Keywords: Deep learning, CAD, mammography.

DOI: 10.1201/9781003527442-26

1. Introduction

The disease that affects women most frequently worldwide is breast cancer. The mortality rate from breast cancer is among the highest among all diseases. The society of medical and biomedical engineering must deal with a serious issue if breast cancer patients are to survive. Patients with breast cancer are saved by early diagnosis of the disease. Biomedical engineers have created a range of screening and computer-aided diagnosis (CAD) devices for the early detection of breast cancer. Actually, there is a larger need for efficient detection methods due to the prevalence and spread of breast cancer. Breast cancer tissue can be identified and categorised using medical image processing. Creation of automated breast cancer screening techniques in the biomedical field using various neural networks. Figure 1 shows mammogram samples with marked malignant tumour.

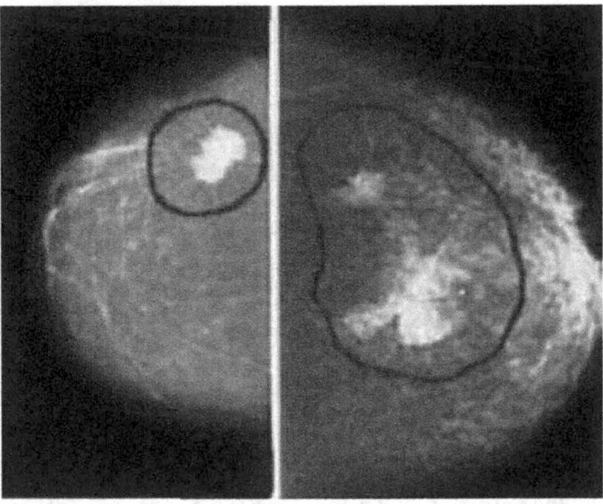

Figure 1: Mammogram samples with marked malignant tumour.

It is challenging to automatically identify the disease at an early stage even using a number of breast cancer screening procedures. Many writers have recently proposed feature optimisation-based breast cancer diagnosis. The texture and boundary value of the mammography image's lower content can be defined as a feature.

2. Literature Survey

We will draw on some of the ideas from the literature study that is presented in the bell mansion below.

Author 1 proposed about a mammogram's normality or abnormality can be automatically determined by the diverse features-based breast cancer detection (DFe BCD) system, which was created by the study. They use four different feature kind sets. Local binary patterns and taxonomy indices are two of these stable statistical features. Using a deep convolution neural network built on the foundation of a highway network, the proposed DFe BCD dynamically extracts the fourth set of characteristics from mammography images. The suggested method extracts dynamic properties by use of a deep Highway-Network author [1]. Authors [2] tested the efficacy of using various data-processing strategies and assessed numerous DL-based AI systems that use diverse approaches and DL models as their backbone using a set of digital mammography images with annotations of pathologically proven breast cancer. By examining several types of algorithms, DL models and data-processing approaches, the study indicates a significant opportunity to enhance the effectiveness of such techniques used to detect breast cancer [2]. Authors [3] as a result, breast cells frequently transform into tumours when they grow out of control, which feel like lumps in the breast. Early breast cancer detection, which has been demonstrated to lower the risk of death, increases the likelihood that the right treatment will be found. The author [4] in order to detect malignant tumours in mammography images, deep learning is widely applied. Since the sensitive nature of the datasets necessitates ongoing accuracy improvement, we introduce segmentation and wavelet transform to improve the key features in the picture scans. The author [5] in order to improve, we started with a standard 8-layer CNN and then incorporated the batch normalisation (BN) and dropout (DO) approaches. Last but not least, rank-based stochastic pooling (RSP) took the place of the conventional max pooling. As a result, CNN, BN, DO and RSP were combined to become BDR-CNN. This BDR-CNN was combined with a two-layer GCN to form our BDR-CNN-GCN model. The author [6] this article suggests a deep convolutional neural network model. The network's initial weight is computed using a variety of techniques, and the model modifies the model's parameters to speed up back propagation learning. We utilised a GPU that supports the parallel computing architecture of Coda to train the model quickly and with the least amount of hardware, in this study, the author [7] uses an enhanced CAD approach to divide breast cancer tumours into benign and malignant classes. This CAD system employs a technique called region growth for segmenting regions. For the purpose of extracting features from the texture and histogram, discrete wavelet transformations are employed. According to the author [8] of this study, background photographs are removed from breast cancer images before Weiner filtering and a contrast-limiting histogram equalisation filter are used to recover the images. The preprocessing stage of the CNN building stage, which was produced by the author [9], includes format unification, noise reduction, picture improvement, ROI extraction, augmentation and image resizing. To learn characteristics and categorise breast cancers in mammography pictures, a completely new model is created throughout the CNN development phase [10]. The creator for the purpose of detecting breast cancer, we have introduced the support vector machine (SVM) method. It creates a set of hyper planes in a high- or infinite-dimensional environment that may be applied to tasks like outlier detection, regression and classification. Compared to ANN, SVM is 85% more accurate at detecting breast cancer. While identifying the exact size and location of breast cancer tumours is crucial for tumour diagnosis, traits gleaned from questionable areas in mammograms aid in the earlier detection of tumours and speedier initiation of treatment. [11] The writer according to previous research findings, transformer-based vision models outperform convolutional models. The self-supervised learning technique known as bootstrap your own latent (BYOL) is presented in this work for the processing of mammography pictures for diagnostic purposes. The deep learning technique used for data analytics is improved by BYOL's ability to handle unlabelled image data.

3. Problem Identification

Breast cancer is a significant condition that impacts a lot of women worldwide. It is feasible to save lives by detecting this cancer in its earliest stages. Mammography has also been regarded as the most popular and uncomplicated method of early cancer screening. Radiologists can anticipate that a mammogram will be more than 90% accurate. However, 10–15% of cancers of the breast may go undetected by radiologists. The possibility of a false-positive outcome may be reduced by double-checking and reviewing the mammography pictures. The same mammogram must be examined by two radiologists twice during reprocessing. However, double checking has been shown to increase proper detection accuracy by 15% in comparison to single checking Breast cancer is a dangerous condition that affects many women globally. Early detection of this cancer allows for the possibility of saving lives. The most popular and straightforward method of detecting cancer in its early stages is mammography, according to research. According to radiologists, a mammography will be accurate more often than 90% of the time. But between 10% and 15% of breast cancer cases may be missed by radiologists. If you verify and interpret the mammography results, a false positive result might be less likely. The same mammography must be viewed by two radiologists twice during reprocessing. It has been demonstrated that double-checking increases the accuracy of proper identification by 15% compared to single checking.

1. Diminished breast cancer symptoms
2. The classifier's training error rate.
3. Choosing the best combination of feature elements for the detection.

4. Objectives

A comprehensive study of breast cancer detection techniques needs to be adjusted and improved if women's lives can be saved by early detection of breast cancer. There is a predetermined objective for the design algorithm and model.

1. To investigate how deep learning and machine learning methods influence breast cancer behaviour.
2. Develop a deep learning model that can recognize breast cancer using variable stage auto-encoding.
3. Create a hybrid stacking-based classifier model to detect breast cancer early.
4. Examine several techniques for improving breast cancer detection.
5. Contrast the suggested approach with the available algorithms for identifying breast cancer.

5. Proposed Methodology

The recommended method for detecting breast cancer employed digitised images from mammograms. In flow diagram here we try to develop protocol of proposed method system broadly divided into two categories. Training dataset and train model in training dataset will try to going all preprocessing activity after that feature processing which involved side and dimension, and find appropriate value of by forwarded propagation, then we used auto encoder deep learning, after getting result we detect where dataset are normal or malignant as shown in figure 2.

as shown in figure 2 , we try........

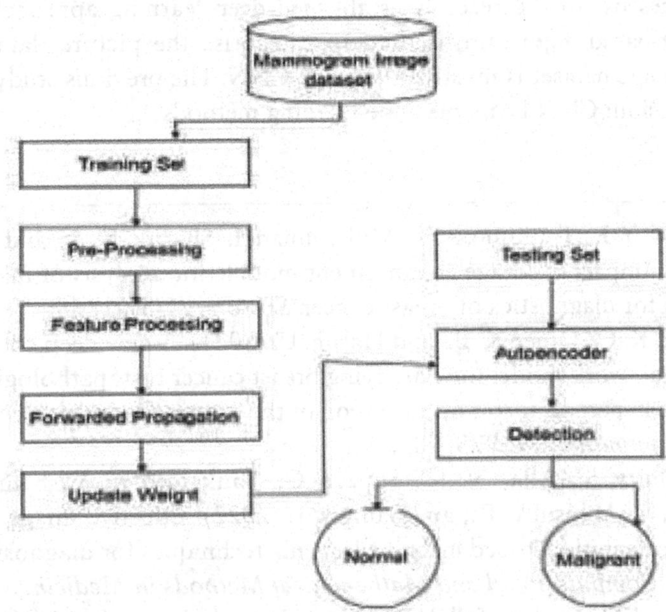

Figure 2: Proposed model of breast cancer detection.

6. Expected Outcomes of the Proposed Work

The intended work will give rise to the following outcomes:

- Boost the efficiency of breast cancer early detection.
- Cuts back on features to speed up prediction processing
- Reduces the fluctuation in mammography scanning image noise.
- Individualised algorithm for detecting breast cancer
- Encourage the centers for disease control and prevention to continue providing primary care.

7. Conclusion

According to this survey, breast cancer is currently the biggest global cause of end of live for women. It is predicted that early diagnosis like the treatment of breast cancer will increase survival rates by reducing the need for surgery. Accurate breast cancer screening and monitoring remain difficult despite technological advancements. To enhance prediction models, it is crucial to combine streams of data from the biological, social and demographic spheres. Every year, there is a declining and worsening chance that someone with breast cancer will recover. For the identification of breast cancer, methods like deep learning and machine learning are frequently applied. Prior studies have shown that machine learning algorithms perform better in their respective sectors. Experiments in the past have shown that the system will fail when the data is in the form of photographs. To overcome the issue, a novel application of machine learning techniques is made. Data science typically makes use of the recently established deep learning approach. CNN, a deep learning-based algorithm, is used to categorise the picture data of breast cancer. The image dataset is mostly utilised by CNN. The previous study reached a similar conclusion: CNN beats machine learning methods.

References

[1] Almalki, Y. E., Toufique, A. S., Muhammad, I., Sharifa, K. A., and Ahmed, A. (2022). Impact of image enhancement module for analysis of mammogram images for diagnostics of breast cancer. *Sensors*, 22(5), 1868.

[2] Burçak, K. C., Ömer, K. B., and Harun, U. (2021). A new deep convolutional neural network model for classifying breast cancer histopathological images and the hyperparameter optimisation of the proposed model. *The Journal of Supercomputing*, 77, 973–989.

[3] Chaudhury, S., Alla, N. K., Suneet, G., Sankaran, K. S., Samiullah, K., Kartik, S., Abhishek, R., and Sammy, F. (2022). Effective image processing and segmentation-based machine learning techniques for diagnosis of breast cancer. *Computational and Mathematical Methods in Medicine.*

[4] Chouhan, N., and Asifullah, K., Shah, J., and Hussnain, M. (2022). Deep convolutional neural network and emotional learning based breast cancer detection using digital mammography. *Computers in Biology and Medicine*, 132, 104318.

[5] Darweesh, M. S., Mostafa, A., Ahmed, A., Omar, F., Ahmed, K., and Mohamed, A. (2022). Early breast cancer diagnostics based on hierarchical machine learning classification for mammography images. *Cogent Engineering*, 8(1), 1968324.

[6] Davoudi, K., and Parimala, T. (2021). Evolving convolutional neural network parameters through the genetic algorithm for the breast cancer classification problem. *Simulation*, 97(8), 511–527.

[7] El Houby, E. M. F., and Yassin, N. I. R. (2021). Malignant and nonmalignant classification of breast lesions in mammograms using convolutional neural networks. *Biomedical Signal Processing and Control*, 70, 102954.

[8] Frazer, H., Qin, A. K., Hong Pan, and Brotchie, P. (2021). Evaluation of deep learning-based artificial intelligence techniques for breast cancer detection

on mammograms: results from a retrospective study using a breast screen Victoria dataset. *Journal of medical imaging and radiation oncology*, 65(5), 529–537.

[9] Fadhil, A. F., and Ornek, H. K. (2021). A computer-aided detection system for breast cancer detection and classification. *Selcuk University Journal of Engineering Sciences*, 1, 23–31.

[10] Ghoneim, O., Soliman, G., Galal, A., and Mahgoub, H. (2021). Breast cancer histological image classification using ensemble convolutional neural network and triplet loss. IOSR- *J. Comput. Eng. Ser*, 23(4), 30–42.

[11] Hanis, T. M., Islam, M. A., and Musa, K. I. (2020). Diagnostic accuracy of machine learning models on mammography in breast cancer classification: a meta-analysis. *Diagnostics*, 12(7), 1643.

[12] Jiang, G., Wei, J., Xu, Y., He, Z., Zeng, H., Wu, J., Qin, G., Chen, W., and Lu, Y. (2021). Synthesis of mammogram from digital breast tomosynthesis using deep convolutional neural network with gradient guided cGANs. *IEEE Transactions on Medical Imaging*, 40(8), 2080–2091.

[13] Kalafi, E. Y., Jodeiri, A., Setarehdan, S. K., Lin, N. W., Rahmat, K., Taib, N. A., Ganggayah, M. D., and Dhillon, S. K. (2021). Classification of breast cancer lesions in ultrasound images by using attention layer and loss ensemble in deep convolutional neural networks. *Diagnostics*, 11(10).

[14] Li, H., Chen, D., Nailon, W. H., Davies, M. E., and Laurenson, D. I. (2021). Dual convolutional neural networks for breast mass segmentation and diagnosis in mammography. *IEEE Transactions on Medical Imaging*, 41(1), 3–13.

[15] Rasheed, A., Younis, M. S., Qadir, J., and Bilal, M. (2021). Use of transfer learning and wavelet transform for breast cancer detection. Ar Xiv preprint arXiv:2103.03602.

[16] Safdar, S., Muhammad, R., Gadekallu, T. R., Javed, A. R., Rahmani, M. K. I., Jawad, K., and Bhatia, S. (2022). Bio-imaging-based machine learning algorithm for breast cancer detection. *Diagnostics*, 12(5), 1134.

[17] Saidnassim, N., Abdikenov, B., Kelesbekov, R., Akhtar, M. T., and Jamwal, P. (2021). Self-supervised visual transformers for breast cancer diagnosis. *In Asia-Pacific Signal and Information Processing Association Annual Summit and Conference (APSIPA ASC)*, 423–427.

[18] Wang, J., Liu, Q., Xie, H., Yang, Z., and Zhou, H. (2021). Boosted efficientNet: Detection of lymph node metastases in breast cancer using convolutional neural networks. *Cancers*, 13(4), 661.

[19] Zhang, Y. D., Suresh, S. S., Guttery, D. S., Górriz, J. M., and Wang, S. H. (2021). Improved breast cancer classification through combining graph convolutional network and convolutional neural network. *Information Processing & Management*, 58(2), 102439.

Chapter 27

Medical Image Multi-View Transformation and Reconstruction with Deep Learning

Vivekanand P. Thakare[1], *Dr.Narendra Chaudhari* [2], *Mr.Amit Nigam*[3], *Ms.Minal Thawakar*[4], *and Mr.Kartik Ingole*[5]

[1]Assistant Professor, Department of Computer Science & Engineering, GWCOE, Nagpur

[2]Associate Professor, Department of Computer Science & Engineering, TGPCET, Nagpur

[3]Assistant Professor, Department of Electronics & Communication Engineering, Narula Institute of Technology, Kolkata

[4]Assistant Professor, Department of Computer Science & Engineering, KDKCE, Nagpur

[5]Assistant Professor, Department of ECE, KDKCE, Nagpur

Abstract: Recent innovations in medical electronics and the widespread availability of high-speed Internet have made it possible to create, store and distribute vast quantities of imaging data. The use of medical pictures equipped with multiple senses has become more important in the diagnosis and treatment of a wide range of illnesses. When diagnosing and treating a patient, doctors often refer to similar past cases for guidance. Because of the sophisticated methods employed in medical diagnosis, treatment planning and response evaluation, enormous databases are generated. X-ray, magnetic resonance imaging and computed tomography are just a few examples of the cutting-edge medical imaging methods that have led to a massive increase of biological data. To effectively retrieve medical images, an approach called the enriched deep residual framework is developed. A module for re-creating images, including a residual encoder and sequential decoder, is included. The characteristics of a picture are encoded via a sequence of residual connections in the EDR framework's encoder module, image reconstruction network (IRNet) framework, which are then sent on to the decoder module. Latent representation for robust input picture reconstruction is provided by the extracted encoded features.

Keywords: Computed tomography, medical electronics, enriched deep residual framework, encoder module, decoder module, image reconstruction network.

DOI: 10.1201/9781003527442-27

1. Introduction

Due to sensor, multimedia, digital electronics and Internet improvements, new information is generated at an unparalleled rate. Processing this data manually is a chore. To ensure the data can be used quickly, a trustworthy data management system is needed. Retrieval systems formerly used human annotation. This image retrieval process relies on metadata. The manual annotation method has two drawbacks when annotating many photos. First, manual annotation takes time and effort, and second, each annotator sees an image differently. Until these two issues are resolved, retrieval may provide a few false positives. Image sensor and multimedia technology made manual annotation impossible in the early 1990s. Physicians have long used X-rays, ultrasound, ultrasonography and CT to observe internal organs, external body parts and other tissues. Various imaging techniques may produce images with low SNR, CNR and picture artifacts (Wang *et al.*, 2016). Image reconstruction was developed to address these challenges and improve image quality for better visual interpretation, comprehension and analysis. Radionics, CAD/CAM and medical image analysis use deep learning (DL; Wang *et al.*, 2020; Wu *et al.*, 2019). Deep learning, or representation learning, has gained popularity in medical picture analysis (Xie *et al.*, 2019). A major advantage of deep learning over standard machine learning is its capacity to learn features from raw input data during training. It learns data abstractions due to its multiple levels (Syben *et al.*, 2019). GPUs and cloud computing platforms have made deep learning more popular in many fields, including medical image reconstruction (Chen *et al.*, 2018). Image reconstruction uses measuring data to rebuild a picture. First, sensor encodes an item in the sensor domain, then reverse the encoding function to create a picture. Due to sensor non-idealities and noise, picture reconstruction requires analytic knowledge of the precise inverse transform, which may not exist a priori. Due to the difficulty of determining the inverse transform, traditional picture reconstruction approaches fail. Inaccurate approximations using chains of carefully tuned signal-processing modules may be needed for noisy data (Xie *et al.*, 2018). Deep learning will revolutionise image reconstruction. Akagi *et al.* (2019) deep-learning algorithms improve medical image reconstruction speed, accuracy and resilience. In particular, this study summarises deep learning in medical picture reconstruction. Open medical imaging research, deep learning datasets and open-source medical image processing software were our focus. Deep learning is widely studied; however medical picture reconstruction is rarely covered. Though it gives a broad overview of deep learning's usage in medical image restoration, it (Zhang *et al.*, 2020) focuses on the mathematical models of many techniques. This article summarises how deep-learning techniques and architectures have been used to rebuild medical images Dar *et al.* (2020)

2. EDR: Enriched Deep Residual Framework

The enriched deep residual framework (EDR) is introduced here. The proposed EDR architecture has a sequential decoder and a residual encoder integrated within the image reconstruction unit. In addition, photos are retrieved from the database using similarity matching followed by retrieval (Ramzi *et al.*, 2020). The EDR framework's encoder block incorporates a chain of residual blocks. After an image's latent characteristics have been encoded, they are sent to a decoder module. Latent information necessary for accurate reconstruction of the input picture is provided by the encoded features. The process of matching and retrieving images also makes use of latent information. Using data from the VIA/IELCAP-CT and the ILD, we can see that the suggested technique is effective. Average accuracy and recall are compared across two datasets for the proposed approach and other state-of-the-art methods (Topal *et al.*, 2020). The suggested architecture has shown superior performance in medical image retrieval experiments and results.

3. Enriched Deep Residual Framework

Image augmentation, super-resolution, reconnection, moving object/object segmentation and depth estimation are just a few of the areas where auto encoder–decoder networks have recently received a lot of attention. Taking cues from the network's impressive technical prowess, we offer an end-to-end enhanced EDR for picture retrieval and reconstruction. For latent feature extraction and reconstruction of input medical pictures, the EDR framework utilises an encoder–decoder module. The picture input is sent into an encoder, which then extracts relevant information. The similarity index and retrieval component receive encoded latent characteristics as input. As can be seen in Figure 1, latent feature estimates are calculated using a series of residual blocks. To aid in the decoder network's capacity to rebuild the abstract representation of the provided medical picture, residual connections are used. Index matching and retrieval of comparable pictures from an image database both rely on these hidden abstract properties. Separate modules for picture indexing and retrieval make up the EDR framework for medical image retrieval.

Step-by-step procedure of EDR method: Input: Medical scan 256 × 256 Features Extracted: 8 × 8 × 2048 Output: Reconstructed Image 256 × 256 as shown in Figure 3 .

1. Medical image of size 256 × 256 × 3 is given as input to Convolution layer.
2. Latent features of input image are encoded using residual blocks.
3. Two sequential convolutions and ReLU followed by maxpooling are performed for feature extraction.
4. Image is down-sampled to extract 2 × 2 × 64 features.
5. Decoder uses up sampler for reconstructing the image by using latent features obtained in step-4.

6. Convolution is done to retain original image size.
7. Features obtained in step-4 are used for comparing query and database images.
8. By using d1 distance, similarity levels are identified for retrieval.

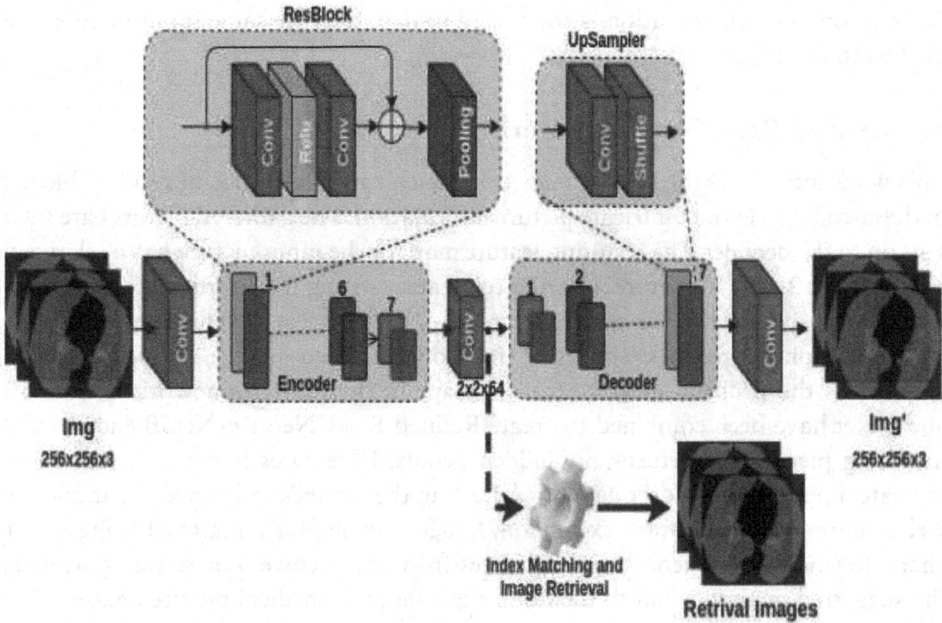

Figure 1: Enriched deep residual framework.

4. Encoder Architecture

In the proposed EDR framework, features are extracted using an encoder. The encoder uses leftover blocks to learn features from the query picture. The primary function of the encoder is to compress the source picture down to smaller dimensions. In the proposed structure, encoder layers are designed with residual blocks in mind. Two convolutions and ReLU come together to form a residual block. The suggested encoder design utilises a sequence of seven residual blocks. When applied to a picture, these leftover blocks create a simplified replica of the original. As can be seen in Figure 1, when residual blocks are used, a picture of size 256 × 256 is reduced to 2 × 2 × 64 features.

5. Decoder Architecture

In order to rebuild the input picture, the suggested framework's decoder is employed to conduct convolution and shuffle on the up-sampled data. The encoder's latent characteristics are sent into the decoder. The seven decoder modules of the proposed EDR system. Convolution and a shuffle operation are carried out in each individual block. The input to the convolution layer of the first decoder block is the 2,264 features received from the encoder. Scaling the lambda layer brings the resolution of these characteristics up to 4 × 4 × 64. Once latent features have been retrieved, they undergo a series of convolution, shuffle and convolution procedures. The picture is up-sampled to 256 × 256 × 64 by shuffling the pixels. After the seventh decoder layer, a convolution operation is conducted to maintain the same input picture size as the rebuilt image.

6. Refined Res-UNet Framework

This work introduces the EDR framework, which uses a sequence of residual blocks in the encoder to extract intricate picture information. The gathered features are then sent on to the decoder. The resulting feature map for the input picture has a relatively small size of 2,264. Therefore, in order to create a strong reconstructed picture, it is necessary to expand the size of the feature map. Refined Res-UNet is a unique image reconstruction framework suggested for content-based medical image retrieval that solves the problem of large feature maps in the EDR framework. ResNet50 and UNet have been combined to create Refined Res-UNet. ResNet50 encodes the incoming picture and returns the hidden details. UNet uses latent information to recreate a picture at the decoder end. Adjusting the model's weights and parameters yields more accurate feature extraction. Images are similarly matched using latent characteristics. After then, we pull pictures from the archive. For testing purposes, the suggested method is put to the test using canonical medical picture datasets.

7. Refined Res-UNet for Image Reconstruction

The suggested system employs a ResNet model variation called ResNet50. One maxpool layer, 48 Convolution layers and an average pool layer make up the ResNet model. It is a popular ResNet model. ResNets were first used in image recognition and other non-computer vision tasks to improve accuracy. UNet has a route that contracts on the left and a path that expands on the right. The convolution, ReLU and maxpooling layers make up the contracting route. As a result of the convolution process removing boundary pixels, cropping is required.

8. IRNET: Image Reconstruction Network

In this study, enhanced Res-UNet uses transfer learning to fix feature map issues. Even with larger feature maps, spatial information is lost while transmitting encoder-level features to decoder layers. Reduce information leakage. This chapter introduces an image reconstruction network (IRNet) feature learning approach to handle this challenge. A bridge network links the encoder and decoder. Skip connections enable feature exchange between encoder and decoder levels. This framework encodes features using an image reconstruction network. The decoder retrieves encoded features. Weight and parameter adjustments did not reduce L1 loss in the planned IRNet. The peak signal-to-noise ratio is determined by ablation with and without a bridge network. Recommended procedure effectiveness is determined by average running time. System performance is assessed utilizing OASIS, ILD and VIA/I-ELCAP medical image repositories.

9. Image Reconstruction Network

The input picture is represented as a feature set in the proposed IRNet. Reconstruction is then performed using the encoded characteristics. The encoded characteristics are seen to function as a simplified representation of the original picture. The input picture is therefore represented by the IRNet's encoder unit. The similarity matching and retrieval module is used to find related pictures after the first feature learning stage.

For the ith node in the network, the characteristics provided to it are denoted by fi. The number of nodes in the network, denoted by N. The suggested network is capable of efficient reconstruction thanks to feature concatenation, which aids the decoder in reconstructing the image's basic edge information. To decode the encoded characteristics and rebuild the input picture, eight decoder blocks are available. Each of these decoder blocks, as shown in Figure 2, processes the encoded abstract features to recreate the 256 × 256-pixel input image.

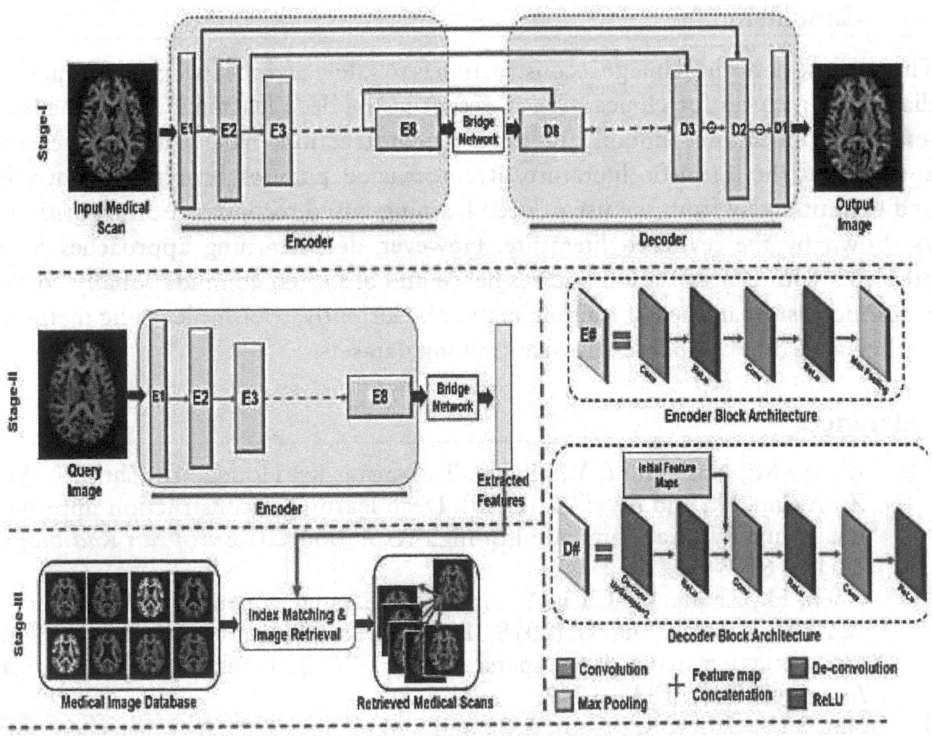

Figure 2: Proposed content-based medical image retrieval framework.

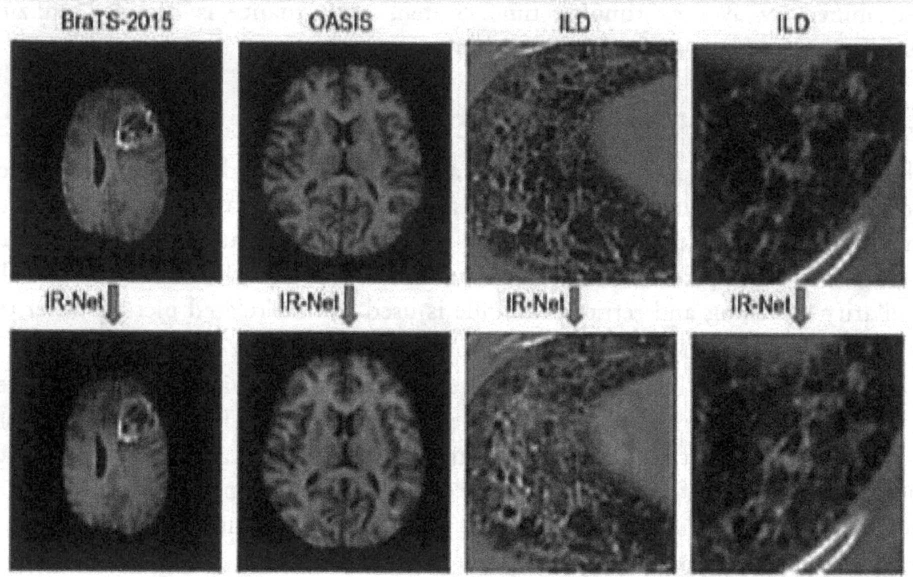

Figure 3: Original images and corresponding reconstructed images.

10. Conclusion

The purpose of medical image reconstruction is to safely and cheaply get high-quality diagnostic pictures for clinical use. In recent years, deep learning and its medical imaging applications, notably in image reconstruction, have gained extensive coverage in the scientific literature. Reconstructed pictures benefit qualitatively and quantitatively from the use of deep learning-based reconstruction algorithms, as shown by the reviewed literature. However, deep learning approaches have problems with generalisation and resilience and are often computationally costly, requiring vast quantities of training material. Currently, transfer learning methods are being used to combat insufficient training datasets.

References

[1] Akagi, M., Nakamura, Y., Higaki, T., Narita, K., Honda, Y., Zhou, J., Yu, Z., Akino, N., and Awai, K. (2019). Deep learning reconstruction improves image quality of abdominal ultra-high-resolution CT. *European Radiology*, 29(11), 6163–6171.

[2] Chen, H., Zhang, Y., Chen, Y., Zhang, J., Zhang, W., Sun, H., Lv, Y., Liao, P., Zhou, J., and Wang, G. (2018). Learn: Learned experts' assessment-based reconstruction network for sparse-data CT. *IEEE Transactions on Medical Imaging*, 37 (6), 1333–1347.

[3] Dar, S.U., Özbey, M., Çatlı, A.B., and Çukur, T. (2020). A transfer-learning approach for accelerated MRI using deep neural networks. *Magnetic Resonance in Medicine*, 84(2), 663–685.

[4] Ramzi, Z., Ciuciu, P., and Starck, J.-L. (2020). Benchmarking MRI reconstruction neural networks on large public datasets. *Applied Sciences*, 10(5), 1816.

[5] Syben, C., Michen, M., Stimpel, B., Seitz, S., Ploner, S., and Maier, A. K. (2019). Technical note: Pyro-nn: Python Reconstruction Operators in neural networks. *Medical Physics*, 46 (11), 5110–5115.

[6] Topal, E., Löffler, M., and Zschech, E. (2020). Deep learning-based inaccuracy compensation in reconstruction of high resolution XCT Data. *Scientific Reports*, 10(1).

[7] Wang, J., Liang, J., Cheng, J., Guo, Y., and Zeng, L. (2020). Deep learning based image reconstruction algorithm for limited-angle translational computed tomography. *PLOS ONE*, 15 (1).

[8] Wang, S., Su, Z., Ying, L., Peng, X., Zhu, S., Liang, F., Feng, D., and Liang, D. (2016). Accelerating magnetic resonance imaging via deep learning. *IEEE 13th International Symposium on Biomedical Imaging* (ISBI).

[9] Wu, Y., Ma, Y., Liu, J., Du, J., and Xing, L. (2019). Self-attention convolutional neural network for improved MR Image reconstruction. *Information Sciences*, 490, 317–328.

[10] Xie, S., Zheng, X., Chen, Y., Xie, L., Liu, J., Zhang, Y., Yan, J., Zhu, H., and Hu, Y. (2018.) Artifact removal using improved googlenet for sparse-view CT reconstruction. *Scientific Reports*, 8 (1).

[11] Xie, H., Shan, H., and Wang, G. (2019). Deep encoder-decoder adversarial reconstruction (DEAR) network for 3D CT from few-view data. *Bioengineering*, 6 (4), 111.

[12] Zhang, H.-M., and Dong, B. (2020). A review on deep learning in medical image reconstruction. *Journal of the Operations Research Society of China*, 8 (2), 311–340.

Chapter 28

Exploring the Effectiveness of Ensemble Learning Techniques for the Classification of Stellar Objects in Gaia DR3

Nived Krishna Prakash[1] *and Shilpee Srivastava*[1]

[1]Department of Mathematics, Chandigarh University, Punjab, India

Abstract: This study explores the effectiveness of ensemble learning algorithms for accurately classifying stellar objects in the Gaia DR3 dataset. We compare various classification algorithms, including random forest, XGBoost and Decision Tree, for both binary and multiclass classification. Our results show that binary classification achieves a maximum accuracy of 87%, while the random forest algorithm outperforms others in multiclass classification, with an accuracy of 79%. This study highlights the potential of ensemble learning methods, particularly the random forest algorithm, for precise classification of stellar objects in Gaia DR3. The findings have implications for research on other stellar objects, providing a foundation for accurate and reliable classification methods.

Keywords: Ensemble learning, gaiadr3, machine learning, stellar object classification.

1. Introduction

Machine learning has unlocked many new opportunities in the field of astrophysics. It enables researchers to quickly and accurately process vast amounts of data and make predictions that that cannot be made without it. Machine learning algorithms have been used to classify galaxies, supernovae and identify young stellar objects in various studies over the years. Some researchers have explored the potential of unsupervised machine learning algorithms for astrophysical data, while others have compared feature engineering and deep learning for astronomical time series classification. Recent studies have successfully classified 1.7 billion stars in Gaia DR2 using Pan-STARRS1 and WISE data.

A 97% accuracy was achieved by applying nearest neighbour instance-based algorithms to analyse celestial objects. The adoption of x-ray data from the ROSAT All-Sky survey (RASS), specifically the RASS Bright Source Catalogue (RBSC) and the RASS Faint Source Catalogue (RFSC), along with optical data from USNO-A2.0 and infrared data from the Two Micron All Sky Survey (2MASS), significantly improved the accuracy of their predictions (Li *et al.*, 2008). The study suggests that

DOI: 10.1201/9781003527442-28

the K-nearest neighbour (KNN) algorithm performs better for predicting celestial objects. RF classifier is relatively superior to other methods in terms of speed and accuracy when applied to Milky Way tomography data (Richards *et al.*, 2011). Furthermore, they introduced tree-ensemble methods for stellar classification. This method was an improvement over existing practices for classification. It was found that the RF classifier was able to identify stars with greater accuracy and faster speed than traditional methods.

Support vector machine classifier (SVM) was successfully used to classify galaxies in WISE/Super Cosmos (Krakowski *et al.*, 2016). To confirm accuracy, they cross-verified results with spectroscopic sources from Sloan Digital Sky Survey (SDSS). A supervised machine learning algorithm that uses Principal Component Analysis (PCA) and an Artificial Neural Network (ANN) was proposed to classify stars into spectral types based on diffraction pattern in images taken with Hubble Space Telescope Advanced Camera for Surveys (HST ACS) and Euclid VIS Imager (Kuntzer *et al.*, 2016). Machine learning algorithms such as K-NN, Naïve Bayes, SVM, BDTs and ANNs were used to classify supernovae based on light curve data (Lochner *et al.*, 2016). K-means clustering algorithm was used on the APOGEE data set and found that unsupervised methods were generally less reliable (Garcia-Dias *et al.*, 2018).

Deep learning representation methods were discovered to be not ideal for astrophysical data (Hinners *et al.*, 2018). The team used Kepler light curve data from MAST and found that classification with feature engineering was more successful for both regression and classification. However, the feature engineering was still laborious and time-consuming. A supervised learning classifier was developed to classify 85 million objects in the Gaia data release 2, using Pan-STARRS 1 and WISE data. These objects were cross-matched with the SIMBAD database, and potential labels for the objects were identified (Bai et al., 2018). This approach was used to create a classification for the objects, consisting of stars (around 98%), galaxies and quasars. The same data catalogue was used in another study (Lam *et al.*, 2018).

Random forest classifier was used to identify young stellar objects in the Gaia DR2 with AllWISE catalogue. This helped for the classification of previously unknown young stellar objects in the Gaia photometric science alerts system (Marton *et al.*, 2019). A new deep learning model for automated planet candidate identification in TESS data showed high accuracy and precision. The model was trained on real TESS data and achieved an average precision of 97.0% and an accuracy of 97.4% in triage mode. In vetting mode, the model achieved an average precision of 69.3% and an accuracy of 97.8%. The authors applied the model on new data from Sector 6 and presented 288 new signals that received the highest scores in triage and vetting and were also identified as planet candidates by human vetters (Yu *et al.*, 2019). Machine learning techniques were used to classify blazars based on their synchrotron peak frequency. Using supervised ML algorithms including random forests, support vector machine, k-nearest neighbours, Gaussian naïve Bayes and the Ludwig auto-ML framework, they found that the support vector machine algorithm achieved a balanced accuracy of 93% (Arsioli and Dedin, 2020). The study demonstrated the potential of machine learning in distinguishing multifrequency spectral characteristics and accurately classifying blazars into low synchrotron peak (LSP) and high synchrotron peak (HSP) categories.

1.1. Gaia Data Release 3

The Gaia mission of the European Space Agency has provided unprecedented amounts of data since its launch in 2013. The third data release of the Gaia mission (DR3) was released in 2022. This included more than 1.7 billion astronomical sources, providing an unprecedented view of our galaxy and its components (Vallenari *et al.*, 2022). This data release also included a catalogue of radial velocity measurements for more than 7 million stars (Seabrok *et al.*, 2021), as well as positions and motion in three dimensions for more than 1.3 billion stars. The data is available at The European Space Agency's online platform for Gaia Program as well as onNASA/IPAC Infrared Science Archive (IRSA). The Gaia Data Release 3 (DR3) release has a lot of variables we can use to get information about stellar objects and for a variety of astronomical studies. These variables include astrometry, photometry and spectroscopy data. Astrometry data provides accurate position, parallax and proper motion measurements for over 1.8 billion stars (Bailer *et al.*, 2021). Photometry data includes flux measurements in different bands, which can be used to study the color and magnitude of stars (Riello *et al.*, 2021). Spectroscopy data provides information about the chemical composition of stars and their radial velocities.

The astrometry data in Gaia DR3 is highly accurate, with a median uncertainty of 0.03 mas in position measurements for bright sources, and 0.1 mas for faint sources. The parallax measurements have a typical uncertainty of 0.03 mas for sources brighter than 17 mag, and up to 0.7 mas for fainter sources. The proper motion measurements are also highly accurate, with a typical uncertainty of 0.03 mas/yr for bright sources, and up to 1 mas/yr for fainter sources. The photometry data in Gaia DR3 includes measurements in three different bands: G, BP and RP. The G-band flux measurements have a typical uncertainty of 1 mmag for sources brighter than 16 mag and up to 10 mmag for fainter sources. The BP and RP-band flux measurements have typical uncertainties of 2 mmag for bright sources, and up to 30 mmag for fainter sources. The spectroscopy data in Gaia DR3 includes radial velocities and chemical abundances for over 7 million stars. The radial velocities have a typical uncertainty of 0.5 km/s for bright sources, and up to 10 km/s for fainter sources. The chemical abundances provide information about the metallicity and alpha-element content of stars, which can be used to study their formation and evolution (Vallenari *et al.*, 2022; Seabrok *et al.*, 2021).

1.2. Data Catalogues

The Set of Identifications, Measurements, and Bibliography for Astronomical Data (SIMBAD) is a widely used reference database for astronomical objects that provides comprehensive information on stars, galaxies and other celestial objects. It has been utilised in a multitude of research studies for object classification, spectral analysis and more. SIMBAD contains data on over 10 million objects, including positional information, magnitudes, proper motions and radial velocities, as well as cross-references to other databases and bibliographic references. In addition, CDXMatch, a software tool developed by the Centre de Donnéesastronomiques de Strasbourg (CDS), enables researchers to cross-match catalogues from various astronomical surveys, facilitating studies of different astrophysical phenomena. CDXMatch is an essential tool for cross-identifying and verifying large datasets and has been extensively used in various research studies. We used SIMBAD to cross-verify object data in the Gaia DR3.

2. Methodology

We used the CDSX match cross-matching tool to extract a subset of the catalogue that contains only sources with high astrometric accuracy and photometric quality. This was a cross-match of objects in Gaia DR3 and SIMBAD catalogue. We collected 6.53 million data points for various stellar objects. This also contains object types for the cross-matched subset as those objects have been studied in previous studies. This subset is then filtered to remove any sources with missing data. Since the dataset was imbalanced, as shown in Figure 1, under sampling using random under sampler function in sklearn library was implemented. We also performed data normalisation to scale the features to a common range. In order to select the most relevant features for the classification task, we employed the k-best feature selection function available in the sklearn library. This approach allowed us to reduce the dimensionality of the pre-processed data while retaining the most important features.

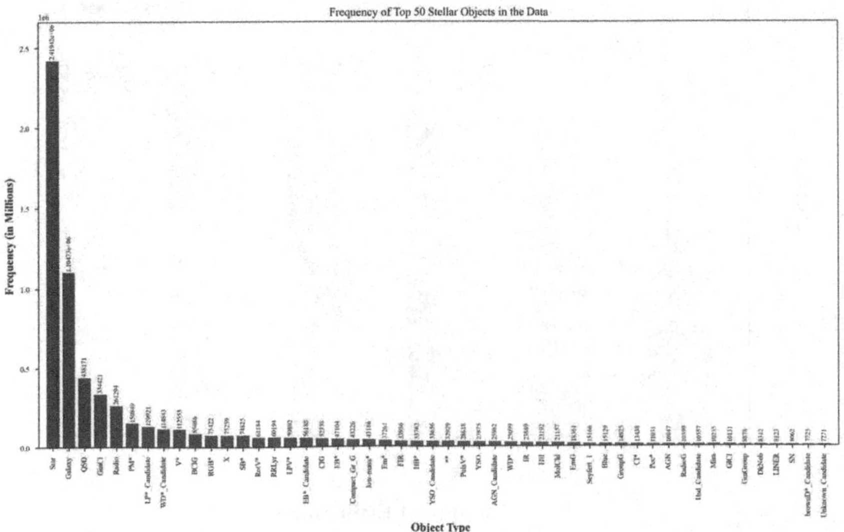

Figure 1: Frequency of Top 50 Objects in the Data

Correlation analysis and mutual information methods were used to compute individual feature scores, which were then used by the k-best function to select the top features based on their scores. It should be noted that this feature selection technique did not remove any entries from the dataset but rather excluded certain variables or features that were deemed less significant. Ultimately, we selected 25 best features out of 67 features excluding the target variable.

We considered a range of models, including K-Nearest Neighbours (K-NN), SVM, RF, ANN, XGBoost and AdaBoost. Dataset with stratified training and testing splits was done before feeding it to a pipeline of models with preprocessing. We evaluated the performance of each model using cross-validation technique and selected the best-performing model among them.

3. Results

We found that binary classification resulted in a maximum accuracy of 87% as shown in Figure 2. However, for multiclass classification, the random forest algorithm yielded the highest accuracy of 79%, shown in Figure 3. Our findings suggest that while binary classification may produce higher accuracy results, multiclass classification can still be effective with the right ensemble learning method.

The random forest algorithm emerged as the most effective method for multiclass classification, outperforming other methods such as XG Boost and decision tree. This highlights the importance of selecting appropriate ensemble learning algorithms for working with large and complex datasets like GaiaDR3.

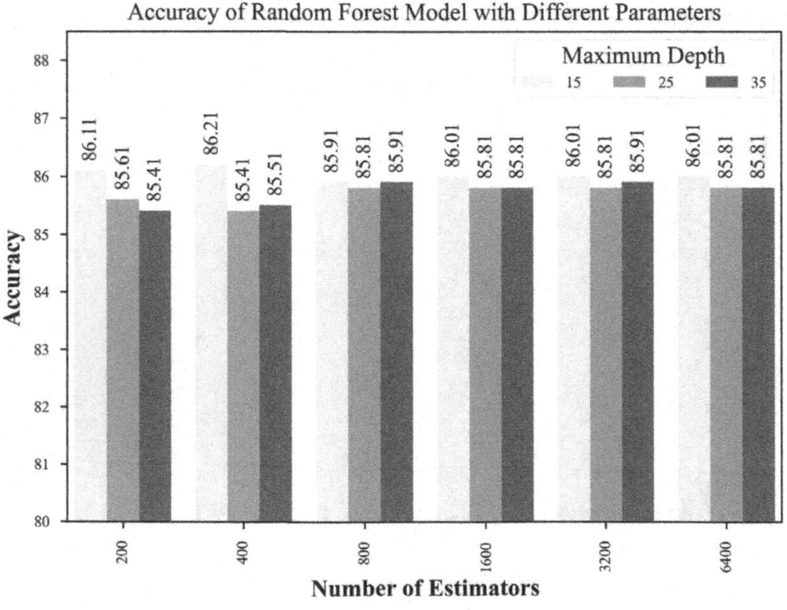

Figure 2: Accuracy of RF Classifier with Different Parameters

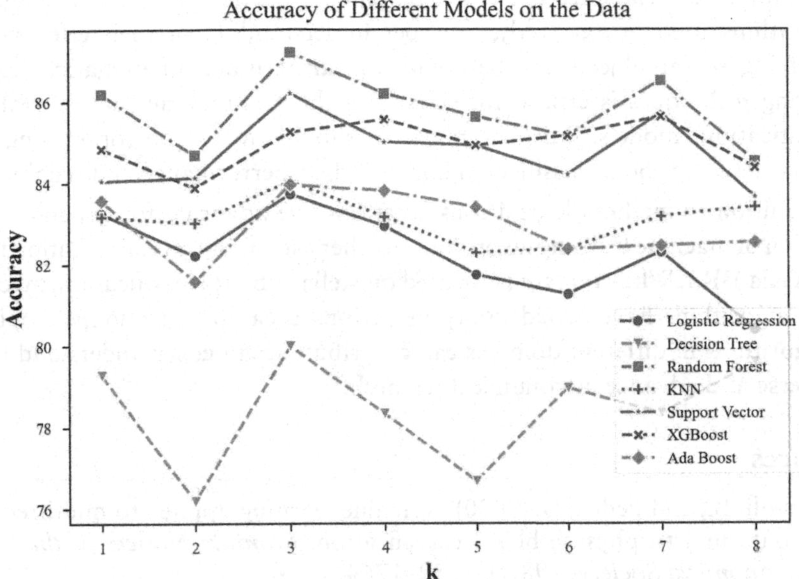

Figure 3: Accuracy of Different Models on the Data

Our study emphasises the potential of ensemble learning methods, particularly the random forest algorithm, for accurately classifying stellar objects in Gaia DR3. These findings may have significant implications for future research on TTau* objects and other stellar objects, providing a foundation for more accurate and reliable classification methods in the field. We discovered that the random forest classifier and XGBoost classifier performed best out of all the models tested. Random forest and XGBoost had an average accuracy of 86% and an F1 score of 0.85 in eight-fold cross-validation. The other models we tested had less than 82% accuracy with F1 scores lower than 0.84.

For our analysis of 1 million objects in the Gaia DR3 data, we selected the random forest classifier with the number of estimators ranging from 200 to 6400 and a maximum tree depth ranging from 15 to 35. During training, random forest classifier showed an 86% accuracy, while during validation, it had an 85% accuracy.

4. Conclusion

Our study highlights the potential of ensemble learning methods, particularly the random forest algorithm, for accurately classifying stellar objects in the Gaia DR3 dataset. We emphasise the importance of selecting appropriate ensemble learning algorithms for large and complex datasets like Gaia DR3. Based on our experiments, we recommend the random forest classifier or XGBoost classifier for classification tasks in Gaia DR3, as it exhibited superior accuracy and F1 scores, which directly confirms the results of previous studies using similar techniques.

We suggest exploring additional learning methods to gain insights into their suitability for classification tasks. By comparing their performance against the algorithms used in our study, a deeper understanding of their effectiveness in classifying stellar objects can be obtained. Furthermore, investigating feature engineering techniques is crucial for enhancing the accuracy and interpretability of classification models. Further research should focus on optimising the hyperparameters of these classifiers to improve their performance and reliability.

In addition to methodological considerations, we advocate for expanding the application of machine learning algorithms to other astronomical classification tasks beyond Gaia DR3. While our study focused on stellar object classification, machine learning algorithms have broad utility in various areas of astronomy. Applying these algorithms in different domains can contribute to a deeper understanding of the universe and advance astronomical research.

References

[1] Arsioli, B., and Pedro, D. (2020). Machine learning applied to multifrequency data in astrophysics: blazar classification. *Monthly Notices of the Royal Astronomical Society*, 498(2), 1750–1764.

[2] Bai, Y., JiFeng, L., Song, W., and Fan, Y. (2018). Machine learning applied to Star–Galaxy–QSO classification and stellar effective temperature regression. *The Astronomical Journal*, 157(1), 9.

[3] Bailer-Jones, C. A. L., Rybizki, J., Fouesneau, M., Demleitner, M., and Andrae, R. (2021). Estimating distances from parallaxes. V. Geometric and photogeometric distances to 1.47 billion stars in Gaia Early Data Release 3. *The Astronomical Journal*,161(3), 147.

[4] Brown, A. G. A., Vallenari, A., Prusti, T., De Bruijne, Babusiaux, C., Biermann, M., Creevey, *et al.* (2021). Gaia early data release 3-summary of the contents and survey properties. *Astronomy & Astrophysics*, 649(A1).

[5] Garcia-Dias, R., Carlos, A. P., Jorge, S. A, and Ignacio, O. P. (2018). Machine learning in APOGEE-Unsupervised spectral classification with K-means. *Astronomy & Astrophysics*, 612(A98).

[6] Hinners, T. A., Kevin, T., and Rachel, T. (2018). Machine learning techniques for stellar light curve classification. *The Astronomical Journal*, 156(1), 7.

[7] Krakowski, T., Małek, K., Bilicki, M., Agnieszka, P., Agnieszka, K., and Magdalena, K. (2016). Machine-learning identification of galaxies in the WISE× SuperCOSMOS all-sky catalogue. *Astronomy & Astrophysics*, 596(A39).

[8] Lam, A., Edward, W., and Matthew, M. (2018). DEIMOS observations of WISE-selected, optically obscured AGNs. *Monthly Notices of the Royal Astronomical Society*, 480(1), 451–466.

[9] Li, L., Yan, X. Z., and Yong, H. Z. (2008). K-nearest neighbours for automated classification of celestial objects. *Science in China Series G: Physics, Mechanics and Astronomy*, 51(7), 916–922.

[10] Lochner, M., Jason, D. E., Hiranya, V. P, Ofer, L., and Max, K. W. (2016). Photometric supernova classification with machine learning. *The Astrophysical Journal Supplement Series*, 225(2), 31.

[11] Marton, G., Péter, A., Elza, S. E., József, V., Mária, K., Ágnes, K., Verebélyi, E. V., *et al.* (2019). Identification of Young Stellar Object candidates in the Gaia DR2 x AllWISE catalogue with machine learning methods. *Monthly Notices of the Royal Astronomical Society*, 487(2), 2522–2537.

[12] Paine, Jennie, Jeremy Darling, and Alexandra Truebenbach. (2018). The Gaia–WISE Extragalactic Astrometric Catalog. *The Astrophysical Journal Supplement Series*, 236(2), 37.

[13] Richards, J. W., Dan, L. S., Nathaniel, R. B., Joshua, S. B., John, M. B., Arien, C. Q., Justin, H., Rachel, K., and Maxime, R. (2011). On machine-learned classification of variable stars with sparse and noisy time-series data. *The Astrophysical Journal*, 733(1), 10.

[14] Riello, M., De Angeli, F., Evans, Montegriffo, P., Carrasco, Busso, G., Palaversa, L., *et al.* (2021). Gaia early data release 3-photometric content and validation. *Astronomy & Astrophysics*, 649(A3).

[15] Seabroke, G. M., Fabricius, C., Teyssier, D., Sartoretti, P., Katz, D., Cropper, M., Antoja, T., *et al.* (2021). Gaia early data release 3-updated radial velocities from Gaia DR2. *Astronomy & Astrophysics*, 653(A160).

[16] Yu, L., Andrew, V., Chelsea, H., Christopher, J. S., Ian, J. C., Gaudi, B. S., Tansu, D., *et al.* (2019). Identifying exoplanets with deep learning. III. Automated triage and vetting of TESS candidates. *The Astronomical Journal*,158(1), 25.

Chapter 29

Secure on Demand Token-Based Binding Update Authentication Mechanism

Bhavana S. Karmore[1], *Shraddha N. Zanjat*[2], *and Vishwajit K. Barbudhe*[3]

[1]Assistant Professor, UG and PG Department of Computer Application, School of Science, G H Raisoni University, Amravati, Maharashtra, India

[2]Assistant Professor, UG and PG Department of Electrical and Electronics Engineering, School of Engineering and Technology, Sandip University, Nashik, Maharashtra, India

[3]Assistant Professor, Department of Artificial Intelligence and Data Science Engineering, Sandip Institute of Technology and Research Center, Nashik, Maharashtra, India

Abstract: Secured request tokens, oriented bindings and updated verification mechanisms were developed to strengthen the safety of Portable IPv6 by defending it against binding-updated attacks. This technique enables CN to verify the MN to make certain this is not an unauthorised node or automatically authorises MN and CN to communicate with one another. It also verifies MN is accessible via HoA and CoA. Encryption interactions among MN and CN to guarantee authenticity and secrecy and shield BU communications from DoS, MITM and hijacked session attacks.

Keywords: Mobile IPv6, security of MIPv6, binding update authentication mechanism, token-based authentication.

1. Introduction

Mobile IPv6 enhances the IPv6 network by allowing mobile devices to roam. The main advantage of the MIPv6 protocol is that nodes that are mobile don't need to alter their IP addresses every time they switch where they connect to the internet in (Adnan and Shamala, 2018). Mobile gadgets may move invisibly using MIPv6 without losing their present connection. Because MIPv6 promotes mobility and uses open airways as a means of communication, it is vulnerable to the numerous security risks that are sent across the MIPv6 system. These threats to safety are among the most important issues that have to be taken into account. There are also several dangers. A lot of attacks are caused by an attacker not understanding the communication nodes' coordinates in (Moravejosharieh *et al.*, 2012). When an MN changes locations in MIPv6, it communicates the HA and CN's current addresses, or CoA, via the BU message. But there are several security risks associated with this BU message, making it possible for an attacker to transmit a bogus BU to divert data for packet collection and eavesdropping or to stop the connection altogether, leading to

DOI: 10.1201/9781003527442-29

a denial of service attack in Antonio and David (2014). Effective identification and encryption of messages are required to defend MIPv6 from such BU attacks. You can use identification to confirm if the alleged node is indeed an approved node. Utilise encryption as well, which enables node-to-node communications to stay private in Rossi (2013). In order to enhance the RR technique, this article suggests using a HA as a token issuance organisation that creates and controls tokens for MN and CN in the MIPv6 system in Barbudhe *et al.* (2013). The suggested approach allows MN and CN to mutually verify one another, enabling each MN to start a BU procedure with CN and build a relationship of confidence with its very own HA or home network router. The data exchange between MN and CN was encrypted using ECC encryption in order to guarantee integrity and secrecy in Barbudhe *et al.* (2014). Redirection-based flooding incidents were also prevented using the CoA test in Barbudhe *et al.* (2015).

2. Proposed Mechanism's Requirements

- Assure the security association between MN and CN.
- Implement efficient and secure authentication mechanisms that protect BU messages and reduce handover latency in Barbudhe *et al.* (2017).
- By CN authenticating the HoA, it is proven that the BU application originated with an approved MN that is the HoA's owner.
- Making sure the CoA sent in the BU response is accurate by having CN confirm the CoA's legitimacy in Perkins (2010).
- In order to protect against a reply attack keys and messages sent in communication are in encrypted form.
- Protect the BU message from the false binding update threat, MITM threat and DOS threat in Kaufman *et al.* (2010).
- Improve the integrity of the BU message in Bo *et al.* (2010).

Network Architecture

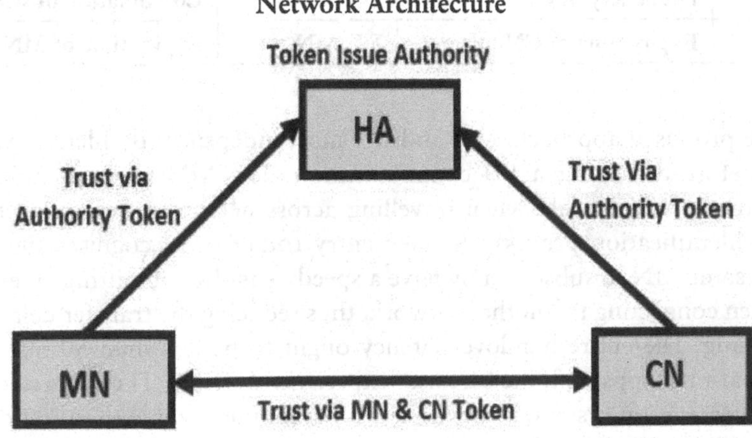

Figure 1: Trust model.

Figure 1 shows the network architecture of the trust model, and the following are several assumptions used in the proposed mechanism:

- It assumes that mobility-related messages exchanged between the MN and the CN are protected using the IPSec ESP tunnel in Chitra (2012).
- HA is a trusted node that verifies the MN's HoA validity. Also, it plays the role of token issue authority and generates and manages tokens for MN and CN.
- HA is in charge of not only token issuing but also ECC public parameter management.
- MN already registered its COA with the HA via a pre-established secure IPv6 tunnel.

Secure On-Demand Token-Based Binding Update Authentication Mechanism

Table 1: Notation table.

Notation	Description	Notation	Description
T_{MN}	Token of MN	ET_MN	Expiry time of MN token.
T_{CN}	Token of CN	K_{BU}	Binding update Key
HoA	ID of MN who keeps MN token	N_{MN}	Nonce generated by MN
HA	ID of Token Issue Authority	N_{CN}	Nonce generated by CN
Sign(HA)	Digital Signature of Token Issue Authority	MN_{TT}	Transfer token of MN
CN	ID of CN	ET_MN_{TT}	Expiry time of MN transfer token
MN_{PUK}	Public Key of MN	$Enc(X_{PUK}(m)$	Encryption of the message m by entity x's public key
CN_{PUK}	Public Key of CN	||	Combination of strings
ET_CN	Expiry time of CN token.	ET_MN	Expiry time of MN token.

Table 1

In the proposed approach, MN and CN must independently identify with one another before delivering a BU communication. The MN must re-authenticate in order to use the internet when travelling across networks. Following the first complete identification process, the new entry router pre-recognises the MN to allow the same MN to subsequently have a speedy pass by submitting an exchange token when connecting to another network, thus reducing the transfer delay during MN roaming. The entire handover latency ought to be less than 50 ms so as to provide real-time apps with the best user experience possible. The suggested token-based authentication system, however, sets a maximum token expiration duration of 1000 milliseconds. The digital token expires after one thousand milliseconds in (Daemon and Vishal, 2017).

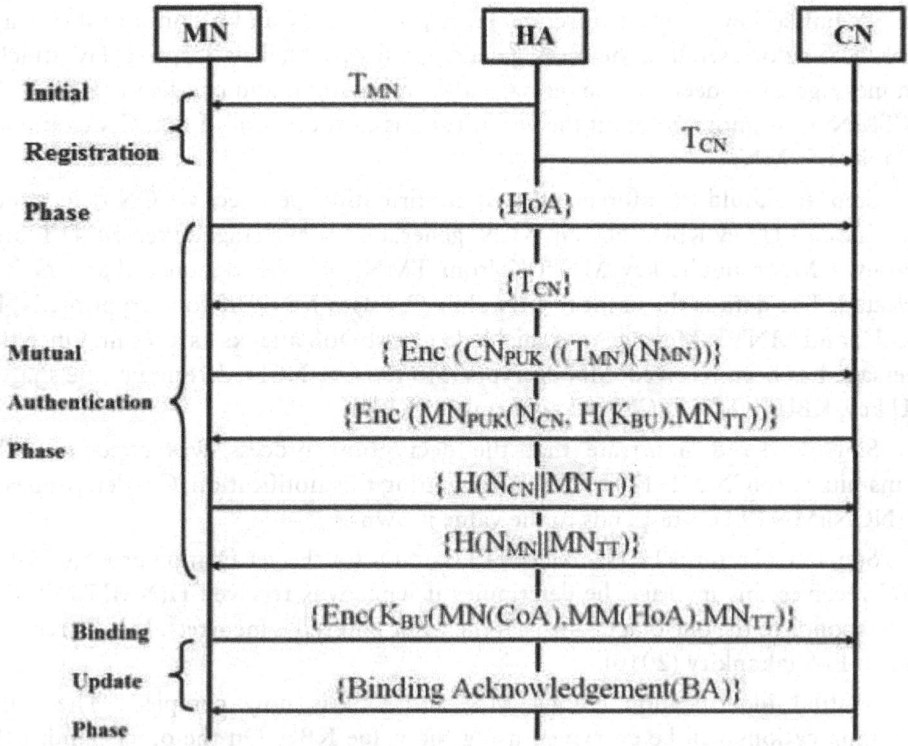

Figure 2: *Phases of the on-demand token-based binding update authentication mechanism.*

The proposed secure on-demand binding update authentication mechanism is composed of three phases: initial registration, mutual authentication and binding update. The nodes that participated in this mechanism are MN, HA and CN. But in the future, after MN performs handover, for BU authentication, only MN and HA participate. Figure 2 depicts phases of the on-demand token-based binding update authentication mechanism.

3. Mutual Authentication Phase

In this section, it is assumed that each of MN and CN has gotten TMN and TCN from HA, correspondingly. The MN and CN are now exchanging tokens for mutual purposes. Here is how processes are broken down into steps:

Table 2 and Table 3 both MN's and CN's validity may................

Step 1: The MN broadcasts a request signal, including its HoA, to the CN as soon as it begins to access the MIPv6 Networks.

Step 2: Next, in order to let MN know of its existence, CN replies to MN using a communication that contains its token TCN. When MN receives TCN, it confirms the Sign (HA) of TCN. MN disregards this TCN if Sign (HA) passes the validation; if it succeeds, MN validates the ET_CN of the TCN in Kavitha *et al.* (2010).

Step 3: Should the aforementioned validation be effective, MN will obtain TCN's public key in order to scramble its token TMN and a nonce NMN using CN PUK before sending the message encrypted to CN. This stops replay attacks on messages. CN decrypts the message after receiving it and checks the Sign (HA) in TMN. CN ignores TMN if the verification is unsuccessful. If not, CN examines TMN's ET_MN.

Step 4: Should the aforementioned confirmation be effective, CN determines the shared BU key KBU =NMN|| NCN, generates an exchange token MNTT, and retrieves MN's public key MNPUK from TMN. NCN is a nonce that CN has selected. The data is then sent to MN after CN uses MNPUK to encrypt NCN, H (KBU) and MNTT. Messages are shielded against DoS attacks as a result. Once the message has been received, MN decrypts it to retrieve NCN, determines the shared BU key, KBU=NMN||NCN and confirms H (KBU).

Step 5: To demonstrate that the decryption process was effective, MN transmits H (NCN|MNTT) to CN. After getting this notification, CN determines if H(NCN||MNTT) corresponds to the value it owns.

Step 6: CN returns H (NMN||MNTT) to MN for the verification process. When MN receives this message, he determines if what was received H(NMN||MNTT) corresponds to the one that he owns. If the value entered is incorrect, MN disregards this in El Shakankiry (2010).

Mutual identification among MN and CN is now complete. Then, BU communications will be encrypted using the value KBU. On the other hand, MN may demonstrate its legitimacy and wander from its home network to a foreign network by using KBU along with its transfer identifier, MNTT in Zhu *et al.* (2011).

4. Binding Update Phase

If the mutual authentication phase is completed successfully, MN sends a BU message to the CN to inform its current location that is CoA. In this MN encrypts the MN's CoA and MNTT by the shared BU key KBU.

MN (CoA)-> CN {Enc(KBU(MN(CoA),MN(HoA),MNTT)

Upon receiving the BU message CN first checks if the message is secured or not. Then decrypts the BU message and verifies its authenticity, confidentiality, integrity and ownership and readability.

- Authenticity
- Confidentiality
- Integrity
- Ownership and Reach ability

5. Mutual Authentication

The MN and CN nodes that are connected check each other's validity during the mutual verification process. MN and CN swap tokens in the suggested immediately token-based bindings, updating authentication technique and cross-verify one another's tokens [20]. Both MN's and CN's validity may be verified using the electronic signature of HA. Additionally, MN encodes its token and data for authentication using CN's public key, making it impossible for anybody other than CN to decode this communication and get the plaintext. As a result, it is challenging for an attacker to gain the identification data (like NMN) and produce the appropriate reply response (like H(NMN). If CN provides the proper H (NMN), then MN can validate CNs authenticity in (Vishwajit and Barbudhe, 2013).

- Forward and Backward Security
- Protection against Man-in-the-middle threat
- Protection against False Binding Update threat
- Protection against Denial of Service Attack
- Performance analysis

Table 2: Comparisons of binding update authentication protocols.

Protocol	RR	CAM	EBU	ECBU	Proposed
Mutual Autho.	No	No	No	Yes	Yes
Message Integrity	No	Yes	No	Yes	Yes
Location Auth.	Yes	Yes	Yes	Yes	Yes
Fast Handoff	No	No	Yes	Yes	Yes
Use of HA	Yes	No	Yes	Yes	Yes
Consideration MN	No	No	No	Yes	Yes
Use Public Key	No	CGA	Credit Autho	DH key	ECC PK

Table 3: Computation time (ms).

No.of Nodes	5	10	15	20	25	30
Return Routability (RR)	5.0471	5.0478	5.0462	5.0538	5.0543	5.0466
Child Proof Authentication (CAM)	5.2329	5.2333	5.2321	5.2331	5.2347	5.232
Early Binding Update (EBU)	5.3319	5.3322	5.3329	5.3337	5.3344	5.3331
Extended Certificate-Based BU	5.7258	5.7263	5.7269	5.7274	5.7279	5.7271
Proposed Mechanism	5.7314	5.7314	5.7315	5.7315	5.7315	5.7315

6. Summary and Discussion

In MIPv6 whenever MN changes its location, it sends a BU message to the CN to inform its CoA. But this BU message suffers from various threats and this is the most important security problem in MIPv6. In this work proposed have analysed the various binding update threats and introduced the feature in order to solve the security issue of the MIPv6. It proposed a secure on demand token-based binding update authentication mechanism. The proposed mechanism shows significant effectiveness in protection against BU threats and also it satisfies all the security requirements.

References

[7] Adnan, J. J., and Shamala, S. (2018). A comprehensive survey of the current trends and extensions for the proxy mobile IPv6 protocol. *IEEE Systems Journal*, 12(1), 1065–1081.

[8] Moravejosharieh, Modares, H., and Salleh, R. (2012). Overview of mobile IPv6Security. *In Proceedings of the 3rd Int'l Conf on Intelligent Systems, Modelling andSimulation* (*ISMS*), 584–587.Antonio, J. J., and David, F. (2014). Lightweight MIPv6 with IPSec support. *International Journal Mobile Information Systems*, 10(1), 37–77.

[9] Rossi, (2013). Secure route optimization for MIPv6 using enhanced CGA and DNSSEC. *IEEE System Journal*, 7, 351–352.

[10] Barbudhe, Vishwajit, B., and Chitra, D. (2013). Mobile IPv6 route optimization Protocol. *International Journal of Computer Trends and Technology* (*IJCTT*), 4(7), 2087–2092.

[11] Barbudhe, Vishwajit, B., and Chitra, D. (2014). Comparative analysis of security mechanisms of mobile IPv6 threats against binding update, Route Optimization and Tunneling. *IEEE 6th International Conference on Adaptive Science & Technology* (*ICAST*), 1–7.

[12] Barbudhe, Vishwajit, B., and Chitra, D. (2015). Comparison of mechanisms for reducing handover latency and packet loss problems of route optimization in MIPv6. *In IEEE International Conference on Computational Intelligence & Communication Technology* (*CICT*), 323–329.

[13] Barbudhe, A., Vishwajit, B., and Chitra, D. (2017). Review: security mechanism of Mobile IPv6 threats against binding update. *International Journal for Research in Applied Science & Engineering Technology* (*IJRASET*), 5(IX), 665–672.

[14] Barbudhe, A., Vishwajit, B., and Chitra, D. (2017). Survey on route optimization mechanism in network mobility. *International Journal of Engineering Development and Research* (*IJEDR*), 5(3), 1051–1055.

[15] Bo, C., Elz, R., and Kamolphiwong, R. (2010). Mobile IPv6 without a home agent. *In Proceedings of Electrical Engineering/Electronics, Computer, Telecommunications and Information Technology* (*ECTI-CON*), 904–908.

[16] Chitra, D. (2012). A robust secured mechanism for mobile IPv6 threats. *International Journal of Engineering Research and Applications* (*IJERA*), 2(6), 918–921.

[17] Daemon, S., and Vishal, S. (2017). Secure and efficient protocol for route optimization in PMIPv6-based smart home IoT networks. *IEEE Access*, 5, 11100–11117.

[18] El Shakankiry, O. (2010). *Securing home and correspondent registrations in mobile IPv6 networks*. University of Manchester, Manchester, England.

[19] Faisal Al Hawi. (2013). Secure framework for the return routability procedure in MIPv6. *In Proceedings of IEEE International Conference on Green Computing and Communications and IEEE Internet of Things and IEEE Cyber, Physical and Social Computing*, 1387–1391.

[20] Kaufman, C., Hoffman, P., Nir, Y., and Eronens, P. (2010). Internet key exchange protocol version 2 (IKEv2). IETF RFC 5996, September.

[21] Kavitha, D., Sreenivasa, M., and Zahoor-ul-Huq, S. (2010). Security analysis of binding update protocols in route optimization of MIPv6. *Paper presented at the International Conference on Recent Trends in Information, Telecommunication, and Computing* (ITC 2010), Kochi, India.

[22] Perkins, (2010). IP mobility support for IPv4: revised. Request for Comments – 5944, Internet Engineering Task Force (IETF), November.

[23] Perkins, Ed., Johnson, D., and Arkko, J. (2011). Mobility support in IPv6. A survey of mobility support on the internet. Request for Comment 6275, Internet Engineering Task Force, July.

[24] Vishwajit, K. B., and Barbudhe, A. (2013). Mobile IPv6: threats and solution. *International Journal of Application or Innovation in Engineering & Management* (*IJAIEM*), 2(6), 265–268.

[25] Zhang, C. (2016). An efficient CGA algorithm against DoS attack on duplicate address detection process. *In Proceedings of IEEE Wireless Communications and Networking*, 1–6.

[26] Zhu, Wakikawa, R., and Zhang, L. (2011). A survey of mobility support in the internet. Request for Comment-6301, Internet Engineering Task Force, July.

Chapter 30

Authentication of Video: Noval Approach for Identifying Video Forgery Using Correlation Coefficients

Mukesh D. Poundekar[1] *and S. P. Deshpande*[2]

[1]Research Scholar, PG Department of Computer Science & Technology, Hanuman Vyayam Prasarak Mandal, Amravati, Maharashtra, India

[2]Professor, PG Department of Computer Science & Technology, Hanuman Vyayam Prasarak Mandal, Amravati, Maharashtra, India

Abstract: Past two decades, the authentication of visual media has emerged as a pivotal area of research, focusing on the development of methods and tools to ascertain the authenticity of digital media. Detecting forged content has garnered significant attention, particularly within the context of videos, given their critical role in scenarios like road accidents and court proceedings. In this paper, an innovative, cost-effective video forgery detection algorithm is introduced. This approach leverages correlation coefficients among video frames and embeds them as encrypted data within each frame. Through experimentation, the final results demonstrate the remarkable effectiveness of the suggested approach based on visual quality PSNR (76.89) and result validation. The algorithm exhibits the capability to identify tampering, even when subjected to simple and subtle attacks with minimal impact.

Keywords: Video authentication, data embedding, video tampering, video forgery.

1. Introduction

The advancement of technology has elevated digital multimedia images, audio and videos to the forefront of data transmission and connection-building across various sources. Due to its ease of collection, delivery and storage, visual data holds the capacity to convey information, emotions, verification and more through a wide array of methods. This has rendered it a highly successful form of interaction, dating back to ancient times. Alongside its application in video monitoring and court evidence, the progress of visual (video) technology encompassing compression, modification, retrieval and transmission has yielded benefits across multiple domains, including social, technological, academic, healthcare, entertainment and industrial sectors suggested by Ramos *et al.* (2020). Multimedia images and videos,

DOI: 10.1201/9781003527442-30

prevalent on diverse video-sharing and social media platforms such as YouTube, TeacherTube, Google Videos, Instagram, etc., are becoming an essential facet of contemporary life, highlighting their unparalleled importance in the present milieu. While visible (video) data finds numerous positive applications across various fields, it also presents some drawbacks, including its potential for misuse. One such aspect is video tampering, wherein an individual manipulates genuine recordings to produce modified versions for illicit purposes such as video-based forgery, altered depiction of illegal activities, defamation and more. Although image manipulation has been practiced since the inception of imaging itself and has led to significant instances of manipulated images, video manipulation is a burgeoning phenomenon. With the availability of advanced video editing tools, forgers can alter authentic videos to create deceptively authentic-looking counterfeit ones by Mandelli *et al.* (2022). In cases where video evidence is pivotal, such as in surveillance footage or legal disputes, the authenticity or originality of the video becomes paramount. This is where video forensic research institutions and experts play a vital role, employing visual examination and scrutiny of potential evidence, clues or traces left behind by the forger during the editing process by Sandoval Orozco *et al.* (2019).

The efficacy of such forensic examinations hinges on the degree of skill with which the forger has tampered with the videos. In instances where the forger has left minimal or no traces, forensic professionals face challenges in verifying the video's authenticity and pinpointing manipulation sites. The field of digital video forensic analysis, a subset of digital media protection, is still evolving and aims to establish standardised techniques for assessing video credibility and identifying manipulation cues suggested by Bourouis *et al.* (2020). Conversely, the realm of anti-forensics, a component of multimodal cybersecurity, empowers forgers to counter current forensic practices and techniques. The primary goal here is to carry out manipulation that visually resembles the original, effectively deceiving even untrained observers. While there are tactics against forensic analysis available in this domain, they often serve as guidance for the forger.

The paper's organisation is structured as follows: Section 2 presents an overview of prior research on video authentication and video tampering detection. Section 3 outlines the comprehensive methodology adopted. In Section 4, an analysis of experimental results for the proposed method is discussed. Finally, Section 5 addresses the conclusion and outlines future directions.

2. Related Work

The anti-forensic approach for source anonymization and forgery in MP4 video was suggested by Sandoval Orozco *et al.* (2019). To measure the performance of the suggested approach for MP4 videos identify the source to make it seem recorded by another model. The author works on video forgery detection but there is no evidence to find the origin of captured video. A complete assessment of the most current works on visual forgery identification was provided by Bourouis *et al.* (2020), along with a straightforward soft categorisation. A more thorough evaluation and future study opportunities are provided after that. This project gives investigators a different way to learn about the current state of the region and helps them improve and assess their video forensics methods. The hybrid steganographic solution designed by Begum *et al.* (2020) satisfies the fundamental design requirements for a range of circumstances compiled and assessed the complexity of several previously presented alternatives after a brief research investigation. The method proposed by Yang *et al.* (2020) is effective, affordable and able to clearly explain its findings. To carry out this operation, the vectorial description of the video container design was done using the machine learning classifier such as a decision tree. The 7,500 digital files, including diverse software-manipulated content and films posted on social media networks, were evaluated by researchers. The recommended approach achieves a 97% accuracy in detecting spotless from modified videos by recognising the editing elements when the video is reduced without re-encoding or resampling to the size of a clip.

According to Zheng *et al.* (2020), an image hash may be produced by merging invariant visual traits with a Bernoulli random space that is represented using a physically unclonable function (PUF) approach. Although it differs from it, the camera's tamper-resistant randomised PUF answer requires a difficult task, that the picture capture timestamp provides. A CMOS image sensor-based PUF modelled using a 180-nm TSMC design as well as a customised CASIA dataset is used to verify the recommended hash. The enormous collection of movies acquired specifically for designing brand identification methods was proposed by Hosler *et al.* (2019) over the video verification and camera recognition dataset. The video collection contains over 11,000 videos from 47 devices made by 37 distinct camera manufacturers. The paper revealed the characteristics, design and collection method of video that incorporates tagged movies for evaluating camera brand identification methods. Additionally, employ deep learning techniques to provide benchmark camera-type detection results on such video evaluation. A trustworthy steganographic technique based on the watermark's texture matching was proposed by Jain *et al.* (2018).

A robust hybrid blind steganography approach for copyright protection of multimedia pictures was presented by Hamidi *et al.* (2018). Such a hybrid technique makes use of the DFT and DCT transform. Bringing these two technologies together will enhance perceptual stability and quality. The Arnold transforms applied to the signature prior to data insertion enhance the indicated security approaches. The research made use of photographs that were realistic and textured. The author suggested data embedding techniques but there is no mechanism suggested for video authentication. It was suggested by Harran *et al.* (2018) to provide a secret key for a video that may be embedded to frame with related metadata. There may be references to the giving organisation in the information contained in the file.

A novel steganographic strategy that integrates the DWT, DCT and single value decomposition was just presented by Assini *et al.* (2018). Using this technique, a clinical image may be included with a hidden image watermark. The covered healthcare picture is divided using the third stage of the DWT coefficients, and then it is transformed using DCT and SVD. The steganographic picture receives an equivalent treatment. The high-frequency sub-bands of the third level DWT of the single value of the cover picture are filled with the single values of the watermark. It, therefore, is possible to increase the durability of the steganography approach without sacrificing the image resolution such as the watermark in specific locations. The experimental results show that the proposed hybrid approach offers a remarkable balance between perceptual quality and dependability against a range of threats when compared to conventional strategies. The DWT, SVD and PSO were used for image steganography suggested by Takore *et al.* (2018). The parts of hosted photographs with the highest pixel intensities are the best places for watermark insertion. The PSO approach produces the best scaling factors, enhancing interpretation and robustness. The experimental results reveal that the proposed technique satisfies the necessary dependability and complexity requirements, making it well-suited for multimedia information.

3. Proposed Methodology

With the purpose of determining the location of inter-frame video tempering, the suggested solution employs passive techniques. In contrast, the collection of tracking linear attributes is retrieved from a proposed model of tube fibre architectural framing building for comparison with its succeeding groups rather than spatial evaluating the entire pixel correlation across every subsequent video frame by Abdulhussain *et al.* (2019) and Mahmmod *et al.* (2020).

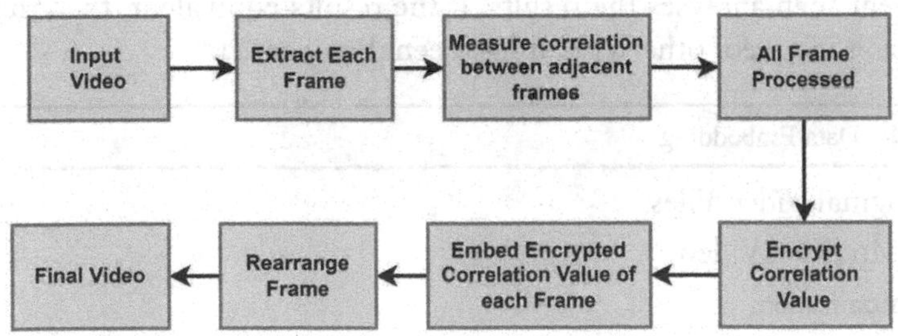

Figure 1: Data embedding process.

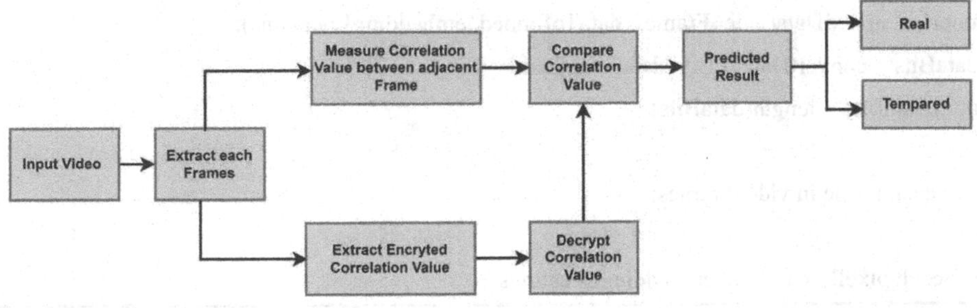

Figure 2: Video tempering detection process.

3.1. Data Embedding Process

The embedding procedure entails three primary parts as shown in Figure 1; computing the correlation values, implementing the AES algorithm to encode the coefficient values in each frame of video and utilising bit substitutions to conceal the encryption's results. Beginning with the subsequent frame, the correlation is computed between each pair of consecutive frames. Take a look at frame (p) in an n-frame video stream. The definition of the frame correlation is:

$$\textbf{Frame}\ (P_i, P_{i+1}) = \frac{\sum\sum(P_i, P_{i+1})}{\sqrt{(\sum\sum(P_i + \overline{P_i})^2)(\sum\sum\sum(P_{i+1} + \overline{P_{i+1}})^2)}}$$

for i = 2 to n-1, and are mean of respectively.

Utilising the AES-128 technique, correlational data for neighboring frames are encoded. The block size for encoding is 128 bits, or 16 bytes for text, and has a randomised 128-bit key size. Following that, encoded correlation scores are inserted into each frame of video using the least significant bits that randomised positions.

3.2. Video Tempering Detection

As seen in Fgure 2, the receiver end determines the correlation values of the received video, retrieves the encoded correlation values from a random point in each frame, and uses the symmetrical key of the AES method to decode these values. The recipient then analyses the results. If the results equivalent, there has been no manipulation in video; otherwise, it has been discovered.

Pseudocode: Data Embedding

Input: Original Video Files

Output: Modified Video

Step 1: Initialization:

- Convert the data to be embedded into a binary format.
- Determine the number of bits needed to embed the data (according to the data size).

Step 2: Embedding Process

```
function embedData(videoFrames, dataToEmbed, embeddingLocations):
 dataBits = convertDataToBits(dataToEmbed)
 numDataBits = length(dataBits)

  for each frame in videoFrames:

 for each pixelLocation in embeddingLocations:
```

```
pixel = frame[pixelLocation]
for bitIndex from 0 to numDataBits –1:
dataBit = dataBits[bitIndex]
replaceLeastSignificantBit(pixel.colorComponent, dataBit)
bitIndex++
frame[pixelLocation] = modifiedPixel

return modifiedVideoFrames
```

Step 3: Modified Frame Generation:

- The resulting modified frame has the embedded data.

4. Experimental Result

Any kind of video extension can be used with the recommended algorithm. The MP4 types of files have been utilised in this research. The effectiveness of the suggested method has been evaluated on five films with varying numbers of frames and frame lengths. The two key factors taken into account in the outcomes were perception quality and resilience tests. To measure the perceptual quality of the video, several evaluation parameters such as MSE, PSNR and SSIM are considered.

Figure 3 and figure 4 shows the measure perceptual quality of video over the MSE, SSIM, and PSNR.

$$MSE = \frac{1}{N}\sum_{i=1}^{N}(X_i - \bar{X}_i)^2$$

where *N* is a total number of frames in video, is the output video frame and is the input video frame.

$$PSNR = 10log_{10}\left(\frac{R^2}{MSE}\right)$$

where *R* is fluctuation in input video frames

$$SSIM\ (X,Y) = \frac{(2\mu_x\mu_y + C_1)(2\sigma_{xy} + C_2)}{(\mu_x^2 + \mu_y^2 + C_1)(\sigma_x^2 + \sigma_y^2 + C_2)}$$
$$C_1 = (k_1, R)^2$$
$$C_2 = (k_2, R)^2$$

where *R* is the range of pixel values in frame. By default, the k1 and k2 values of 0.01 and 0.03, respectively, are assigned.

Table 1: Measure the perceptual quality of video.

Videos Files	Frames in Video	Size	MSE	PSNR	SSIM
1	135	720*480*0.5	0.0015	76.89	0.9973
2	140	1920*1080*0.5	0.0321	65.84	0.9974
3	150	1920*1080*0.3	0.0114	68.45	1.00
4	400	1280*720*0.3	0.1459	70.63	0.9945
5	750	720*480*0.5	0.0094	72.94	0.9958
6	800	1920*1080*0.5	0.0789	67.99	0.9973
7	850	480*550*0.5	0.0985	69.78	0.9912

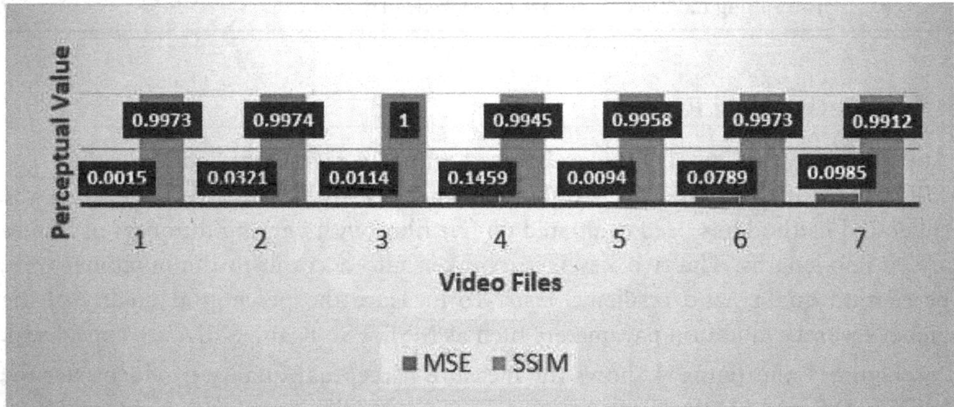

Figure 3: Perceptual quality of videos in term of MSE and SSIM.

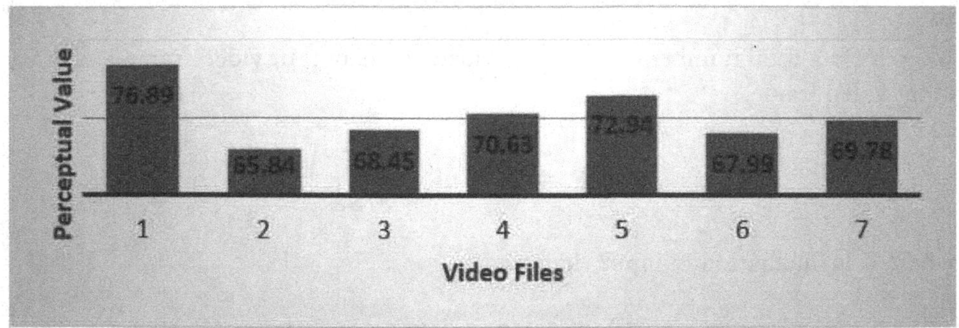

Figure 4: Perceptual quality of videos in terms of PSNR.

4.1. Result Validation

To validate the proposed model, apply various attacks such as insertion, deletion and modification to the video. Perceptual quality metrics such as MSE, PSNR and SSIM are employed to assess the impact of data embedding on the visual fidelity of the video. These metrics provide quantifiable measures of distortion and degradation caused by the embedding process. Comparing the perceptual quality scores of the original and embedded videos can evaluate the extent to which the embedded data affects the viewer's visual experience. Result validation not only helps optimise embedding techniques but also guides the selection of appropriate embedding strategies that minimise perceptual distortions while successfully concealing the data within the video frames.

Table 2: Validation of result.

Video Files	Type of Attack	MSE	PSNR
	Insertion	**0.0756**	**69.12**
1	**Deletion**	**0.0742**	**70.41**
	Modification	**0.0765**	**71.56**
	Insertion	**0.0713**	**71.78**
2	**Deletion**	**0.0725**	**68.45**
	Modification	**0.0731**	**69.98**

Figure 5: *Comparative analysis of MSE and PSNR after applying attack.*

Figure 5 depicts a comparative analysis between the original video files and the attacked video files in terms of MSE and PSNR values. It is evident that the MSE values increase, and the PSNR values decrease after applying the attacks to the original video files.

5. Conclusion

This study introduces an innovative video tampering detection algorithm, which revolves around determining correlation coefficients among each frame and subsequently embedding them into randomly selected positions within the initial frame. To bolster security, these correlation coefficients are encrypted through the utilisation of the AES algorithm prior to embedding. The experimental findings underscore the algorithm's robust detection capabilities, even when exposed to diverse attacks such as frame insertion, deletion and modification, all while maintaining a commendable level of visual frame quality. The future direction endeavours are centered on refining the precision of tamper localisation, whether it pertains to specific regions within frames or spans across the entire video stream.

References

[1] Abdulhussain, S., Mahmmod, B., Saripan, M., Al-Haddad, S., and Jassim, W. (2019). A new hybrid form of krawtchouk and tchebichef polynomials: Design and application. *J. Math. Imaging Vis.*, 61(4), 555–570.

[2] Assini, A., Badri, A., Badri, K., Safi, A., and Sahel A., (2018). A robust hybrid watermarking technique for securing medical image. *International Journal of Intelligent Engineering and Systems*, 11(3), 169–176.

[3] Begum, M., and Shorif Uddin, M. (2020). Analysis of digital image watermarking techniques through hybrid methods. *Advances in Multimedia*, 12. https://doi.org/10.1155/2020/7912690.

[4] Bourouis, S., Alroobaea, R., Abdullah, M., Andejany, M., and Rubaiee S. (2020). Recent advances in digital multimedia tampering detection for forensic analysis. *Symmetry*, 12(11), 1811. https://doi.org/ 10.3390/sym12111811.

[5] Hamidi, M., Haziti, M., Cherifi, H., and Hassouni M., (2018). Hybrid blind robust image watermarking technique based on DFT-DCT and Arnold transform. *Multimedia Tools and Applications*, 77(20), 27181–27214.

[6] Harran M., Farrelly W., and Curran K., (2018). A method for verifying integrity & authenticating digital media. *Applied Computing and Informatics*, 14(2), 145–158.

[7] Hosler, B., Zhao, X., Mayer, O., Chen, C., Shackleford, J., and Stamm, M., (2019). The video authentication and camera identification database: a new database for video forensics. *In IEEE*, 7, 76937–76948. https://doi.org/10.1109/ACCESS.2019.2922145.

[8] Mahmmod, B., Abdul-Hadi, A., Abdulhussain, S., and Hussien, A., (2020). On computational aspects of Krawtchouk polynomials for high orders. *J. Imaging*, 6(8), 81.

[9] Mandelli, S., Bonettini, N., and Bestagini, P., (2022). Source camera model identification. In: Sencar H., Verdoliva L., and Memon N., (eds) multimedia forensics. *Advances in Computer Vision and Pattern Recognition*. Springer, Singapore. https://doi.org/10.1007/978-981-16-7621-5_7.

[10] López R., Luengo E., Orozco A. L., and Villalba L. J. (2020). Digital video source identification based on container's structure analysis. *In IEEE*, 8, 36363–36375. https://doi.org/10.1109/ACCESS.2020.2971785.8., , and,vsaftbpsne.*I*–

[11] Takore, T., Kumar, P., and Devi, G., (2018). A new robust and imperceptible image watermarking scheme based on hybrid transform and PSO. *International Journal of Intelligent Systems and Applications*, 10(11), 50–63.

[12] Yang, P., Baracchi, D., Luliani, M., Shullani, D., Ni, R., Zhao, Y., and Piva, A. (2020). Efficient video integrity analysis through container characterization. *In IEEE Journal of Selected Topics in Signal Processing*, 14(5), 947–954. https://doi.org/10.1109/JSTSP.2020.3008088.

[13] Zheng, Y., Cao, Y., and Chang, C., (2020). A PUF-based data-device hash for tampered image detection and source camera identification. *In IEEE Transactions on Information Forensics and Security*, 15, 620–634. https://doi.org/10.1109/TIFS.2019.2926777.

Chapter 31

Overspeed Alert System for Vehicles to Avoid E-Challan on Highways using Machine Learning Algorithm

Yogesh N. Thakare[1], *Anushka Khandelwal*[1], *Muskan Kumar*[1], *Muskan Kumar*[1], *and Jyotiraditya Purohit*[1]

[1]Department of Computer Science and Engineering (AIML), Shri Ramdeobaba College of Engineering and Management, Nagpur, India

Abstract: The overspeed warning system is a cutting-edge endeavour aimed at improving traffic safety and reducing accidents caused by speeding vehicles. The system uses computer vision techniques and intelligent algorithms to identify, recognise and warn drivers to traffic speed signs and speed limits. A broad collection of traffic speed signs is collected and annotated as the system's first stage. The dataset is then utilised to train a deep-learning model for sign recognition and identification. The technology detects and locates traffic speed signs in real time using cutting-edge convolutional neural networks. The technology uses powerful image processing methods to extract speed restriction information from the observed signs after successfully detecting the speed indicators. The vehicle's current speed, as assessed by a speed detection module, is then integrated with this data to determine whether the speed limit has been exceeded. The proposed system is intended to be adaptive, scalable and suitable to a wide range of traffic scenarios. It can be installed as a standalone device in vehicles or integrated into existing infrastructure. Extensive tests and evaluations are utilised to establish the system's robustness across a variety of weather situations, lighting configurations and road kinds. Furthermore, the system can be enhanced with new functions such as pedestrian detection, traffic signal recognition and congestion monitoring, all of which would improve overall road safety.

Keywords: speed limit, accident, road safety, deep learning, traffic control.

1. Introduction

1.1. Overview

The rapid increase in traffic congestion, combined with the worrisome rise in road accidents, has highlighted the critical need for improved traffic management

DOI: 10.1201/9781003527442-31

systems in recent years. Speeding is a major contributor to accidents, injuries and fatalities on the road, making it a vital concern around the world (Kaur *et al.*, 2019). To address this problem, reliable detection and recognition of traffic speed signs is critical in assisting drivers in adhering to speed restrictions and avoiding potential fines.

Manual patrols and radar-based speed cameras, for example, have limits in terms of coverage, efficacy and the capacity to deliver real-time feedback to drivers. Furthermore, because these systems rely significantly on human intervention, they are prone to errors, inconsistencies and delays. As a result, there is an increasing demand for automated and intelligent systems that can detect and understand speed signs in real time, giving drivers with timely information and alerts to ensure compliance with traffic regulations.

1.2. Motivation

Road accidents caused by fast automobiles continue to be a major concern around the world, resulting in serious injuries and fatalities. The failure of drivers to recognise and obey traffic speed signs contributes considerably to this problem. Manual patrols and radar-based speed cameras, for example, have limits in terms of coverage,

efficacy and the capacity to deliver real-time feedback to drivers (Sermanet *et al.*, 2011). As a result, there is an urgent need for an automatic and intelligent system that can detect and understand speed signs in real time, allowing drivers to get timely alerts and follow traffic laws.

The goal of this proposed system is to use advances in computer vision and machine learning techniques to provide an innovative solution that tackles the shortcomings of existing speed sign detection and recognition systems. We can acquire high-quality images or videos of the road environment by putting a high-definition camera within the car and integrating the necessary technology, allowing for more accurate and reliable speed sign identification. This, in turn, allows drivers to get real-time speed restriction alerts, promoting safer driving habits and improving overall road safety.

1.3. Objective

The primary goal of the proposed system is to create a reliable and efficient traffic speed sign detection, recognition and alarm model designed exclusively for four-wheelers. The model will use high-definition video footage to accurately recognize and understand speed signs. By incorporating this model into automobiles, drivers will receive real-time speed limit alerts, assuring compliance and contributing to improved road safety (Kaur *et al.*, 2019). Several critical components must be addressed in order to attain this goal. First, we must create an algorithm capable of detecting speed signs in a variety of lighting and weather circumstances. The system should be able to deal with issues like occlusions, blur and distortion that are frequent in real-world driving circumstances. Second, a speed recognition module must be created to appropriately interpret detected signs and extract necessary information, such as the prescribed speed limit. This module should be able to handle many sorts of speed signs, including shapes, colours and text formatting (Wu *et al.*, 2020). Furthermore, to achieve real-time performance, the model must be smoothly connected with the vehicle's hardware. This integration entails capturing and analysing camera footage, conducting the computations required for speed sign detection and recognition and generating timely alerts for the driver. Finally, to assure the model's dependability and effectiveness, it must be rigorously tested and evaluated under various road situations and driving conditions.

2. Literature Survey

In recent years, the pressing need for enhanced traffic management systems has become increasingly evident due to the escalating challenges posed by traffic congestion and road accidents. This issue is not only a significant concern for urban planners and transportation authorities but also a matter of great importance to everyday commuters and motorists. Among the various factors contributing to the complexity of traffic management, one of the most critical and persistent problems is speeding, which remains a leading cause of accidents and fatalities on roadways around the world. Recognising the gravity of this issue, researchers and engineers have been tirelessly working to develop innovative solutions to mitigate its impact and create safer road environments for all.

2.1. Speeding – A Persistent and Deadly Problem

Speeding is a multifaceted problem that encompasses not only the excessive speed of vehicles but also the failure to adhere to posted speed limits and the inability to adjust speed according to changing road and weather conditions. It is a well-established fact that higher speeds reduce a driver's reaction time and increase the severity of accidents. The consequences of speeding can be catastrophic, leading to not only a higher likelihood of collisions but also more severe injuries and fatalities. In response to this pressing issue, the development of intelligent traffic management systems has become a paramount concern for governments and organisations worldwide.

2.2. Recognition of Traffic Speed Signs

One of the key elements in addressing speeding is the accurate and timely recognition of traffic speed signs. This technology is fundamental to informing drivers of the speed limits in a particular area and ensuring that they comply with these limits. Over the years, significant strides have been made in the field of computer vision and image recognition, leading to the development of advanced systems capable of automatically detecting and deciphering traffic speed signs (Li *et al.*, 2019).

Computer vision algorithms, often powered by deep learning techniques, have made substantial progress in the accurate detection and interpretation of speed limit signs. These systems can analyse real-time video footage from cameras placed at strategic locations along roadways. By processing this visual data, they can identify speed limit signs, extract relevant information and communicate it to drivers through various means, such as dashboard displays, heads-up displays (HUDs) or even smartphone apps. The ability to promptly recognise and relay this critical information to drivers is a significant step towards encouraging compliance with speed limits and reducing the incidence of speeding-related accidents.

2.3. *Development of Intelligent Systems for Alerting Drivers*

In addition to recognising traffic speed signs, the development of intelligent systems for alerting drivers about speed restrictions is another essential aspect of mitigating speeding-related issues. These systems leverage a combination of hardware and software solutions to provide real-time feedback to drivers, helping them stay within the prescribed speed limits. Advanced driver assistance systems (ADAS) have gained prominence in recent years, offering features

such as adaptive cruise control, lane-keeping assistance and speed limit recognition. ADAS can use GPS data, onboard cameras and sensors to monitor a vehicle's speed and location relative to the road network. When a driver exceeds the speed limit, these systems can issue visual and auditory warnings, prompting the driver to slow down and comply with the speed restriction.

Moreover, the integration of artificial intelligence (AI) into these systems enables more sophisticated functionalities, such as predictive speed limit adjustments based on traffic conditions, weather and road characteristics. These AI- driven systems can provide dynamic speed recommendations to drivers, considering factors like congestion, accidents and construction zones.

Kaur *et al.* (2019) conducted an extensive review of traffic sign detection and recognition techniques. They explored the application of various algorithms and machine learning approaches, emphasising the significance of robust feature extraction and classification methods. Their study shed light on the challenges of real-world scenarios, such as varying lighting conditions and occlusions, and the need for adaptable models that perform effectively across diverse environments.

Sermanet *et al.* (2011) proposed a breakthrough in traffic sign recognition through multiscale convolutional networks. This pioneering work demonstrated the potential of deep learning in detecting and interpreting traffic signs. The authors highlighted the advantages of convolutional neural networks (CNNs) in capturing hierarchical features, which significantly improved recognition accuracy.

Li *et al.* (2019) delved into the synergy between deep learning and visual saliency for traffic sign detection and recognition. Their approach leveraged the human visual attention mechanism to prioritise regions of interest within images. By combining deep learning techniques with visual saliency, the authors achieved higher accuracy in identifying traffic signs and their associated meanings.

Wang *et al.* (2019) contributed to the field by enhancing the Faster R-CNN architecture for traffic sign detection. Their research focused on refining the region proposal network to improve localisation accuracy. The study highlighted the importance of addressing localisation precision, especially in complex road scenarios.

Zhang *et al.* (2018) investigated traffic sign detection using the popular YOLOv3 architecture. Their work showcased the efficiency of YOLO's object detection framework in real-time scenarios. The authors demonstrated how the YOLO architecture could effectively identify and classify traffic signs within cluttered scenes.

Ding and Lin (2017) made significant strides by implementing deep convolutional neural networks for traffic sign detection and recognition. Their work emphasised the necessity of well-designed architecture to handle the intricacies of traffic sign diversity. The study contributed insights into network design considerations and training strategies.

Eric. M. Masatu *et al.* in their paper titled "Development and Testing of Road Signs Alert System Using a Smart Mobile Phone" published by "Nelson Mandela African Institution of Science and Technology, Arusha, Tanzania" in the journal "Hindawi" discussed about the development and testing of a system for alerting drivers about road signs using a smartphone. The Haversine formula was used to measure and estimate the distance between two pairs of coordinates using the smartphone-based navigation application, Google Map. The application provides a voice alert to a needed action that enhances the driver's attention. The system didn't have the speed tracker which alerts the driver when the vehicle exceeds the speed limit.

Ms. Nethravathi *et al.* in their paper titled "Traffic Sign Recognition System Using Machine Learning" published by Visvesvaraya Institute of Technology, Bengaluru in IRJETS used CNN and SVM to predict the traffic sign. The model build was real time along with shape and color detection to find the target region. The paper though only focussed on classification and not the alert system and speed tracking.

Adonis Santos *et al.* in their paper titled "Real-Time Traffic Sign Detection and Recognition System for Assistive Driving" published by Ateneo de Manila University in ASTESJ concluded that shadow and highlight invariant method is the superior pre-processing method. Also, it stated that CNN is the best algorithm for classification, in terms of speed and accuracy. This paper compared different algorithms and described each one on surface level rather than going in-depth with one algorithm.

3. Methodology

A traffic sign recognition system captures pictures or video frames including traffic signs using cameras. To recognise and analyse these indications, it follows a set of steps:

- **Image Acquisition:** Cameras capture images or video frames of road scenes/ sign board on road side.
- **Preprocessing:** Images may be enhanced for better quality and analysis.
- **Detection:** Object detection algorithms locate potential traffic signs.
- **Feature Extraction:** Relevant characteristics like shape and color are extracted.
- **Classification:** A trained model identifies the type of each sign.
- **Post-processing:** Refinement of results by removing duplicates and false positives.
- **Interpretation:** The recognised signs' meanings are understood in the road context.
- **Alert:** Systems provide alert to drivers or autonomous vehicles for appropriate actions.

In addition to superior traffic sign recognition, our integrated technology raises the bar for road safety by seamlessly merging real-time vehicle speed monitoring. This novel function not only improves overall driving safety but also encourages strict respect to traffic laws. The system works by continuously tracking the vehicle's speed using cutting-edge sensor technology strategically placed along routes. This continuous speed monitoring procedure ensures that the vehicle's speed is constantly monitored, allowing for immediate response to any deviations from the authorised speed restrictions (Maldonado-Bascón *et al.*, 2018).

When the system comes across a speed limit traffic sign, its functionality changes to incorporate speed limit enforcement. The technology commences complete speed monitoring for the vehicle the moment it detects a relevant speed restriction sign. This entails comparing the vehicle's speed in real time against the indicated speed limit. If the vehicle's speed exceeds the established speed restriction, the system immediately activates an alert mechanism (Ding and Lin, 2017). This message is effortlessly transmitted to the driver, clearly expressing the need to slow down and conform to the current speed limits. The technology guarantees that the driver is completely aware of their speed by offering rapid and proactive warnings, reducing the likelihood of accidents and promoting responsible driving behavior (Pan and Zhang, 2020).

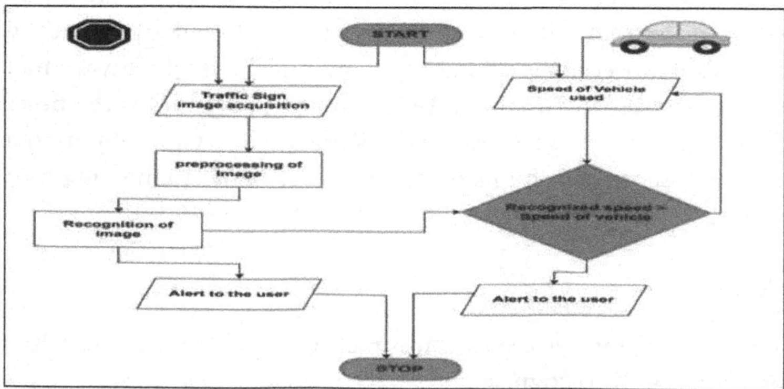

Figure 1: Flow representation of how the system works.

4. Implementation

4.1. Dataset Preparation:

Obtained diverse traffic sign images dataset with corresponding labels prepared by the University of Copenhagen containing more than 11,000+ images.

- Data Augmentation: Generated augmented training data through random transformations using Keras' Image DataGenerator for augmentation.
- Applied rotations, zooming, shifting, and flipping to enhance model's generalization.
- The dataset was partitioned into distinct subsets for various purposes: a training set comprising 7,500 images, a testing set encompassing 2,300 images and a validation set consisting of 1,800 images.

4.2. Data Pre-processing:

- Converted the images from BGR colour space (commonly used in OpenCV) to grayscale using cv2.cvtColor.Applied histogram equalisation to the grayscale image using cv2.equalizeHist.
- Histogram equalisation enhances the contrast of the image, making it more visually informative.
- Finally reshaped all the images to 32 × 32 format for better model compatibility.

4.3. Model Architecture:

- The model starts with two sets of Conv2D layers, each followed by ReLU activation, to extract features from the input images.
- MaxPooling2D layers reduce spatial dimensions and help in capturing important features.
- Dropout layers are used for regularisation, reducing overfitting by randomly deactivating neurons during training.
- The flatten layer converts the 2D feature maps into a 1D vector.
- The dense layers add fully connected neural network components for classification.
- The output layer uses softmax activation for multiclass classification:

Models summary:

Convolutional Layers: 6

MaxPooling2D Layers: 2

Dropout Layers: 2

Fully Connected (Dense) Layers: 2 Total params: 360,989

Trainable params: 360,989

4.4. Model Training:

- A batch-based model training approach was employed to efficiently optimise the neural network's performance with the following parameters:
- batch_size_val: Each training iteration involved 50 images per batch.
- steps_per_epoch_val: A total of 150 batches were processed in each epoch, effectively covering a substantial portion of the training dataset.
- epochs_val: The model underwent 10 complete passes over the training dataset, facilitating the learning process.
- The training procedure involved iteratively feeding batches of preprocessed images (X_train) and corresponding labels (y_train) through the network. Additionally, the model's performance was continuously monitored using a validation set (X_validation, y_validation) to monitor overfitting. The shuffle parameter ensured that the training data was randomly reorganised in each epoch, further enhancing model robustness.
- The training process was aimed at minimising the categorical cross-entropy loss by adjusting the model's internal weights through the backpropagation algorithm. The resulting history object captured valuable insights into the training progression, including loss and accuracy metrics, which were instrumental in evaluating the model's efficacy for traffic sign detection.

4.5. Model Evaluation:

The model was primed for assessment using the Adam optimizer with a learning rate of 0.001, thereby enabling effective weight updates during training. The primary evaluation criterion was the categorical cross-entropy loss, which quantified the disparity between predicted class probabilities and ground-truth labels. In conjunction, the evaluation encompassed the computation of the "accuracy" metric, which gauged the model's ability to correctly classify traffic sign images.

4.6. Result:

The final model achieved a training accuracy of 91.79%, while the validation accuracy reached 97.72%. Furthermore, during the final testing phase, the model exhibited an impressive accuracy of 98.13%.

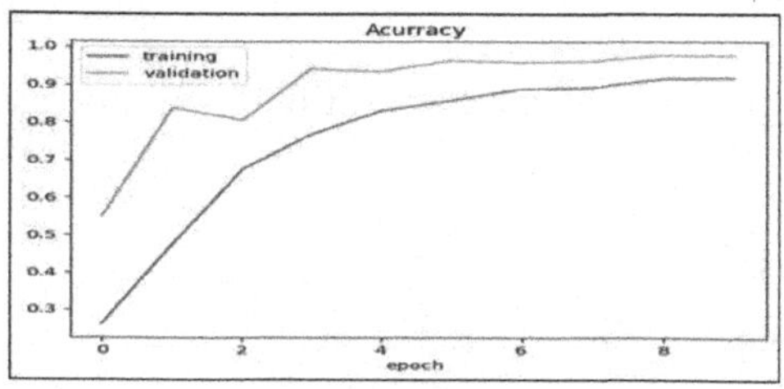

Test Accuracy: 0.9813242554664612

Figure 2: Model accuracy using machine learning.

Figure 3: Detection and alert of the speed using mobile.

4. Conclusions

To sum up, our study has been effective in creating a reliable model for spotting, identifying and warning four-wheel vehicle drivers about traffic speed signs. By assisting drivers in obeying speed limits and lowering accidents brought on by over speeding, the implemented sign detecting module exhibited substantial progress in enhancing road safety. The system handled difficulties like shifting lighting and occlusions with ease in real-world conditions. While some restrictions do exist, such as the difficulties with broken signage, they present chances for advancement in the future. The study also demonstrated the possible combination of hardware and high-resolution cameras to improve real-time performance and image quality. Making a speed recognition module, real-time detection and interaction with vehicle displays and alarms are the next phases. These developments will improve notifications' timeliness and accuracy even more. In conclusion, this research lays a solid foundation for a thorough traffic speed sign system, employing cutting-edge computer vision and clever algorithms to encourage safe driving. The model has the potential to be extended to additional traffic signs, and with continued development and practical use, it might have a substantial impact on road safety for all users.

References

[1] Ding, X. and Lin, Y. (2017). Traffic sign detection and recognition using deep convolutional neural networks. *IET Intelligent Transport Systems*, 11(4), 225–232.

[2] Kaur, M. *et al.* (2019). Traffic sign detection and recognition: a comprehensive review. *IEEE Transactions on Intelligent Transportation Systems*, 20(3), 1016–1033.

[3] Li, Y. *et al.* (2019). Traffic sign detection and recognition using deep learning and visual saliency. *IEEE Transactions on Intelligent Transportation Systems*, 20(1), 348–357.

[4] Maldonado-Bascón, S., *et al.* (2018). Traffic sign detection and recognition in the wild: A survey. *IEEE Transactions on Intelligent Transportation Systems*, 19(2), 631–643.

[5] Pan, L. and Zhang, K. (2020). Traffic sign detection based on convolutional neural network with attention mechanism. *IEEE*, (Access, 8) 89846-89856.

[6] Sermanet, P. *et al.* (2011). Traffic sign recognition with multiscale convolutional networks. *In Proceedings of the International Joint Conference on Neural Networks* (*IJCNN*), 2809–2813.

[7] Wu, W. *et al.* (2020). Traffic sign detection and recognition using YOLOv3. *In Proceedings of the International Conference on Computer Science, Information Technology and Telecommunications* (*CSITT*), 1–5.

[8] Wang, Z., *et al.* (2019). Traffic sign detection based on deep learning and improved Faster R-CNN. *IEEE* (Access, 7), 170308-170317.

[9] Xu, J., *et al.* (2020). Traffic sign recognition based on improved YOLOv3. *In Proceedings of the International Conference on Artificial Intelligence and Big Data* (*ICAIBD*), 1–6.

[10] Zhang, T. *et al.* (2018). Traffic sign detection and recognition using convolutional neural networks. *Sensors*, 18(11), 3746.

Chapter 32

Detection of Nodule like Objects Using Local Contrast Thresholding

Dnyaneshwar Kanade[1], Jagdish Helonde[2], Mangesh Nikose[2], and Prakash Burade[2]

[1]Research Scholar, School of Engineering and Technology, Sandip University

[2]Professor, School of Engineering and Technology, Sandip University

Abstract: Lung cancer is one of the leading causes of cancer deaths worldwide, per the WHO. Using image processing techniques and computer aided diagnosis (CAD) systems, the detection of lung nodules has vastly improved. A unique strategy for extracting a candidate nodule from postero-anterior chest radiographs based on their local features is provided. Chest radiographs with varying degrees of subtlety are used to test the algorithm. For the initial test, 40 chest radiograph pictures with varying degrees of subtlety are utilised. The experimental results of candidate nodule extraction via local contrast thresholding indicate that actual positive nodules are accurately found in 90% of images and reduce the false positive rate by 3.6%.

Keywords: Lung cancer, chest radiograph, local contrast thresholding.

1. Introduction

Lung cancer is among the leading causes of cancer-related mortality worldwide. Radiography, computed tomography (CT) and magnetic resonance imaging (MRI) are the most common imaging modalities utilised for earlylung cancer identification. Despite the increased accuracy and sensitivity of CT and MRI, radiography remains the technique of choice for the early detection and diagnosis of lung cancer due to its non-invasive nature, radiation doseand cost-effectiveness. Before classifying the lungs based on their features, it is required to extract lung segments from the chest radiograph and identify potential nodules within the segmented lungs. Numerous researchers have proposed various methods for detecting lung nodule-like structures on chest radiographs. Inaccurate segmentation of the nodule from the chest radiograph may have an impact on its local features. Using SWLCT, the proposed research examines the detection of nodule-like structures in segmented lungs. The SWLCT is based on the Bernsen Local Contrast thresholding approach and uses a square window. The square window's dimensions are equivalent to diameter of thecircular nodule. The active shape model (ASM) (Cootes, 2000; Lee

DOI: 10.1201/9781003527442-32

et al., 2009; Van Ginneken *et al.* 2002), the most robust segmentation technique, is used to extract the lung masks. Chest x-rays are the most used diagnostic test for cardiopulmonary problems. However, bone features such as ribs and clavicles can mask small anomalies, leading to diagnostic mistakes. A deep learning (DL)-based (Rajaraman *et al.*, 2021) bone suppression model was suggested that recognises and removes these obstructing bony structures in frontal chest radiographs. The authors presented a convolutional neural filter (CNF) (Matsubara *et al.*, 2019) for bone suppression, which is based on a convolutional neural network that has outstanding image processing performance and is widely used in the medical area. Using CNF with six convolutional layers, the system achieved a bone suppression rate of 89.2%.

Using (Horváth *et al.*, 2013; Juhász *et al.*, 2010) dynamic programming, pictures are segmented to remove shadows on the rib and clavicle. In distinct space, the divided shadows disappear. The cleaned pictures are processed using a hybrid lesion detector based on gradient convergence, contrast and intensity statistics. A SVM eliminates erroneous conclusions. Approximately 80% of bone shadows are removable. The suggested approach employs the bone shadow removed dataset from the Japanese Society of Radiological Technology (JSRT) (Juhász *et al.*, 2010; Junji *et al.*, 2000). Several lung nodule segmentation strategies based on classical methodologies have been presented at the present level of technological development. Implementation of the Gaussian filter for nodule detection (Soleymanpour *et al.*, 2011) with the premise that nodules are round. Smoothed pictures from a multi-scale Gaussian filter were added, resulting in the nodule's conspicuous appearance. For all 247 photos with a sensitivity of 0.96, this technique generated a substantial number of false-positive nodules. Extending prior work (Campadelli *et al.*, 2006) for minimising false positive nodules, the SVM classifier is utilised and a sensitivity of 0.71 for 1.5 false positive nodules/image is reported. Using directional Laplacian of Gaussian (LoG), a method was presented for improving (Shi *et al.*, 2010) dot-like patches. The primary flaw of the strategy (Soleymanpour *et al.*, 2011; Shi *et al.*, 2010) is that it is predicated on the assumption of equal complexity level. Using global threshold value of (Patil and Udupi, 2010) nodules were segmented from chest radiographs and identified based on their geometrical and textural characteristics. The forward stepwise selection approach (Wei *et al.*, 2002) was used to choose the optimal set of 210 features. The ideal feature set achieves an 80% true positive detection rate and an average of 5.4 false positives per image. Otsu's thresholding (Mya *et al.*, 2014) was used to CT scan images to detect nodules. Since CT scan images are more distinct than X-Ray images, the identical thresholding procedure is ineffective for X-Ray. Statistical and geometrical features were calculated, and the ANN classifier was used to detect the cancer stage. A computationally intensive CAD system (Al Gindi *et al.*, 2014) for lung nodule characterisation and classification of nodules as benign or malignant utilizing multiscale wavelet transform yielded notable results. The MTANN technique is used to construct a method for suppressing (Chen and

Suzuki, 2013) clavicles and ribs in the virtual dual-energy (VDE) image, hence enhancing the visibility of nodules obscured by clavicles and ribs. The system lacks an extraction procedure for nodules. Three distinct techniques (Savithri and Chaya, 2016) for extracting nodules from chest radiographs were proposed. All three methods require images to have same levels of subtlety. A linear SVM classifier was used to categorise the ROI as either nodule or non-nodule based on the features extracted from the patch (region of interest) using constrained Boltzmann. Using Lindeberg's multiscale blob detector (Schilham *et al.*, 2003), the potential nodules are detected on the chest radiograph. It had a sensitivity greater than 50% for two false positives per image and 70% for 10 false positives per image. Principal deficiency of the system is the assumption that all images have the same level of subtlety. Image binarisation commonly employs thresholding techniques (Mehmet and Bulent, 2004). Clustering-based methods, histogram shape-based methods, entropy-based methods, object attribute-based methods, spatial methods and local methods are broad classifications for thresholding approaches (Singh *et al.*, 2011b; Nie *et al.*, 2013). These methods modify the threshold value for each pixel based on the local characteristics of the image. Comparative research of (Chaya and Savithri, 2018) distinct algorithm techniques for rib suppression from chest radiographs, lung segmentation algorithms, candidate nodule detection, feature estimation and classification as nodule or non-nodule was conducted. A comparative evaluation of alternative strategies for segmenting lung nodules was proposed (Zheng and Lei, 2018). The proposed system segmented nodules using active contour, differential operator, and region growing approaches. Proposed is a deep learning (Liang *et al.*, 2020) CAD system for detecting pulmonary nodules from chest radiographs. The proposed system utilised four algorithms: the algorithm for aberrant probability, the algorithm for nodule probability, the algorithm for heat maps and the algorithm for mass probability. This system proposal established a comparative analysis of the performance of these algorithms. Among the four algorithms, the mass probability algorithm performs the best with a sensitivity of 76.6% and a specificity of 88.68% while analysing a total of 100 cases. Combining CNN with a visual geometry group, a neural network hybrid technique (Bharati *et al.*, 2020) was proposed to address CNN's shortcomings. A novel approach (Annangi *et al.*, 2010) to segmenting lungs from digital chest x-ray images, overcoming challenges such as local minima and strong edges due to the rib cage and clavicle. The algorithm was tested on over a thousand clinical images, with promising results. A new fully automated approach (Sun *et al.*, 2012) for segmenting lungs with high-density pathologies, which delivers statistically significant better segmentation results compared to two commercially available lung segmentation approaches.

1.1. Locally Variable Thresholding

Threshold $Tn(x, y)$ is a value in locally adaptive thresholding and corresponds to the parameters x and y.

$$b_n(x,y) = \begin{cases} 0 & if\ (x,y) \le T(x,y) \\ 1 & otherwise \end{cases} \qquad (1)$$

When binarised image is represented by bn(*x*, *y*) and pixel intensity of image I is represented by I (*x*, *y*). Local adaptive approach determines a threshold for each pixel based on factors such as range, variance or surface-fitting parameters of neighboring pixels (Al Gindi *et al.*, 2014). Background subtraction, the mean value of pixel values, the standard deviation, the local picture contrast and the water flow model are the most prominently distinct local adaptive technique techniques. The primary disadvantages of local adaptive thresholding algorithms are that they depend on region size, image properties and are time-intensive. Some scholars (Schilham *et al.*, 2003) proposed a mix of global and local thresholding methods coupled with the usage of morphological operators. Niblack (Mehmet and Bulent, 2004), Sauvola and Pietaksinen (Singh *et al.*, 2011a) and Bernsen (Nie *et al.*, 2013) employed the midrange value within the local block, whereas Niblack (Mehmet and Bulent, 2004), Sauvola and Pietaksinen (Singh *et al.*, 2011a).

2. Proposed Method

2.1. Assumptions

Prior to using the suggested approach for candidate nodule extraction from chest radiograph utilising local contrast thresholding, the following assumptions are made.

1) The dataset used for the proposed method consists of digital chest radiograph pictures without bone shadow from the Japanese Society of Radiological Technology (Matsubara *et al.*, 2019; Rajaraman *et al.*, 2021). The dataset consists of 154 chest radiographs of lung nodules and 93 chest radiographs without nodules.
2) Each chest radiograph in the JSRT dataset including lung nodules contains a single nodule.
3) Images with subtlety levels ranging from 5 (Nodule is obvious) to 2 (Nodule is very subtle) are utilised for analysis.The proposed method does not use images with a subtlety level of 1 (Extremely subtle) because it is extremely difficult to extract nodules from such images due to their low contrast, small size or overlap with normal structures.

2.2 Algorithm

The proposed method is divided into sub processes namely (i) pre-processing of the image, (ii) nodule segmentation using local contrast thresholding, (iii) postprocessing and (iv) nodule feature extraction and classification.

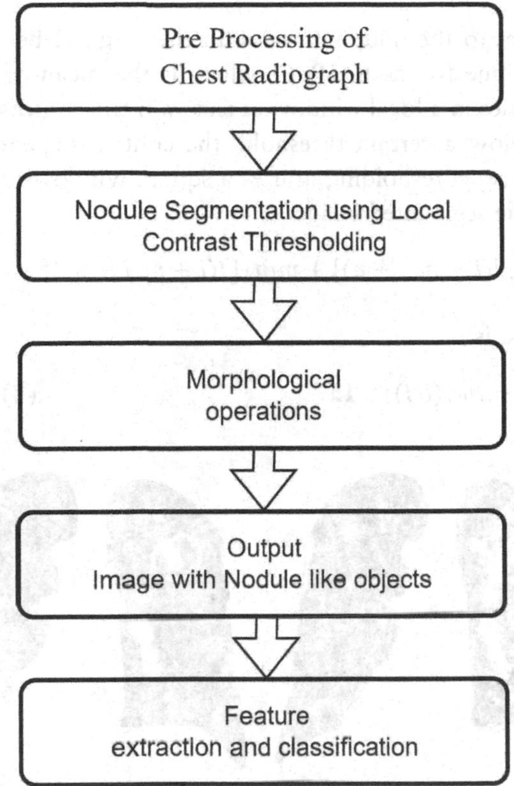

Figure 1: System flow chart.

2.2.1. Preprocessing

In the preprocessing, the original chest radiograph images are contrast-enhanced as shown in Figure 2 and down sampled from 2048 × 2048 to 512 × 512 before being used for lung segmentation with Active Shape Model (Cootes, 2000). Figure 1b shows the generated lung mask using active shape model.

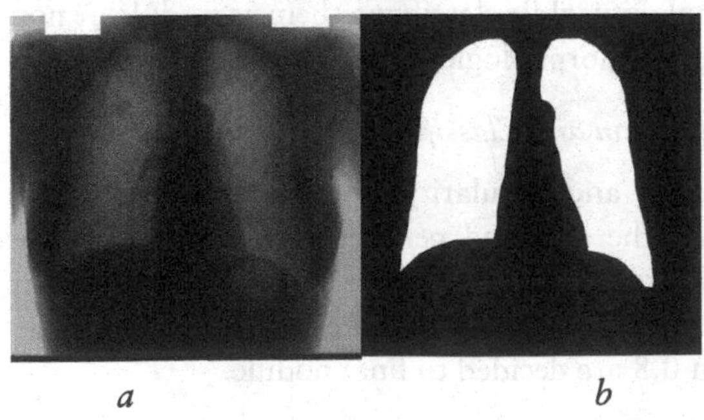

Figure 2: Lung segmentation using ASM (a) Original chest radiograph (b) Lung mask using ASM.

2.2.2. Nodule Segmentation using Local Contrast Thresholding

The threshold is set to the midpoint value in the original Bernsen local contrast thresholding technique (Bernsen, 1986), which is the mean of the minimum and maximum grey values in a local window of size w. If the contrast C(p, q) = Ihigh(p, q) Ilow(p, q) is below a certain threshold, the context-dependently on T (p, q). Bernsen local contrast thresholding utilises a square window of size w x w. Figure 3a shows the nodule segmented image.

$$(i,j) = 0.5\{max[l(i+m,j+n)] + min_w[l(i+m,j+n)]\} \quad (2)$$

where local contrast is

$$(i,j) = I_{hig}(i,j) - I_{low}(i,j) \geq 15 \quad (3)$$

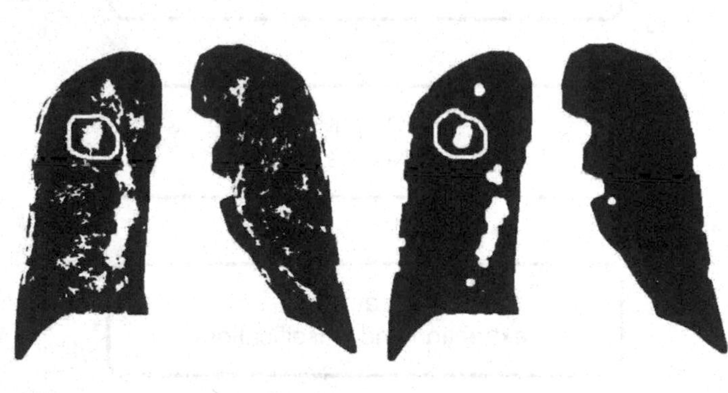

Figure 3: Nodule segmentation using local contrast thresholding (a) local contrast thresholded image (b) image after morphological operations.

2.2.3. Postprocessing

In the postprocessing the morphological operations are employed on the nodule segmented images using local contrast thresholding. This operation minimises the noise and the complexity while classifying them as nodule or non-nodule. Figure 3b shows the image after morphological operations.

2.2.4. *Feature Extraction and Classification*

Using area, perimeter and circularity index as a features, final nodule or actual nodule is detected. The area and perimeter range are developed using the JSRT Dataset clinical information. Most of the actual nodules are circular in shape or closer to circular shape. Based on this knowledge the nodule having circularity index greater than 0.8 are decided to final nodule.

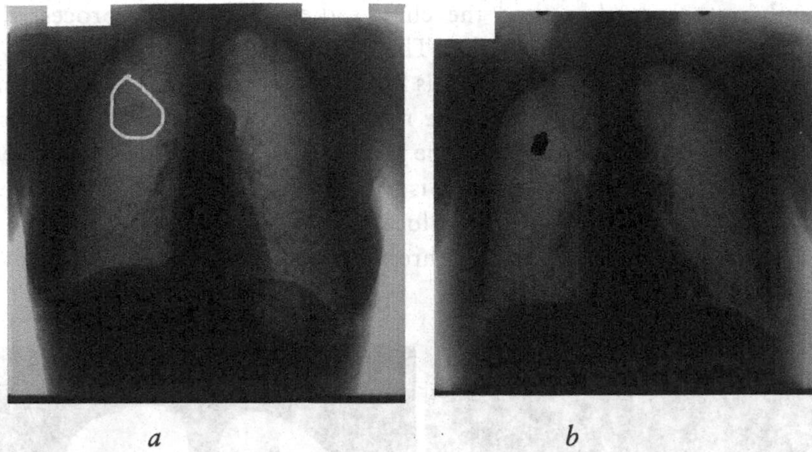

Figure 4: Final nodule detection based on their geometrical features (a) original image (b) final nodule detection.

3. Experiments and Results

3.1. Experimental Test

As shown in the flow chart Figure 5, 50 images with subtlety levels 5–2 are selected and tested with Square Window Based Bernsen Local Contrast Thresholding.

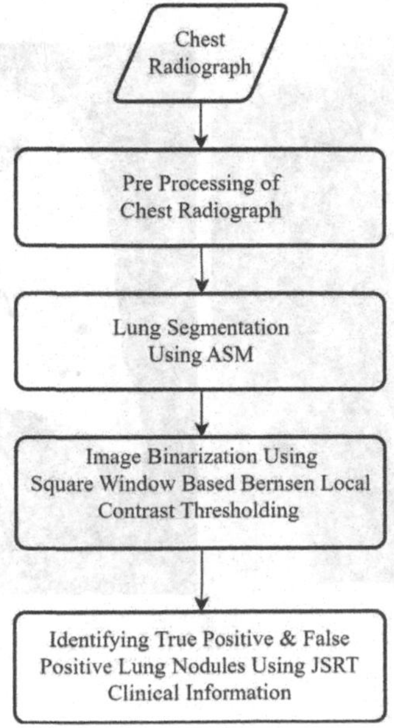

Figure 5: System flow chart for Test-I.

Prior applying proposed method the chest radiographs are preprocessed for enhancing nodule like regions visibility. Then lungs are segmented using ASM. In this test, proposed method tested and it is found that the 90% of the true positive nodules extracted are part of candidate nodules. The true positive nodules are the nodules having been mentioned in the JSRT clinical information. The average rate of false positive nodules per image is 5 fp/image. Figure 6 shows the results of candidate nodules extraction using local contrast thresholding with square window based bernsen local contrast thresholding for the images with subtlety level 5–2.

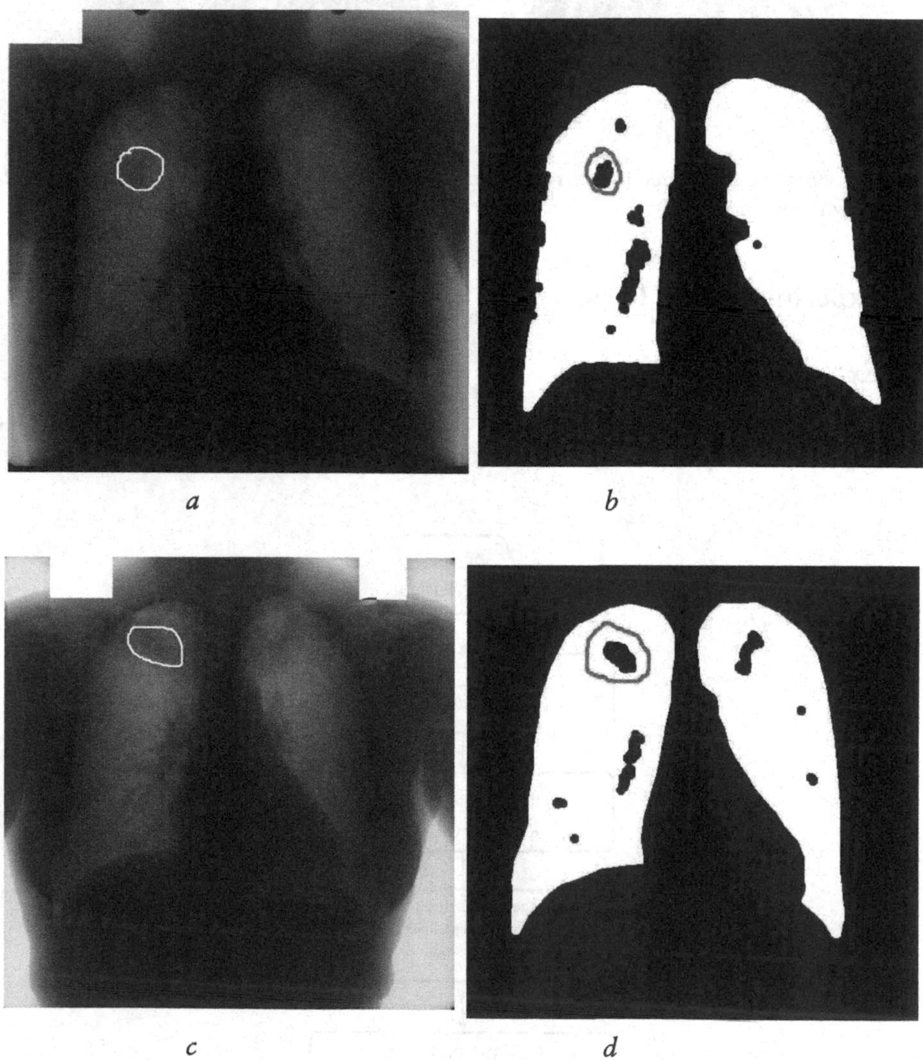

a *b*

c *d*

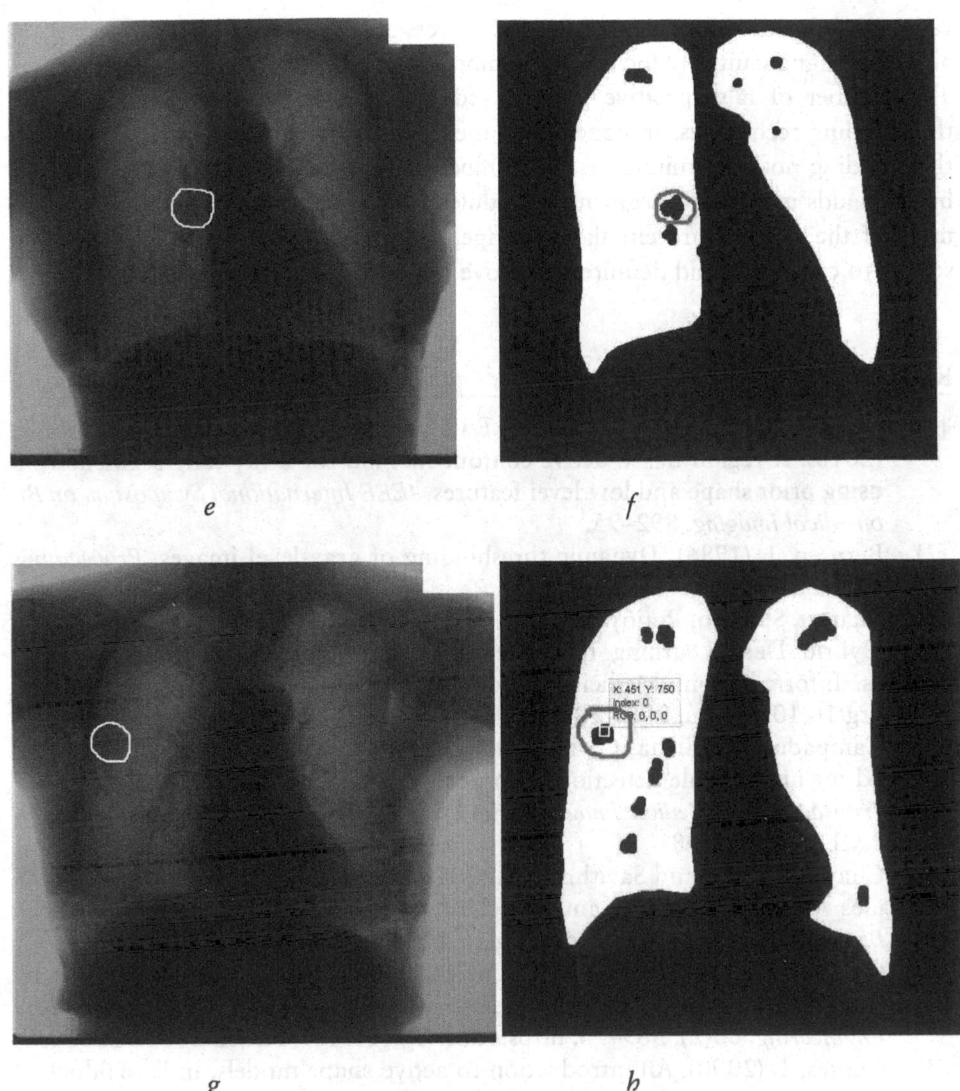

Figure 6: Some of the results obtained -Original Images with subtlety levels 5, 4, 3 & 2 sequentially (Left Column-(a),(c), (e), (g)), results of candidate nodule segmentation (circled-True positive nodules) using local contrast thresholding(Right column-(b), (d), (f), (h)).

As shown in the Figure 6, the sample images with subtlety levels from 5 to subtlety level 2 are used for testing the proposed algorithm. Ten images from each subtlety level are chosen and the proposed method is applied.

4. Conclusion and Future Scope

The proposed approach of local contrast thresholding satisfies the objective of improving the precision of nodule detection. The experimental setup with chest radiograph images of subtlety levels 5–2 detects actual positive nodules with 90% accuracy, significantly reducing the number of false positive nodules per image. The number of false positive nodules reduces by 3.6% as compared to other thresholding techniques. In conclusion, the application of Bernsen local contrast thresholding not only minimizes the number of false-positive nodules per picture but also aids in precisely segmenting nodules from chest radiograph images. Since most of the nodules are circular in shape, modification in window shape from square to circular would definitely improve the nodule detection accuracy.

References

[1] Annangi, P., Thiruvenkadam, S., Raja, A., Xu, H., Sun, X., and Mao, L. (2010). A region based active contour method for x-ray lung segmentation using prior shape and low level features. *IEEE International Symposium on Biomedical Imaging*, 892–95.

[2] Bernsen, J. (1986). Dynamic thresholding of gray-level images. *Proceedings International Conference on Pattern Recognition*, 1251–1255.

[3] Bharati, Subrato, Prajoy Podder, and M. Rubaiyat Hossain Mondal. (2020). Hybrid Deep Learning for Detecting Lung Diseases from X-Ray Images. Informatics in Medicine Unlocked 20 (January): 100391. https://doi.org/10.1016/j.imu.2020.100391.

[4] Campadelli, P., Elena, C., and Diana, A. (2006). A fully automated method for lung nodule detection from postero-anterior chest radiographs. *IEEE Transactions on Medical Imaging*, 25(12), 1588–1603. https://doi.org/10.1109/TMI.2006.884198.

[5] Chaya, D. S. K., and Savithri, T. S. (2018). Review: on segmentation of nodules from posterior and anterior chest radiographs. *International Journal of Biomedical Imaging*. https://doi.org/10.1155/2018/9752638.

[6] Chen, S., and Suzuki, K. (2013). Computerized detection of lung nodules by means of "virtual dual-energy" radiography. *IEEE Transactions on Biomedical Engineering*, 60(2), 369–78. https://doi.org/10.1109/TBME.2012.2226583.

[7] Cootes, T. (2000). An introduction to active shape models. In R. Baldock et al. (Eds.), *Image Processing and Analysis*, 223–48, Oxford University Press.

[8] Gindi, A. M. A., Essam, A. R., and Mostafa, M. S. M. (2014). Development and evaluation of a computer-aided diagnostic algorithm for lung nodule characterization and classification in chest radiographs using multiscale wavelet transform. *Journal of American Science*, 10(2), 1545–1003. http://www.jofamericanscience.orghttp://www.jofamericanscience.org.14.

[9] Ginneken, B. V., Frangi, A. F., Staal, J. J., Romeny, B. M. T. H., and Viergever, M. A. (2002). Active shape model segmentation with optimal features. *IEEE Transactions on Medical Imaging* 21(8), 924–33. https://doi.org/10.1109/TMI.2002.803121.

[10] Horváth, A., Gergely, O., Ákos, H., and Gábor, H. (2013). An x-ray CAD system with ribcagesuppression for improved detection of lung lesions. *Periodica Polytechnica Electrical Engineering and Computer Science*, 57(1), 19–33. https://doi.org/10.3311/PPee.2079.

[11] Juhász, S, Horváth, A., Nikházy, L., and Horváth, G. (2010). Segmentation of anatomical structures on chest radiographs. MEDICON 2010, *IFMBE Proceedings*, 29, 359–62. Junji, S., Katsuragawa, S., Ikezoe, J., Matsumoto, T., Kobayashi, T., Komatsu, K., Matsui, M., Fujita, H., Kodera, Y., and Doi, K. (2000). A development of a digital image database for chest radiographs with and without a lung nodule: receiver operating characteristic analysisof radiologists' detection of pulmonary nodules. *American Journal of Roentgenology*, 174, 71–74. www.macnet.or.jp/jsrt2/.

[12] Lee, J., Wu, H., and Yuan, M. Z. (2009). Lung segmentation for chest radiograph by using adaptive active shape models. *Fifth International Conference on Information Assurance and Security*. IEEE Computer Society.

[13] Liang, C. H., Liu, Wu,., Garcia-Castro, F., Alberich-Bayarri, A., and Wu, (2020). Identifying pulmonary nodules or masses on chest radiography using deep learning: external validation and strategiesto improve clinical practice. *Clinical Radiology*, 75 (1), 38–45. https://doi.org/10.1016/j.crad.2019.08.005.

[14] Matsubara, N., Atsushi, T., Kuniaki, S., and Hiroshi, F. (2019). Bone suppression for chest x-ray image using a convolutional neural filter. *Australasian Physical and Engineering Sciences in Medicine*, 43. https://doi.org/10.1007/s13246-019-00822-w.

[15] Mehmet, S., and Sankur, B. (2004). Survey over image thresholding techniques and quantitative performance evaluation. *Journal of Electronic Imaging*, 13(1), 146–65. https://doi.org/10.1117/1.1631316.

[16] Mya, K., Mya, T., and Aung, S. K. (2014). Feature extraction and classification of lung cancer nodule using image processing techniques. *International Journal of Engineering Research & Technology*, 3(3). www.ijert.org.

[17] Nie, F., Yonglin, W., Meisen, P., Guanghan, P., and Pingfeng, Z. (2013). Two-dimensional extension of variance-based thresholding for image segmentation. *Multidimensional Systems and Signal Processing*, 24(3), 485–501. https://doi.org/10.1007/s11045-012-0174-7.

[18] Patil, S. A., and Udupi, V. R. (2010). Chest x-ray features extraction for lung cancer classification. *Journal ofScientific & Industrial Research*, 69, 271–77.

[19] Rajaraman, S., Ghada, Z., Les, F., Philip, A., and Sameer, A. (2021). Chest x-ray bone suppression for improving classification of tuberculosis-consistent findings. *Diagnostics* (Basel)11(5). https://doi.org/10.3390/diagnostics11050840.

[20] Savithri, T., and Chaya, D. (2016). Nodule detection from posterior and anterior chest radiographs with different methods. *2016 Future Technologies Conference* (FTC), San Francisco, 504–15.

[21] Schilham, A. M. R., Bram, V. G., and Marco, L. (2003). Multi-scale nodule detection in chest radiographs. *International Conference on Medical Image Computing and Computer-Assisted Intervention*,602–9. http://www.isi.uu.nl.

Shi, Z., Jun, B., Kenji, S., Lifeng, H., Quanzhu, Y., and Tsuyoshi, N. (2010). A method forenhancing dot-like regions in chest x-rays based on directional scale log filter. *Journal of Information & Computational Science*, 7(8), 1689–1696. http://www.joics.com.

[22] Singh, T. R., Sudipta, R., Singh, O. I., Sinam, T., and Singh, K. M. (2011). A new localadaptive thresholding technique in binarization. *International Journal of Computer Science*, 8(6). www.IJCSI.org.

[23] Soleymanpour, E., Hamid, R. P., Emad, A., and Mehri, S. Y. (2011). Fully automatic lung segmentation and rib suppression methods to improve nodule detection in chest radiographs. *Journal of Medical Signals and Sensor*, 1, 191–99.

[24] Sun, S., Christian, B., and Reinhard, B. (2012). Automated 3-D segmentation of lungs with lungcancer in CT data using a novel robust active shape model approach. *IEEE Transactions on Medical Imaging*, 31(2), 449–60. https://doi.org/10.1109/TMI.2011.2171357.

[25] Wei, J., Yoshihiro, H., Akinobu, S., and Hidefumi, K. (2002). Optimal image feature set for detecting lung nodules on chest x-ray images. *Computer Assisted Radiology and Surgery*, Springer. https://doi.org/10.1007/978-3-642-56168-9_118.

[26] Zheng, L., and Yiran, L. (2018). A review of image segmentation methods for lung nodule detection basedon computed tomography images. MATEC Web of Conferences, 232, EDP Sciences. https://doi.org/10.1051/matecconf/201823202001.

Chapter 33

Efficient Set-D4: A Deep Learning Approach to Identify Human Handwritten Signatures

Shailesh Chandrakant Sahu[1], *Sakshi Aniket Khamankar*[1], *and Amit Kumar Tripathi* [2]

[1]Assistant Professor, Department of Computer Engineering, Cummins College of Engineering for Women, Nagpur, Maharashtra, India

[2]Assistant Professor, Department of Electronics and Telecommunication, Cummins College of Engineering for Women, Nagpur, Maharashtra, India

Abstract: Identification of handwritten signatures has experienced extensive usage in the world of information handling. Likewise, there are numerous distinctions in people's styles of writing; it can be difficult to correctly recognise those individuals from photographs. This process is also complicated by the existence of numerous visual artifacts, such as vibration, deformation and intensity variations. This paper proposed a deep learning (DL)-based approach to categorising integers termed EfficientDet-D4 using the recommended method in an attempt to overcome those restrictions. To clearly show the region of fascination, the input signature images are first accurately annotated. These photos are used to train the Efficient Set-D4 algorithms to detect and categorise their unique signatures. Utilising the MNIST dataset evaluate the detection accuracy of the suggested model and succeeded in obtaining a 99.83% total accuracy rating. Furthermore conducted the cross-dataset analysis on the USPS datasets and received a dependability score of 99.10%. The graphical outputs and the real-world results show that system can accurately identify handwritten signatures compared to pictures regardless of how their writing style differs, as well as when there are many testing artifacts there, including motion, deception, chrominance, status fluctuations and dimension shifts of numerals.

Keywords: Handwritten signatures, deep learning, MNIST dataset, efficientDet-D4.

1. Introduction

The identification of characters in a variety of languages has been thoroughly researched by researchers (Singh *et al.*, 2022) but handwritten signature recognition (HSR) is a particularly significant area. Considering the vast quantity of information accessible in the form of written material or photos, HSR serves a critical role in processing information (Kumar *et al.*, 2022). Furthermore, analysing online data

DOI: 10.1201/9781003527442-33

can be less expensive than manually manipulating data on printed pages. The purpose of HSR techniques is to identify textual words as well as numbers and translate them into machine-understandable representations. Researchers' interest in HSR has grown recently as a result of its many applications.

These structures may be used to read text from paper pages and allow researchers to look at crucial information kept in ancient documents and files that are hidden from the human retina (Misgar *et al.*, 2023). Additionally, HSR techniques are essential for the digital transformation of any organisation or institution. Computerised HSR structures can be useful in several ways, including the automatic recognition of signatures in online and offline modes (Al-wajih *et al.*, 2020) and the recognition of the signature printed on healthcare invoices, which may be beneficial to pharmacists, customers and staff. Psychiatrists can also use HSR techniques to assess a person's character (Abdulrazzaq *et al.*, 2018.). In the aforementioned domains, there are enormous amounts of data; nonetheless, automatic HSR solutions must be effective and successful with rapid processing rates and solid conclusions. Many HSR solutions are now available to help in a range of situations. Those systems must exhibit better identity categorisation and recognition efficiency more frequently (Shamim *et al.*, 2018). Given the diverse handwriting characteristics, scripts and patterns that are susceptible to underwriting and underwriting inconsistencies, automated recognition of prehistoric written handwriting is a difficult process (Ahlawat and Choudhary, 2020). Additionally, it is more difficult to distinguish written signatures on digital records and photos since their arrangement, size and spacing vary depending on the file limits and breadth (Albahli *et al.*, 2021). Because they are designed for specific purposes, handwriting technologies now offer more diversity, better digit identification and better classification. Performance of the system as a whole (Wang *et al.*, 2020).

In the discipline of HSR, pattern-based picture processing and recognition are essential for memorising sequences. To identify a written signature, word and digit in handwriting, many HSR frameworks have just been presented. The following are the key phases of a typical HSR structure: data preparation, segmentation, computation of key points and categorisation. Rapid developments in HSR show improvements in learning algorithms and access to vast databases. Many theories that employ homemade keys and are based on point techniques, dense network approaches, and other methods have been put forth by the scientific community (Masood *et al.*, 2021).

Boundary attributes, projecting, FFT explanations, contoured alignment figures, sequential deciphering and unchangeable seconds represent a few of the most widely utilised handmade qualities in recognising images. These basic elements can be merged to provide a trustworthy set of features for teaching a system of classification. The effectiveness of categorisation can be significantly impacted by the computation of attributes. Several categorising approaches are still in use right now, including supporting vector models. A proper list of characteristics should include all literary characteristics distinctive to a certain group and be as distinct as possible from all others (Ali *et al.*, 2020). Customised characteristic-based solutions, on the other hand, have repeatedly been demonstrated to be unsuccessful owing to the enormous duration and work necessary for data collection. Furthermore, these approaches are incapable of detecting letters and numbers in deformed images (Uzair *et al.*, 2020). Because of the biased main calculation and categorisation of property, DL has expanded into a fast-expanding topic that incorporates machine-learning (ML) methodologies for improving accuracy in the fields of pattern identification and recognising texts. Because there are more hidden components and links in deep learning than in other DL frameworks, it can be more time-consuming.

CNN techniques developed by the DL category have increased in favour for the analysis of images since they use more concealed levels as DNNs as well as a smaller number of variables. These techniques are also capable of computing location-invariant critical points quickly because of their outstanding pattern recognition skills. Additionally, CNNs are skilled at employing time sampling to map input data to output data and are thus unaffected by aberrations or straightforward geometrical changes like movement, translation, pressing or scaling. Because a written signature, word and digit can appear in a variety of formats and contexts, researchers have carefully investigated a wide range of CNN techniques to address challenges in the area of automated HSR (Uzair *et al.*, 2021).

Although earlier studies achieved exceptional accuracy, automatic HSR may still be improved in terms of speed and precision efficiency. For the successful and effective identification of numbers using images, a lot of algorithms have been presented in the past, with the bulk of articles depending on DL techniques. The primary drawback of DL-based HSR methods is how inefficient and time-consuming it is to handle data. Additionally, a lot of research in the scientific literature on identifying written and printed text focuses only on standard characters without any distinctive properties or noise (Aly *et al.*, 2020; Tan *et al.*, 2020). Handwritten letters can be damaged or deformed in a word as well as in a number of real-world situations, including contrasts, blurriness from movement, vibration and varying lighting conditions.

Dani *et al.* (Malik and Roy, 2019) suggested an extensive CNN approach using Alex-Net to differentiate signs with fragmentary and overcrowded features. According to the model's specifications, 92% of the dataset created artificially is accurate. Similar to Nawaz *et al.* (2022) written digits were identified without noise or distortion using an image detection framework. This method results in comparatively better identification outcomes at the cost of more computational work.

The proposed DL architecture can recognise printed numerals even in the absence of distorted backgrounds, despite the fact that it uses many scales of feature combining throughout feature computation. The practical outcomes demonstrate the recommended approach's effectiveness in dealing with complex situations, including variations in dimensions, introductions, handwriting styles and approaches, as well as gigantic fitting between various signature, word and digit layouts. The proposed system can also successfully distinguish signature, word and digit while minimising implementation time because of its shallow architecture, which also makes it immune to noise and distortion in input instances (Pashine *et al.*, 2019).

The following are the main characteristics of the recommended strategy:

(1) For precise handwritten signature, word and digit identification and classification, researchers developed a top-to-bottom deep learning system based on the Efficient Set-D4 algorithm (Ge *et al.*, 2019).
(2) The suggested approach makes use of the EfficientNet-B4 compact base to provide reliable and biased points that raise overall HSR achievement while requiring less training and time to implement (Beikmohammadi *et al.*, 2021).
(3) A detailed statistical and subjective evaluation of the suggested approach was performed using typical databases to demonstrate the utility of the structure that was provided. The results show that the recommended CNN method works better at identifying percentages than prior network-based strategies (Agrawal *et al.*, 2021).

2. Related Work

In recent years, a lot of work has gone into developing computerised HSR solutions. Boukharouba and Bennia proposed a handmade HSR approach based on characteristics. Utilising the pixel's transitional information across the upward and horizontal directions, a separate key variable set is constructed using the chained encoding dispersion approach. Following that, the SVM classifiers underwent training to categorise the acquired key features. This approach can correctly distinguish handwritten signatures by word and digit, but it needs a lot of practice on a larger database. A combination method for categorisation with increased precision was provided by the author (Singh *et al.*, 2022).

The findings showed better identification efficiency; nevertheless, vibration, deformation or an uncommon style of writing degraded efficiency. Sheikh *et al.* suggested a novel characteristic computation method based on architecture and analytical methodologies. Preparation was first undertaken to binarize agricultural products and normalise the raw data. Then, utilising cavity, zoning, Frei chain encoding (FFC) and profile projections, four separate sets of features were generated. Finally, KNN was used to categorise crucial points. Using FFC on the MNIST database, this method achieved a recognition precision of 95% (Kumar *et al.*, 2023; Misgar, 2023).

Although it might not function effectively in difficult datasets. In recognising handwritten signatures by word and digit, Hou and Zhao used a mixture of handmade and deep characteristics. A Gaussian pattern of features approach was used for calculating unique handcrafted areas of attention. The calculated points of interest were used for developing the CNN categorisation algorithm. This method enhanced the precision of signatures, phrases and number identification;

nonetheless, it had a greater computing expenditure. To improve RNN resilience for HSR, Pham *et al.* presented an abandoned regularisation strategy. This approach significantly enhances RNN efficiency by lowering letter and phrase mistake rates (Al-wajih *et al.*, 2020).

Then, CNN was trained to classify the observed digits using the recovered attributes. According to CNN, having fewer layers offers better identification reliability for smaller datasets. For finding signatures, word and numbers given sample inputs, the deep convolution autonomous mapping architecture is presented in. What happens is described in. This method works well for signature, digit classification when interference is present; however, it might not work well when rotation fluctuations are present. By merging DL with the Q-learning-based reinforcement training technique, Hafiz and Bhat showed an outstanding hybrid classification. This study shows that the precision of digit categorisation is better when rotational adjustments are present, albeit at a higher cost.

Islam *et al.* showed handwritten variety identification using a CNN and additional educational modules. For collecting biased important indications, an improved CNN built upon the LeNet model (Al-wajih *et al.*, 2020) with five levels of suppression and an ELM encoder for basic classification was built. Although this method increases categorisation precision, the resulting model is prone to excessive fitting. The recognising researchers also examined the impact of adding concealed levels on model efficiency and observed that doing so affects modelling efficiency for HSR. Similarly, the researchers proposed an ANN- and ELM-based approach for recognising handwritten letters.

This paper studied various automatic ways for properly identifying handwritten integers; nonetheless, performances both in terms of precision and computation expense can still be enhanced. Current methods either require substantial preliminary processing or are useless when confronting damaged backdrops or complicated situations, including size, introductions, handwriting forms, methods and noise. Additionally, such approaches need extensive training and are prone to modelling excessively, leading to poor effectiveness with unreliable data that may be rectified.

3. Proposed Methodology

PCA is employed in this research to analyse interval information features with the goal of minimising the size of a design area to uncover additional prominent qualities. Then, in the PCA calculation, we made an intelligent selection of features by picking certain additional characteristics. During the classification stage, collected characteristics obtained from the extraction of features as shown in Figure 1 and selection phases are integrated into a signature.

Extracting features and screening PCA is a prominent statistical technique for the extraction of features, reduction of dimension and presentation of information in recognition of patterns and machine vision. To translate multivariate distribution into smaller dimensions with minimal loss of essential data. This is accomplished by significant associations with factors into the fresh domain with independent parameters (Beikmohammadi *et al.*, 2021). The generated elements are used as features to describe the information.

As mentioned in Section 2, every signature example in the SIGMA dataset consists of time-varying signals (x, y, and p), leading to attribute vectors with a large dimension in our chosen signature subset. However, while selecting features, we discuss six key procedures for calculating PCA convey features for smaller dimensions. The following are the reduced procedure stages:

Step 1: Considering (1), calculate the average value of the database X for every parameter (x, y, p):

$$\overline{x} = \frac{\sum_{i=1}^{n} X_i}{N}$$

during which N is the number of specimens provided.

Table 1: It shows the number of samples collected for each user throughout education and assessment.

True signatures	20
Sign created with ability	10
Non-skill-forged	10
Number users	200
Maximum number of specimens	8000

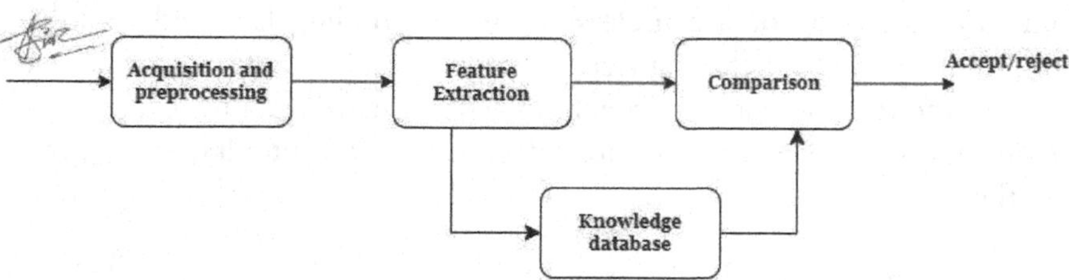

Figure 1: *Design of an online sign validation platform.*

Second step: Remove the average value (X) with every sample's result (X), as stated in the equation that follows, to get an additional matrix of a single,

$$M(N x M) : \Phi i = Xi - X$$

Third step: Applying (3) to the preceding matrix (N M), calculate the correlation for each of the pairs separately.

$$Con(M)=\sum_{i=1}^{n}\left(X_i-\bar{X}\right)\left(Y_i-\bar{Y}\right)/\left(N-1\right)$$

Fourth step: Applying the formula below, get the eigenvalue of the correlation: lM Il = 0. (4)

Fifth step: Also, a formula that follows the solution to compute the eigenvalues from the correlation matrix: $(M\ jI)ej=0.(5)$

Sixth step: Finally, keep K as the primary elements in relation to the Eigenvalues.

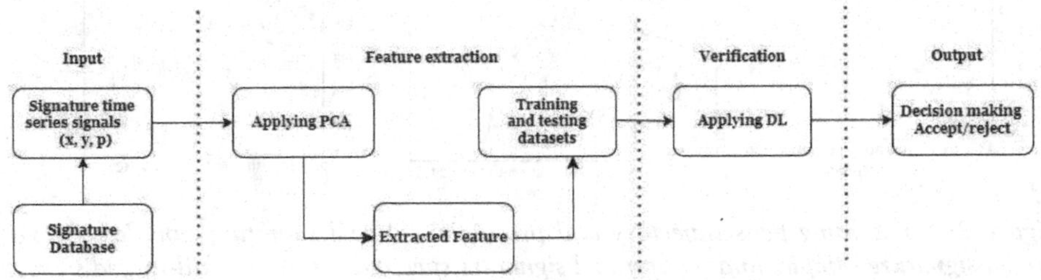

***Figure 2:** A representation of the proposed online sign validation mechanism.*

To categorise the provided sample into signature, word and digit classifications, precise and effective extraction of features is required. Challenging due to several factors, including the computation of an enormous include set, causing the framework to be biased, and the calculation of a smaller include collection, which leads to the models failing to acquire certain significant image features, such as size, colour and appearance. As a result, it is critical to use an automated characteristic extract rather than handmade methods of feature extraction that can predict the more representative characteristics using the provided photos. Because of many parameters such as location, form, colour, chrominance and so on, handmade feature-estimation approaches are ineffective in correctly localising and categorising the signature, word and digits (Pashine *et al.*, 2021).

To address such issues, we applied the efficient detection DL approach, which extracts characteristics from the pictures under evaluation. The term convolution filtering refers to generating the characteristics of an image by looking at the physical makeup of an item. Various object-based diagnosis approaches were developed for the identification of various illnesses. These approaches are classified as one-stage (YOLO, CornerNet, CenterNet and so on) and two-phase (Fast-RCNN, Faster-RCNN, Mask-R CNN, EfficientDet and so on). This choice of efficient design across one-stage strategies is because these approaches combine to enhance recognition accuracy by exposing an insignificant quantity of time dedicated to classification, whereas two-phase approaches show more effective signatures and digit-detection efficiency; however, those techniques are logically complex due to the two-stage design of networks (Beikmohammadi *et al.*, 2021).

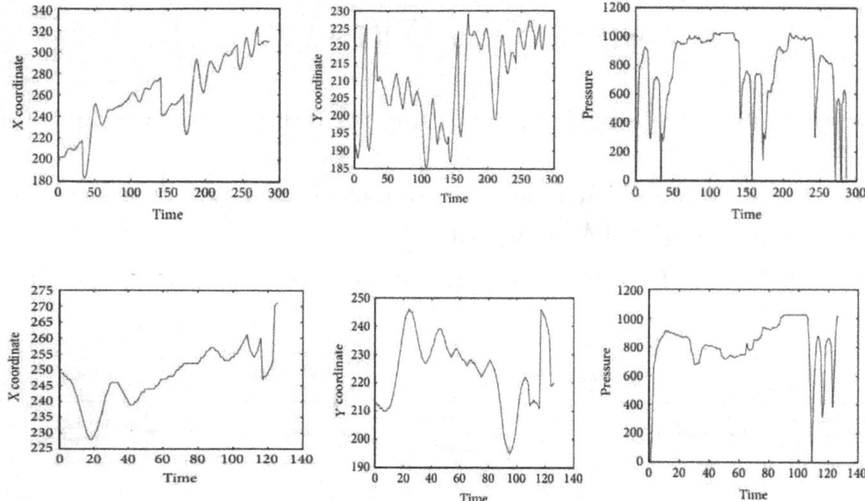

Figure 3: DB signing pens trajectory and pressures: (a) real signature samples, (b) skill-forged signature samples and (c) user real signature specimen as a non-skill-forged sign.

$$component = score\ latent + residual$$

Although dataset area is three-dimensional in nature, containing factors, which formed by all three of them as features. These elements allowed us to reassemble our initial data. The remainder is information that cannot be clarified through the parts included in the initial information. The number of components is determined by the remaining detail value. As a consequence, any of the three parts as shown in Figure 3 that are produced can be used to convey the initial written observation. The numbers in the rating matrices are sorted in a decreasing sequence depending on their variation, which correlates to the ordering of the main component (Malik *et al.*, 2019).

In comparison to the other two elements, the initial component possesses the biggest variation in terms of its rating. Similarly, the subsequent element has the most volatility, whereas the final element has the lowest variation.

Verification: Its categorisation technique in the present instance is deep learning, which relies on an unsupervised training technique called back propagation. Networked development is an ongoing process. For calculating concealed level mistakes, weighting variables (w) on the neuron are adjusted in every round based on the resulting mistake communicated through the subsequent layer on input to the highest layer.

Measurements (w) are configured as minuscule integers spanning between 0 and a single at the beginning of the instructions, and the result for every neuron is utilised to feed the following buried level. The retraining procedure in a reverse-propagating system is described in depth in the next images. The result (*Y*) is a calculated nonlinear combination of the components, whereby *i* represents the input sampling index, l is the neuron's directory and *N* is the total number of input repetitions (Misgar *et al.*, 2023) namely:

$$Y = \sum_{l=1}^{N} w_{il} X_i + w_{in+1}$$

Its output (*y*) is subsequently matched to the intended production, yielding an error (e). The resultant error is determined as follows, wherein *tl* are the intended variables and *ol* are the resulting quantities (Singh *et al.*, 2022):

$$E = \frac{1}{2} \sum_{l=1}^{N} (tl - ol)^2$$

Therefore, the error (e) for every neuron is employed to change the weight in order to achieve the intended output; the mistake is sent downstream to determine the error number of any layer (e.g., layer j) in lesser concealed levels depending on its upper layer (K) error:

$$\partial j = oj(1 - oj) \sum_k w_{kj} \partial_k$$

At last, the error in every single neuron is utilised to modify the neuron parameters so as to minimise the entire error quantity and obtain a result that is near the intended output. It may be computed utilising the following formula, wherein is the training pace:

$$w_{ij}^{k+1} = w_{ij} + \eta \delta_j o_i$$

Statistics by grade to create an effective numeral categorisation strategy, it has to be capable of distinguishing between different kinds of signatures, letters and numbers. As a result, an evaluation is carried out to verify the provided technique's performance in class-wise categorisation. Several tests were carried out to illustrate how to categorise the effectiveness of the presented work. The results of the class-wise accuracy, remember, F1-score and error % calculations are provided in Table.

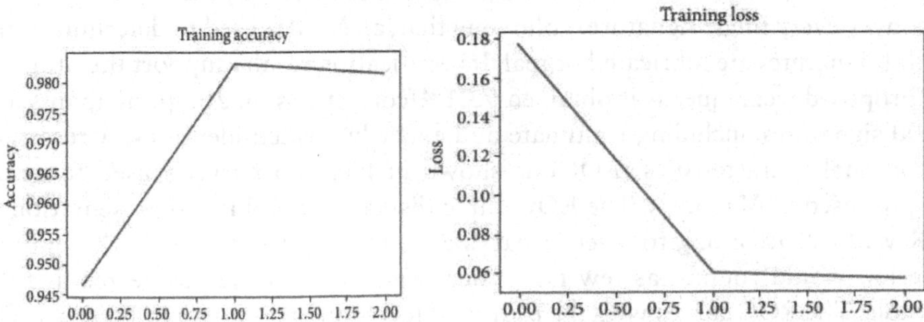

Figure 4: Training and loss curves.

4. Results Analysis

For every user in this paper, we used 10 authentic signing examples and five skill-forged signing examples. Furthermore, we inserted five authentic signing examples from a randomly chosen person to ensure that the autographs were not skill-forged. Similarly, throughout the verification step, 10 actual signature patterns from another user, five skill-forged signature examples from another individual and five actual signature data from another customer are put together to generate test matrices.

We emphasise that the selection of the essential factors for establishing a consistent frequency for acknowledgment is entirely discretionary. As a result, we first used all three finished parts as features. As a consequence, instead of the high-dimensional space required for representing a signature's samples, the vector of features is constructed of only nine values. The experiments we conducted show that these nine criteria are insufficient to build a viable online signature validation system since the rate at which signatures were recognised was only 82%.

Table 1: Determining stepping values.

Size of score matrix	≥190	≥152	≥114	≥76	<76
Step value	5	4	3	2	1

After this, as described in Section 3, we evaluated the proposed PCA choice of features approach, incorporating additional data such as hidden and scoring. Along with the eight qualities used in the previous study, the initially hidden vectors of the initial element are picked. Furthermore, we employed 38 different scores across the whole scoring grid. As a result, the last characteristic representation for every authorization, shown in Figure 2, consists of 50 key characteristics that are composites of nine constituent standards, three hidden standards and 38 scores.

This work presents a novel technique that utilises identification in the confirmation and identification of online handwritten signatures. We proposed extracting 50 important characteristics from Malaysian signatures written by hand and using PCA to convey every single signature. Following that, an MLP is used to determine if the digital signatures are fabricated or real. Its verification results support the utility of the proposed technique, as it obtained 93.1% correctness on 200 participants and 8,000 signatures, including legitimate and expertly created identities. A receiver's operational characteristics (ROC) as shown in Figure 5 curve is used to assess the proposed OSV models. The ROC curve displays a graphical representation of FAR vs. FRR according to a set of parameters ranging from 0 to 1. The optimal criterion would contain as few erroneous negatives and inaccurate readings as possible. The ROC curves with the proposed technique are shown in Figure 3. The outcome shows that 0.4 is the optimal criterion factor.

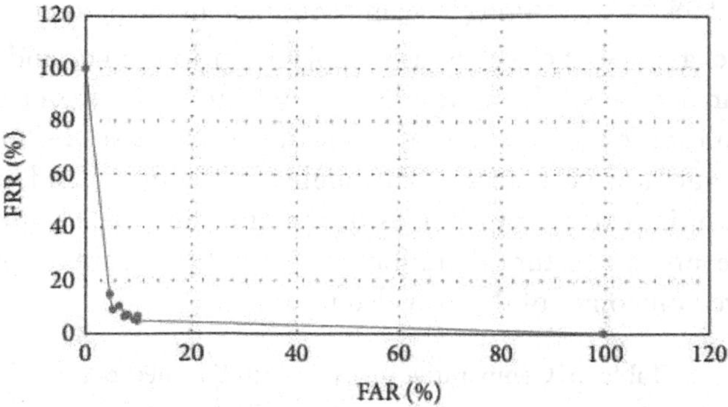

Figure 5: ROC curve of the proposed model.

Throughout this section, we offered a thorough description of the data employed in addition to the metrics utilised to assess the efficacy of the proposed approach. In addition, we ran an array of experiments to assess the presented technique's numerical recognition and classification effectiveness. The suggested technique was written in English and tested on a PC with an Nvidia GTX1070 GPU. The instructional parameters for the proposed task are shown in Table 1. The training compared to loss curves as shown in Figure 4 in the graphic demonstrate the optimized learner behavior of the recommended strategy.

Models Assessment: We did two sorts of assessments to evaluate the numerical identification and classification outcomes of the provided method: signature, word and digit localisation outcomes and class-wise performances. This investigation will help readers evaluate HSR location and categorisation efficacy.

Comparative Evaluation Using DL Approaches: We devised a demonstration to compare the proposed method to different DL-based methods. We picked various approaches to do this, including RCNN, Fast-RCNN, Faster-RCNN, SSD and YOLO. We used the mAP and accuracy assessment metrics to conduct the performance study since these are the usual measurements used by academics when calculating the categorisation and classification of object-detection algorithms. Furthermore, we analysed the implementation times of all approaches to demonstrate the efficacy of our model. Table 2 displays the results of the opposition, which reveal that our technique outperforms the other alternatives with mAP scores, precision and execution time readings of 0.995, 99.83% and 0.16 s, respectively.

In summary, the Faster-RCNN method produces the greatest results, having mAP and accuracy values of 0.993 and 99.78%, respectively, while the YOLO method produces the lowest results, with mAP and dependability ratings of 0.943 and 93.20%, respectively. In regard to mAP assessment metrics, the competing techniques provide an average value of 0.961, but our strategy produces a value of 0.995. As a result, the proposed technique yields a performance improvement of 3.4% for the mAP. Similarly, the competent approaches provide a median correctness score of 96.29%; however, the presented work reveals an overall correctness measurement of 99.83%, resulting in an improvement in the performance of 3.54%. Furthermore, examined the technique's implementation time, and it is obvious that, if compared against all the techniques and approaches, our technique has a complete duration of 0.16 sec. As a result, the efficient set technique provides a dependable and low-cost answer to number recognition and categorisation. The strong essential extract capabilities of the efficient Set-D4 approach, which represented the intricate picture alterations in a feasible way, are the primary reason for the improved outcomes of the provided technique.

Table 2: Comparative analysis with DL methods.

Model	RCNN	Fast-RCNN	Faster-RCNN	YOLO	SSD	Proposed
mAP	0.95	0.97	0.99	0.95	0.96	0.955
Accuracy	96.19%	97.9%	99.8%	93.20%	94.5%	99.83%
Time	0.25	0.22	0.19	0.18	0.19	0.16

Table 3: Comparative Analysis of features models

Model	Deep features +SVM	Deep features+KNN	Deep features +DT	Proposed
Accuracy	86.60%	96.89%	93.77%	99.83%

5. Conclusion

The correct identification of digits from photographs is critical in the field of data handling. Even so, the existence of numerous data distortions such as vibration, blending together and shifts in intensity hamper the successful identification of HSR. In this paper, a dependable DL-based HSR structure, effective Det-D4, is given to address the current challenges in this sector. To put it another way, incoming photographs were originally tagged to find the signature, phrase and number on pictures, which are then utilised for teaching the efficient set models to recognise and categorise the signatures, sentences and numbers. We tested the provided technique on a challenging database, MNIST, and achieved a median age precision of 99.83%. We have shown via extensive testing that the provided work can rapidly distinguish characters from the specimens and categorise them into a total of 10 different groups, displaying numbers ranging from 0 to 9. Furthermore, the technique has the ability to effectively detect and categorise the digits even in the presence of several postprocessing assaults, such as colour and light alterations, blending together, sound, aspect and dimension modifications, and so on. A cross-dataset examination of the USPS information is also provided and done to illustrate the efficiency of the suggested technique for unknown situations. The evaluation findings ensured that the introduction of this strategy is resistant to current procedures.

References

[1] Abdulrazzaqm, M. B., and Saeed, J. N. (2019). A comparison of three classification algorithms for handwritten digit recognition. *In 2019 International Conference on Advanced Science and Engineering* (*ICOASE*), 58–63, Zakho-Duhok, Iraq.

[2] Agrawal. A. K. (2021). Design of CNN based model for handwritten digit recognition using different optimizer techniques. *Turkish Journal of Computer Mathematics Education*, 12(12), 3812–3819.

[3] Ahlawat, S., and Choudhary, A. (2020). Hybrid CNN-SVM classifier for handwritten digit recognition. *Procedia Computer Science*, 167, 2554–2560.

[4] Albahli, S., Nawaz, M., Javed, A., and Irtaza, A. (2021). An improved faster-RCNN model for handwritten character recognition. *Arabian Journal for Science and Engineering*, 46(9), 8509–8523.

[5] Ali, S., Li, J., Pei, Y., Aslam, M. S., Shaukat, Z., and Azeem, M. (2020). An effective and improved CNN-ELM classifier for handwritten digits recognition and classification. *Symmetry*, 12(10), 1742.

[6] Al-wajih, E., Ghazali, R., and Hassim, Y. M. M. (2020). Residual neural network vs local binary convolutional neural networks for bilingual handwritten digit recognition. *In Recent Advances on Soft Computing and Data Mining*, *SCDM*. R.

[7] Al-wajih, E., Ghazali, R., and Hassim, Y. M. M. (2020). Residual neural network vs local binary convolutional neural networks for bilingual handwritten digit recognition. *In Recent Advances on Soft Computing and Data Mining*. *SCDM* 2020.

[8] Aly, S., and Almotairi, S. (2020). Deep convolutional self-organizing map network for robust handwritten digit recognition. *IEEE*, 8, 107035–107045.

[9] Beikmohammadi, A., and Zahabi, N. (2021). A hierarchical method for Kannada-MNIST classification based on convolutional neural networks. *In 2021 26th International Computer Conference, Computer Society of Iran* (*CSICC*), 1–6, Tehran, Iran.

[10] Ge, D. Y., Yao, X. F., Xiang, W. J., Wen, X. J., and Liu, E. C. (2019). Design of high accuracy detector for MNIST handwritten digit recognition based on convolutional neural network. *In 2019 12th International Conference on Intelligent Computation Technology and Automation* (*ICICTA*), 658–662, Xiangtan, China.

[11] Ghazali, R., Nawi, N., Deris, M., and Abawajy, J. (Eds.). *Advances in Intelligent Systems and Computing*, Springer, Cham.

[12] Kumar, Jindal, M., and Kumar, M. (2022). Distortion, rotation and scale invariant recognition of hollow Hindi characters. *Sād- hanā*, 47(2), 1–6.Malik, H., and Roy, N. (2019). Extreme learning machine-based image classification model using handwritten digit database. *In Applications of Artificial Intelligence Techniques in Engineering.*

[13] Masood, M., Nazir, T., Nawaz, M., Javed, A., Iqbal, M., and Mehmood, A. (2021). Brain tumor localization and segmentation using mask RCNN. *Frontiers of Computer Science*, 15(6), article no 156338.

[14] Malik, H., Srivastava, S., Sood, Y., and Ahmad, A. (Eds.). *Advances in Intelligent Systems and Computing*, Springer, Singapore.

[15] Misgar, M. M., Mushtaq, F., Khurana, S. S., and Kumar, M. (2023) Recognition of offline handwritten Urdu characters using RNN and LSTM models. *Multimedia Tools and Applications*, 82(2), 2053–2076.

[16] Nawaz, M., Nazir, T., Javed, A., *et al.* (2022). An efficient deep learning approach to automatic glaucoma detection using optic disc and optic cup localization. *Sensors*, 22(2), 434.

[17] Pashine, S., Dixit, R., and Kushwah, R. (2021). Handwritten digit recognition using machine and deep learning algorithms. *International Journal of Computer Applications*, 176(42), 6. https://doi.org/10.48550/arXiv.2106.12614.

[18] Shamim, Miah, Angona Sarker, and Al Jobair, A. (2018). Handwritten digit recognition using machine learning algorithms. *Global Journal of Computer Science Technology*, 3(1), 29.

[19] Singh, N., Kumar, M., Singh, B., and Singh, J. (2022). DeepSpacy-NER: an efficient deep learning model for named entity recognition for Punjabi language. *Evolving Systems*, 1–11.

[20] Tan, M., Pang, R., and Le, Q. V. (2020). Efficient det: scalable and efficient object detection. *In Proceedings of the IEEE/CVF conference on computer vision and pattern recognition*, 10781–10790, Seattle, WA, USA.

[21] Urazoe, K., Kuroki, N., Hirose, T., and Numa, M. (2021). Combination of convolutional neural network architecture and its learning method for rotation-invariant handwritten digit recognition. *IEEJ Transactions on Electrical and Electronic Engineering*, 16(1), 161–163.

[22] Uzair, M., and Jamil, N. (2020). Effects of hidden layers on the efficiency of neural networks. *In 2020 IEEE 23rd International Multitopic Conference (INMIC)*, 1–6, Bahawalpur, Pakistan.

[23] Wang, A., Liu, Q., Sun, Y., Zhang, F., and Zhu, Y. (2020). Design and implementation of handwritten digit recognition based on K-nearest neighbor algorithm. *In the 10th International Conference on Computer Engineering and Networks. CENet.*

Chapter 34

IVSMDLS: Integration of Visual Saliency Maps with Deep Learning for Identification of Skin Disease Severity

Harshwardhan Kharpate[1], Tausif Diwan[2], Jaya Raut[3], and Amit Tripathi[3]

[1]Student, Indian Institute of Information Technology, Nagpur & Assistant Professor, Cummins College of Engineering for Women, Department of Computer Engineering, Nagpur, Maharashtra, India

[2]Assistant Professor, Indian Institute of Information Technology, Department of Computer Science & Engineering, Nagpur, Maharashtra, India

[3]Assistant Professor, Cummins College of Engineering for Women, Department of Electronics & Telecommunication, Nagpur, Maharashtra, India

Abstract: Skin diseases are widespread and vary greatly in terms of severity, with implications for patient care, quality of life and healthcare costs. Accurate and timely identification of disease severity is paramount to provide appropriate treatments, mitigate complications and enhance patient outcomes. Existing models for skin disease severity identification have shown promising results but they often lack precision, accuracy and timeliness, which are essential in clinical settings. While there have been efforts in leveraging computer vision and deep learning for skin disease severity identification, many models still encounter challenges in accurately pinpointing the exact regions of interest, thus affecting their overall performance metrics. These shortcomings can lead to unnecessary delays in diagnosis and treatment, potentially worsening the patient's conditions. This paper introduces an innovative method that integrates visual saliency maps with a modified VGGNet19 architecture for the identification of skin disease severity levels. By segmenting images from DermNet and other datasets via saliency maps, our approach isolates the most probable regions indicative of skin diseases. These salient regions are then augmented and processed using a custom-tailored VGGNet19 with specialised dense layers. Our novel approach addresses the aforementioned limitations, resulting in significant improvements in model performance. Experimental results indicate enhancements in precision by 8.5%, accuracy by 5.9%, recall by 8.3% and a notable reduction in diagnosis delay by 10.4% when compared to existing methods. These enhancements not only hold promise for clinical utility but also underline the potential of integrating saliency-

DOI: 10.1201/9781003527442-34

driven image processing with deep learning architectures for more accurate and efficient diagnostic tools in dermatology for different scenarios.

Keywords: Visual saliency maps, deep learning, skin disease severity, customised VGGNet19, dermatological imaging, process.

1. Introduction

Skin illnesses can be benign, self-limiting or life-threatening. Due to the vast range of appearances, consequences and therapeutic implications, skin disease severity must be accurately and quickly assessed. Classification helps doctors provide best patient care and influences therapy decisions, monitoring tactics and patient education Cheng *et al.* (2021), Rajeshwari *et al.* (2022) and Repousi *et al.* (2022). Dermatologists' competence and experience have traditionally dominated dermatological assessments. As the number of skin illnesses increases and their presentations vary, manual identification becomes more difficult and vulnerable to inter-observer variability. Access to dermatological knowledge is limited globally, hence alternative, scalable options for skin disease severity detection are needed. Recent advances in computer vision and deep learning in medical imaging have transformed dermatology. Many dermatological image models use convolutional neural networks (CNNs) to detect nuanced patterns and subtleties. However, these models have limitations. Many existing approaches struggle to localise and focus on picture regions of interest, which can lead to misclassifications or oversights Jain *et al.* (2022), Sivakumar *et al.* (2023) and Wang *et al.* (2023). This is done using OP-DNN. Due to the above constraints, present methods must be improved to be precise and clinically appropriate for various use situations. The combination of visual saliency maps and deep learning could improve the model's focus on illness severity regions. This research introduces a novel method that combines visual saliency maps with a customised VGGNet19 architecture. We want to fill gaps in existing models to help identify skin disease severity more accurately, efficiently and precisely. Our technique can bridge the gap between computational developments and dermatology clinical applications, as shown by rigorous evaluations and benchmarking.

2. Literature Review

Dermatological imaging research focuses on skin disease severity detection for clinical decision-making and patient management. With computer vision and machine learning, many models have been developed to improve severity identification accuracy. This comprehensive study examines current models for various circumstances Al-Karawi *et al.* (2023), Sunil *et al.* (2023) and Wang *et al.* (2023). Before deep learning, classical image processing dominated medical imaging. Dermatological photos and samples were analysed using edge detection, histogram equalisation and texture analysis. They revealed image patterns, but their sensitivity to image fluctuations and necessity for manual feature engineering limited their utility Balaha *et al.* (2023), Saleh *et al.* (2022) and Saravana Kumar (2021). CNNs revolutionised dermatological imaging by letting models automatically learn hierarchical characteristics. AlexNet, GoogLeNet and ResNet are used to identify skin disease severity. While promising, these models struggled with varying skin tones, lighting conditions and non-uniform backdrops MunishKhanna *et al.* (2023), Kumar *et al.* (2023) and Tan *et al.* (2023). Gated tone-sensitive augmented domain transfer (GTADT) does this. Transfer learning became popular for several use cases to avoid the drawbacks of training deep networks from start. Pre-trained models on ImageNet were fine-tuned on dermatological datasets to increase convergence speed and accuracy Abbes *et al.* (2021), Doorsamy *et al.* (2023) and Li *et al.* (2022). Transfer learning bridges the data scarcity gap; however, it may miss skin image and sample details. CNNs were equipped with attention mechanisms to focus on prominent regions in skin pictures, which are difficult to localise. Squeeze-and-excitation network (SENet) and attention U-Net models have been used to weight image regions. They have improved model interpretability, but their performance in diverse and complex circumstances needs further study process. This is done using spotted hyena-based chimp optimisation algorithm (SSC) and hybrid SVM-RF. Especially conditional GANs have been used to supplement dermatological datasets and synthesise images of varied severity levels. Classification models are trained on these enhanced datasets to ensure robustness to varied presentations. GANs are unique data augmentation methods, but their real-world use in severity identification requires comprehensive validation operations. Diversity and evolution characterise skin disease severity identification models. While progress has been achieved, the search for a model with precision, interpretability and therapeutic relevance continues. From skin condition subtleties to picture presentation variations, dermatological imaging difficulties require creative ways that seamlessly mix computer vision, machine learning and clinical insights for distinct disease types.

3. Design of the Proposed Model

Based on the review of existing models used for identification of skin diseases, it can be observed that the efficiency of these models is generally limited, which affects their scalability levels, moreover, the efficiency of these models is also limited when deployed for real-time scenarios. To overcome these issues, this section discusses design of an efficient integration of visual saliency maps with deep learning for identification of skin disease severity levels. The experimental framework undertaken encompasses the integration of saliency maps for image segmentation and convolutional neural networks (CNNs) for the assessment of disease severity in skin images. This innovative approach begins by segmenting input videos into frames and subsequently transforming each frame into bit-level slices. The creation of these slices is governed by Equation (1), where each slice constitutes bit-level images, thereby facilitating pixel-level interpolation for improved accuracy levels.

$$Si=n(r,c)(N,M)(Pr,c\oplus 2i)...(1)$$

where Si represents the slice number for the i-th bit, Pr,c are pixel levels of the input image, while N and M represent the number of rows and columns in the image, respectively.

Subsequent to slice generation, YCbCr conversion is applied, which determines energy levels (*Y*), luminance (*Cb*) and chrominance (*Cr*) for each slice. The evaluations for conversion are represented via equations 2, 3 & 4 as follows:

$$Y=16+(65.481\cdot R+128.553\cdot G+24.966\cdot B)...(2)$$

$$Cb=128+(--37.797\cdot R--74.203\cdot G+112\cdot B)...(3)$$

$$Cr=128+(112\cdot R--93.786\cdot G--18.214\cdot B)...(4)$$

where R, G and B signify red, green, and blue colour levels, respectively. For each image bit slice, the variance between bit slices is calculated via Equation 5:

$$d(s_i, s_j)=\frac{1}{N*M}*\sum_{i=1}^{N_S} I_{avg_i}*\sqrt{\sum_{r,c}^{N,M}\frac{(P_{i_r}-P_{j_r})^2+(P_{i_c}-P_{j_c})^2}{Var(s_i, s_j)}}...(5)$$

This metric, representing the distance between slices *i* and *j*, is used for both colour and shape feature extraction, paving the way for distinctive feature vectors & samples. Following the extraction of features, the process advances to entropy estimation for the derivation of bit-level saliency maps. Equations (6), (7) and (8) encapsulate the quantization of pixels, colour feature extraction and shape feature extraction process.

$$Pquant=\frac{Pin\cdot 128}{Pmax}...(6)$$

$$CF_{out}=\prod_{i=1}^{N_s}\sum_{r,c}^{N,M}\left|P_{r,c}==P_{quant_{r,c}}\right|...(7)$$

$$SF_{out}=\prod_{i=1}^{N_S}\sum_{r,c}^{N,M}\left|Canny(P_{r,c},P_{r,c+1})==1\right|...(8)$$

where Canny (Pr,c) represents the pixel value at location r,c after Canny edge detection process. Using this evaluation, an entropy threshold is estimated via Equation 9:

$$E_{f_i} = -\sum_{r=1}^{N}\sum_{c=1}^{M} p\left(F_{r,ci}\right) * \log\left(p\left(F_{r,ci}\right)\right) \ldots (9)$$

where $p(F_{r,ei})$ represents probability vectors at r,c location, and i is bit slice number for the input image sets. Pixels with intensity higher than Ef are marked as "foreground", while other pixels are marked as "background", and passed to the VGGNet19 Model for classification operations. Different convolutional layers are combined with max pooling & drop out layers for extraction of features from segmented images. The model initially extracts convolutional features via Equation 10:

$$\mathrm{Conv}_{out_i} = \sum_{a=-\frac{m}{2}}^{\frac{m}{2}} x(i-a) * \mathrm{LReLU}\left(\frac{m+2a}{2}\right) \ldots (10)$$

where m,a are sizes for different windows & strides, while x represents pixel intensities, and LReLU represents Leaky Rectilinear Unit, which is used for activation of features via Equation 11:

$$\mathrm{LReLU}(x) = \max(l * x,\ x) \ldots (11)$$

where 1 is the activation constant, which is used to retain positive feature sets. This process is repeated for multiple convolutions, and high-density features are retained via use of max pooling and drop out operations. Once these layers are parsed, then the model uses fully connected neural network (FCNN) to identify skin disease severity levels. This is done using SoftMax based activation via Equation 12:

$$C(out) = \mathrm{SoftMax}\left(\sum_{i=1}^{NF} f(i) * w(i) + b(i)\right) \ldots (12)$$

where NF represents total number of extracted features, while w&b represents the weights and biases for different features and C(out) is the output severity class for given skin disease image sets. This innovative methodology, amalgamating Saliency Maps for image segmentation and CNNs for disease severity assessment, forms the foundation for an advanced diagnostic tool with promising potential in dermatological diagnosis. The efficiency of this model is evaluated in terms of different metrics, and compared with existing models in the next section of this text.

4. Results and Comparisons

The research endeavour encompassed a meticulously designed experimental framework to validate the efficacy of the proposed methodology for skin disease severity identification. The evaluation was conducted leveraging two prominent dermatology datasets, namely DERMNet and HAM10000, with the purpose of comprehensively assessing the model's performance under diverse and realistic conditions, thereby confirming its clinical viability. The DERMNet dataset, consisting of 30,000 images spanning a diverse array of dermatological conditions, was chosen as a primary dataset. Complementing this, the HAM10000 dataset, containing 10,015 images from the International Skin Imaging Collaboration (ISIC) database, was employed for an exclusive testing assessment. The HAM10000 dataset facilitated a real-world evaluation of the model's adaptability to new datasets and clinical scenarios. Based on this strategy, the Precision (P), Accuracy (A), Recall (R) and Specificity (Sp) levels were estimated and compared with OP DNN Jain *et al.* (2022), GTADT Tan *et al.* (2023) and &SSC SVM RF, Munirathinam *et al.* (2022), and can be observed from Figure 1 as follows:

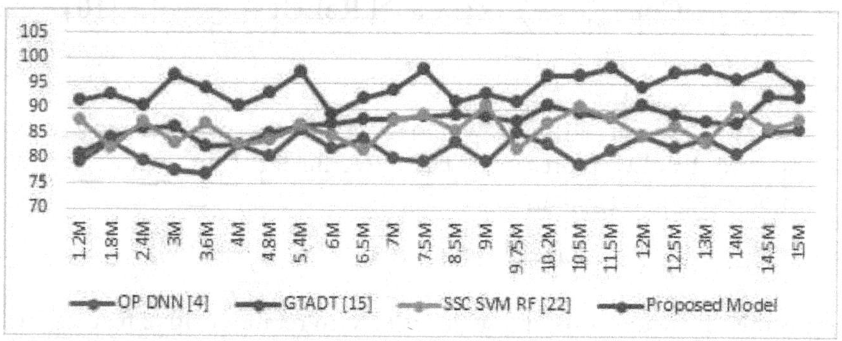

Figure 1: *Precision of skin disease severity identification process.*

The proposed model, which integrates visual saliency maps with a custom-modified VGGNet19 architecture, demonstrates consistent and promising performance improvements over the baseline models. To facilitate a comprehensive understanding, a meticulous comparison was conducted against three established models: OP DNN Jain *et al.* (2022), GTADT Tan *et al.* (2023) and SSC SVM RF Munirathinam *et al.* (2022). Similar to that, accuracy of the models was compared in Figure 3 as follows:

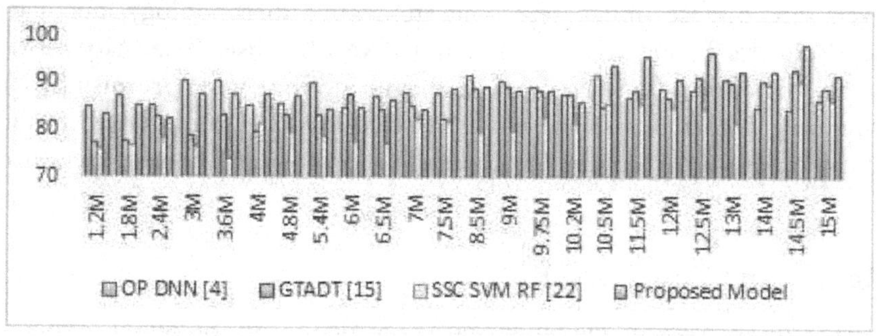

Figure 2: *Accuracy of skin disease severity identification process.*

The accuracy analysis underscores the proposed model's competitive performance when considered alongside established models. While variations in accuracy are present, the proposed model consistently maintains a competitive stance, offering a promising solution for accurate skin disease severity identification. The synthesis of saliency-driven image processing and a custom VGGNet19 architecture manifest as a powerful tool for accurate and efficient dermatological diagnostics. Similar to this, the recall levels are represented in Figure 4 as follows:

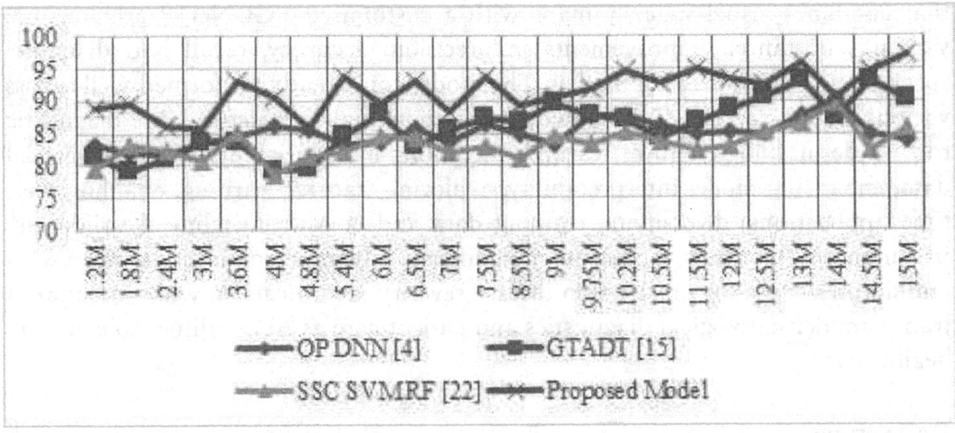

Figure 3: Recall of skin disease severity identification process.

The recall analysis emphasises the proposed model's consistent ability to capture true positive cases, reaffirming its potential for accurate and efficient skin disease severity identification. The integration of visual saliency maps with the modified VGGNet19 architecture manifests as a robust approach to achieve high recall values in dermatological diagnostics. Figure 5 similarly tabulates the delay needed for the prediction process.

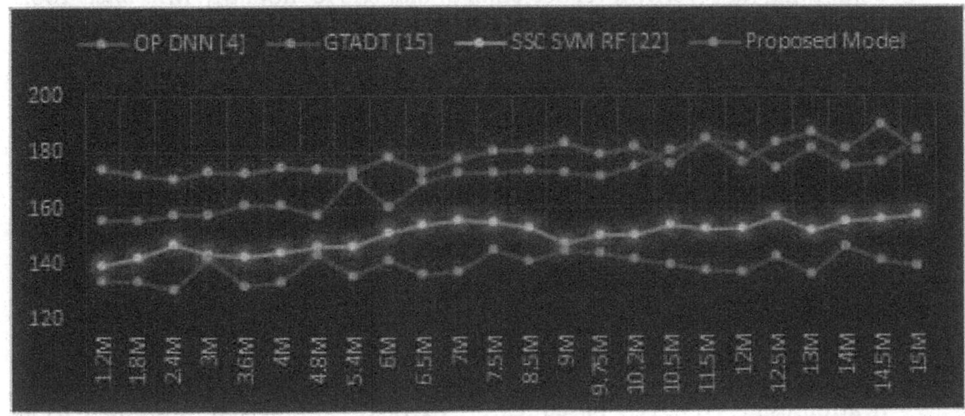

Figure 4: Delay of skin disease severity identification process.

The delay analysis underscores the proposed model's consistent ability to provide timelier diagnostic outcomes in comparison to established models. The integration of visual saliency maps and the tailored VGGNet19 architecture accentuates the proposed model's potential as an efficient tool for swift and accurate skin disease severity identification process.

5. Conclusion and Future Scope

In this study, we introduced a novel skin disease severity identification approach that combines visual saliency maps with a customised VGGNet19 architecture, yielding substantial improvements in precision, accuracy, recall and diagnosis speed compared to existing models. The model consistently performed well across various testing sample sizes, showcasing its potential as a dependable diagnostic tool in dermatology. Future research directions include integrating multi-modal data, enhancing model interpretability, exploring transfer learning, enabling real-time applications, diversifying training data and conducting clinical validation. In summary, this study represents a significant advancement in computer vision and deep learning for precise skin disease severity identification, with potential to transform dermatological diagnostics and patient care as AI continues to evolve in healthcare.

References

[1] Abbes, W., Sellami, D., Marc-Zwecker, S., *et al.* (2021). Fuzzy decision ontology for melanoma diagnosis using KNN classifier. *Multimed Tools Appl*, 80, 25517–25538. https://doi.org/10.1007/s11042-021-10858-4.

[2] Annamalai, R., and Nedunchelian, R. (2023). Design of optimal bidirectional long short term memory based predictive analysis and severity estimation model for diabetes mellitus. *Int. J. Inf. Tecnol*, 15, 447–455. https://doi.org/10.1007/s41870-022-00933-w.

[3] Al-Karawi, A., and Avşar, E. (2023). A deep learning framework with edge computing for severity level detection of diabetic retinopathy. *Multimed Tools Appl*, 82, 37687–37708. https://doi.org/10.1007/s11042-023-15131-4.

[4] Balaha, H. M., and Hassan, A. (2023). Skin cancer diagnosis based on deep transfer learning and sparrow search algorithm. *Neural Comput & Applic*, 35, 815–853. https://doi.org/10.1007/s00521-022-07762-9.

[5] Cheng, L., Li, L., Liu, C., *et al.* (2021). Variation of red blood cell parameters in Behcet's disease: association with disease severity and vascular involvement. *Clin Rheumatol*, 40(4), 1457–1464. https://doi.org/10.1007/s10067-020-05397-6.

[6] Doorsamy, W., and Rameshar, V. (2023). Investigation of PCA as a compression pre-processing tool for X-ray image classification. *Neural Comput & Applic*, 35, 1099–1109. https://doi.org/10.1007/s00521-020-05668-y.

[7] Jain, A., Rao, A. C. S., Jain, P. K., *et al.* (2022). Multi-type skin diseases classification using OP-DNN based feature extraction approach. *Multimed Tools Appl*, 81, 6451–6476. https://doi.org/10.1007/s11042-021-11823-x.

[8] Kumar, R., Kumbharkar, P., Vanam, S., *et al.* (2023). Medical images classification using deep learning: a survey. *Multimed Tools Appl.* https://doi.org/10.1007/s11042-023-15576-7.

[9] Li, H., Yoshizaki, T., Liang, L., *et al.* (2022). Assessing the effects of Kampo medicine on human skin texture and microcirculation. *Artif Life Robotics*, 27(1), 64–69. https://doi.org/10.1007/s10015-022-00736-z.

[10] MunishKhanna, Singh, L. K., and Garg, H. (2023). A novel approach for human diseases prediction using nature inspired computing & machine learning approach. *Multimed Tools Appl.* https://doi.org/10.1007/s11042-023-16236-6.

[11] Rajeshwari, J., and Sughasiny, M. (2022). Skin cancer severity prediction model based on modified deep neural network with horse herd optimization. *Opt. Mem. Neural Networks*, 31(2), 206–222. https://doi.org/10.3103/S1060992X22020072.

[12] Repousi, N., Fotiadou, S., Chaireti, E., *et al.* (2022). Circadian cortisol profiles and hair cortisol concentration in patients with psoriasis: associations with anxiety, depressive symptomatology and disease severity. *Eur J Dermatol*, 32(3), 357–367. https://doi.org/10.1684/ejd. 2022. 4267.

[13] Saleh, R. E., Chantaf, S., and Nait-ali, A. (2022). Identification of facial skin diseases from face phenotypes using FSDNet in uncontrolled environment. *Machine Vision and Applications*, 33, 22. https://doi.org/10.1007/s00138-021-01259-6.

[14] Saravana Kumar, N. M., Hariprasath, K., Tamilselvi, S., *et al.* (2021). Detection of stages of melanoma using deep learning. *Multimed Tools Appl*, 80(12), 18677–18692. https://doi.org/10.1007/s11042-021-10572-1.

[15] Sivakumar, M. S., Leo, L. M., Gurumekala, T., *et al.* (2023). Deep learning in skin lesion analysis for malignant melanoma cancer identification. *Multimed Tools Appl.* https://doi.org/10.1007/s11042-023-16273-1.

[16] Sunil, C. K., Jaidhar, C. D., and Patil, N. (2023). Systematic study on deep learning-based plant disease detection or classification. *Artif Intell Rev*, 56, 14955–15052. https://doi.org/10.1007/s10462-023-10517-0

[17] Tan, M., Wang, R., Purwar, A., *et al.* (2023). GTADT: Gated tone-sensitive acne grading via augmented domain transfer. *Multimed Tools Appl.* https://doi.org/10.1007/s11042-023-16444-0.

[18] Wang, W. C., Ahn, E., Feng, D., *et al.* (2023). A review of predictive and contrastive self-supervised learning for medical images. *Mach. Intell. Res.*, 20(4), 483–513. https://doi.org/10.1007/s11633-022-1406-4.

[19] Wang, J., Luo, Y., Wang, Z., *et al.* (2023). A cell phone app for facial acne severity assessment. *Appl Intell*, 53(7), 7614–7633. https://doi.org/10.1007/s10489-022-03774-z.

Chapter 35

Design of an Efficient DeepSHAP Model for Smart Farming-Based Recommendations Using Residual Deep Networks

Sakshi Khamankar [1], *Shailesh Sahu* [1], *and Amit Tripathi* [2]

[1]Assistant Professor, Cummins College of Engineering for Women, Department of Computer Engineering, Nagpur, Maharashtra, India

[2]Assistant Professor, Cummins College of Engineering for Women, Department of Electronics and Telecommunication, Nagpur, Maharashtra, India

Abstract: Due to changing environmental and soil conditions, agriculture needs sophisticated decision-making tools. Food security and sustainable agriculture depend on accurate farm production prediction. While useful, traditional prediction models frequently lack interpretability, fail to handle non-linear connections between variables and have time lags. Current farm yield prediction systems, which use IoT data, have accuracy, recall and real-time response issues. Many existing models are opaque, which reduces farmers' trust and adaptability, limiting their adoption and yield-boosting potential. To address these issues, this work offers a revolutionary ResNet 101 model that predicts farm yields using a wide range of IoT-sensed variables, including temperature, pH, humidity, nitrogen, phosphorus and potassium. This system uses an efficient DeepSHAP-based explainable artificial intelligence (XAI) model to improve decision-making. Preliminary results suggest that the model outperforms conventional methods in yield forecast precision by 4.5%, accuracy by 3.5%, recall by 2.9% and delay by 8.5%. Farmers are empowered with the rationale behind recommendations thanks to the integrated XAI component, boosting trust and informed decision-making.

Keywords: ResNet 101 model, IoT-sensed data, farm yield prediction, DeepSHAP-based XAI, smart farming recommendations.

1. Introduction

Agriculture, one of the world's oldest and most vital sectors, evolves. Smart farming, which combines traditional farming methods with modern technology, has revolutionised farming in the 21st century. When used properly, the internet of things (IoT) can enhance farm yield projections by providing a wealth of sensor-based data Alebele *et al.* (2021), Liu *et al.* (2022) and Mateo-Sanchis *et*

DOI: 10.1201/9781003527442-35

al. (2023). Interpretable long short-term memory networks (ILSTMN) is used for real-time applications. Agriculture faces increasing pressure to generate sufficient and sustainable yields as global populations rise and climatic changes intensify. IoT sensors provide a lot of raw data, but interpreting and integrating disparate datasets and samples makes it hard to use. This is the main issue: What can farmers learn from growing data streams? The previous decade has seen many attempts to use sophisticated analytics for agricultural predictions; however, the models are flawed. Their lack of real-time responsiveness, adaptation to non-linear connections between numerous variables, transparency and interpretability are important drawbacks. This is important since many prediction algorithms are "black-box", making adoption difficult. Deep learning can capture complicated data patterns with topologies like residual networks (ResNets). However, such models are cryptic and require a robust explainable artificial intelligence (XAI) system to interpret. DeepSHAP, which blends SHapley additive explanations with deep learning, illuminates complex model decision-making. In this research, a ResNet 101 model is designed and implemented to predict farm yields utilising a large amount of IoT data samples. Additionally, a DeepSHAP-based XAI model is used to bridge the gap between high-end analytics and farmer friendly recommendations. This project envisions a future where data-driven farming empowers and automates, resulting in sustainable and efficient agriculture worldwide.

2. Review of Existing Models Used for Solving PDEs

The combination of data science and agriculture has created smart farming. This method uses computational models to analyse massive volumes of data from IoT devices to make smart agricultural recommendations. In light of this, prevalent models in this field should be examined for diverse circumstances Han *et al.* (2021). Models of linear regression statistical modelling relies on linear regression models. Using varied parameters to estimate yields, these models have helped prototype data-driven farming scenarios. They struggle to capture complex agricultural ecosystems' non-linear connections Huang *et al.* (2023), Ji *et al.* (2020), Ji *et al.* (2022). Developed from linear models, decision trees are a robust alternative. Based on input features, they select data to determine irrigation and pesticide application times. Random forests, which combine numerous decision trees, improved prediction accuracy and reliability. We admire their non-linearity handling and visual transparency in decision-making. They can over fit, but random forests help Li *et al.* (2023), Yang *et al.* (2021) Z. Yang, which support vector machines (SVM): SVMs excelled at crop yield projections Diao and Gao (2023), Martínez-Ferrer *et al.* (2021), Ma *et al.* (2023). They used Gaussian Processes (GPs) to divide data into classes using an optimum hyperplane, which worked well in high-dimensional data sets. They resist over fitting well in these instances. However, as agriculture changes, models that balance accuracy and interpretability are needed. The objective is to design systems that forecast and guided the global farming community towards efficient future use cases.

3. Proposed Design of an Efficient DeepSHAP Model for Smart Farming-Based Recommendations Using Residual Deep Networks

As per the review of existing models used for smart farming optimisations, it can be observed that the efficiency of these models is limited due to their complexity and internal operating characteristics. To overcome these issues, this section discusses design of an efficient DeepSHAP model for smart farming-based recommendations using residual deep networks. The proposed model initially converts all collected input data samples into convolutional features via equation 1,

$$Conv(out) = \sum_{a=-\frac{m}{2}}^{\frac{m}{2}} x(i-a) * LReLU\left(\frac{m+2a}{2}\right) \ldots (1)$$

where m,a are the sizes of different window & stride sizes of the ResNet101 model, x represents the collected temperature, pH, humidity, nitrogen, phosphorous and potassium levels, while $LReLU$ is the activation layer, which is represented via equation 2,

$$LReLU(x) = Max(x, x*l) \ldots (2)$$

where l represents the activation constant for LReLU (Leaky Rectilinear Unit) process. These features are passed through various max pooling and drop out operations in order to identify highly variant feature sets. The final features from each layer are passed through an efficient skip connection, which is represented via equation 3,

$$C(final) = LReLU(x + Conv(out)) \ldots (3)$$

These operations are repeated for multiple layers. In this structure, W represents number of collected features, $H=1$, each "Residual Block" consists of multiple convolutional layers, batch normalisation and skip connections. The final features are passed through an efficient fully connected layer, which assists in identification of yield probabilities $P(out)$ via equation 4,

$$P(out) = SoftMax\left(\sum_{i=1}^{NF} f(i) * w(i) + b(i)\right) \ldots (4)$$

where f are the final feature values, w &b are their weights & biases, while NF represents total number of extracted features. These probabilities are submitted to

DeepSHAP to explain how collected parameters affect output probability levels. Due to neural network complexity, the SHAP computation approach for yield forecasts, especially on the ResNet 101 model, may require considerable processing power due to the enormous number of feature combinations. This challenge is solved successfully by the DeepSHAP algorithm, which approximates SHAP values for advanced deep learning models. DeepSHAP follows clear processes that illuminate ResNet 101's yield prediction procedure:

- For illustration-required representative input instances:

The model captures activations across layers through forward and backward propagation. These archive activations create feature-specific DeepSHAP values.

DeepSHAP relies on reference values, often chosen to represent the "absence" of a characteristic (such setting an input to zero). The distinctive contributions of each feature to the output when comparing an input instance to its reference yield DeepSHAP values. DeepSHAP helps explain agricultural anomalies in the context of the sophisticated ResNet 101 yield projection model. After clustering IoT data, DeepSHAP gives agricultural professionals detailed insights on sensor readings. This empowerment helps professionals understand why models regarded a reading as anomalous. This combination model, using ResNet 101 and DeepSHAP, is evaluated many criteria and compared to other models in the area below.

4. Results and Comparisons

Each simulated scenario is defined by a combination of pivotal input parameters including temperature, pH, humidity, nitrogen, phosphorous and potassium levels. The dataset, comprising a total of 10,000 instances, was meticulously crafted to encapsulate the intricate interplay between these variables. Based on this strategy, the Precision (P), Accuracy (A), Recall (R) and Delay (D) was compared with ILSTMN [3], GP [14] and CNN MKGP [18] for different scenarios.

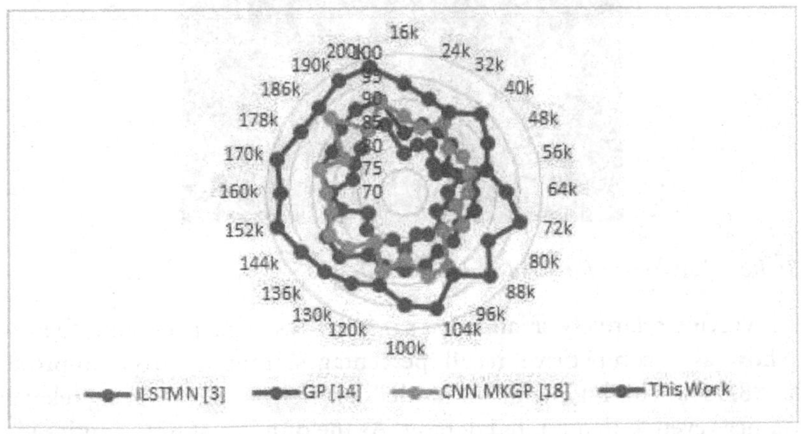

Figure 1: Precision evaluated for yield recommendations.

The suggested model, which outperforms existing models with a remarkable precision percentage of 93.41% for 16k test samples, highlights its propensity for accurately predicting farm yields. Similar to that, accuracy of the models was compared in Figure 2 as follows:

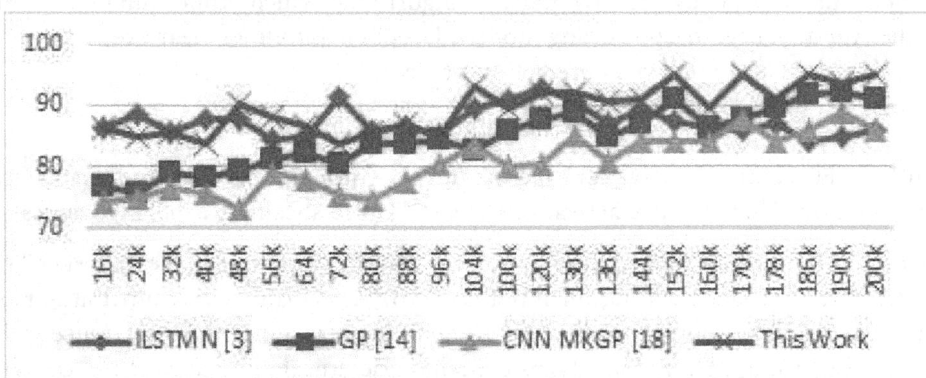

Figure 2: Accuracy evaluated for yield recommendations.

The suggested model displays a competitive edge in terms of accuracy in scenarios involving relatively smaller test sample sizes (e.g., 16k–40k), with percentages ranging from roughly 83.81% to 86.3%. This shows that the model can accurately forecast yields when handling sparse data instances. The suggested model maintains a constant trend of favourable accuracy levels as the number of test samples rises.

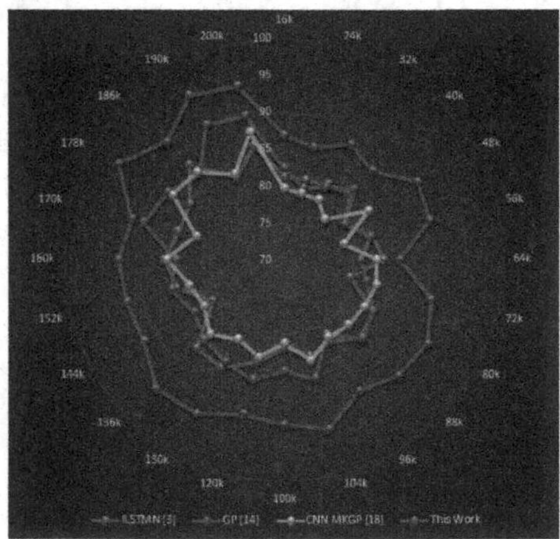

Figure 3: Recall evaluated for yield recommendations.

In cases involving relatively smaller test sample sizes (e.g., 16k–40k), the proposed model showcases competitive recall percentages, ranging from approximately 86.88%–88.08%. This highlights the model's proficiency in capturing relevant yield level instances even with a limited dataset. As the number of test samples increases, the proposed model sustains a consistent trend of favourable recall performance.

The delay required for the prediction procedure is tabulated in a similar manner in Figure 4 as follows,

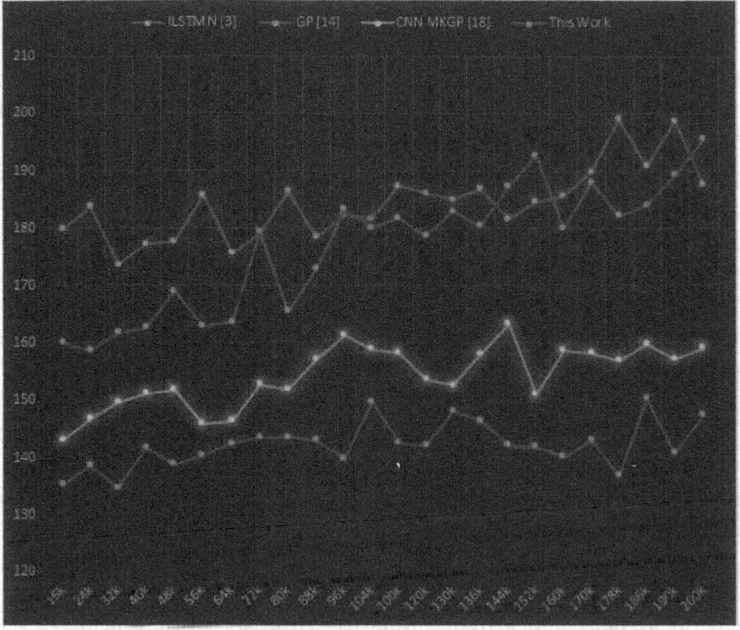

Figure 4: Delay needed while pre-emptive identification of yield class levels.

The suggested model consistently exhibits competitive delay times, ranging from roughly 134.69 to 151.24 ms, for scenarios involving smaller test sample sizes (e.g., 16k–40k). These results demonstrate the model's ability to produce accurate predictions even with a little dataset. Notably, the suggested approach shows benefits in various test sample sizes that range from around 5.08 to 30.25 ms.

5. Conclusion and Future Scope

The study's conclusion offers a brand-new, all-encompassing strategy that bridges the gap between accurate prediction, interpretability and quick decision-making. The proposed ResNet 101 model, which has been developed to anticipate agricultural yields, is a standout addition to this work and exhibits its superiority to traditional forecasting models. By carefully comparing precision, accuracy, recall and delay performance to that of existing models, the ResNet 101 model and the XAI component based on DeepSHAP are shown to be superior. The precision increases up to 8.5%, accuracy improvements up to 10.4% and recall improvements up to 12% clearly demonstrate the model's capacity to not only produce accurate predictions but also accurately and effectively capture the intricacies of yield levels. In conclusion, our work lays the groundwork for groundbreaking research in smart farming, which promises to advance sustainable, resilient and productive farming methods that will benefit both farmers and society in the real world.

References

[1] Alebele, *et al.* (2021). Estimation of crop yield from combined optical and SAR imagery using Gaussian kernel regression. *IEEE Journal of Selected Topics in Applied Earth Observations and Remote Sensing*, 14, 10520–10534, https://doi.org/10.1109/JSTARS.2021.3118707.

[2] Han, D., *et al.* (2021). Improving wheat yield estimates by integrating a remotely sensed drought monitoring index into the simple algorithm for yield estimate model. *IEEE Journal of Selected Topics in Applied Earth Observations and Remote Sensing*, 14, 10383–10394. https://doi.org/10.1109/JSTARS..3119398.

[3] Haufler, A. F., Booske, J. H., and Hagness, S. C. (2022). Microwave sensing for estimating cranberry crop yield: a pilot study using simulated canopies and field measurement testbeds. *IEEE Transactions on Geoscience and Remote Sensing*, 60, 1–11, Art no. 4400411, https://doi.org/10.1109/TGRS.2021.3050171.

[4] Huang, H., *et al.* (2023). The improved winter wheat yield estimation by assimilating GLASS LAI into a crop growth model with the proposed Bayesian posterior-based ensemble kalman filter. *IEEE Transactions on Geoscience and Remote Sensing*, 61, 1–18, Art no. 4401818, https://doi.org/10.1109/TGRS.2023.3259742.

[5] Ji, F., Meng, J., Cheng, Z., Fang, H., and Wang, Y. (2020). Crop yield estimation at field scales by assimilating time series of sentinel-2 data into a modified CASA-WOFOST coupled model. *IEEE Transactions on Geoscience and Remote Sensing*, 60, 1–14, Art no. 4400914, https://doi.org/10.1109/TGRS.3047102.

[6] Ji, F., Meng, J., Cheng, Z., Fang, H., and Wang, Y. (2022). Crop yield estimation at field scales by assimilating time series of sentinel-2 data into a modified CASA-WOFOST coupled model. *IEEE Transactions on Geoscience and Remote Sensing*, 60, 1–14, Art no. 4400914, https://doi.org/10.1109/TGRS.2020.3047102.

[7] Li, X., Dong, Y., Zhu, Y., and Huang, W. (2023). Enhanced leaf area index estimation with CROP-DualGAN network. *IEEE Transactions on Geoscience and Remote Sensing*, 61, 1–10, Art no. 5514610, https://doi.org/10.1109/TGRS.3230354.

[8] Liu, Yu, Q., Zhou, Q., Wang, C., Bellingrath-Kimura, and Wu, W. (2022). Mapping the complex crop rotation systems in Southern China considering cropping intensity, crop diversity, and their seasonal dynamics. *IEEE Journal of Selected Topics in Applied Earth Observations and Remote Sensing*, 15, 9584–9598. https://doi.org/10.1109/JSTARS.2022.3218881.

[9] Mateo-Sanchis, Adsuara, Piles, M., Munoz-Marí, J., Perez-Suay, A., and Camps-Valls, G. (2023). Interpretable long short-term memory networks for crop yield estimation. *IEEE Geoscience and Remote Sensing Letters*, 20, 1–5, Art no. 2501105. https://doi.org/10.1109/LGRS.2023.3244064.

[10] Martínez-Ferrer, L., Piles, M., and Camps-Valls, G. (2021). Crop yield estimation and interpretability with Gaussian processes. *IEEE Geoscience and Remote Sensing Letters*, 18(12), 2043–2047. https://doi.org/10.1109/LGRS.2020.3016140.

[11] Ma, Y., Yang, Z., and Zhang, Z. (2023). Multisource maximum predictor discrepancy for unsupervised domain adaptation on corn yield prediction. *IEEE Transactions on Geoscience and Remote Sensing*, 61, 1–15, Art no. 4401315. https://doi.org/10.1109/TGRS.2023.3247343.

[12] Qiao, M., *et al.* (2021). Exploiting hierarchical features for crop yield prediction based on 3-D convolutional neural networks and multikernel Gaussian process. *IEEE Journal of Selected Topics in Applied Earth Observations and Remote Sensing*, 14, 4476–4489. https://doi.org/10.1109/JSTARS.2021.3073149.

[13] Reyana, Kautish, S., Karthik, Al-Baltah, Jasser, and Mohamed, (2023). Accelerating crop yield: multisensory data fusion and machine learning for agriculture text classification. *IEEE Access*, 11, 20795–20805, 2023, https://doi.org/10.1109/ACCESS.3249205.

[14] Wu, S., Ren, J., Chen, Z., Yang, P., Li, H., and Liu, J. (2021). Evaluation of winter wheat yield simulation based on assimilating LAI. Retrieved from networked optical and SAR remotely sensed images into the WOFOST model. *IEEE Transactions on Geoscience and Remote Sensing*, 59(11), 9071–9085. https://doi.org/10.1109/TGRS.2020.3038205.

[15] Yang, *et al.* (2021). Integration of crop growth model and random forest for winter wheat yield estimation from UAV hyperspectral imagery. *IEEE Journal of Selected Topics in Applied Earth Observations and Remote Sensing*, 14, 6253–6269, https://doi.org/10.1109/JSTARS.2021.3089203.

[16] Yang, Diao, C., and Gao, F. (2023). Towards scalable within-season crop mapping with phenology normalization and deep learning. *IEEE Journal of Selected Topics in Applied Earth Observations and Remote Sensing*, 16, 1390–1402, https://doi.org/10.1109/JSTARS.2023.3237500.

[17] Zhang, *et al.* (2023). Enhanced feature extraction from assimilated VTCI and LAI with a particle filter for wheat yield estimation using cross-wavelet transform. *IEEE Journal of Selected Topics in Applied Earth Observations and Remote Sensing*, 16, 5115–5127, https://doi.org/10.1109/JSTARS.2023.3283240.

Chapter 36

Design of an Efficient Mutable Blockchain for QoS-Aware IoT Deployments

Amit Tripathi[1], Kanchan Wagh[2], Pravin Pokle[3], Sakshi Khamankar[4], Shailesh Sahu[4], Harshwardhan Kharpate[4], and Pravin Gorantiwar[5]

[1]Student, Indian Institute of Information and Technology, Nagpur, Assistant Professor, Cummins College of Engineering for Women, Department of Electronics and Telecommunication, Nagpur, Maharashtra, India

[2]Assistant Professor, Cummins College of Engineering for Women, Department of Electronics and Telecommunication, Nagpur, Maharashtra, India

[3]Associate Professor, Priyadarshini J.L. College of Engineering, Department of Electronics and Telecommunication, Nagpur, Maharashtra, India

[4]Assistant Professor, Cummins College of Engineering for Women, Department of Computer Engineering, Nagpur, Maharashtra, India

[5]Assistant Professor, Cummins College of Engineering for Women, Department of Allied Science, Nagpur, Maharashtra, India

Abstract: The relentless proliferation of internet of things (IoT) applications necessitates blockchain frameworks that can aptly address the dynamic demands of vast device networks. Traditional blockchain deployments, while offering unparalleled security, often grapple with bottlenecks in terms of delay, energy consumption, throughput and packet delivery ratio. These limitations pose severe challenges in ensuring optimal quality of service (QoS) for IoT deployments, especially in contexts requiring high adaptability and responsiveness. In light of these constraints, this paper presents a novel design for a mutable blockchain tailored for QoS-aware IoT deployments. Pivotal to our approach is the deployment of the Ant Lion Optimizer (ALO), which strategically identifies mutable segments of individual blocks. Contrary to a one-size-fits-all format, each block in this schema possesses a unique, context-specific format, thereby amplifying its security. This mutable characteristic not only fortifies the blockchain against security threats but also paves the way for a more adaptive and responsive system, especially beneficial for varied IoT contexts. Empirical results demonstrate the prowess of our model over conventional systems. Specifically, our model reported a 4.5% decrease in delay, 3.9% reduction in energy consumption, 8.3% surge in throughput and an impressive 4.9% elevation in the packet delivery ratio. Collectively, these enhancements underline the capability of our proposed blockchain design in meeting

DOI: 10.1201/9781003527442-36

the rigorous QoS demands of contemporary IoT frameworks, providing a robust solution to the impediments faced by existing models.

Keywords: Blockchain, mutable, Ant Lion, optimisation, contextual, security, scenarios.

1. Introduction

The advent of IoT has transformed the digital landscape, spawning applications ranging from smart cities to complex industrial automation systems. By 2021, there were approximately 35 billion IoT devices worldwide, with exponential growth expected. While these devices offer convenience and efficiency, they also present complex challenges, particularly in terms of security, scalability and quality of service (QoS). Blockchain technology, known for its decentralisation and cryptographic security, emerged as a promising solution to address IoT network challenges. It ensures data integrity and transparency, mitigating many IoT security threats. However, traditional blockchains have limitations (Buttar *et al.*, 2023; Rodrigues *et al.*, 2021; Zhou *et al.*, 2022). Despite their security, they struggle to meet IoT's dynamic demands, resulting in issues like latency, high-energy consumption, limited throughput and subpar packet delivery levels (Ren *et al.*, 2022; Xu *et al.*, 2022; Wang *et al.*, 2023). Furthermore, IoT device designs vary widely, necessitating a blockchain system that can adapt to each deployment's specific needs and contexts. Fixed structures hinder adaptability, making it challenging to achieve optimal QoS in diverse IoT environments. This paper delves into these challenges, emphasising the need for a blockchain design that is not only secure but also adaptable to IoT's multifaceted demands. We introduce a novel blockchain model that uses the Ant Lion Optimizer (ALO) to provide context-specific block structures, enhancing security and QoS. Through a comprehensive exploration of our proposed model and its advantages over existing solutions, this paper charts a path forward for the next generation of blockchain implementations in IoT deployments.

2. Review of Existing Models Used for Blockchain Optimisations

IoT applications have drawn attention to blockchain technology's potential to disrupt several sectors decentralisation and cryptography promise security. As blockchain integration with IoT devices has grown, researchers have struggled to achieve acceptable QoS. This section examines current approaches and their usefulness in enhancing blockchain QoS. Techniques for Sharding (Akrasi-Mensah *et al.*, 2023; Qian *et al.*, 2023): Sharding splits the blockchain into shards, each managing its own transactions and smart contracts. Distributed approaches improve scalability and throughput. Despite its faster validation and higher transaction throughput, it poses security risks including "single-shard take over" attacks. Solution off-chain (Qiu *et al.*, 2021; Vaiyapuri *et al.*, 2023; Zhou *et al.*, 2022): Before registering transactions on the main chain, off-chain solutions execute them on a sidechain. This method may speed up and lower transaction costs. But off-chain solutions may not be as secure as the main chain. PoS and its variants (Kumar *et al.*, 2021; Kalapaaking *et al.*, 2023; Wei *et al.*, 2023) with blockchain-based federated learning – BFL:PoS selects validators to produce new blocks based on the bitcoin they "stake" or lock up. It uses less energy than PoW and speeds transactions. Centralisation is a danger because individuals with more stakes may have more power. Directed acyclic graph (DAG) using MLEB (Liao *et al.*, 2022; Sanghami *et al.*, 2023; Xu *et al.*, 2023) DAG processes several transactions simultaneously by organising them in a graph rather than a straight sequence. Some methods provide zero transaction costs and significant scalability. It is still young, and certain surroundings have security problems. DPoS (Baucas *et al.*, 2023; Rawlins *et al.*, 2022): In DPoS, coin holders elect representatives to verify transactions and produce blocks. It speeds transaction confirmations and reduces energy use but risks centralisation and top delegate collusion (Abegaz *et al.*, 2023; Riahi *et al.*, 2023): Hybrid models incorporate blockchain or consensus algorithm characteristics. They combine performance and security from numerous models. Their sophisticated implementation may bring unexpected weaknesses when integrating techniques. The graphs (Rahman *et al.*, 2023; Zhang *et al.*, 2021) IOTA's DAG graphs combine users and validators. Validating two transactions is required to issue an augmented set of transactions. This micro transaction-focused IoT strategy is very scalable for different use cases. However, a real-time hostile majority adversary raises security issues for real-time scenarios.

3. Proposed Model

As per the review of existing models used for optimising blockchain performance, it can be observed that the efficiency of these models is generally limited when applied to real-time deployments; moreover, these models have higher computational complexity, which limits their scalability levels.

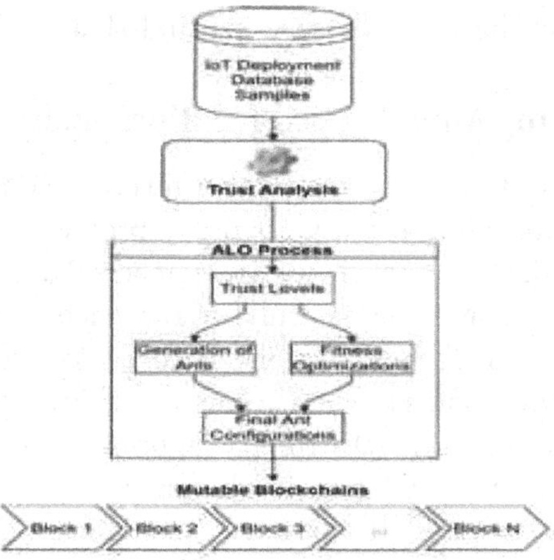

Figure 1: Design of the blockchain optimisation process.

To overcome these issues, this section discusses design of an efficient ALO to create mutable blockchains. As per Figure 1, the proposed ALO model initially collects spatial & temporal node parameters and converts them into node trust levels (NTL) via equation 1:

$$NTL = \frac{1}{NC}\sum_{i=1}^{NC}\frac{PDR(i)*THR(i)}{D(i)*E(i)}\ldots(1)$$

where NC represents total number of communications, which were performed by the nodes, while PDR, THR, D&E are the QoS metrics obtained during these communications, which represents packet delivery ratio (PDR), throughput, delay and energy consumption levels, which are estimated via equations 2, 3, 4, and 5 as follows:

$$PDR = \frac{Rx(P)}{Tx(P)}\ldots(2)$$

where Rx & Tx are the number of packets received and transmitted by these nodes.

$$THR = \frac{Rx(P)}{D}\ldots(3)$$

$$D = ts(complete) - ts(start)\ldots(4)$$

where ts represents the timestamp for completion and starting of these communications.

$$E = e(start) - e(complete)\ldots(5)$$

where e is the residual energy of these nodes.

Based on this trust level, the proposed model selects miners with higher trust values to add new blocks to the chains. Before adding these blocks, they are passed through an ALO process, which works as follows:

Assuming the block has NE entities (including source, destination, timestamp, etc.), the ALO model initially generates NA Ants, where each Ant selects a different set of entities to mutate (EM) via equation 6:

$$EM = STOCH(1,\ NE*LA)\ldots(6)$$

where LA represents learning rate for the Ants. After removing the mutable entries from hashing process, the model hashes remainder of entities and performs ND dummy communications with the current blockchain configurations. In each of these communications, around 15% of packets are sent as Sybil, Finney, Masquerading and DDoS Attacks. After these communications are completed, then Ant Fitness is estimated via equation 7,

$$fa = \frac{1}{ND}\sum_{i=1}^{ND}\frac{NTL(i)}{Max(NTL)}\ldots(7)$$

After generation of NA Ants, an iterative Ant fitness threshold is evaluated via equation 8,

$$fth = \frac{1}{NA}\sum_{i=1}^{NA} fa(i)*LA\ldots(8)$$

Ants with *fa>fth* are passed to next iteration, while others are discarded and replaced with new configurations via equations 6 & 7, which assists in identification of new mutable configurations. This process is repeated for *NI* Iterations, and new mutable configurations are generated and analysed during this process. Once all iterations are completed, then Ants with maximum fitness levels are identified, and their mutable configuration is used to add new blocks. Due to which, the selected blockchain configuration has higher QoS and higher security performance under real-time scenarios. This performance was evaluated for multiple configurations and compared with existing models in the next section of this text.

4. Results and Comparisons

We carried out a comprehensive experimental study to quantify improvements in delay, energy consumption, throughput and PDR across a range of communication scenarios in a simulated IoT environment in order to empirically validate the effectiveness of our proposed mutable blockchain design tailored for QoS-aware IoT deployments. Based on this configuration, the frequency of network attacks was modified from 1% to 20%, and parameters for different performance metrics, such as throughput (T), delay (d), PDR and energy (E) levels, were evaluated. The throughput levels were compared with regard to this analysis. Figure 2 shows the total number of communications (TNC) with BLB [5], BFL [15] and MLEB [18] as follows:

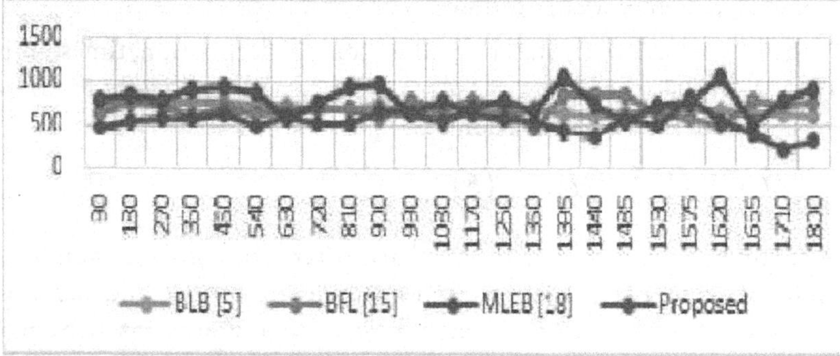

***Figure 2:** Throughput for different models under attacks.*

Upon examination, the table yields intriguing data pertaining to the performance of the models under investigation. It is noteworthy to observe that the "Proposed" model consistently exhibits competitive or superior throughput values when compared to the other models that were considered throughout a range of communication counts. Upon considering the specific attributes of the proposed model procedure, the repetitive nature of this pattern becomes more apparent. Similarly, Figure 3 depicts the delay required for these communications.

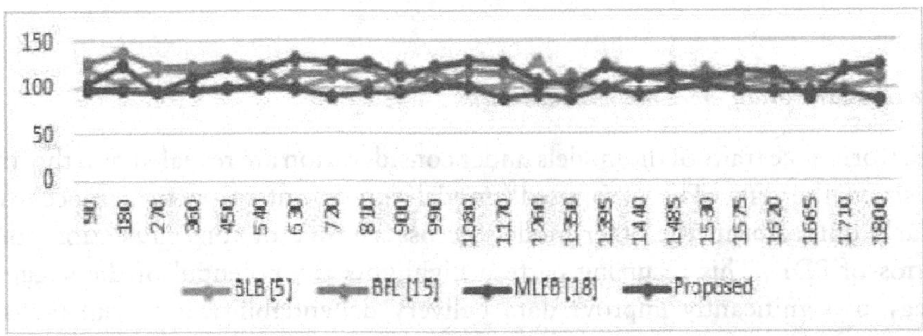

Figure 3: Communication delay for different models under attacks.

Investigating the data reveals interesting patterns regarding the delay properties of the tested models. The "Proposed" model consistently maintains lower delay values than the other models throughout a range of communication counts, which is noteworthy. The potential for this suggested strategy to drastically reduce communication scenario delay times is highlighted by these trends. Similar to that, Figure 4 shows the energy needed for these evaluations.

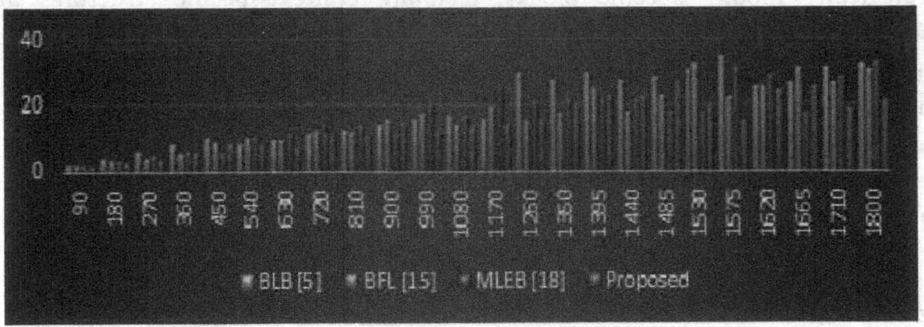

Figure 4: Energy consumptions for different models under attacks.

Examining the energy consumption traits demonstrated by the models under consideration reveals some interesting tendencies. The "Proposed" model stands out in particular since it consistently outperforms the other models under consideration in terms of energy efficiency throughout a range of communication counts. This pattern highlights the model's potential to make a major contribution to lowering energy use, an important factor in scenarios with limited resources. Similar to that, Figure 5 shows the PDR observed for these evaluations.

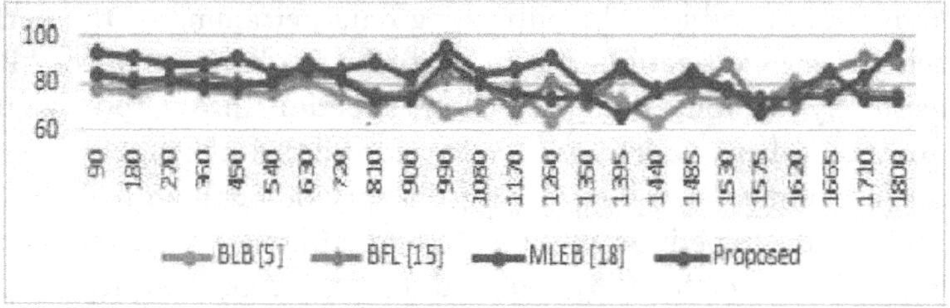

Figure 5: PDR for different models under attacks.

The performance traits of the models under consideration are revealed by a thorough analysis of the data. The "Proposed" model stands out in particular because it regularly outperforms the other models across a range of communication counts in terms of PDR. This recurring pattern highlights the potential of the suggested strategy to significantly improve data delivery dependability, a crucial factor in communication systems.

5. Conclusion and Future Scope

In this work, we introduce a customisable blockchain design tailored for QoS-aware IoT deployments to address challenges posed by the expanding array of IoT applications. Conventional blockchain frameworks struggle with the dynamic demands of large device networks, resulting in delays, energy consumption, throughput and packet delivery issues. To address this, we utilise the ALO to strategically modify specific blocks, enabling context-specific structures and enhancing security, flexibility and responsiveness.

Empirical testing demonstrates our model's superiority, with key performance improvements: 4.5% reduction in delay, 3.9% lower energy consumption, 8.3% higher throughput and an impressive 4.9% increase in packet delivery ratio. This design meets strict QoS requirements for modern IoT systems, surpassing previous approaches.

6. Future Scope

Scalability and Network Size: Extend the model to handle larger IoT networks with higher device density. Security Considerations: Explore potential weaknesses and solutions. Dynamic Block Creation: Investigate adaptive block creation techniques. Consensus Processes: Compare our architecture with different consensus methods. Real-world Deployment: Validate in actual IoT setups. Integration with IoT Protocols: Assess integration with existing IoT communication protocols. Our research enables the integration of adaptable blockchain designs with optimisation strategies, offering potential solutions for QoS-aware IoT deployments. Further research and development will refine these solutions for the evolving IoT landscape.

References

[1] Akrasi-Mensah, N. K., *et al.* (2023). Adaptive storage optimization scheme for blockchain-IIoT applications using deep reinforcement learning. *In IEEE Access*, 11, 1372–1385. https://doi.org/10.1109/ACCESS.2022.3233474.

[2] Baucas, M. J., and Spachos, P. (2023). Federated kalman filter for secure IoT-based device monitoring services. *In IEEE Networking Letters*, 5(2), 91–94. https://doi.org/10.1109/LNET.2023.3264622.

[3] Buttar, H. M., Aman, W., Rahman, M. M. U., and Abbasi, Q. H. (2023). Countering active Attacks on RAFT-based IoT blockchain networks. *In IEEE Sensors Journal*, 23(13), 14691–14699. https://doi.org/10.1109/JSEN.2023.3274687.

[4] Kalapaaking, A. P., Khalil, I., Rahman, M. S., Atiquzzaman, M., Yi, X., and Almashor, M. (2023). Blockchain-based federated learning with secure aggregation in trusted execution environment for internet-of-things. *In IEEE Transactions on Industrial Informatics*, 19(2), 1703–1714.

[5] Kumar, P., *et al.* (2021). PPSF: A privacy-preserving and secure framework using blockchain-based machine-learning for IoT-driven smart cities. *In IEEE Transactions on Network Science and Engineering*, 8(3), 2326–2341. https://doi.org/10.1109/TNSE.2021.3089435.

[6] aqLiao, H., *et al.* (2022). Blockchain and semi-distributed learning-based secure and low-latency computation offloading in space-air-ground-integrated power IoT. *In IEEE Journal of Selected Topics in Signal Processing*, 16(3), 381–394. https://doi.org/10.1109/JSTSP.2021.3135751.

[7] Qian, K., Liu, Y., Shu, C., Sun, Y., and Wang, K. (2023). Fine-grained benchmarking and targeted optimization: enabling green IoT-oriented blockchain in the 6G Era. *In IEEE Transactions on Green Communications and Networking*, 7(2), 1036–1051. https://doi.org/10.1109/TGCN.2022.3185610.

[8] Qiu, C., Wang, X., Yao, H., Du, J., Yu, F. R., and Guo, S. (2021). Networking integrated cloud–edge–end in IoT: a blockchain-assisted collective Q-learning approach. *In IEEE Internet of Things Journal*, 8(16), 12694–12704. https://doi.org/10.1109/JIOT.2020.3007650.

[9] Rawlins, C. C., and Jagannathan, S. (2022). An intelligent distributed ledger construction algorithm for IoT. *In IEEE Access*, 10, 10838–10851. https://doi.org/10.1109/ACCESS.2022.3146343.

[10] Riahi, A., Mohamed, A., and Erbad, A. (2023). RL-based federated learning framework over blockchain (RL-FL-BC). *In IEEE Transactions on Network and Service Management*, 20(2), 1587–1599. https://doi.org/10.1109/TNSM.2023.3241437.

[11] Rodrigues, D. S., and Rocha, V. (2021). Towards blockchain for suitable efficiency and data integrity of IoT ecosystem transactions. *In IEEE Latin America Transactions*, 19(7), 1199–1206. https://doi.org/10.1109/TLA.2021.9461849.

[12] Ren, Li, J., Liu, H., and Qin, T. (2022) Task offloading strategy with emergency handling and blockchain security in SDN-empowered and fog-assisted healthcare IoT. *In Tsinghua Science and Technology*, 27(4), 760–776. https://doi.org/10.26599/TST.2021.9010046.

[13] Xu, Qu, Y., Luan, P., Eklund, W., Xiang, Y., and Gao, L. (2022). A lightweight and attack-proof bidirectional blockchain paradigm for internet of things. *In IEEE Internet of Things Journal*, 9(6), 4371–4384. https://doi.org/10.1109/JIOT.2021.3103275.

[14] Wang, Chen, J., Jiao, Y., Kang, J., Dai, W., and Xu, Y. (2023). Connectivity-aware contract for incentivizing IoT devices in complex wireless blockchain. *In IEEE Internet of Things Journal*, 10(12), 10413–10425.

[15] Ray, Dash, D., Salah, K., and Kumar, N. (2021). Blockchain for IoT-based healthcare: background, consensus, platforms, and use cases. *In IEEE Systems Journal*, 15(1), 85–94. https://doi.org/10.1109/JSYST.2020.2963840.

[16] Sanghami, S. V., Lee, J. J., and Hu, Q. (2023). Machine-learning-enhanced blockchain consensus with transaction prioritization for smart cities. *In IEEE Internet of Things Journal*, 10(8), 6661–6672. https://doi.org/10.1109/JIOT.2022.3175208.

[17] Vaiyapuri, Shankar, K., Rajendran, S., Kumar, S., Acharya, S., and Kim, H. (2023). Blockchain assisted data edge verification with consensus algorithm for machine learning assisted IoT. *In IEEE Access*, 11, 55370–55379. https://doi.org/10.1109/ACCESS.2023.3280798.

[18] Wei, An, Z., Leng, S., and Yang, K. (2023). Evolved PoW: integrating the matrix computation in machine learning into blockchain mining. *In IEEE Internet of Things Journal*, 10(8), 6689–6702. https://doi.org/10.1109/JIOT.2022.3165973.

[19] Xu, *et al.* (2023). BESIFL: Blockchain-empowered secure and incentive federated learning paradigm in IoT. *In IEEE Internet of Things Journal*, 10(8), 6561–6573. https://doi.org/10.1109/JIOT.2021.3138693.

[20] Zhou, J., Feng, G., and Wang, Y. (2022). Optimal deployment mechanism of blockchain in resource-constrained IoT systems. *In IEEE Latin America Transactions*, 19(7), 1199–1206.

[21] Zhou, Q., Zheng, K., Zhang, K., Hou, L., and Wang, X. (2022). Vulnerability analysis of smart contract for blockchain-based IoT applications: a machine learning approach. *In IEEE Internet of Things Journal*, 9(24), 24695–24707. https://doi.org/10.1109/JIOT.2022.3196269.

Chapter 37

Experimental Investigation of Segmentation Methods for Segmenting Fruit Images in Natural Environment

Amit R. Welekar[1] *and Dr. Manoj Eknath Patil*[2]

[1]Research Scholar, Department of Computer Science & Engineering, Dr. A. P. J. Abdul Kalam University, Indore (M.P), India

[2]Research Guide, Department of Computer Science & Engineering, Dr. A. P. J. Abdul Kalam University, Indore (M.P), India

Abstract: In this paper we are focusing on problem of segmenting irregular shape fruit images in natural light through different segmentation algorithms and trying to provide best possible solution. Fruits are the integral part of our day-to-day life; it provides various nutritional values to keep our body fit and healthy. Segmenting fruit images to classify its features are extremely popular these days. Many researchers tried original and hybrid segmentation techniques to work on various aspect of fruit image. Fruits are present in different shapes; regular shape and irregular shape. These fruits images give distinct results when captured in natural light and artificial light environment. Some algorithms work better for regular- and irregular-shaped fruit images but does not provide accurate result when it is captured under natural light. Researchers worked hard to attend these problems but could not guarantee a one solution and one segmentation algorithm which could address all problems. So in this paper, we will put efforts to investigate different segmentation algorithms which can be applied for fruit segmentation and will see if get one algorithm which could address all the problems by adding some or more features to it.

Keywords: image processing, segmentation algorithms, ripening of fruit, artificial ripening, natural ripening, fuzzy c-means.

1. Introduction

Segmenting image is a practice of grouping together objects with comparable characteristics. Segmentation breaks photo into diverse components and come up with all of the necessary facts required for evaluation of a digital image. Image segmentation method is a key step in image analysis in which the outcome affects the entire process. Only the entity of interest was processed during the analysis phase since the selected component was segmented. However, the segmentation

DOI: 10.1201/9781003527442-37

procedure is complicated by the photos' intricate backgrounds and variable light. As a result, it is critical to have a sophisticated segmentation approach that can properly and rapidly separate a picture into foreground and background in the presence of both natural and artificial lighting. Image segmentation has become a critical stage in automating the analysis of many objects. It provides wide application in the field of agriculture, industries, medical and in recognition tasks. Image segmentation is primarily used in agriculture, notably in identifying damaged fruit parts (Dubey *et al.*, 2013), categorising fruits (Prabha and Kumar, 2012), managing fruit quality (Ramprabhu and Nandhini, 2015), recognising on-plant fruit (Yamamoto *et al.*, 2014) and estimating crop (Payne and Walsh, 2013). Also used in fruit ripeness detection (Amarasinghe and Sonnadara, 2009). Quality control is critical in agriculture and grading in food companies in order to export high-quality crops. Because manual examination of agricultural goods is time consuming and unpredictable, it is vital to replace it with an automated and computerised process.

Many segmentation techniques are available like thresholding (Prabha and Kumar, 2012), (Chaudhary and Prajapati, 2014), (Riyadi *et al.*, 2007), (Abdullaha *et al.*, 2012); clustering, Fuzzy-C means (Dubey *et al.*, 2013), (Ramprabhu and Nandhini, 2015), (Amarasinghe and Sonnadara, 2009), (Kaur *et al.*, 2014), (Abdullah *et al.*, 2013), and (Ghabousian and Shamsi, 2012). Successful picture segmentation technique includes tools and a comprehensive environment for data analysis, conceptualisation and algorithm development.

The researchers placed a strong emphasis on threshold-based segmentation. For photos with significant contrast levels, the thresholding approach partitioned the background and foreground, whereas the clustering-based methodology separated apart groupings of objects.

Clustering k-means is the most popular approach since it is relatively easy and can simply sort out datasets or pixels over a specified number of clusters. Aside from these improved segmentation approaches, they are incompatible with photos acquired under uniform light.

Thresholding (Dubey *et al.*, 2013), (Dahapute and Welekar, 2016), (Dahapute and Welekar, 2013), (Kaur *et al.*, 2014); Colour-based segmentation (Prabha and Kumar, 2012) that is K-means clustering (Feng *et al.*, 2008), (Ramprabhu and Nandhini, 2015), (Dhillon and Bathla, 2014), (Yamamoto *et al.*, 2014), (Singh and Chadha, 2013) and Fuzzy C-Means clustering (Chaudhary and Prajapati, 2014); region-based segmentation (Ghabousian, 2012); edge-based segmentation (Amarasinghe and Sonnadara, 2009) that is watershed algorithm (Slamet Riyadi, 2007), texture-based segmentation that is. texture filter (Hambali *et al.*, 2014) are the most common segmentation techniques. Some of the aforementioned approaches (Silwal *et al.*, 2014), (Ghabousian, 2015) and (Hambali *et al.*, 2014) are borrowed to achieve more precise segmentation and eliminate the limitations of ongoing methods.

2. Discussion on Segmentation Methods

As discussed in our earlier paper (Welekar and Patil, 2022), fuzzy C-Means is an effective segmentation method to segment a regular-shaped fruit images if we use L/*a/*b colour space method with fuzzy C-Means (Mendoza and Aguilera, 2004). So over to Otsu thresholding, K-Means clustering, region based, edge based and some enhanced segmentation methods based on original methods, Fuzzy C-Means clustering is best solution which can worked on regular-shaped fruit images.

But fuzzy C-Means clustering has a major problem of defining initial cluster value. It also provides indistinct values for the fruit images captured under natural environment. So if you are using fuzzy C-Means clustering for segmenting fruit images, you have to go through uniform light source. Rule-based segmentation (Hambali and Abdullah, 2014) could solve this problem and provides better result for fruit images captured in natural light. Processing time is directly proportional to number of iteration in fuzzy C-Means clustering method, so processing time will increase with increasing number of iteration. To work on segmentation processing time if we use K-Means along with graph-based technique, we could minimise the processing time of segmentation. So above method is effective than using K-Means and graph-based technique separately.

Another problem of segmenting irregular-shaped image can be overcome if we could use gray-scale segmentation and K-means adaptive segmentation (Sundari, 2017).

So from above discussion we can conclude that not a single method or enhanced method is capable of segmenting an irregular-shaped fruit image captured in natural light. Each segmentation technique restricts itself to particular shaped fruit along with provision of light source.

Fuzzy C-Means could be best solution than K-Means for irregular-shaped fruit image if we could work on processing time and its inability to produce accurate results for fruit images captured in natural light source.

Most of the researchers used K-Means and Fuzzy C-Means for segmenting the fruit images. They tried to address the problem and enhanced these two methods as per their requirement. So we keep our focus on K-means and Fuzzy C-Means clustering methods and will try to find out its pros and cons for segmentation of random shape fruit image captured in outdoor environment. We have experimented K-Means (Dahapute and Welekar, 2016) and Fuzzy C-Means clustering on various shaped fruit images. So below we will first discuss working principle of K-Means and Fuzzy C-Means clustering and then will try to quote our findings.

3. Experimental Analysis of K-Means Clustering

We tried to apply K-Means clustering on different irregular and regular shape fruit images. As we know, data items are sorted into numerous classes in K-Means clustering based on their inherent distance from each other.

Let us see the working principle of K-Means.

1) Select "C" number of clusters
2) Select random "C" point which is different from input data objects
3) Now assign these points to their nearby centroid, which helps to form new "C" clusters
4) Again calculate centroids for these new clusters
5) Repeat step number 3 until any new reassignments occurs
6) If any reassignment occurs then go-to step-4
7) Finish

For our experiment we have used apple, banana, mango and pineapple. RGB values are stored (Dahapute and Welekar, 2016).

- RGB value found of apple [40, 70, 27]
- RGB value found of banana [100,88, 20]
- RGB value found of mango [100, 76, 14] and
- RGB value found of pineapple [33, 23, 5]

To change a picture, colour space conversion is used. The colour space L*a*b provides a more conceptually consistent colour space.

Euclidean distance matrix is used to measure the difference between two colours

*a and*b.

The results we found are, K-Means clustering works well for regular and irregular shape fruit images but it produces inaccurate results when these regular- and irregular-shaped fruit images are captured in natural light source.

4. Analysis of Fuzzy C-Means Clustering

Fuzzy C-Means clustering is a type of soft clustering in which each data object belongs to many clusters. Each data objects are assigned with membership grades and these grades indicate the degree of data object belong to each cluster.

So, Fuzzy C-Means work in basically four steps:

1) Defining a number of clusters
2) Assigning the membership degree to each data object to be part of the cluster.
3) Calculating the centroid of cluster
4) Repeat the above step until the sensitivity threshold is achieved.

4.1. Calculating Centroid of Each Cluster

Calculate the coefficients for each data object in the cluster.

In above steps, processing time increases with increased number of iteration. So processing time is directional to number of iteration

$P_t \alpha N_i$

where P_t is the processing time and Ni is the number of iteration.

Fuzzy clustering algorithm is based on strategy that every data point present is a part of every cluster by means of some degree. And this makes Fuzzy C-Means more applicable in applications where one data object belongs to more than one clusters.

In K-Means Clustering each data point is assigned with values 0 and 1, and in Fuzzy C-Means values are between 0 and 1

Let us assume that,

$S = \{s_1, s_2 \dots s_n\}$ be the number of data points, and

$K = \{k_1, k_2, \dots k_m\}$ are the number of cluster

So,

	k_1	k_2	k_m	
X_1	0.1	0.3	0.6	= 1
X_2	0.2	0.4	0.4	= 1
.			.	
			.	
X_n	0.6	0.1	0.3	= 1

where mean of all weights should be equal to one.

The fundamental limitation of Fuzzy C-Means Clustering is that the initial number of clusters must be defined before iteration can begin. Fuzzy membership degree of each data point is calculated as,

$$\mu_{i,j} = 1/\sum_{k=1}^{c}(d_{i,j}/d_{i,k})^{(2/m-1)} \qquad 1$$

where "m" is the fuzziness parameter

Fuzzy centres are calculate by V_j:

$$V_j = ((\mu_{i,j})^m x_i)/(\sum_{i=1}^{n}(\mu_{i,j})^m), \forall_j = 1, 2, \dots, c \quad 2$$

Equations 1 and 2 are repeated until the objective function j is minimise or $\|U^{(k+1)} - U^{(k)}\| < \beta$

where

"β" is termination criteria.

Overall performance of Fuzzy C-Means is affected by this β.

Fuzzy C-Means' overall performance varies based on a Fuzzy parameter "m". Hence initialising a fuzziness parameter and selecting its appropriate value can change the game.

5. Final Discussion on Performance of K-Means and Fuzzy C-Means Clustering

From above discussion we can clearly say that K-Means clustering has good performance as compared to Fuzzy C-Means clustering. Fuzzy C-Means performance is affected by two parameters, "β" and "m". In K-Means clustering each data point is part of only one cluster at a moment, but in real-life scenario applications data available can be part of more than one segment. So, while K-Means clustering outperforms Fuzzy C-Means in some situations, it falls short in others.

In fruit segmentation like applications, fruit is of random shaped where each pixel could not a part of only one segment. In Fruit like banana K-Means clustering gives the inaccurate results. In Fuzzy C-Means clustering each data point can be a part of different segments based on its distance from cluster centres. So, in segmentation of random size fruit Fuzzy C-Means could perform well and give better results as compared to K-Means clustering.

When we captured fruit image in natural light or in outdoor environment, the intensity values of every pixel are affected. To achieve good results, colour and texture features of fruit need be accurately analysed. To extract exact colour feature, the original intensity values of each pixels is to be maintained.

Colour space conversion techniques like Hue Saturation Value (HSV), XYZ and L*a*b could solve this problem of over brightness of colour pixels.

6. Conclusion

After analysing and discussing various segmentation algorithms, we can clearly state that some modified and composite approaches had also overcome the illumination effect, but even more effort should be made to develop a method that can separate photos of several irregularly shaped fruits under natural lighting.

Based on above experimental analysis of K-means clustering and discussion of Fuzzy C-Means clustering we can say that these two methods are mostly suitable for applications like segmenting a fruit images. Based on above discussion we can conclude that K-Means clustering works good for regular shape fruit images, running time performance is also good. But when we captured images of random shape fruit in natural light source it gives inappropriate results. Fuzzy C-Means is a soft clustering method so it could be a better solution for segmentation of irregular fruit images. But running time of fuzzy C-Means is directly proportional to number iteration "β" and fuzziness parameter. So if we could work on "β" termination criteria and fuzzy parameter "m", and its uncertainty for images captured in natural light, Fuzzy C- Means will be a best solution for segmenting regular and irregular shape fruit images captured in natural environment.

References

[1] Abdullaha, S. L. S., Hambalia, H. A., and Nursuriati, B. (2012). Segmentation of natural images using an improved thresholding-based technique. Jamilc International Symposium on Robotics Aand Intelligent Sensors.

[2] Abdullah, S. L. S., Hambali, H. A., and Sarawak, N. J. (2013). Adaptive K-Means method for segmenting images under natural environment. *Proceedings Oof Tthe 4th International Conference on Computing Aand Informatics*, ICOCI, 28–30.

[3] Amarasinghe, D. I., and Sonnadara, D. U. J. (2009). Surface color variation of papaya fruits with maturity. *Proceedings of the Technical Sessions*, 25, 21–28, Institute of Physics, Sri Lanka.

[4] Chaudhary, S., and Prajapati, B. (2014). Quality analysis and classification of bananas. *International Journal of Advanced Research in Computer Science and Software Engineering*, 4(1), ISSN: 2277 128X.

[5] Dahapute, M., and Welekar, A. (2016). K-mean clustering for segmentation of irregular shape fruit images under various illumination. *International Conference on Modern Trends in Engineering Science and Technology*, ISSN: 2454-4248, 2(5).

[6] Dahapute, M. R., and Welekar, A. (2016). A review: segmentation approaches for fruit images. *2nd International Conference Technosoft* 16, TGPCET, Nagpur, Special Issue, ISSN: 2454-1958.

[7] Dahapute, M. R., and Welekar, A. (2016). Improved K-means clustering with colour classification for segmentation of fruit images. *International Journal of Research*, 3(5), ISSN: 2348-6848.

[8] Dubey, S. R., Dixit, P., Singh, N., Gupta, J. P., and Mathura. (2013). Infected fruit part detection using k-means clustering segmentation Technique. *International Journal of Artificial Intelligence Aand Interactive Multimedia*, 2 (2). Ddoi: 0.9781/Ijimai.

[9] Dhillon, S., and Bathla, E. A. K. (2014). Detecting guava quality using gradient function histogram plotting. *International Journal of Engineering and Technical Research* (*IJETR*), ISSN: 2321-0869, 2(9).

[10] Feng, G., Qixin, C., and Masateru, N. (2008). Fruit detachment and classification method for strawberry harvesting robot. *International Journal of Advanced Robotic Systems*, 5, (1), ISSN 1729–8806, 41–48.

[11] Ghabousian, A., & Shamsi, M. (2012). Segmentation of apple color images utilizing fuzzy clustering algorithms. *Advances in Digital Multimedia*, 1(1), 59–63.

[12] Hambali, H. A., Hazaruddinharun, Abdullah, S. L. S., and Jamil, N. (2014). A rule-based segmentation method for fruit images under natural illumination. *IEEE*, 978-1-4799-4575.

[13] Hambali, H. A., Abdullah, L. S., and Jamil, N., Harun, H. (2014). A rule based segmentation method for fruit images under natural Illumination. *In:*

IEEE International Conference on Computer, Control, Informatics and its application, Bandung, Indonesia, 21-23, 13–18.

[14] Kaur, M., Kaur, N., and Singh, H. (2014). Adaptive k-means clustering techniques for data clustering. *International Journal of Innovative Research in Science, Engineering and Technology*, ISSN, 3(9), 2319–8753.

[15] Leemans, V., Magein, H., and Destain, M. F. (2002). On-line fruit grading according to their external quality using machine vision. *Biosystem Engineering*, 83(4):, 397–404.

[16] Mendoza, F., and Aguilera, J. M. (2004). Application of image analysis for classification of ripening bananas. *Journal of Food Science*, 69 (9), 415–423.

[17] Payne, A. B., Walsh, K. B., Subedi, P. P., and Jarvis, D. (2013). Estimation of mango crop yield using image analysis -segmentation method. *Computers Aand Electronics In Agriculture*, 91, 57–64.

[18] Prabha, D. S., and Kumar. J. S. (2012). A study on image processing methods for fruit classification. *Proc. Int. Conf. on Computational Intelligence and Information Technology*, CIIT.

[19] Ramprabhu, J., and Nandhini, S. (2015). Embedded system based fruit quality management using PIC micro controller. *International Journal for Research in Applied Science & Engineering Technology* (IJRASET), 3(I), ISSN: 2321–9653.

[20] Riyadi, S., Rahni, A. A. A., Mustafa, M. M., and Hussain, A. (2007). Shape characteristics analysis for papaya size classification. *The 5th Student Conference on Research and Development – SCOReD*, 11–12, Malaysia.

[21] Silwal, A., Gongal, A., and Karkee, M. (2014). Identification of red apples in field environment with over the row machine vision system. *Agric Eng Int: CIGR Journal*, 16(4).

[22] Singh, P., and Chadha, R. S. (2013). A novel approach to image segmentation. *International Journal of Advanced Research in Computer Science and Software Engineering*, 3(4).

[23] Sundari, T. K. (2017). Maturity detection of fruits and vegetables using k-means clustering technique. *International Journal of Engineering Technology Science and Research*, ISSN, 4 (11), 2394–3386.

[24] Wang, Y., Kan, J., Li, W., and Zhan, C. (2013). Image segmentation and maturity recognition algorithm based on color features of Lingwu Long Jujube. *Advance Journal of Food Science Aand Technology*, 5(12), 1625–1631, Issn: 2042-4868; E- Issn: 2042-4876.

[25] Welekar, A. R., and Patil, M. (2022). Review paper on segmentation algorithms for segmentation of irregular shaped fruit image captured in natural light. *Dickensian Journal*, ISSN NO: 0012-2440, 22(6).

[26] Yamamoto, K., Guo, W., Yoshioka, Y., and Ninomiya, S. (2014). On plant detection of intact tomato fruits using image analysis and machine learning methods. *Sensors*, 14, 12191–12206. doi:10.3390/s140712191.

C. Emerging Technologies in Mechanical Engineering

DOI: 10.1201/9781003527442-38

Chapter 38

Influence on the Air Flow due to Boundary Layer

V. CVS Phaneendra[1] *and M. V. Mallikarjuna*[2]

[1]Research Scholar, Department of Mechanical Engineering, JNTUK, A.P., India
[2]Professor & Principal, Department of Mechanical Engineering. NIT Raichur, Karnataka, India

Abstract: Engine designers today are working with an objective to achieve lowest possible emission levels and best performance. The air admitted into the engine cylinder characterizes the efficiency and the level of emission. To maximize the air mass induced inside the engine cylinder during the intake stroke, the design of inlet manifold plays an important part and has to be improved. The in-cylinder movement of the air decides the extent of combustion within the engine cylinder. The plane or even inner wall surface makes the air to flow easily inside the inlet manifold with least drop in pressures.

With improved velocity of air stream, air admitted into the engine cylinder will increase, improving the volumetric efficiency of the engine. But the stream velocity at and near the wall is reduced and while calculating volumetric efficiency the effective diameter of the flow field is to be reduced for obtaining the exact inducting capacity and velocity gradient was analyzed in the present work by means of pressure gradient near the wall surface which inversely varies the velocity. The hollow pins were inserted in the manifold about the cross-section to predict the pressure variation due to suction (using manometer) about the diameter at several points (0-3 mm from wall). The pressure near the wall surface was found to be more, and pressure away from the wall surface was quite low when compared with near the surface i.e., the velocity of air flow is greater at the center of the manifold and less near the wall surface.

Keywords: boundary layer, in-cylinder movement, inlet flow field, intake manifold, velocity profile, volumetric efficiency.

1. Introduction

One of the key components of an automobile is the engine. Technology for internal combustion engines is advancing quickly to be more potent that is necessary to

DOI: 10.1201/9781003527442-38

increase engine efficiency and enable the use of alternative fuels. Any engine's power production is the result of an air–fuel mixture that is properly burned in the combustion chamber. The two biggest challenges are lowering global carbon emissions and oil usage. The air entrance is essential for the ignition and burning of the fuel for any engine

Intake and exhaust manifolds are a feature of both compression ignition and spark ignition engines. The intake manifold's design is one among several elements that influence the engine's efficiency. The intake manifold's job is to evenly distribute a consistent air quantity to every cylinder. The air flow through the intake manifold is being improved by a number of designers, researchers and engine manufacturers since the volumetric efficiency and output of the engine are significantly influenced by the amount of air supplied. When the cylinder is filled with a lot of air, more fuel may be pumped into the combustion chamber, which will increase the peak pressure and enhance the engine's overall performance. The Bernoulli's equation, which was derived from the energy conservation law, predicts that when fluid velocity (kinetic energy) increases, the associated pressure energy (energy per unit volume) decreases.

2. Literature Review

According to Loong *et al.* (2013), the design of intake manifold has substantial influence over engine's performance since the air intake can boost combustion. A balanced distribution is necessary for improved performance and maximum effectiveness.

The research on the intake airflow is currently the most key fields in engine's development. A well-tuned manifold pipe can notably improve an engine's performance. Therefore, for better engine performance, a well-designed intake manifold is crucial. Some of the considered parameters while designing the intake manifold are the length, diameter, plenum volume, smoothness of the joints and shape of the runners.

In fact, more air can boost the engine's power and torque. Small changes to the intake pipe's design could improve performance by reducing flow resistance and changing the pressure fluctuations brought on by the piston's cyclic intake strokes. The common requirements to be considered while designing the intake manifolds are runner must be kept as short as feasible, avoid too rough or jagged wall surface and to maintain good diameter to combustion chamber relation.

An analytical method was created by Mashkour (2012) to determine the intake manifold pipe's effective diameter (which impacts the effective air flow rate). The effect of engine's knocking was addressed relating to an increase in intake pressure. The increased intake pressure raises engine output pressure, which raised the engine knocking.

According to research tests conducted by Hushim *et al.* (2015) the air flow behaviour varies with various intake manifold angles. The bend's angle limits the airflow resistance. This demonstrates that the intake manifold pipe's maximum angle of bend can speed up air intake while lowering potential constraints. When the intake manifold was bent, the velocity profile, the pipe's internal pressure and flow pattern will vary with respect to the bending. Low air velocity was observed at the outer bending wall, while high air velocity was at the inner bending wall. Air restriction can cause the engine to produce less power so the least amount of air resistance is required. In this instance, the pipe's bend acts as a barrier to the airflow. The ideal intake manifold for removing air resistance is the straight manifold. This demonstrates that the intake manifold's maximum angle is the best angle for producing higher performance because it has the least amount of airflow resistance. An intake manifold with a high angle will result in increased air velocity and increases brake power, brake torque, volumetric efficiency and brake mean effective pressure of the engine. Although intake manifolds have a straightforward design, the processes taking place inside them are overly complicated because the airflow is pulsing rather than constant. With reference to the pressure and velocity contours when the pressure is low the velocity at that location is found to be high, which justifies the Bernoulli's principle that is the pressure is inversely proportional to the velocity.

The influence of the intake valve opening on flow profiles was demonstrated by Krishna *et al.* (2010), lower intake valve apertures were seen to be less smooth, and it is possible that this is related to the air intake pipe's diameter. The airflow velocity also reduces at lower intake valve openings as a result of air friction brought on by the contact between the flowing fluid and the wall surface. The main flow constraints in the intake system are the inlet valve and ports, which causes a significant pressure drop over the intake valve. These issues are less severe with bigger intake valve openings, which produce smooth velocity profiles.

A smaller diameter intake manifold will improve engine response and increase torque at low speed. However, performance and reaction at lower RPMs will suffer with a diameter that is too large. When researching the impact of air intake pipes, Shannak *et al.* (2005) exclusively considered engine exhaust emissions and found that at small intake pipe sizes (20 mm) significant hydrocarbon and carbon monoxide concentrations were observed at speeds of about 1000 rpm (revolutions per minute). At speeds of about 4000 rpm, the HC and CO were found to be lowest at bigger pipe diameters (60 mm), and at different operating conditions, the values of CO and HC remain consistent.

Air restriction can cause the engine to produce less power. The least amount of air resistance is required. In this instance, the pipe's bend acts as a barrier to the airflow. The ideal intake manifold for removing air resistance is the straight manifold. This indicates that the biggest manifold angle has the least airflow resistance, which is the ideal angle for providing higher performance.

The bends and connections will limit the pipe's flow. Aziz *et al.* (2018) has recreated the venturi effect in the inlet manifold. The venturi tube shape concept on the pipe was illustrated by Norizan *et al.* (2017) as a potential means of improving airflow. When a fluid moves through a surface area that is contracting, air velocity and pressure increase and the venturi effect is created.

According to Ceviz *et al.* (2010), changing the plenum length enhances engine performance quality in terms of specific fuel consumption. According to Hasan (2013), experimental analysis, the length of the inlet pipe has an impact on the airflow into the combustion chamber. Long pipes can improve volumetric efficiency, which is thought to be a critical component in engine combustion.

According to Singla *et al.* (2015), altering the length of the intake pipe results in an improved engine. To enable faster pressure wave propagation and air intake at high-volume flow rate, the pipe length must be increased at low engine speeds while it must be decreased at high engine speeds.

Narendran *et al.* (2020) reviewed various works relating to the inlet flow field that is geometry, surface finish and the bend angle of inlet manifold. The airflow pattern can be influenced by the optimum tuned inlet flow geometry which can improve the engine performance.

In essence, the distribution of pressure and velocity along the pipe is calculated using the Bernoulli's number. The pressure wave will occasionally change because of how the engine is running. Due to the disruption caused by the pulsing pressure, the intake air velocity will be impacted by this as well. Studying the sort of airflow inside the pipe using the Reynolds number will also be crucial. Maharudrayya *et al.* (2004) investigations claim that the Reynolds number and geometrical characteristics of the pipes impact how much pressure is lost.

According to Bordjane *et al.* (2010), the volumetric, scavenging and trapping efficiencies indicated power, emissions and flow field inside the engine cylinder are all significantly influenced by the intake and exhaust manifold's flows. Friction, pressure and inertial forces exist in these systems when the air or the gas moves in an unpredictable manner. The pipe geometry can be adjusted to alter the pressure and velocity behaviors of the flowing air in accordance with the pipe design. The diameter and length of the pipe can both affect the Reynolds number; thereby significantly affects the engine performance. Reynolds numbers have an impact on performance since they can be used to establish the flow profile.

3. Experimental Procedure

In any fluid flow there comes a common resisting force called the boundary layer which is caused by the friction between the flowing fluid and the surface. The boundary layer may majorly or marginally affect the flow leading to reducing the effective flow of the fluid flowing. Here, the test was performed on an engine with discharge rate of 1.4 cubic meter per minute and the modified manifold was attached for the pressure readings. The influence of the boundary layer on the flow cross-section is analysed. In order to precede with the work, an intake manifold with tubes arranged along the periphery of inlet pipe was used. When the air is flowing in the pipe the pressure gradient along the diameter was calibrated. The tubings were arranged in such a way to obtain the pressure at an appropriate diameter during the flow. Based on the pressure gradients with the ambient the pressure variations will be determined and further the velocity of the flow at that diameter can be found. By determining the variations of the flow, the impact of boundary layer can be known and to what extent the boundary layer is influencing the flow geometry leading to reduced mass flow rate or volumetric efficiency.

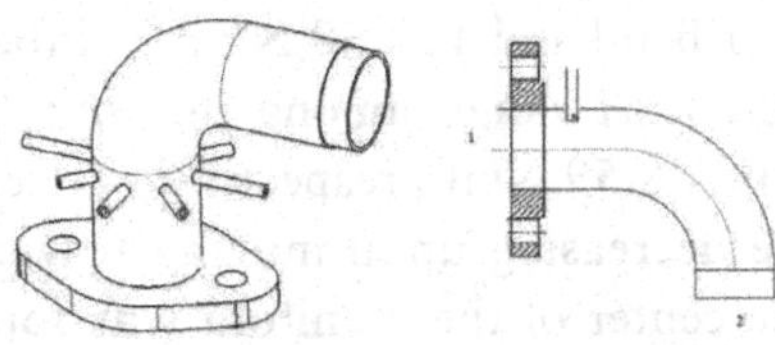

***Figure 1:** (a) Arrangement of pins. (b) pins opening orientation.*

Table 1: Readings of the manometer.

Distance from wall Surface (mm)	Manometer difference h_1 (cm)	Manometer difference h_2 (cm)	Pressure P_1 $(\rho_l-\rho_g)gh_1$ N/m^2	Pressure P_2 $(\rho_l-\rho_g)gh_2$ N/m^2
0	2.1	1.9	205.76	186.17
1	1.5	1.7	146.97	166.57
2	1.4	1.3	137.18	127.38
3	1.3	0.7	127.38	68.59
15	0.7		68.59	

Note: h_1 represents the manometer difference in the pin located towards the outer bend

h_2 represents the manometer difference in the pin located towards the inner bend

ρ_l represents the density of the liquid used in manometer(water)

ρ_g represents the density of the fluid whose pressure is to be determined (air)

formulae should be Pressure = P1 = (ρl-ρg)gh1 N/m and Pressure = P2 = (ρl-ρg) gh2 N/m2

As shown in the Figure 1 the hollow pins are inserted in the manifold about the curved portion in such a way that the variation of the suction pressure about the diameter is measured. The opening of the pin (1-mm thick and width of half the pin circumference) is faced away from engine towards the air filter end and the openings are maintained at 0 mm, 1 mm, 2 mm and 3 mm from the manifold inner wall surface. The pressure variations at these diameters will be measured using U-tube manometer.

4. Results and Discussions

From the obtained readings, the pressure variation with the atmospheric pressure was determined at various diameters (0, 1, 2 and 3 mm from the manifold wall surface) of the inlet manifold. The pressure at the 0 mm from wall gave higher value and was found to be 205.76 N/m^2 and 186.17 N/m^2 in the outer and inner bends, respectively. The pressure at 1 mm from wall was 146.97 N/m^2 in outer bend and 166.57 N/m^2 in inner bend. The pressure at 2 mm from wall was 137.18 N/m^2 in outer bend and 127.38 N/m^2 in inner bend. The pressure at 3 mm from wall gave the least values among the four in both outer and inner bends of 127.38 N/m^2 and 68.59 N/m^2, respectively (refer Table 1). The pressure values were found to be decreasing upon moving towards the center from the wall. The pressure at the center of the manifold was found equal to that at the 3 mm from wall surface which implies that the influence of boundary layer is up to 3 mm from the surface. The pressure values also showed the variations as obtained (Hushim *et al.*, 2015) at outer and inner bends of the inlet manifold. The pressure values were decreasing from the wall surface towards the center and from the Bernoulli's Principle states that the pressure inside a moving fluid falls as the fluid"s speed increases; higher pressure will result in decreased velocity that is the velocity near the wall will be less when compared with near the center, which represents that the restriction caused by the boundary layer affects the volumetric efficiency significantly and the variation of effective flow diameter has to be calibrated.

5. Conclusion

The velocity was found increasing from the manifold surface towards the center up to 3 mm. The influence of the boundary layer is up to 3 mm from the wall surface and the overall influence of it is 6 mm about the diameter. The effective diameter during calculation of volumetric efficiency will be around in between 24 and 30 mm and is always less than that of the core diameter.

6. Future Scope

The work can further be carried at different lengths of the flow field and at different diameters. The work can be utilised to obtain a relation among flow quantity, flow diameter based on pressure drop. This work can be analysed with different surface profiles for analysing the boundary layer for different contours.

Acknowledgments

The work reported here has been carried out as supporting evidence for a research work and the authors are thankful to the members who helped to carry out the experiment.

References

[1] Aziz, S., Amin, N. A. M., Rahman, M. T. A., Rahman, A., Syayuthi, A. R. A., Majid, M. S. A., and Suhaimi, S. (2018). Design and analysis of an operative inlet. *IOP Conference Series: Materials Science and Engineering*, 429(1), IOP Publishing.

[2] Bordjane, M., Chalet, D., Abidat, M., and Chesse, P. (2010). Inertial effects on fluid flow through manifolds of internal combustion engines. *Proc. IMechE Journal of Power and Energy*, 225.

[3] Ceviz, M. A., and Akin, M. (2010). Design of a new SI engine intake manifold with variable length plenum. *Energy Conversion and Management*, 51(11), 2239–2244.

[4] Hasan, M. R. (2013). Experiment and analysis of intake pipe for single cylinder engine. PhD dissertation, UMP.

[5] Hushim, M. F., Alimin, A. J., Razali, M. A., Mohammed, A. N., Sapit, A., and Carvajal, J. C. M. (2015). Air flow behavior on different intake manifold angles for small 4-stroke PFI retrofit kit system. *ARPN Journal of Engineering and Applied Sciences*, 1–7.

[6] Krishna, B. M., and Mallikarjuna, J. M., (2010). Characterization of flow through the intake

[7] valve of a single cylinder engine using particle image velocimetry. *Journal of Applied Fluid Mechanics*, 3(2), 23–32. ISSN 1735–3645.

[8] Lim, J. W., Narendran, N., Chai, C. E., Maarof, M. I. N. (2020). A review on the air flow behaviour in the intake pipe. *IOP Conf. Series: Materials Science and Engineering.*

[9] Loong, Y. K., and Salim, S. M. (2013). Experimentation and simulation on the design of intake manifold port on engine performance. *EURECA.*

[10] Maharudrayya, S., Jayanti, S., and Deshpande, A. P. (2004). Pressure losses in laminar flow through serpentine channels in fuel cell stacks. *Journal of Power Sources*, 138, 1–13.

[11] Mashkour, M. A. (2012). Naturally aspirated internal combustion engine. *Journal of Applied Sciences*, 12(2), 161–167.

[12] Norizan, A., Rahman, M. T. A., Amin, N. A. M., Basha, M. H., Ismail, M. H. N., and Hamid, A. F. A. (2017). Study of intake manifold for University Malaysia Perlis automotive racing team formula student race car. *In Journal of Physics Conference Series*, 908(1), IOP Publishing.

[13] Shannak, B., Damseh, R., and Alhusein, M. (2006). Influence of air intake pipe on engine exhaust emission. *Forsch Ingenieurwes*, 70, 128–132.

[14] Singla, S., Sharma, M. S., and Gangacharyulu, D. (2015). Study of design improvement of intake manifold of internal combustion engine. *International Journal of Engineering Technology, Management and Applied Sciences*, 3, 234–242.

Chapter 39

Design and Development of a Semi-Automated Pneumatic System for Production of Washers

J. Om Prasaad[1], Chetan Aade[1], Pranav Ghanvat[1], Biradar Mustaffa[1], Praseed Kumar[1], and Shamim Pathan[1]

[1]Department of Mechanical Engineering, Fr. C. Rodrigues Institute of Technology, Vashi, Navi Mumbai, India

Abstract: Sheet metal forming is a field in product manufacturing in which many sheet metal forming operations are coordinated to enhance the formability of three-dimensional forming involving good quality and accuracy. Punching is a common sheet metal forming process carried out in small- and medium-scale industries. To perform this operation, such industries incorporate machinery which is operated manually which in turn does not promise accuracy and precision. Thus, such industries require an economical option for performing punching operation that provides better accuracy than the current systems used. The main objective of this project is to design, develop and fabricate a pneumatic sheet metal punching system that can punch sheets successively to produce washers. Pneumatic cylinders, actuated by an air compressor, would be employed to punch the metal sheets successively. The sheets would be fed using a system of automated rollers, to avoid manual input. After the washers get manufactured, primary packaging is performed, by collecting a specific number of washers within boxes, over an automated conveyor belt. This project, fabricated in association with S-Tech Engineering, Navi Mumbai, aids small- and medium-scale industries to produce washers, by decreasing costs incurred while providing accurate products, which ultimately results in profits for the company.

Keywords: Automation, forming, pneumatics, punching, solenoid valve.

1. Introduction

Punching is a common sheet metal forming operation incorporated to produce and manufacture washers of variable sizes. The most common method implemented for this purpose is punching through mechanical systems, with a standard die and punch unit. Small and medium-scale industries find such mechanical systems to be expensive to purchase and implement. Thus, they are on the lookout for economical

DOI: 10.1201/9781003527442-39

and easy alternatives which would promise the same results as their mechanical counterparts, with minimal human interference.

The proposed system is a possible alternative to current mechanical systems, to produce washers. The system tends to manufacture washers using a pneumatic system, which consists of two pneumatic cylinders, along with their respective punch and die units, to carry out the punching operation. The sheet supply unit is automated so that the sheets are fed automatically for sequential punching. The system is complemented by a conveyor unit, which collects the washers produced in containers placed over the conveyor, and conveys them forward, in accordance with the batch production protocol. The washers produced are counted using an IR sensor, which enables batch production. During the entire operation, human involvement is minimal and is limited to include only the replenishment of a fresh sheet, regularly.

2. Literature Review

Arunkumar *et al.* (2021) focus on the requirements of various varieties of mechanical presses used for riveting or punching in industries. The author used pneumatic presses to increase efficiency and reduce workload. The machine, powered pneumatically, consists of a five-way directional control solenoid valve with a single port for input, two ports for outputs and two exhausts. Ghatge *et al.* (2017) display the importance of the shearing operation. Typically, the clearance between the two is 5–10% of the material's thickness. The machine developed can cut bars of mild steel material by the shearing operation. Shearing between punch and die can deform the metal plastically in the die.

Air is used as a working fluid which is available free of charge. Raman *et al.* (2019) focus on alternative applications for working on an automated sheet metal cutting unit powered pneumatically, to create a low-cost machine for laboratory use. The machine proposed was successfully assembled after designing in solid works and optimising blade design. The unit uses a metal cutter, with a double-acting cylinder and a two-way system. After fabrication, the machine is activated through solar energy, which can be used in small-scale industries, as an unconventional source. Prajapati *et al.* (2018) developed a machine for the sheet metal industry, and they can be customised. The pressure exerted during the punching operation results in the plastic deformation of the metal. Low clearance between the die and the punch ensures guaranteed deformation. The sheet metal cutting machine is actuated using a double-acting cylinder. The machine is compact, making it easy for transport. The pneumatic cylinder actuates the machine, to perform cutting and punching. The piston is linked to the cutting tool in motion. The timer control unit circuit alters the speed of the cutting and releasing strokes. Sharma (2015) fabricated an automatic pneumatic hole punching machine. Upon opening the exhaust valve, the compressed air exerts pressure over the cylinder's piston forcing it downward, ultimately pushing the punch downward. Thus, the punch applies an impact load over the sheet placed on the die, effectively punching it in the process. The die is attached to the machine's base. The power supply for the whole machine is provided using solar energy. This technology could be used in small-scale industries, eliminating the need for electricity, and saving money on energy bills.

3. Scope of Research

The machine is designed to manufacture circular washers of aluminium material. The following specifications signify the dimensions of the sheet that can be fed into the system per pass for optimum punching (which means after each pass in the punching cycle the same sheet can be adjusted, so as to utilise the available material, for punching in subsequent cycles, until the sheet is completely exhausted and scrap is minimised) and the specifications of the final washer produced,

Dimensions of sheet supplied per pass

Length 60–100 cm

Breadth 10–15 cm

Thickness 1 mm

Dimensions of washer produced

Inner Diameter 5 mm Outer Diameter 10 mm Thickness 1 mm

4. Theoretical Analysis

A theoretical analysis is performed to calculate and analyse the sufficient pressure, required to effectively punch an aluminium sheet followed by the selection of suitable pneumatic cylinders based on the results obtained.

4.1. *Pressure Calculations*

For the successful completion of the theoretical analysis, certain assumptions must be considered to obtain the subsequent values. The respective assumptions are as follows:

- The outer diameter and inner diameter of the desired washer are 10 mm and 5 mm, respectively.
- The shear strength of an aluminium sheet (Tmax) is found to be 103.86 N/mm^2.
- The machine can punch a sheet thick up to 1 mm.
- The operating pressure of each pneumatic cylinder is 6 bars.

The bore diameter of the pneumatic cylinder for punching the 10-mm hole is calculated as follows:

Punching Force

FC = L × t3 × Tmax (1)

The Eqn. (1) is used to evaluate the force necessary to punch the aluminium sheet. FC – force required to punch metal sheet

L – perimeter to be punched = ∏ × d = 31.41mm t3 – sheet thickness = 1 mm

Tmax – shear strength of aluminium sheet = 103.86 N/mm^2 FC = 31.41 × 1 × 103.86

FC = 3262.24 N

Stripping Force

FS = 0.15 × FC (2)

The Eqn. (2) is used to evaluate the force necessary to strip the aluminium sheet. FS – stripping force

FC – force required to punch metal sheet = 3262.24 N FS = 0.15 × 3262.24

FS = 489.336 N

Total Force

FT = FC + FS (3)

The Eqn. (3) is used to evaluate the total force necessary to shear the aluminium sheet. FT – total shear force

FC – punching force = 3262.85 N

FS – stripping force = 489.336 N FT = 3262.85 + 489.336

FT = 3752.18 N

Reduced Force

FR = 0.37 × FT (4)

The Eqn. (4) is used to evaluate the total force necessary to shear the aluminium sheet, after considering reduction.

FR – reduced force

FT – total shear force = 3752.18 N FR = 0.37 × 3752.18

FR = 1388.306 N

Bore of Pneumatic cylinder

PC = (F/A) (5)

The Eqn. (5) is implemented to calculate the bore of the cylinder. PC – punching pressure = 6 bar = 0.6 N/mm^2

F – actual shear force = 1388.306 N

A – bore area =

d – bore of rod d = 54.27 mm

A =

π d2

4

Thus, a suitable bore w.r.t a standard pneumatic cylinder is d = 70 mm

Thus, a double-acting cylinder of a bore diameter of 70mm is suitable to punch a 10mm hole in the Aluminium sheet.

The bore diameter of the pneumatic cylinder for punching the 5-mm hole, is calculated as follows,

Punching Force

FC = L × t3 × Tmax (6)

The Eqn. (6) is used to evaluate the force necessary to punch the aluminium sheet. FC – force required to punch metal sheet

L – perimeter to be punched = ∏ × d = 15.707mm t3 – sheet thickness = 1 mm

Tmax – shear strength of aluminium sheet = 103.86 N/mm^2 FC = 15.707 × 1 × 103.86

FC = 1631.33 N

Stripping Force

FS = 0.15 × FC (7)

The Eqn. (7) is used to evaluate the force necessary to strip the aluminium sheet. FS – stripping force

FC – force required to punch metal sheet = 1631.33 N FS = 0.15 × 1631.33

FS = 244.69 N

Total Force

FT = FC + FS (8)

The Eqn. (8) is used to evaluate the total force necessary to shear the aluminium sheet. FT – total shear force

FC – punching force = 1631.33 N FS – stripping force = 244.69 N FT = 1631.33 + 244.69

FT = 1876.02 N

Reduced Force

FR = 0.37 x FT (9)

The Eqn. (9) is used to evaluate the total force necessary to shear the aluminium sheet, after considering reduction.

FR – reduced force

FT – total shear force = 1876.02 N FR = 0.37 × 1876.02

FR = 694.12 N

Bore of Pneumatic cylinder

PC = (F/A) (10)

The Eqn. (10) is implemented to calculate the bore of the cylinder. PC – punching pressure = 6 bar = 0.6 N/mm^2

F – actual shear force = 694.12 N

A – bore area =

d – bore of rod d = 38.37 mm

$A =$

π $d2$
4

Thus, a suitable bore w.r.t a standard pneumatic cylinder is d = 40 mm

Thus, a double-acting cylinder of a bore diameter of 40 mm is suitable to punch a 5-mm hole in the aluminium sheet.

4.2. Proposed Design of the Machine

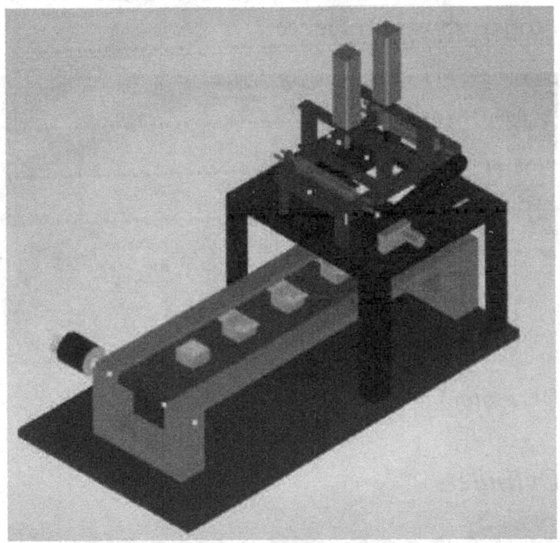

Figure 1: *CAD model of design.*

The proposed system is designed to manufacture washers through a semi-automated pneumatic system. Two pneumatic cylinders, with their punches, are positioned vertically over their respective die units, held over a frame. A roller unit feeds the sheet inside, upon the activation of the system, after subsequent punching is performed. An automated conveyor unit is positioned below the punching unit. Once a washer is produced, it gets collected in a box placed over the conveyor unit. After a certain quantity of washers is produced, the conveyor unit conveys the next box forward, for the collection of the next set of washers. The design is illustrated in Figure 1.

4.3. *Punching Unit*

The punching unit consists of a pneumatic unit, with two cylinders of specifications 40-mm bore, 100-mm stroke and 70-mm bore 125-mm stroke, for punching holes of 5 mm and 10 mm, respectively. The punches with punching diameters of 5 mm and 10 mm are attached through a threaded joint, to their respective cylinders. The cylinders are held over a base frame of dimensions, 350 mm × 200 mm × 100 mm, of MS material. A pair of rolling units, consisting of two rollers in each unit, is positioned on either end across the width of the base frame, which facilitates easy feeding of the sheet. The driver roll is actuated using a belt drive unit, which consists of a driver and a driven pulley, and is driven using a DC motor. The entire unit is held over a table, which houses a hopper below the 10-mm die, to guide the washer produced. 5/2 solenoid valves are equipped to actuate the cylinders individually, by regulating the air they receive from the air compressor.

4.4. *DC Geared Motor (12V, 60 RPM)*

This DC Motor has an RPM of 60, which operates at 12V. It has a shaft of 8-mm diameter. It provides a torque of 20 kg^{-cm}, which is achieved using a gearbox unit, attached to the motor, as shown in Figure 2.

Figure 2: 60 RPM DC motor.

4.5. *Pneumatic Cylinders*

Figure 3: Pneumatic cylinders.

The cylinders used are of the double-acting type, with 70-mm bore 125-mm stroke and 40-mm bore 100-m stroke, respectively. The pistons are provided threading at their ends, to attach the punches to the cylinder. The operable pressure range is 1–10 Bars, as shown in Figure 3.

4.6. Punches

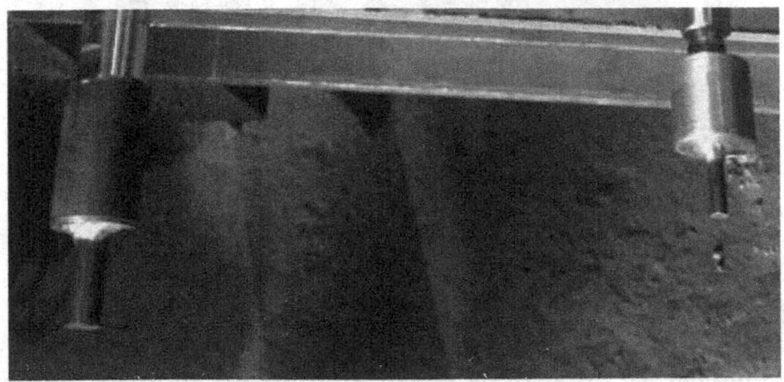

Figure 4: Punches.

The two punches have outer diameters 5 mm and 10 mm individually and are held by an encasement. They are composed of Hard Steel material, optimum for performing punching operations. The encasement has a threading along its inner diameter, for hassle-free attachment to the piston of the pneumatic cylinder, as shown in Figure 4.

4.7. Dies

The dies used are of dimensions 200 mm × 70 mm × 10 mm. A clearance of 0.06 mm that is 6% of the thickness of the sheet that would be punched, is provided to both the die holes, for optimum punching, as shown in Figure 5.

Figure 5: Dies.

4.8. Pulleys

Figure 6: Pulleys.

The driver pulley has an outer diameter of 50 mm and an inner diameter of 8 mm to accommodate the motor shaft. The driven pulley has an outer diameter of 50 mm and an inner diameter of 10 mm to accommodate the roll shaft, as shown in Figure 6.

4.9. 5/2 Solenoid Valve

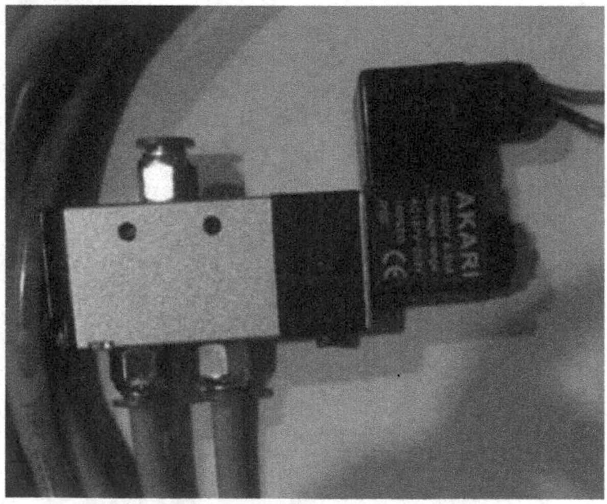

Figure 7: 5/2 Solenoid valve.

Two solenoid valves are used to operate upon two pneumatic cylinders individually. Both the solenoid valves are single-type valves, which work under a pressure range of 6–8 Bars. They are powered using a 220V AC supply. Air hoses are connected to its ports, to receive and regulate air as per requirements, as shown in Figure 7.

4.10. Belt

Figure 8: *Belt.*

The belt used in the belt drive is of Nylon material and has a centre-to-centre distance of 350 mm and a thickness of 2 mm, as shown in Figure 8.

4.11. Belt Conveyor

The belt conveyor is positioned below the punching unit, to collect the washers produced after each consequent punching operation. The assembly consists of a base frame, which supports the legs along the sides. A pair of rolls is positioned at either end of the belt conveyor frame, along its length. Roller contact bearings are positioned over the shafts of the rolls, so that the rolls rotate smoothly, with minimal slippage. The rolls support the belt, along its ends. The conveyor is designed in such a way that the tension along the belt can be varied, based on desired capacity, using a slot mechanism. The DC motor is coupled to the driver roll using a coupling. Once the system gets activated, after the desired number of washers gets manufactured and counted (by the counting unit), the DC motor transfers its rotary motion to the roller shaft, axially attached to it, through the coupling, which leads to the rotation of the conveyor belt.

4.12. DC Geared Motor (12V, 45 RPM)

Figure 9: *45 RPM DC Motor.*

This DC Motor has an RPM of 45, which operates at 12V. It has a shaft of 10-mm diameter. It provides a torque of 80 kg-cm, which is achieved using a gearbox unit, attached to the motor, as shown in Figure 9.

4.13. Belt

Figure 10: Conveyor belt.

The PVC belt used has a total length of 900 mm and a thickness of 1 mm, as shown in Figure 10.

4.14. Shaft Coupling

The shaft coupling has an outer diameter of 22 mm. One end has an inner diameter of 10 mm, and the other end has an inner diameter of 15 mm. The material used is mild steel, as shown in Figure

Figure 11: Shaft coupling.

11. Two holes are provided at either ends along an axis, perpendicular to the axis of the shaft to accommodate grub screws, which would then be used to fix the roller support from the conveyor roller and the motor shaft, along the same axis.

4.15. Rolls

Figure 12: Rolls.

The rolls used are of MS material. The rolls are of outer diameter 20 mm. The shaft along its sides has an outer diameter of 15 mm, as shown in Figure 12.

4.16. *Bearings*

***Figure 13:** Bearing unit.*

The bearings used are of roller contact type and have an inner diameter of 15 mm, as shown in Figure 13. They ensure that the rollers in the belt conveyor rotate freely, without any obstructions, which ensure the consistent delivery of the manufactured washers, without any slipping.

5. Simulation Study of Pneumatic System

The punching unit mainly comprises the punching system, which performs the punching operation. It comprises two double-acting pneumatic cylinders, controlled using 5/2 solenoid valves. A study is conducted on this system, through a simulation analysis using FluidSim.

The pneumatic system is modelled schematically in the FluidSim environment. It consists of a pneumatic cylinder, controlled using a 5/2 solenoid valve, which receives air from a compressor. An FRL unit is used between the compressor and the solenoid valve, as shown in Figure 14.

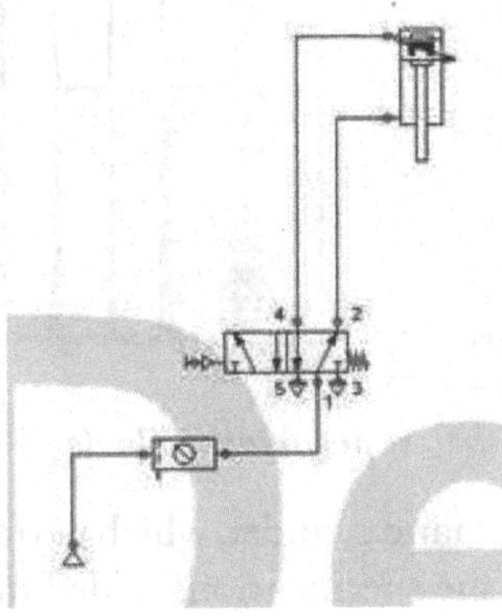

***Figure 14:** Schematic of the pneumatic system.*

In the case of the first pneumatic cylinder, which performs the punching operation to produce the 5-mm hole, the specifications added into the software are a stroke length of 100 mm, piston diameter of 40 mm, piston rod diameter of 16 mm and an added mass (punch) of 0.75 Kg. From the simulation performed, it was inferred that the piston gets actuated forward with a mean velocity of about 0.22 m/s and an acceleration of 42 m/s^2, as per Figure 15. These results promise a quick punching action with minimal delays.

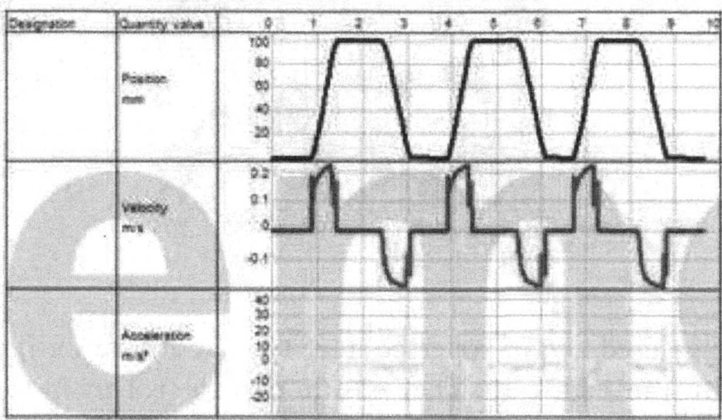

***Figure 15:** Simulated velocity and acceleration of the 1st pneumatic cylinder.*

From the simulation performed for determining the force exerted by this pneumatic cylinder, it was inferred that the piston is capable of exerting a force of 700N on the aluminium sheet, through the punch, which complements the theoretical analysis performed for calculating the desired punching force, as per Figure 16. Thus, a pneumatic cylinder of the aforementioned specifications is optimal for ensuring desired results.

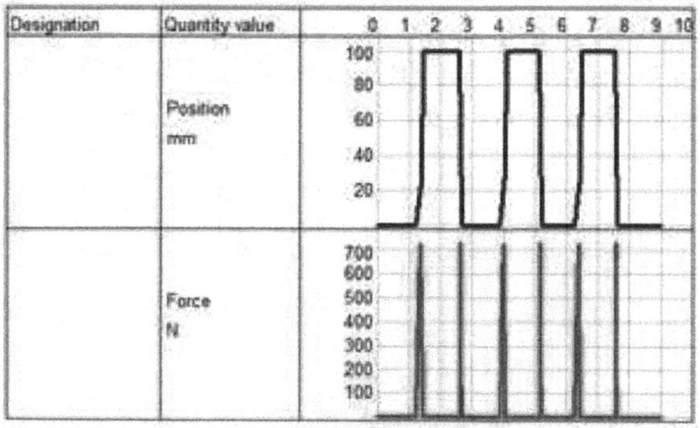

***Figure 16:** Simulated forces of the 1st pneumatic cylinder.*

In the case of the second pneumatic cylinder, which performs the punching operation to produce the 10-mm hole, the specifications added into the software are a stroke length of 125 mm, Piston diameter of 70 mm, piston rod diameter of 20 mm and an added mass (punch) of 1 Kg. From the simulation performed, it was inferred that the piston gets actuated forward with a mean velocity of about 0.17 m/s and maximum acceleration of 50 m/s^2, as per Figure 17. These results promise a quick punching action with minimal delays.

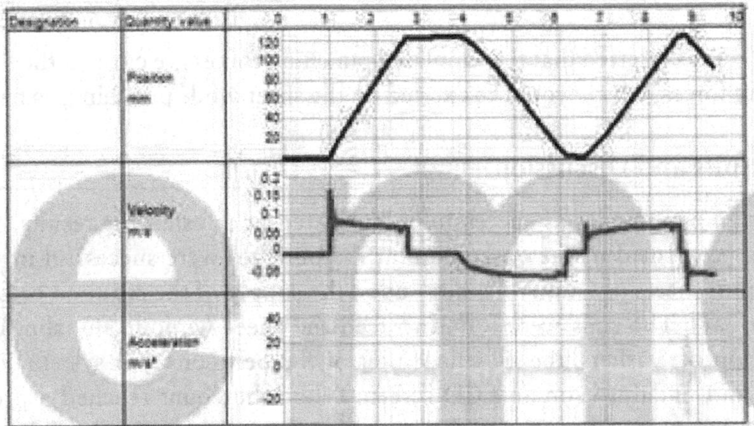

Figure 17: Simulated velocity and acceleration of the 2nd pneumatic cylinder.

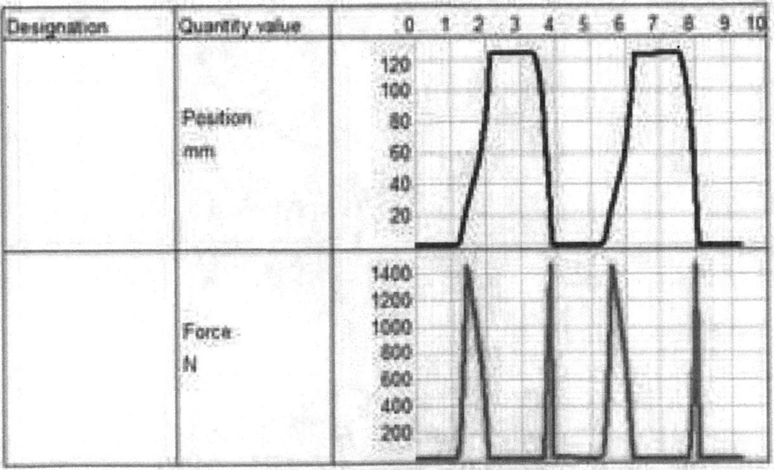

Figure 18: Simulated forces of the 2nd pneumatic cylinder.

From the simulation performed for determining the force exerted by this pneumatic cylinder, it was inferred that the piston is capable of exerting a force of 1400N, on the aluminium sheet, through the punch, which complements the theoretical analysis performed for calculating the desired punching force, as per Figure 18. Thus, a pneumatic cylinder of the aforementioned specifications is optimal for ensuring desired results.

Table 1: Summary of theoretical and simulated punching forces.

Cylinder No.	Hole Diameter (mm)	Theoretical Force (N)	Simulated Force (N)
1	5	694.12	700
2	10	1388.306	1400

Thus, it can be inferred that the simulated punching forces are close to the theoretical punching forces, which would be exerted on the sheet while punching, as per Table 1.

6. Results and Discussion

The design proposed was successfully fabricated and tested for results. Based on the tests performed it was observed that the punches were successful in punching Aluminium sheets of 1-mm thickness, upon the supply of the required pressure that is seven bars. The rolls were found to feed the sheet without any slip. Upon the production of washers, the IR sensor unit of the belt conveyor system counts the washer and updates it on an LCD screen. When the count reached a pre-defined limit, the DC motor was activated and conveyed the belt forward, so that the next box in line can start collecting the next lot of washers. The washers produced were found to be free of sharp edges, which can cause skin injuries. The final assembly of the model is shown in Figure 19.

Figure 19: *Front view of the assembly.*

Figure 20: Application of washers produced

7. Conclusion

The design proposed in this paper aims at introducing a safe and efficient way to manufacture washers while making sure that the cost incurred during purchase and maintenance is minimum. Based on the results obtained from tests, it can be inferred that the machine was successful in accomplishing the desired tasks, with assured safety, when operated at the optimum pressure and the washers produced were found to be feasible for application, as shown in Figure 20. The pneumatic system was thus found to be a feasible substitute to conventional mechanisms, as it is economical comparatively.

The design was thus fabricated as per the established scope and can be updated upon further usage, based on desired applications and possible improvements, as a part of future scope.

References

[1] Achchagam, K. (2016). *Design data-data book for engineers*. Coimbatore, TN: PSG College of Technology.

[2] Arun, S., Rajendra, S., and Bongale, V. (2014). Automatic punching machine: a low cost approach. *International Journal of Advanced Mechanical Engineering*. ISSN, 2250–3234.

[3] Arunkumar, G., Antonio, J. T., and Pon, V. (2021). Design and development of autofeed pneumatic punching and riveting machine. *Journal of Physics: Conference Series*. 2054 No. 1, p. 012025. IOP Publishing.

[4] Bhandari, V. B. (2017). *Design of machine elements*. Chennai, TN: McGraw Hill.

[5] Dagar, R. R. M. G. K., and Kaushik, A. (2019). Some investigation on design analysis and fabrication of automated metal sheet cutting machine. *IJSRD-International Journal for Scientific Research & Development*, 1000–1003.

[6] Ghatge, D. A., Birje, C., and Yadav, P. S. (2017). Use of shearing operation for MS bar cutting by pneumatic bar cutting machine. *Young*, 11(11.3), 10–18.

[7] Kelaginamane, S., and Sridhar, D. R. (2015). PLC based pneumatic punching machine. *Journal of Mechanical Engineering and Automation*, 5(3B), 76–80.

[8] Polapragada, A. A., and Varsha, K. S., (2012). Pneumatic auto feed punching and riveting machine. *International Journal of Engineering Research & Technology* (*IJERT*), 1(7), 2278–0181.

[9] Prajapati, P. A., Patel, M. G., Patel, M. R., Lalit, M., and Patel, D. (2018). Design and development of pneumatic metal sheet cutting machine. 4(2), 340–344.

[10] Sharma, U. (2015). Design of automatic pneumatic hole punching machine. *International Research Journal of Engineering and Technology* (*IRJET*), 2(09).

Chapter 40

Design and Development of Cycloidal Gear Box

Soham Ravindra Nevgi[1], Parth Vinod Trivedi[1], Hardik Mahendra Thasale[1], Siddhesh Nagesh Padate[1], and Bipin Mashilkar[1]

[1]Department of Mechanical Engineering, Fr. Conceicao Rodrigues Institute of Technology, Vashi, University of Mumbai, Navi Mumbai, India

Abstract: Gears and gearboxes are used in daily life in many mechanical devices. One of the gears used is cycloidal gear or drive. It is a gear which reduces speed ratio and increase torque in transmission devices. They are highly efficient and are used mostly for high-speed reduction. They have wide applications where precise output and large drive payloads are needed. Just like cycloidal gears there are many gears such as Harmonic drives, REFLEX Torque Amplifier, Archimedes Drive, NuGear, Bilateral Drive, Gear Bearing Drive and Galaxie Drive. One of the research papers reviewed for this project was on how to build a cycloidal gear with SOLIDWORKS. It is really helpful in designing of the different components of the gearbox. Through the transmission ratio provided, the output torque obtained from the input torque can be calculated. The total torque transmission ratio of the proposed design of cycloidal gearbox is 20:1. The project provides the CAD model and detailed drawing of the components of the cycloidal gearbox. Through ANSYS software we have compared the structural and modal analysis of normal gearbox and cycloidal gearbox. The proposed design of cycloidal gearbox has a Factor of Safety of 15 which tells that it is quite stable. Through this gearbox, the weight lifted is quite large as compared to that of normal gearbox. Even though current design of the cycloidal gear is quite useful, some slight modifications can be made so that it can be operated at low input torque and give high output speed. The current design can be used to get high output torque from low input torque, high precision, speed reducing and to lift heavy objects. The cycloidal gearbox can be used in JCB, excavation equipment, milling machine, turning machine, CNC machine, servo motors, vehicles, lifts, etc. So, this project provides us how cycloidal gearbox is better than normal gearbox and how it can be implemented in day-to-day chores.

Keywords: Cycloidal drive, CNC machine, SOLIDWORKS.

DOI: 10.1201/9781003527442-40

1. Introduction

Cycloidal gear or drive is a gear which is used to reduce ratio motion and increase torque in transmission devices. In geometry, a cycloid is the curve traced by a point on a circle as it rolls along a straight line without slipping. A cycloid is a specific form of trochoid and is an example of roulette, a curve generated by a curve rolling on another curve. The input shaft drives an eccentric bearing that in turn drives the cycloidal disc in an eccentric, cycloidal motion. The perimeter of this disc is geared to a stationary ring gear and has a series of output shaft pins or rollers placed through the face of the disc.

Advantages of the cycloidal gear over normal gear are as follows:

- More robust construction
- Virtually backlash-free for the whole lifetime longer life
- Higher rigidity
- More compact design
- Lower weight
- Higher overload protection
- No additional pre-stages

1.1. Aim and Objective

The aim of this work is to design a cycloidal gearbox for construction pulley. The objective is to design and develop one stage cycloidal drive and analysis of the cycloidal gearbox for the high output torque requirement.

2. Methodology

The methodology followed is described below.

Identification of a need: This was based on rising accuracy demands and the opportunity for greater efficiency and safety through automation.

Literature Review: Conducted an extensive literature review to better grasp the present gear methods and technologies. Mechanical de-husking techniques, automation concepts, safety considerations and past attempts at developing de-husking machines were all included in this review.

Conceptual Design: Based on the literature review, create an early conceptual design for the cycloidal gearbox. Factors such as mechanism type, power source, components, materials and so on were considered. Different design possibilities were sketched, and their viability in addressing the indicated needs was evaluated.

CAD Modeling: Identified suitable computer-aided design (CAD) software and generated a comprehensive 3D model of the selected conceptual design. The CAD model involves all the components, mechanisms and assemblies that were decided upon throughout the conceptual design stage. Dassault systems SolidWorks was used to model this project.

Design Calculations: Performed design calculations to guarantee that the chosen components and mechanisms can be withstanding the mechanical forces and stresses involved. Material availability as well as cost were essential considerations for material selection.

Material Purchasing: A list of materials needed to manufacture the machine was prepared based on the finished CAD model and design calculations. I obtained the essential materials from trustworthy vendors and confirmed they met quality standards.

Manufacturing: Manufacturing involved the construction of the prototype of a cycloidal gearbox based on the CAD model. It included additive manufacturing process 3D printing.

Testing: Tested the cycloidal gearbox once it was manufactured. Assembled the entire gearbox and verified the results by adding different weights.

Performance Optimization: Analysed the results of testing and identified areas for improvement. To achieve optimum performance of the gearbox, changes are made accordingly.

3. Literature Review

Younis (2014) showed how to design a cycloidal drive using SOLIDWORKS. Cycloidal drive consists of five main parts the rotor, the output disk, the input disk, the rollers and the main housing. In Figure1 we can see that the rotor is the most complex one to make, it is made in SOLIDWORKS using the Parametric Equations from Equation Driven Curve Tool.

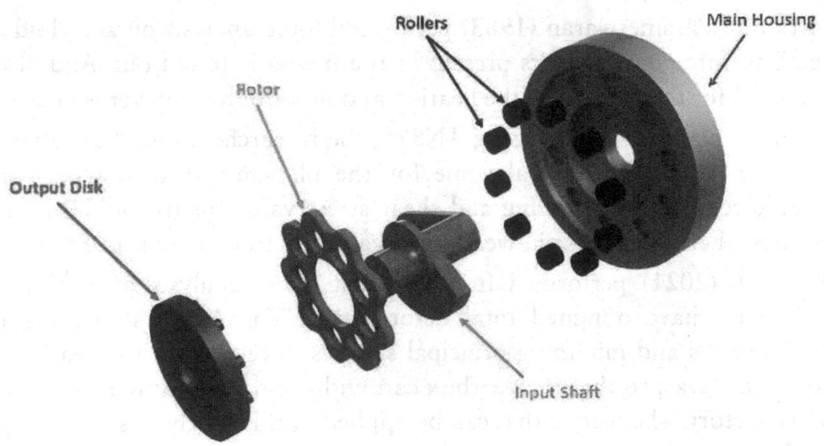

Figure 1: Design of cycloidal drive [1].

The two equations that define the Rotor's shape are the following:

$$X = R\cos(\Theta) - R_r\cos(\Theta - \psi) - E\cos(N\Theta) \quad (1)$$

$$Y = -R\sin(\Theta) + R_r\sin(\Theta - \psi) + E\sin(N\Theta) \quad (2)$$

Where

$$\psi = -\tan^{-1}\left[\frac{\sin((1-N)\Theta)}{(R/EN) - \cos((1-N)\Theta)}\right] \quad (0° \leq \Theta \leq 360°) \quad (3)$$

$$X = (R*\cos(t))-(R_r*\cos(t+\arctan(\sin((1-N)*t)/((R/EN)-\cos(((1-N)*t))))))-(E*\cos(N*t)) \quad (4)$$

$$Y = (-R*\sin(t))+(R_r*\sin(t+\arctan(\sin((1-N)*t)/((R/EN)-\cos(((1-N)*t))))))+(E*\sin(N*t)) \quad (5)$$

Garcia *et al.* (2020) experimented on Strain Wave Gearbox invented by Musser in 1995, and it is commonly referred to as Harmonic Drive. They have zero-backlash and are a lightweight gear. The strain wave trains show large load-independent losses and no-load staring torque particularly in back-driving conditions. Worth noticing fact in the strain wave gearbox is that they have large latent power ratio, multiple teeth engagement, lost motions below 1 arcmin and also improved stiffness linearity.

Lin *et al.* (2014) researched on the coordinate system for the tooth contact analysis of a cycloidal drive. And also came up with a mock-up design for the two-stage cycloidal gear drive. The Tooth Contact Analysis (TCA) can provide information about the surface contact condition at every instant and simulate the transmission errors caused by the profile modifications. The suitable model design for the two-stage cycloidal drive is also previewed.

Malhotra and Parameswaran (1983) performed force analysis on a cycloid speed reducer. The force on all shafts present in the model is found out. And also, the formula used for the torque for the bearing and input/output power is determined.

Yogesh and Kalamkar (2011) using ANSYS, the researchers found out that value where nearer to the theoretical value for the planetary gear system. There is appreciable reduction in bending and shear stress value for trochoidal root filler design. Also, there is increase in wear stress value for trochoidal root filler design.

Saritas *et al.* (2021) performed finite element stress analysis on a three-stage gearbox. They have obtained total deformation, Von Mises stress, maximum principal stresses and minimum principal stresses developed in the gearbox. It is done in such a way, so that the gearbox can withstand the targeted torque with a 1.5 safety factory. The torque that can be applied in this gearbox is in the range of 43–301 Nm.

Li and Yuanzhe (2012) have designed the first stage of the RV reducer, as well as the related components. The details contain design of input shaft, planetary gears, output shaft, common bearings and eccentric bearings. The fatigue analysis is mostly used in the calculation process because the fatigue failures are frequent in this type of rotation machine. In the same time, the general bearings designs are based on the SKF General Catalogue and the eccentric bearings design are based on the Chinese standard.

Blagojevic *et al.* (2011) designed a two-stage cycloidal speed reducer that is presented in this paper. A traditional two-stage cycloidal speed reducer is obtained by the simple combination of single-stage cycloidal speed reducers. A single-stage reducer engages two identical cycloid discs in order to balance dynamical loads and to obtain uniform load distribution. Consequently, the traditional two-stage reducer has four cycloid discs, in total, which means that it is rather compact. Due to its specific concept, this reducer is characterised by good load distribution and dynamic balance, and this is described in the paper. Stress state analysis of cycloidal speed reducer elements was also realised, using the finite elements method (FEM), for the most critical cases of conjugate gear action (one, two or three pairs of teeth in contact). The results showed that cycloid discs are rather uniformly loaded, justifying the design solution presented here. Experimental analysis of the stress state for cycloid discs was realised, using the strain gauges method. It is easy to conclude, based on the obtained results, that even for the most critical case (one pair of teeth in contact) stresses on cycloid discs are in the allowed limits, thus providing normal functioning of the reducer for its anticipated lifetime.

3.1. *Construction*

Figure 2: Assembly of gearbox. *Figure 3: Individual components of gearbox.*

As shown in Figure 3, the gearbox consists of two cycloidal discs which are 180° out of phase. These discs have six holes present in them and 20 lobes on the exterior side of the discs. The case also has 21 lobes on interior side of the case. The output disc has 6 pins which meshes with the holes in cycloidal discs. The eccentric shaft has two discs with an eccentricity of 3 mm. The working of the final assembly is discussed below.

3.2. *Working*

As the working of the machine is concerned, it has a 0.36-watt single phase AC motor as power source whose operating speed is 100 rpm. As per our application,

100 rpm is the most optimum speed. The gearbox reduces speed of the motor with a ratio of 20:1. The motor and the gearbox are coupled together. The output of the gearbox will be connected to a pulley. The motor power is transmitted to the eccentric shaft, from eccentric shaft to two cycloidal discs. Then from two discs to the output disc via the six pins.

3.3. Theoretical Analysis

Some theoretical calculations have been made for the design of the system as described below

Torque Calculations:

Here, Input Torque = 3.43 Nm

n = No. of Lobes on the Cycloidal Gear = 20

N = No. of lobes on the Case = 21

The Torque Transmission ratio for the cycloidal gear is given as:

$$\text{Output Torque} = \frac{n}{N-n} \times \text{Input Torque}$$

For Cycloidal Gear,

$$\text{Output Torque} = \frac{20}{21-20} \times 3.43$$

Output Torque = 68.6 Nm

Diameter of the shaft using Torque will be given as:

2 × Torque = Force × Diameter

2 × 3.43 = 343 × Diameter

Diameter = 0.02 m

Bearing Calculations:

Here, SKF deep groove ball bearings for smooth working of the gearbox. Suitable bearings as per the shaft diameters are:

Table 1: Bearings used.

Bearing No.	Shaft Diameter	DGBB
1	12	6801
2	20	6804
3	20	6804
4	12	6801
5	20	6804

(Source: Author's compilation)

Input Torque (T_i) = 3.43 Nm

Output Torque (T_o) = 64.6 Nm

Shaft Speed (N) = 200 rpm

Service Factor (S) = 1.2

Service of bearing = 12 hrs a day for a year

L_{hr} = 12 x 365 = 4380 hrs

Life of the bearings in million revolutions will be: -

$$[L_{mr}]_{required} = \frac{L_{hr} \times x60 \times x\, N}{10^6}$$

$$[L_{mr}]_{required} = \frac{12 \times x365 \times x60 \times x200}{10^6}$$

$$[L_{mr}]_{required} = 52.56 \text{ mr}$$

Radial load for all the bearings is given as

$$F_r = \frac{\text{K} \times \text{x T} \times \text{x f}}{\text{Y}}$$

where F_r = Radial Load

K = Load Coefficient

T = Torque Applied

f = Service Factor for Load

Y = Effective Radius

Table 2: Radial force on bearing.

DGBB	K	T	f	Y	F_r
6801	1.25	3.43	1	0.006	714.58
6804	1.25	3.43	1	0.01	428.75
6804	1.25	3.43	1	0.01	428.75
6801	1.25	3.43	1	0.006	714.58
6804	1.25	3.43	1	0.01	428.75

(Source: Author's compilation)

Axial load for all the bearings is given as

$$F_a = \frac{T}{k \times \text{x}\, D}$$

where F_a = Axial Load

k = Torque Coefficient

T = Torque Applied

D = Diameter

Table 3: Axial force on bearing.

DGBB	k	T	D	F_a
6801	0.2	3.43	0.012	1429.16
6804	0.2	3.43	0.02	857.5
6804	0.2	3.43	0.02	857.5
6801	0.2	3.43	0.012	1429.16
6804	0.2	3.43	0.02	857.5

(Source: Author's compilation)

Life for every bearing will be as follows: -

Table 4: Life of bearing.

DGBB	$\frac{F_a}{C_o}$	$\frac{F_a}{F_r}$	e	X	Y	P	LIFE
6801	1.37	2	0.44	0.56	1.0	2195.2	83.636
6804	0.34	2	0.37	0.56	1.2	1522.9	61.616
6804	6.94	2	0.44	0.56	1.0	26342.4	55.117

(Source: Author's compilation)

Life of every bearing is greater than the required L_{mr}. Hence, all the bearings are suitable.

3.4. Design

Figure 4: *Exploded view of cycloidal gearbox.*

Shown above Figure 4 is the exploded view of the cycloidal gearbox. Here we can see that it consists of a case, two cycloidal discs, an eccentric shaft, an output disc, a lid, three DGBB 6804 bearing, two DGBB 6801 bearings and seven M4 bolts. The output disc translates the output of the revolving cycloidal disc to an output shaft. The eccentricity of the shaft allows the discs to be in internal meshing with the gearbox casing. This while changing the direction of rotation of the output shaft provides a large increase of torque in small and compact casing. There are gear boxes which have one disc and the eccentric shaft has a cam which balances the offset forces of the disc. But we have opted to using a two-disc system which makes it easier to maintain, repair and manufacture this gearbox. This allows us to reduce the load on the disc and have balanced forces on the casing of the gearbox which reduces the number of vibrations generated while spinning.

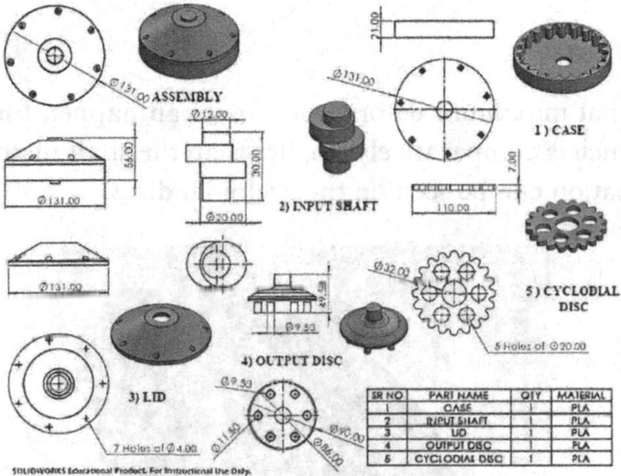

Figure 5: *Detailed drawing for the cycloidal gearbox.*

Figure 5 shows the detailed drawing of the components of the cycloidal gearbox. Case has diameter of 131 mm having 21-mm thickness and we have made a hole of 12-mm diameter in the middle. On the upper surface of the case a profile of cycloidal disc with 21 lobes has been cut out for 18-mm thickness. Output disc has diameter of 90 mm and a shaft of 20 mm and 15-mm height. On the lower surface six pins of 9.5-mm diameter and 14 mm in length are attached. Cycloidal Disc is modeled using the Parametric Feature of the Equation Driven Tool. Six holes of 20-mm diameter and a hole of 32 mm in middle are made on the cycloidal disc. Last component to be designed was eccentric shaft. It is made up of four discs, out of which two discs have diameter of 20 mm and thickness of 7.5 mm. Other disc has diameter of 12 mm and thickness of 55 mm and last disc has dia. of 12 mm but height of 10 mm. The center of first cycloidal disc is at 3-mm eccentricity from the center of the second cycloidal disc. Distance between the two cycloidal discs are 1 mm. Lid also has diameter of 131 mm, height of 31 mm with 30° angle chamfer. A hole of 22-mm diameter in the center and seven holes of 4-mm diameter near the end. These holes fit seven M4 bolts.

4. Results and Discussion

Static Structural Analysis of Cycloidal Gearbox:

Figure 6: Total deformation.

Figure 6 shows that maximum deformation that can happen for the gearbox will be .000175 m which is comparatively smaller than the shaft diameter at that point. Also, less deformation can be seen on the cycloidal discs.

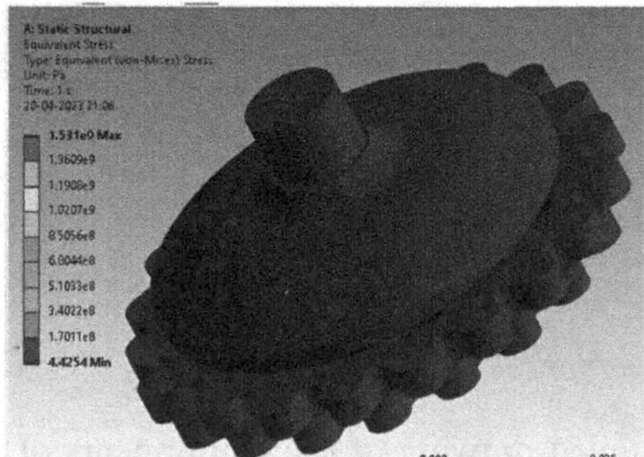

Figure 7: Von-Mises stress.

Figure 7 depicts Von Mises Stress, maximum stress induced is 1.531×10^9 Pa at the edge of the shaft of output disc. Induced stress in the discs are very small 4.4254 Pa.

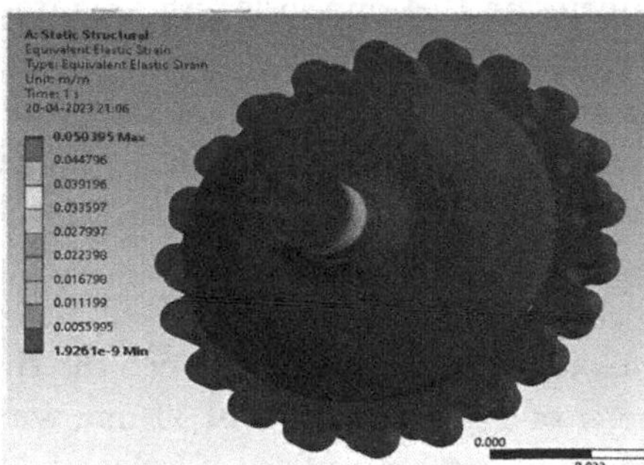

Figure 8: Equivalent strain.

Figure 8 depicts the equivalent strain developed in the gearbox. Maximum strain developed is .05395 m/m and it is located at the middle of shaft of output disc. Strain developed in the discs are negligible.

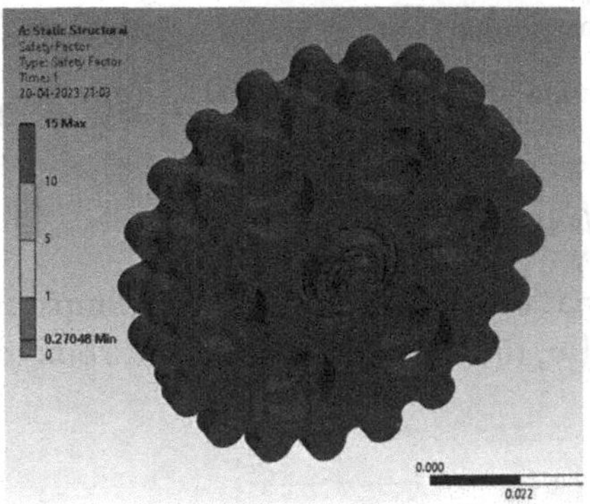

Figure 9: Factor of safety.

Figure 9 shows factor of safety for designed gearbox, it can be seen that most regions of the gearbox have 15 as safety factor but only at edges the safety factor is low as there is high load at that point which is around 1.

Comparison of Cycloidal Gearbox with Normal Gearbox

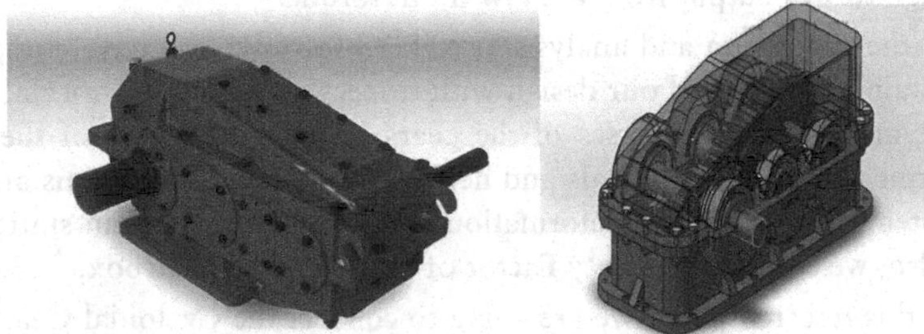

Figure 10: Different types of gearboxes used in construction pulleys.

Figure 10 depicts normal gearbox which is used in the construction pulleys. As shown in the figure the size of the gearbox is quite big. And also due to its size the weight of the gearbox is also large. When this gearbox is compared to designed cycloidal gearbox, it is seen that:

Table 5: Comparison between normal gearbox and designed cycloidal gearbox.

Parameters	Normal Gearbox	Cycloidal Gearbox
Input Speed	Minimum needed is 250 rpm	Minimum needed is 100 rpm
Output Speed	Maximum up to 900 rpm	Maximum up to 1000 rpm
Input Power	Requires 200 W	Requires 20 W

Output Power	**Delivers 132 kW**	**Delivers 100 W**
Input Torque	**Requires up to 350 Nm**	**Requires up to 3 Nm**
Output Torque	**Delivers up to 18000 Nm**	**Delivers up to 20 Nm**
Weight	**45 to 190 kg**	**450 gm**
Size	**Nominal is 500 × 300 × 300 mm**	**132 × 132 × 56 mm**

(Source: Author's Compilation)

From Table 5 we get that the designed gearbox is comparatively more useful than the normal gearbox. Also, the designed cycloidal gearbox is far more efficient than the normal gearbox.

5. Conclusion

From this project we get to know that there are many forms of cycloidal gears available for to design a gearbox, but the design proposed in this project is most efficient, beneficial as well as profitable to be designed for a gearbox. Using SOLIDWORKS it is quite easy to design the outline of the gear needed to be used in the gearbox. Also, the case has been designed in such a way that it can provide the gearbox with an external covering as well as give us rollers needed by us so that we can get high torque output from the low-input torque.

From the simulation and analysis in ANSYS software, we have compared the results obtained by FEM of our design with respect to that of the normal gearbox. The static and dynamic analyses of the gears, shafts and casing of the gearbox are performed. After the materials and necessary boundary conditions are applied to the system, we can get the deformation and Von Mises stress in static analysis results. Also, we can get the Safety Factor of the Cycloidal Gearbox.

As in this research paper we are going to connect the Cycloidal Gearbox with a construction pulley or crane, we can also find the maximum amount of mass it can lift for the given amount of the torque. When it is compared to mass lifted by the normal gearbox it is seen that it is quite large and it can be very beneficial to use it instead of a normal gearbox. It is true the mass lifted by Cycloidal Gearbox is dependent on many various factors such as the yield strength of the arm of the crane, the Young's modulus of the wire used in the crane and the integrity of the components to be transported.

6. **Future Scope**

Cycloidal gear was firstly used in mechanical clocks and then found their way into robotics. Similarly, we can make use of this concept to implement the usage of cycloidal gears in our day-to-day life. Using the current design available, it can be used where we require high-output torque from low-input torque, for high precision, for reducing speed and where we have to lift very heavy masses to a certain height. In future slight modifications can be done in the cycloidal gear so that it can be provide us with a design which has low input and high-speed output. This can have many day-to-day applications around us.

It can be used in JCBs and excavation tools as it can provide very high-output torque and can lift heavier masses. And in sports cars to increase torque as it can provide quite good control over the car. It can be used in milling, turning, facing and the CNC machines to provide us with precisely measured products. It can used in servo motor used for the toy airplanes and in various other devices to give them very high precision and give us total control over our device. In a car cycloidal gear can be used in unison with some contraption to slow down the speed of the car and ensure us the safety of the passenger inside. It can provide the people with a very good control over the vehicle they are driving.

References

[1] Blagojevic, M., Marjanovic, N., Djordjevic, Z., Stojanovic, B., and Disic, A., (2011). A new design of a two-stage cycloidal speed reducer. 085001.

[2] García, L., Crispel, S., Saerens, E., Verstraten, T., and Lefeber D. (2020). Compact gearboxes for modern robotics: a review. *Frontiers in Robotics and AI*, (7), 103.

[3] Li, Y., and Yuanzhe, W. (2012). Design of a cycloid reducer: planetary stage design, shaft design, bearing design, bearing selection, and design of shaft related parts.

[4] Lin, W., Shih, Yi., and Lee, J. (2014). Design of a two-stage cycloidal gear reducer with tooth modifications. *Mechanism and Machine Theory*, (79), 184–197.

[5] Malhotra, S. K., and Parameswaran M. A. (1983). Analysis of a cycloid speed reducer. *Mechanism and Machine Theory*, 18(6), 491–499.

[6] Yogesh H., and Kalamkar V. (2011). Analysis of stresses and deflection of sun gear by theoretical and ANSYS method. *Modern mechanical engineering*, 1(02), 56.

[7] Saritas, M., Gölbol, Ö., and Yayla, P. (2021). Finite element stress analysis of three-stage gear box. *Niğde Ömer Halisdemir Üniversitesi Mühendislik Bilimleri Dergisi*, 10(2), 784–790.

[8] Younis, O. Building a cycloidal drive with SOLIDWORKS. https://blogs.solidworks.com/teacher/wp-content/uploads/sites/3/Building-a-Cycloidal-Drive-with SOLIDWORKS.pdf

Chapter 41

Design, Development and Manufacturing of Coconut De-husking Machine

Suyash Bare[1], Jebin Samuel[1], Pradnya Bokade[1], Yash Mekade[1], and Dhananjay Panchagade[1]

[1]Department of Mechanical Engineering, Fr. Conceicao Rodrigues Institute of Technology, Vashi, University of Mumbai, India

Abstract: Coconut is widely used in our day-to-day life. The coconut fruit is used for various reasons across the globe. India is one of the largest producers of coconut. The coconut's inner part contributes for its healthy juice, copra to extract oil, whereas the husk (outer shell) is used for making coir, potting soil, etc. The demand for its husk as well as the soft edible meat is always high. Different methods are used to de-husk the coconut, manual operation with machetes is one of them; however, its major drawbacks are safety issues, lethargy of workers and less production. There are numerous machines in the market to overcome this problem but they have many drawbacks, the machine that de-husk the coconut of any shape and size quickly is very costly whereas another machine is economical but is limited to a particular shape and size. This paper presents the design and manufacturing of coconut de-husking machine whose objective is to de-husk the coconut of any shape and size safely, effortlessly, quickly and economically without partial de-husking or crushing the coconut. The machine is designed to effortlessly remove the outer shell of the coconut, thereby increasing productivity along with safety. To achieve this, various components like electric motor, gearbox, rollers and power transmission drive works together. The paper outlines the design, testing and manufacturing phase of the project along with the methodology followed. As the testing was based on a trial-and-error method, the paper briefly explains all the changes made in the initial design to achieve the aim. As coconut exists in various shapes, a unique adjuster mechanism is designed to de-husk coconut. The result discusses various changes made in the design parameters during testing and their effects on the performance of the machine so that the aim and objectives are fulfilled. The paper concludes by representing the final data obtained from the testing which leads to the successful working of the machine.

Keywords: De-husk, coconut, machine, adjuster mechanism.

DOI: 10.1201/9781003527442-41

1. Introduction

Coconut is produced in across the 93 countries of the world, with the production of 54.9 billion tons per year across the world (Ingle, *et al.*, 2022). India stands as a significant global contributor to coconut production. India contributes to 15.30 % of the global area and 19.49 % of global production, and is the largest single market for coconut, consuming almost its entire production of 12.6 billion nuts (Sakhare *et al.*, 2014). This versatile crop offers nourishment, edible oil, industrial applications and a healthful beverage to humanity. Every component of the coconut tree holds value, impacting the socio-economic well-being of countless farming communities. Coconut oil, classified within the edible-industrial category, serves as both a cooking medium and finds applications in hair care, massages and industrial processes (Ramadurai *et al.*, 2019). Coconuts are available in different forms, such as dry coconut and tender coconut, each with its own unique applications and demand. The coconut fruit is composed of outer layer, fibrous husk, seed coat, meat and water of coconut that is consumed by both humans and animals (Yohanes *et al.*, 2016). In contrast, dry coconuts are in high demand in Indian cuisine. Additionally, the coconut husk has a traditional use in cleaning utensils, where its fibers are separated and combined with charcoal powder and lemon juice to create effective scrubbing pads for cleaning dishes. In India, a sport known as Jallikattu involves releasing bulls into a crowd, where individuals attempt to subdue them either by riding or restraining them. For safety reasons during this event, the ground is covered with dry coconut husks, providing a soft and cushioning surface to prevent accidents.

The list of things goes on but to achieve it, coconut and its husk are required. A large amount of husk is required for this. There are various methods available to de-husk the coconuts. De-husking is the process of removing the husk from the nut (Gaikwad *et al.*, 2017). Traditionally sharp machetes are used to de-husk the coconut and are still used in the modern era. Manual coconut peeler is also used to de-husk the coconut but both of the method mentioned above has some drawback or the other. The de-husking of a coconut is regarded as the most time consuming, tiring, and difficult operation to perform and involves much human drudgery (Amal *et al.*, 2018). Limited efficiency is observed in manual coconut peeler as compared to traditional machetes; however, for the separation of husk the operator has to apply a force of 52.77 N (Prajwal *et al.*, 2021) and to de-husk one single coconut, the operator has to apply this separation force at least four times.

These methods require skilled labour and are tiring to use. Efforts invested in the advancement of de-husking tools have yielded only limited success, falling short of fully replacing traditional manual techniques. The shortcomings attributed to these tools encompass inadequate and incomplete husk removal, the risk of coconut shell breakage during the de-husking process, loss of valuable coir fibers and the requirement of greater exertion compared to manual methods. (Kadam *et al.*, 2023).

After identifying the problems present in the market, this machine was designed and fabricated in such a way that it overcomes all the issues discussed above. This machine doesn't compromise production rate, safety or cost as it overcomes all the major drawbacks of the earlier machines and can also de-husk coconuts of varying sizes and shapes without partial de-husking or crushing the coconut. This machine has a combination of splined rollers and spiked rollers which de-husk the coconut. The splined roller penetrates and pushes the coconut to the spikes and then the spikes pierce and tear the husk, making it effortless and safe. The machine is designed using ergonomic considerations, making it easy and comfortable to use. A special adjuster mechanism is designed and implemented which is discussed further in the paper.

2. Aim and Objective

The aim of this work is to design and manufacture coconut de-husking machine. The objective is to design and manufacture an innovative coconut de-husking machine that prioritises safety, cost-effectiveness and versatility to effortlessly and rapidly de-husk coconuts of varying sizes without any of the problems mentioned above.

3. Methodology

The methodology followed is described below.

3.1. Identification of Need

This was based on factors such as the labour-intensive nature of traditional de-husking methods, increasing demand for processed coconuts, and the potential for increased efficiency and safety through automation.

3.2. Literature Review

Conducted an in-depth literature review to understand the existing methods and technologies for coconut de-husking. This review comprised of mechanical de-husking techniques, automation concepts, safety considerations and any previous attempts at designing de-husking machines.

3.3. Conceptual Design

Based on the literature review developed an initial conceptual design for the coconut de-husking machine. Considered factors such as the type of mechanism, power source, components, materials etc. Sketched out different design options and assessed their feasibility in meeting the identified needs.

3.4. CAD Modelling

Identified a suitable computer-aided design (CAD) software and created a detailed 3D model of the selected conceptual design. The CAD model includes all components, mechanisms and assembly which was decided in conceptual design. For modelling this project Auto Desk Inventor 2024 was used.

3.5. Design Calculations

Performed design calculations to ensure that the chosen components and mechanisms can handle the mechanical forces and stresses involved in the de-husking process. Validated these calculations through physical testing. For material selection availability and cost were considered.

3.6. Material Procurement

Based on the finalised CAD model and design calculations, a list of materials was created which was required for building the machine. Secured the necessary materials from reputable suppliers and ensured that they meet quality standards.

3.7. Manufacturing

Manufacturing involved the actual construction of the coconut de-husking machine based on the CAD model. It included processes such as machining, welding, fabrication, painting, etc.

3.8. Testing

Thoroughly tested the coconut de-husking machine once it was manufactured. Assembled the entire machine and ran it through various shapes and sizes of coconuts.

3.9. Performance Optimisation

Analysed the results of testing and identified areas for improvement. To achieve optimum performance of the machine, changes are made accordingly.

4. Construction

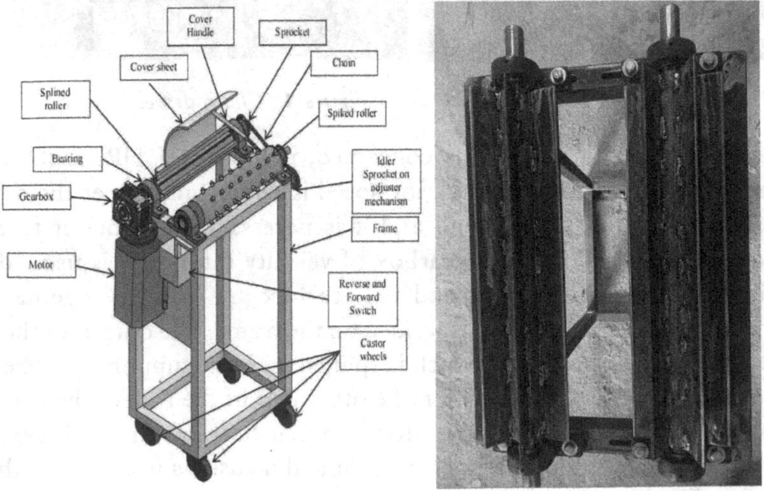

Figure 1: CAD model of final assembly. *Figure 2: Initial design.*

As shown in Figure 1, the machine consists of primary components such as motor, gearbox, driver roller, driven roller, chain drive, bearings, reverse forward starter switch, adjuster mechanism with idler sprocket, cover sheet, castor wheels and frame. The machine also consists of secondary components such as slotted members, supporting plate, nuts and bolts, etc. The CAD model shows the final working assembly of the machine after successful testing which de-husk the coconut in the most optimum way and fulfil all the objectives mentioned above. Initially, the design consisted of all the primary and secondary components; however, the driven and driver were splined rollers as shown in Figure 2. After testing it was observed that the initial design assembly was a failure. The results obtained from the initial design assembly were not satisfactory as desired. To achieve optimum and satisfactory results several changes were made in the design of the machine. The details of all the changes that were made in the initial design to final design during testing period are discussed further in the paper. The working of the final assembly is discussed below.

5. Working

Figure 3: Adjuster mechanism.

Figure 4: Chain drive.

As the working of the machine is concerned, it has a 0.5 HP single phase AC motor as power source whose operating speed is 1440 rpm. As per the de-husking application, 1440 rpm is very high and it is necessary to reduce it to a certain limit. For the speed reduction, a gearbox of velocity ratio 40:1 is used, due to its availability and cost. The motor and the gearbox are coupled together and are bolted on supporting plate which is welded on the frame. The output of the gearbox is connected to the driver roller which is splined and it is supported by bearings at both the ends. A sprocket is keyed on the other end of the roller. The driven roller which is larger and spiked is supported between bearings and the bearings are bolted on the frame. Another sprocket of same dimensions is keyed on the end of the driven roller.

To change center-to-center distance, a unique adjuster mechanism was designed and manufactured and an idler sprocket was mounted on it which has different dimensions than the other two. The reason for smaller idler sprocket is that it is easy to tighten and loosen the chain, with the help of adjuster mechanism. The adjuster mechanism shown in Figure 3 is easy to operate as it can be operated by hand. A chain is passed over the sprockets in a special manner as shown in the Figure 4, which allows both the rollers to rotate in different directions. The output of gearbox is 35 rpm. The driven roller which is connected to the driver through chain starts to rotate in opposite direction with same speed as the size of sprockets is same. Before starting the machine, a coconut is placed in between the two rollers and the cover sheet is pressed downwards against the coconut. After securing the coconut between two rollers and cover sheet, the machine is started. Optimum center-to-center distance between the rollers was found during the testing which is discussed further in the paper. As the roller rotates the splined roller penetrates and pushes the coconut towards the spiked roller and the spikes tears the husk. The cover sheet prevents the coconut from escaping the clutch of two rollers. The selection of components was based on design calculations.

6. Design

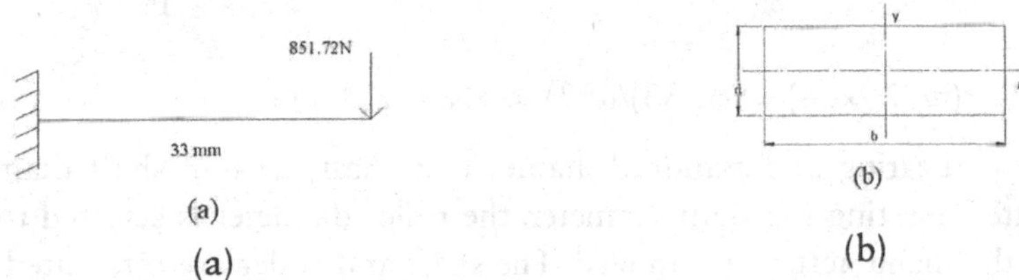

Figure 5 (a): Spline as cantilever beam. (b) Cross section of splines.

6.1. Selection of electric motor & gearbox

Shear load required to de-husk dry coconuts is 420 N (Venkataramanan *et al.*, 2014). A single-phase AC electric motor of 0.5 HP of 1440 rpm for this application is selected. Reasons for selection are:

as a single-phase AC motor is easy to connect anywhere. It does not require a DC source or a special three-phase supply which is used in industries. As the rating of the motor is 0.5 HP, the power consumption is less. One of the main reasons for selecting this motor is its availability, cost and maintenance. A reverse forward starter switch is connected to the motor which enables it to rotate in both directions.

P = 0.5 HP, N_1= 1440 rpm, 1 HP = 746 Watts, 0.5 HP = 746 × 0.5, P = 373 Watts

Torque given by motor as: $P = \frac{2\pi NT}{60}$ => $T = 2.47$ Nm.

The torque of the motor can be increased in many ways, the gear is selected to increase the torque and to the transmit power. Reasons for selecting the gearbox are:

it is a positive drive which gives constant velocity ratio. It is better than belt drive, as it does not undergo creep, has no slip, has long life, low maintenance and repair, and is also compact. Due to the availability of gearbox, gearbox of velocity ratio of 40:1 is selected for the application.

$\frac{N1}{N2} = \frac{Torque2}{Torque1}$ = 40/1, N2 = 36 rpm, Torque 2 = 98.8 Nm.

6.2. Design of Shaft and Roller

Torque (maximum) = 98.8 Nm = 98.8 × 10^3 Nmm. Mild steel is considered as shaft material as it is economical and easily available. S_{yt} = 250 MPa. According to maximum shear stress theory: τ = 250/2 = 125 MPa, FOS = 1.5, τ (allowable) = 83.33 MPa

Assumptions for design of shaft:

The shaft is subjected to pure torsion only. Material is homogeneous & isotropic. Weight of the shaft is neglected. Pure Torsion equation, $\frac{T}{J} = \frac{\tau_{allowable}}{R}$

(98.8 × 103)/ ((π/32) ×d4) = (83.33)/(d/2) => d= 18.20 mm.

As per size of bearing and standard diameter of shaft, 25-mm shaft diameter is selected. After selecting the shaft diameter, the roller diameter is selected to be 50 mm as per the manufacturing standard. The shaft and roller are integrated in one component.

6.3. Design of Splines

Diameter of roller = 50 mm, radius = 25 mm, penetration required = 33 mm, 25 + 33 = 58 mm, tangential force = T/R = (98.8 × 103)/ 58 = 1703.44 N. As two rollers are used with same specification and dimensions: 1703.44 N/2 = 851.72 N, 851.72 N > 520 N.

So, the required force is obtained to de-husk the coconut. For the design of splines, assumptions are: strength of weld is considered to be equal to strength of original material, hence high factor of safety (FOS) is considered. Splines are subjected to pure bending only and for analysis, the splines are considered to be a cantilever beam. Material is assumed to be isotropic and homogeneous.

BM = 33 mm × 851.72 N = 28.10 × 10^3 N.mm.

Mild steel is considered as spline material as it is economical and easily available. S_{yt} = 250 MPa, FOS = 4, S_{yt} (allowable) = 62.5 MPa. *b* = 472 mm, *d* = 2.5 mm (*b* and *d* are assumed).

$I_{XX} = \frac{BD^3}{12}$, $I_{XX} = 614.58\ mm^4$

y= d/2 = 2.5/2 = 1.25 mm, Z= 491.67 mm^3

σ = M/Z = 28.10 × 103/491.67 = 57.15 N/mm . Hence, design and assumptions are safe.

6.4. Chain Drive

The two rollers driven and driver has a sprocket with 18 number of teeth on it. As it is desired to have same velocity ratio throughout and same speed, same number of teeth is chosen. Another sprocket which is idler sprocket as to move so that adjustment can be made has a smaller number of teeth which is 17. The size of chain was determined by trial-and-error method and the chain is of industrial grade.

7. Manufacturing

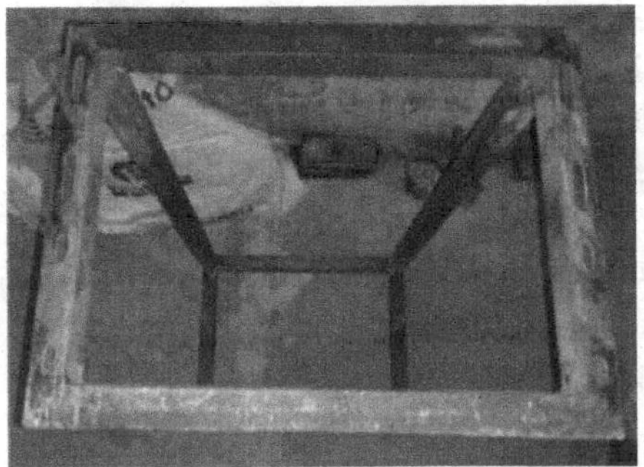

Figure 6: Frame.

Figure 7: Splined roller.

7.1. Manufacturing of the Frame

A square pipe of mild steel has been used in this project of dimensions 560 mm × 410 mm × 900 mm to support the entire assembly of the machine. A square pipe of 38 mm × 38 mm with thickness of 3 mm at bottom and thickness of 2 mm for overall frame is used. This varying thickness of 3 mm at bottom is used to give extra strength to the frame and making it powerful to withstand shocks and vibration produced by the machine. The upper side of the frame has slots so that the bolts of the bearing can slide inside it; thus, center-to-center distance between the rollers can be varied according to the size of the coconut. The slots of width 10 mm were manufactured by end milling operation. The square pipes were cut at 45 degrees with a chop saw machine at its ends and were welded together by arc welding. First, the bottom square of 3-mm thickness was manufactured followed by top square of 2-mm thickness. Vertical members were welded between the top and bottom squares as shown in Figure 6.

7.2. *Manufacturing of Splined Roller*

A raw material of mild steel bright bar of diameter 50 mm and length 642 mm was used for shaft and roller. The raw material underwent machining in a lathe and the shaft, and roller was manufactured as a single component. The shaft end was machined to a diameter of 25 mm and a length of 88 mm, while the roller was machined to a length of 466 mm. Slots of dimensions 2 mm × 5 mm were manufactured by end milling operations on the surface of roller throughout its length to hold the splines while welding and to give it extra support. Splines of dimensions 466 mm × 5 mm × 38 mm were manufactured out of mild steel strips. Six of these splines were manufactured and then arc-welded onto the surface of the roller within the slots, as illustrated in Figure 7. Grinding was done on splines to make them sharp.

7.3. *Manufacturing of Spiked Roller*

A hollow pipe with an outer diameter of 110 mm and an inner diameter of 100 mm was selected for the spiked roller. To enclose the ends of the hollow pipe, a 7-mm thick plate was welded to it. On the plate, a 30-mm rod was welded to serve as the shaft, which was then turned to a 25-mm diameter. The total length of the roller, including the shaft, is 602 mm, with the roller itself measuring 466 mm in length. The spikes were manufactured from mild steel bright bar of 13-mm diameter. The rod was cut to length of 15 mm, which was taper turned to be shaped into pointed spikes. Numerous such spikes were produced and subsequently subjected to a hardening process to enhance their strength and hardness. Holes were drilled on the surface of the roller and the spikes were press fitted into the holes.

8. Testing

During testing the performance of the machine was evaluated, whether it achieves the aim and objectives of the project or not. It was done by de-husking multiple coconuts of various sizes. After the manufacturing of the initial design assembly shown in Figure 2, the machine was started to check the proper functioning of motor, gearbox and chain drive. After checking the drive and other components, the machine was stopped and a coconut was placed in between the two splined rollers and the machine was started. It was observed that the coconut instead of getting de-husk got stuck in between the two rollers hence stopping the machine completely as shown in Figure 8. After the first failed attempt the center-to-center distance between the two splined roller was increased and the machine was run again. It was observed that the coconut passed through the gap between the two rollers and fell on the ground. By changing the center-to-center distance continuously and testing it again and again, it was concluded that the initial design assembly (two splined rollers) was not able to de-husk the coconut and was a failure. After the failure of the machine, it was decided to make changes in it. All the changes were made in iterations that is one at a time and testing of the new assembly was conducted. All the changes made in the initial design assembly and their effects on the performance of the machine will be discussed below. The changes in design are categorised in three designs as there were three main changes.

Figure 8: Coconut stuck between rollers. Figure 9: Partial de-husking.

8.1. Design 1 (Initial Design)

The initial design Figure 2 has two identical splined rollers with the total diameter of 122 mm. Six splines of height 38 mm were welded on its surface in the slots throughout its length in such way that the height of spline above the surface of roller is 36 mm. After the failure of this design few changes were made. First was to reduce the height of the spline by trial-and-error method. The height during this reduction was not measure as it was not certain that this height is perfect or not. Testing was carried out again and slight improvement was observed in terms of de-husking. Earlier without modification, the coconut used to get stuck in between the rollers but after modification it was observed that the splines started to penetrate inside the shell of coconut but after a certain length of penetration it got stuck again. After this observation, the height of the spline was further reduced and testing was continued. At some point it was observed that there was no significant change in the machine performance.

8.2. Design 2

Next was the major change in the design of the machine and it was to replace the driven splined roller with larger diameter roller. The new roller has a diameter of 110 mm and was placed in the exact same way as the previous splined roller. Testing was conducted on the new assembly and slight improvement was observed. Proper de-husking of coconut was initiated in this assembly. However, it was observed that the effort required to initiate the de-husking was extremely large. The operator had to put the entire weight on the handle to initiate the de-husking. Even though de-husking was initiated, it was not desired to put so much effort as it did not fulfil the objectives of the project. To resolve this problem center-to-center distance between the rollers was varied and testing was continued but the performance remained the same. To bring the roller closer, the pedestal bearing block base was cut using end milling operation and the center-to-center distance was reduced further. After the following modifications, partial de-husking was observed as shown in Figure 9.

8.3. Final Design

Figure 10: First successful de-husking.

Figure 11: Optimum de-husking.

After the failure of the second design which consisted of driver splined roller and a big driven roller, third major change in the design was made. Spiked roller was then used as the driven roller instead of normal plain roller. After the installation of the spiked roller testing was resumed. Significant improvement in the performance of the machine was observed. It was observed that the coconut was getting de-husk, with the application of very less effort as compared to the previous design. After the observation, few more changes such as reducing the spline height and varying the center-to-center were done. After all the changes mentioned above, first coconut was finally de-husked successfully with minimum effort shown in Figure 10.

After observing successful de-husking, spline height was measured to be 25 mm and center-to-center was measured to be 145 mm. After resuming the testing, it was observed that some of the coconuts were getting properly de-husked while some were getting crushed in between the rollers. To address this issue, a reverse-forward starter switch was employed which can change the direction of rotation of rollers. The optimum design was achieved by de-husking multiple coconuts with varying sizes and shapes. The optimum center-to-center distance between the rollers was found out to be 137 mm and the spline height was found out to be 20 mm. Since the bearing block base was cut, the force exerted in de-husking operation used to push the roller backward. To resolve this, extra-length nuts were welded on the top surface of the frame and bolts were fastened so that the bolts can withstand the backward force while de-husking operation.

During testing it was observed that an unskilled person takes about 30 seconds on an average to de-husk the coconut completely. While de-husking coconuts again and again, the safety of the machine was demonstrated. The cover plate and handle play an important role regarding safety. Because of the cover plate while de-husking, the rollers are completely covered and our hands cannot touch the rollers when the plate is lowered nor the coconut being de-husk can rebound and come back at the operator as the plate keeps its secured between the rollers. As long cover plate and handle is provided, effort required to de-husk the coconut has also reduced. As mentioned in Design 3, spiked roller also plays an important role to reduce the effort applied by the operator. Finally, to give mobility to the machine, castor wheels were arc welded to the bottom of the frame and final painting was completed. Figure 12 shows the final working assembly of the machine which can de-husk the coconut effortlessly, safely and quickly.

Figure 12: Assembly of machine.

9. Results and Discussion

After testing the machine with multiple number of coconuts, the optimum center-to-center distance which can de-husk coconut of varying size and shape was found out to be 137 mm. This distance is optimum only for coconuts of length up to 450 mm and diameter from 80 to 200 mm. If the size of the coconut does not fall under the mentioned dimensions, then the adjuster mechanism which is hand operated can be used to increase or to decrease the center-to-center distance according to the size of coconut. The maximum and minimum center-to-center distance than can be achieved is 220 and 107 mm, respectively. The optimum height of spline was found out to be 20 mm. As per design the diameter of the splined roller which was assumed to be 50 mm was optimum and the diameter of the spiked roller was also found to be optimum which was 110 mm. Average time required for an unskilled operator to de-husk the coconut is around 20–25 seconds. After using the machine again and again this time can be reduced depending upon the operator. Optimum length of the handle was found out to be 170 mm.

The safety of the machine was also evaluated during testing as it was observed that the cover plate when lowered covers the two rollers in such a way that even if the operator tries to intentionally touch the rollers, he will not be able to do so. A reverse forward starter switch is also provided in case some mishap occurs or when the coconut gets stuck between the rollers. The force applied to de-husk the coconut was observed to be very less as the operator has to hold the cover plate in the position and the de-husking takes place automatically. After conducting successful testing of the machine, it was observed that the performance of the machine was good. The machine can be used by small-scale farmers who cannot afford a fully automatic de-husking machine as the cost of such machines are very high. This machine has a potential to solve the problems of small-scale farmers as its de-husk coconut quickly without crushing them with ease. The main problems faced by the farmers are fatigue, cost and requirement for high production rate. If we consider the operator to get accustomed of the machine and could de-husk in 20 seconds then in 1 minute, he can de-husk 3 coconuts and 180 in an hour. This rate of de-husking can be increased by further developing the machine. The effective length of the roller is 466 mm which can de-husk one coconut so if the length is increased by two times, then two coconuts can be de-husked in a single shot and so on. Many modifications such as addition of an extra roller, pedal operation can be made in the design to increase the productivity.

10. Conclusion

The goals were completely met with the design, construction and testing of a coconut de-husking machine. The optimum design has been achieved in terms of center-to-center distance, length of splines, diameter and size of the driven and driving roller, adjuster mechanism, etc. after extensive testing with more than 35 dried coconuts. With the aforementioned specification, the machine can effortlessly de-husk coconuts of all shapes and sizes without crushing or only partially de-husking them and in a faster way. The majority of the issues were addressed and successfully resolved as the machine was constructed by taking into account the issues related to previously established de-husking machines.

References

[1] Amal, P., Sebastian, S., Babu, A., Saibu, A., and Kuriakose, S. (2018). Design and fabrication of coconut dehusking machine. *International Research Journal of Engineering and Technology*, 5(4), 4485–4489.

[2] Gaikwad, R., Bagadi, P., Kamble, B., and Gadkari, J. (2017). Design and development of automatic coconut de-husking and de-shelling machine. *International Journal of Engineering Development and Research (IJEDR)*, 5(2), 1–6.

[3] Ingle, A., Damahe, A., Rahangdale, A., Chahande, A., Wasnik, B., and Tembhare, B. (2022). Numerical analysis and fabrication of coconut peeling machine. *Journal of Emerging Technologies and Innovative Research (JETIR)*, 9(6), 1–5.

[4] Kadam, S., Dasuri, S., Motghare, A., Ranade, A., Rajesh, P., and Shekapure, R. (2023). Design and fabrication of coconut dehusking machine. *International Journal of Recent Research in Thesis and Dissertation (IJRRTD)*, 4(1), 42–53.

[5] Prajwal, R., Pradhan, P., Mohanty, S., Mishra, J., and Behera, D. (2021). Ergonomic evaluation of manually operated coconut de-husker. *Journal of ergonomics*, 11(4), 1–9.

[6] Ramadurai, K., Mohamed, N., Kutty, I., and Balaji, R. (2019). Coconut de-husking machine. *International Journal of Engineering Research and Technology (IJERT)*, 7(11), 1–10.

[7] Sakhare, V., Tonpe, K., and Sakhale, C. (2014). Design and development of coconut de-husking machine. *International Journal of Engineering Research & Technology (IJERT)*, 3(7), 670–674.

[8] Venkataramanan, S., Ram, A., and Raj, R., (2014). Design and development of automated coconut dehusking and crown removal machine. *International Journal of Sciences: Basic and Applied Research (IJSBAR)*, 13(2), 183–219.

[9] Yohanes, N., Satriardi, H., Susilawati, A., and Arief, D. (2016). Design of coconut de-husking machine using quality function deployment method. *Proceeding of Ocean, Mechanical and Aerospace*, 3, 506–510.

Chapter 42

Application of Low-Cost Automation for Corrugated Box Flaps Twisting to Improve Productivity

Sumit Patil[1], *Laxman Waghmode*[2], *Sanjaykumar Gawade*[3], *Yuvaraj Ballal*[1], *Pradip Patil*[1], *and Ajit Mane*[1]

[1]Assistant Professor

[2]Professor, Department of Mechanical Engineering, Annasaheb Dange College of Engineering and Technology, Ashta, Maharashtra, India

[3]Professor, Department of Mechanical Engineering, Rajarambapu Institute of Technology, Uran Islampur, Maharashtra, India

Abstract: Corrugated boxes play an essential role in packaging dry grapes, ensuring their safe and secure transport. While manually packing dry grapes in the corrugated box the flaps of each box is folded inversely. The act of folding the box inversely leads to the tearing of the box at its creased positions. The tearing of boxes can be avoided by twisting each box at the creased positions. This twisting needed to perform when boxes are in wet/damp condition due to the application of liquid gum. This is the exceptional requirement of corrugated boxes made from recycled paper and used for packing of dry grapes. Currently, these boxes are twisted manually which has several drawbacks. To overcome the identified problems, a low-cost automation approach is applied for twisting boxes at creased positions. Three different mechanisms are developed to perform stopping, clamping and twisting of boxes. The developed mechanisms are integrated, coordinated and automated by using mechatronics technology. It is observed that by using a prepared low-cost automation solution the boxes can be twisted in lesser time with less number of workers along with reduced muscular efforts. The knowledge gained from this endeavour can serve as a foundation for the complete automation of this device.

Keywords: Low-cost automation, mechatronics, productivity improvement.

1. Introduction

In the English language, the term "raisin" is commonly employed to refer to grapes that have been dried. India contributes 1.20% to the global share of exported dried grapes. A combined effort from the states of Maharashtra and Karnataka results

DOI: 10.1201/9781003527442-42

in the production of 96.13% of grapes in India, which are subsequently used for the production of dry grapes (Kerutagi, 2018). The dry grapes are packed in the corrugated boxes with polythene bag inside the box. During manual packing of dry grapes polythene bag is placed in the empty box; the box are flaps folded inversely. Then the open end of polythene bag is hanged in folded flaps and the dry grapes are poured in it. Once the desired weight of dry grapes is attained, the open end of the polythene bag is extracted from the folded flaps. Subsequently, a knot is tied to securely seal the bag. At the last flaps of the box are closed and sealed. When the flaps of corrugated box are folded inversely for hanging polythene bag, the box get tear along creased positions. The corrugated boxes are prepared from paper sheets. These paper sheets consist of multiple paper layers that are bonded together using liquid adhesive. The application of liquid gum during the gluing operation to bond together various layers of paper causes the paper sheets to become moist or wet. It is required to twist each box two or three times at predetermined creased positions while the boxes are still slightly damp. This twisting action while the boxes are in their wet state preventing potential tearing along these creased lines as the boxes eventually dry out. The twisting activity at the creased positions applied exclusively to boxes designed for the manual packing of dry grapes. This practice adopted by numerous small- and medium-scale enterprises engaged in the packing, marketing and distribution of dry grapes. Figure 1 illustrates the issue of box tearing.

Figure 1: Tearing of box at creased positions.

An alternative approach to prevent tearing involves opting non-recycled paper possessing a higher burst factor (Adamopoulos, 2007). Again using recycled paper for manufacturing boxes is green production which is important for sustainable development (Strzelczak, 2017). To address economic considerations, corrugated box manufacturers opt for utilising standard recycled paper. Additionally, they implement a manual twisting process, particularly when the boxes are damp or wet. Low-cost automation (LCA) empowers industrialists to enhance their manufacturing techniques and efficiency while avoiding the need for expensive machinery investments (Patil 2021; Samuel 2009). The LCA successfully applied to address a productivity challenge within a fruit packing (Khire, 2012). An innovative carton-folding device has been created specifically for packaging export-quality grapes, utilising corrugated paper cartons. In the research, Akella (2000) documented the utilisation of a SCARA (selective compliance articulated robot arm) robot to facilitate the folding of cartons. For folding cartons in the confectionery industry reconfigurable technology was investigated (Yao, 2010). A case study of a reconfigurable demonstration system in the confectionery industry is outlined (Dai, 2013). Balkcom (2008) developed four degrees of freedom adept SCARA robot arm to make simple folds. Rotating wheels or spiral bars are used to fold carton panels at their creases (Marschke, 1989). While LCA has been applied to address numerous challenges in various industries, there appears to be a noticeable gap in the application of LCA to the process of corrugated box twisting.

2. Manual Box Twisting Method

The interactions with 23 distinct semi-automated small-scale corrugated box manufacturing industries located in Maharashtra and Karnataka are conducted. It is observed that in all of these industries, the boxes are twisted manually which causes excessive labour cost, less productivity due to more production time, need of supervision and engagement of man power in repetitive, monotonous and fatiguing work. Based on questionnaire survey and responded answers, it was concluded that there is a need to develop a LCA solution which is capable to perform this twisting work automatically. The results of 29 study trails are presented in Table 1.

Table 1: Average time required to twist one box.

Activity No.	Description of Activity	Activity Wise Average Time
1	Pick box from a stack and put it on work the table	6.67
2	Twist flaps on one side (by two workers)	8.33
3	Turn over the box	4.92
4	Twist flaps on the other side (by two workers)	8.08
5	Pick box from the work table and put in a stack	6.08
Average actual total time required to twist one box		34.08

The average actual total numbers of boxes that are manually twisted during a single shift has been calculated based on data collected over a period of 10 days documented in Table 2.

Table 2: Average actual total numbers of boxes twisted in one shift.

Day	1	2	3	4	5	6	7	8	9	10	x
Shift I	430	441	431	422	417	421	412	422	418	421	425.5
Shift II	427	409	398	412	407	394	395	389	397	402	402.4
Average	428.5	425	414.5	417	412	407.5	403.5	405.5	407.5	411.5	413.9

***x- Average actual total numbers of boxes twisted in one shift**

The percentage efficiency of the manual box twisting method is determined by considering the average actual total number of boxes twisted within a single shift as shown in Table 3 is 55.81 % which is considerably poor.

Table 3: Efficiency of conventional manual box twisting method.

Number of seconds available in one shift (considering 7 working hours per shift)	25200
Average actual total time required to twist one box in seconds	34.08
Number of boxes expected to be twisted in one shift (considering zero downtime) [25000/34]	739.43 ≈ 740
Average actual total numbers of boxes twisted per shift (considering 15 days data)	413.9 ≈ 413
Total less number of boxes twisted per shift due to non-performance of workers (considering all valid and invalid downtime reasons) [740-413]	327
The efficiency of the conventional manual box twisting method	55.81%

3. Methodology for Implementing Low-Cost Automation

The proposed methodology for implementing LCA solution for considered problem consists of four steps – first step is to perform a QFD, second step is design and development of mechanisms and third step is mechanisation to automation. The final step is performance confirmation.

3.1. Mechanisation

The envisioned system comprises three distinct primary mechanisms. These mechanisms are discussed in details in the next part of the paper.

3.1.1 Box Twisting Mechanism

An innovative box twisting mechanism was designed to simultaneously twist both sides' flaps, effectively minimising the time required for the twisting process. Figure 2 illustrates the wireframe model of the newly devised box twisting mechanism. Through a single work cycle, the box twisting mechanism facilitates the simultaneous twisting of both sides' flaps, achieved by its alternating clockwise and anticlockwise motions. The power is conveyed from the prime mover to the driving shaft, which then transfers it to the four mobile arms via timing belts and timing pulleys. The actual developed mechanism can be seen in Figure 5.

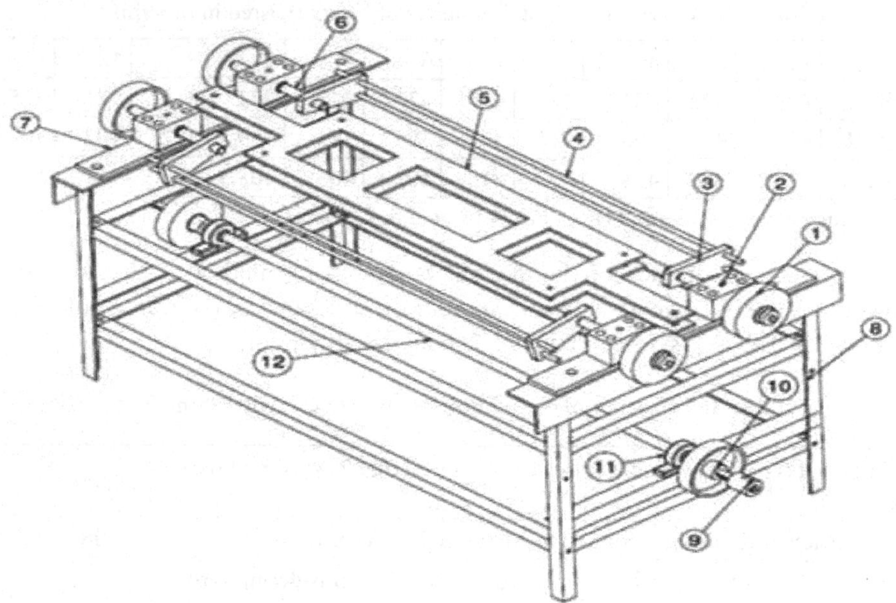

Figure 2: Wire frame CAD model of box twisting mechanism.

1. Timing pulley, 2. bush block, 3. moving arm, 4. bars, 5. bracket, 6. arm shaft, 7. base plate, 8. table 9. hub, 10. pulley spacer, 11. pedestal bearing, 12. driving shaft

3.1.2 Box Stopper Mechanism

After inserting the box within the twisting mechanism and advancing it, it becomes necessary to halt its movement at a specific point. This ensures that the creased position on the box aligns appropriately with the twisting mechanism, enabling the box to twist precisely along the creased positions. To achieve this, two box stopper mechanisms were designed and integrated onto the twisting mechanism. The motion of the box is restrained by a limit switch mounted on the stopper mechanism, effectively controlling its positioning.

3.1.3 Box Clamping De-Clamping Mechanism

During the clockwise and anticlockwise motion of the twisting mechanism, the box's flaps are twisted in predefined directions at specific angles. Ensuring the box's stability during this flap-twisting action is crucial. To address this, a mechanism for securely holding and releasing the box was developed. This box clamping and de-clamping process is achieved by controlling the rotary direction of two power screws.

3.2. Mechanisation to LCA

To minimise the physical exertion of the human operators, incorporating actuators becomes necessary during the automation process. Controlling the angle, speed and direction of rotation are the prime requirements in all of three proposed mechanisms for their proper functioning. Hence, DC motors having inbuilt gear box was used as an actuator to run these mechanisms. Consequently, the total torque needed to drive the twisting mechanism was 20 N^{-mm}. For this purpose, a standard geared motor with a torque capacity of 22 N^{-m} was selected. While the motor operates at a voltage of 12 volts, the necessary amperage adjusts according to torque demands, with fluctuations reaching up to 10 amps. To cater to this requirement, a sophisticated switched mode power supply (SMPS) was furnished. For coordinating and integrating three mechanisms, Arduino Mega 2560 board is used as a controller. The integration of the box stopping mechanism, clamping and de-clamping mechanism, and twisting mechanism is achieved through the utilisation of diverse electronic components as illustrated in the Figure 3.

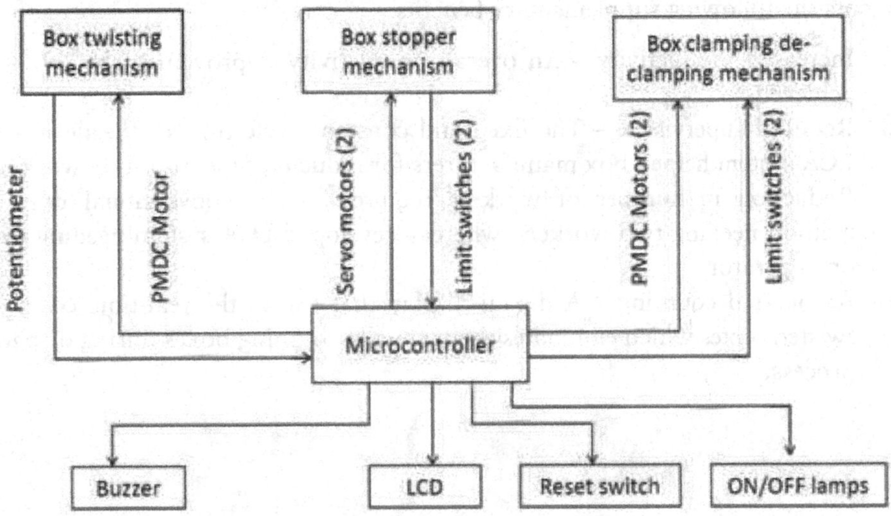

Figure 3: Block diagram of system.

4. Results and Discussions

In the developed automation system worker is required to just load and unload the box; all physical efforts required to twist the box is offered by the system. The manual process involves a total of five activities to twist the box, whereas the automated process achieves the same task with just three activities. The initial activity involves manually loading the box, followed by the automated simultaneous twisting of both sides' flaps as the second step. Finally, the third activity encompasses the manual unloading of the twisted box. In Figure 4, the graph visually represents the variation in cycle time for each activity, both prior to and after implementation of automation.

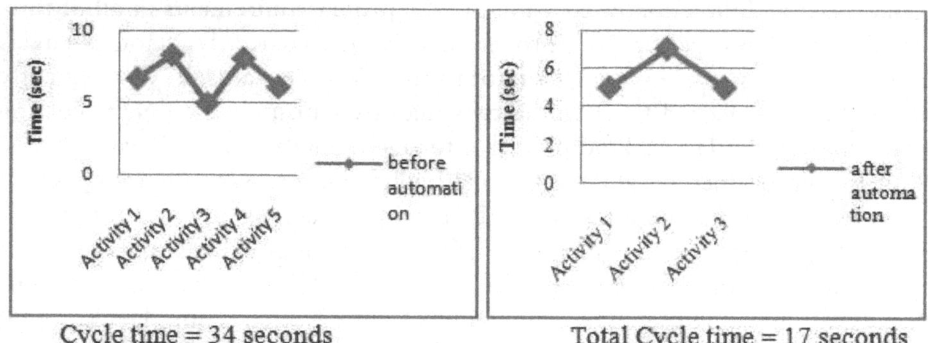

Figure 4: Variation in cycle time for box twisting.

Considering the preceding discussions, it can be deduced that the proposed system presents the following supplementary benefits:

1. Increased productivity – An overall productivity improvement by 50 % is achieved.
2. Reduced supervision – The fixed and constant cycle time of the developed LCA system helped box manufacturers for reducing supervision on workers.
3. Reduction in number of workers required – The conventional twisting method needing two workers, whereas developed LCA system needing only one operator.
4. Automated counting – A digital display that shows the real-time count of twisted boxes which eliminates the manually counting boxes during dispatch process.

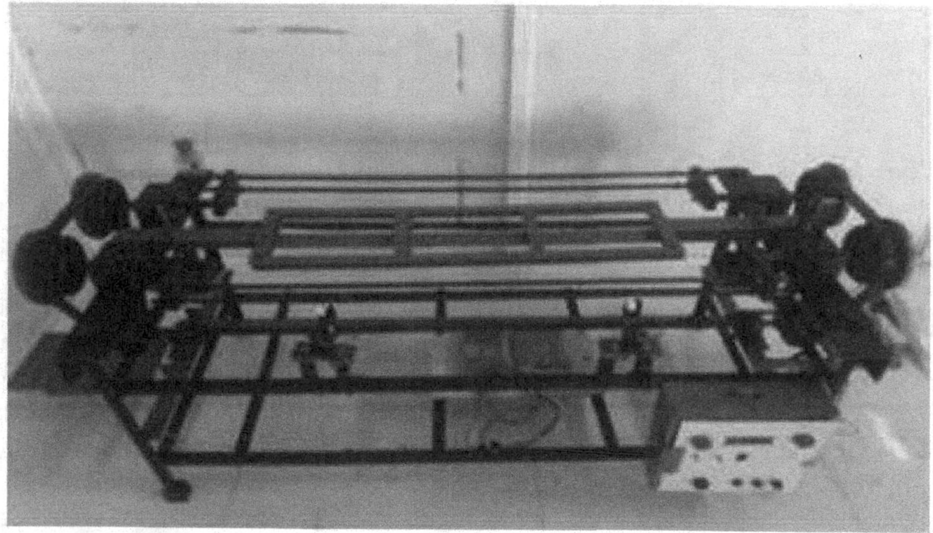

Figure 5: Photograph of developed LCA solution.

5. Conclusion

A successful implementation of a LCA solution has effectively addressed the challenges associated with the manual box twisting method which are the excessive time requirement, numbers of operator required, labour-dependent productivity and muscular efforts needed to apply. The cost required for implementing the proposed system is notably economical, rendering it a viable solution for industrial adoption. By considering total investment, running cost and savings in labour cost, payback period or return on investment period is of two years. Automated loading and unloading of boxes can be facilitated by integrating conveyors and other appropriate mechanisms, yet its implementation must be evaluated based on economic viability and justification.

References

[1] Adamopoulos, S. (2007). Characterization of packaging grade papers from recycled raw materials through the study of fibre morphology and composition. *Global NEST Journal*, 9(1), 20–28.

[2] Akella, S. (2000). Folding cartons with fixtures: a motion planning approach. IEEE *Transactions on Robotics and Automation*, 16(4), 346–356.

[3] Balkcom, D., and Mason, T. (2008). Robotic origami folding. *The International Journal of Robotics Research*, 27(5), 613–27.

[4] Dai, J. (2013). Robotics and automation for packaging in the confectionery industry. *Robotics and Automation in the Food Industry*, 13(2), 401–419.

[5] Kerutagi, M. G. (2018). Production and marketing of dry grapes in Vijayapura District – an economic analysis. *International Journal of Current Microbiology and Applied Sciences*, 7(9), 658–671.

[6] Khire, M., and Madnaik, S. (2001). Folding cartons using lowcost automation – a case study. *Assembly Automation*, 21(3), 210–212.

[7] Marschke C. (1989). Folding of paperboard sheets and the like, US Patent 4834696.

[8] Patil, S. V. (2021). Effect of low-cost automation on labor productivity and labor fatigue for corrugated boxes flaps twisting. *Work – A Journal of Prevention, Assessment & Rehabilitation*, 73,143–155.

[9] Samuel, G. L. (2009). Design and implementation of lowcost automation system in heavy vehicle brakes assembly line. *International Journal of Productivity and Quality Management*, 4(2), 199–211.

[10] Strzelczak, S. (2017). Integrated assembly of 'Green – Lean' production. *International Journal of Automation Technology*, 11(5), 815–828.Yao, W. (2010). A reconfigurable robotic folding system for confectionery industry. *Industrial Robot: An International Journal*, 37(6), 542–551.

Chapter 43

Heat Energy Storage in LiBr and Phase Change Material using Solar Parabolic Dish Collector

B. Kumar[1]*, M. S. Pervez*[2]*, and A. M. Achari*[3]

[1]Associate Professor, Vignan Institute of Information Technology, Visakhapatnam, Andhra Pradesh, India

[2]Associate Professor, Anjuman College of Engineering and Technology, Nagpur, Maharashtra, India

[3]Associate Professor, Institute of Technology, Nirma University, India

Abstract: A comprehensive investigation into latent heat energy storage using phase change materials (PCMs), with a specific focus on exploring their energy storage capabilities within the medium and high-temperature ranges. The study employs a sophisticated three-dimensional numerical model based on enthalpy to facilitate this examination. The simulations are carried out on a 1 kW system, subjecting the lower surface of the container to a constant heat flux. The outcomes of the simulations reveal intriguing insights into the energy storage potential of various PCM options, including NaOH, LiBr and NaCl. Notably, the highest total energy storage is observed when the PCM is fully melted, as opposed to cases where the PCM remains either unmelted or only partially melted. This finding underscores the significance of complete phase change in optimising energy storage capacity. The utilisation of a three-dimensional numerical model adds depth and accuracy to the exploration, enhancing our ability to harness the benefits of latent heat energy storage across diverse temperature ranges. Ultimately, this research contributes valuable insights that could inform the development and deployment of more efficient and effective energy storage solutions.

Keywords: Energy density, thermal energy storage, parabolic dish collector, heat flux, phase change material.

1. Introduction

Energy stands as a fundamental necessity within our society. Among the myriad energy sources available, solar energy reigns supreme. It harbors the potential to satiate both domestic and industrial energy needs. To address this, thermal energy storage emerges as a solution, enabling the utilisation of solar energy in the absence of direct solar radiation.

DOI: 10.1201/9781003527442-43

This technique has been successfully employed for an extended period exceeding two decades reported by Crespo *et al.* (2019). Phase change materials (PCMs) have captured significant interest within the realm of energy storage for their exceptional capacity to store and release thermal energy amidst phase shifts. This unique characteristic has led to considerable exploration and study in the energy storage domain. Substantial focus has been directed towards PCMs in the sphere of energy storage, driven by their impressive capability to absorb and release thermal energy during phase transitions. The exceptional capacity of PCMs to store and release thermal energy during changes in phases has garnered considerable attention in the context of energy storage. One prominent PCM of interest is lithium bromide (LiBr), which exhibits favourable thermal properties suitable for energy storage applications. LiBr PCM has been extensively studied for its potential in various domains, including solar energy systems, building heating and cooling and waste heat recovery. The distribution of heat flux at different surfaces of the PCM container significantly influences the phase change process and overall energy storage performance. In this context, the focus on the bottom surface heat flux adds an additional layer of complexity to the study, as it plays a pivotal role in influencing the temperature distribution and phase change dynamics within the PCM, presented in Table 1.

Table 1: Research conducted on latent heat storage by different researchers.

Approaches	Reference
1D Enthalpy (Gauss-Seidel)	Zivkovic and Fujji (2001)
1D Enthalpy (Finite difference)	Silva *et al.* (2002)
1D & 2D Enthalpy (Finite Difference)	Velraj *et al.* (1999)
2D Energy (Finite Element)	Gong and Mujumdar (1997)
2D Enthalpy (Finite Volume)	Li *et al.* (2019)
2D Conjugate Heat Transfer Enthalpy (Finite Volume)	Ismail *et al.* (2003)
2D Fixed-Grid Enthalpy (Gauss-Seidel)	Sarı *et al.* (2002)

The integration of solar thermal technologies into industrial processes holds the promise of diminishing both fossil fuel consumption and CO_2 emissions. Despite this potential, the practical application of solar energy storage within the industrial sector faces challenges attributed to spatial constraints. However, these challenges find resolution through the utilisation of PCMs, which not only alleviate the spatial requirements but also ensure a consistent and sustained supply of heat energy for thermal applications.

The rationale behind the adoption of PCMs resides in their exceptional energy storage capacity and their ability to furnish heat at a consistent temperature. Notably, the existing literature in the realm of three-dimension numerical simulations focusing on conduction and radiation in latent heat energy storage remains relatively scarce. To address this gap, the present study undertakes an exploration of the total energy stored within a constant volume of various PCMs across the medium to high-temperature spectrum. This endeavour is accomplished through the application of an enthalpy-based model, contributing to an enhanced understanding of latent heat energy storage.

2. Mathematical Model and Governing Equation

The heat transfer processes taking place within both the container and the PCM are recognised as transient and characterised by three-dimensional attributes. In this analysis, any influences stemming from body forces and convection have been intentionally disregarded. The interface shared by the container and PCM maintains a uniform characteristic. Throughout all domains, the mechanisms solely encompass conductive and radiative heat transfer processes. To facilitate this modelling approach, a three-dimensional heat transfer methodology is employed, specifically the three-dimensional enthalpy-based model as outlined by Kumar *et al.* (2020 and 2022).

2.1. *Simulation and Validation*

The computational model employing a three-dimensional enthalpy-based approach is computed using the COMSOL Multiphysics® software. To conduct the simulation, three distinct sets of grids are generated through the mesh generation module. Specifically, these encompass a first set with 34,080 grids, a second set comprising 70,386 grids and a third set involving 145,592 grids. To validate the accuracy of the current computational model, a comparison is carried out against well-established benchmarks provided by Zivkovic and Fujii (2001), as well as Siyabi *et al.* (2018) shown in Figure 1 with a maximum error percentage is 2.81% which is within the permissible limits. Grid independence test and validation are taken and compared with reference to Kumar *et al.* (2020) and (2021).

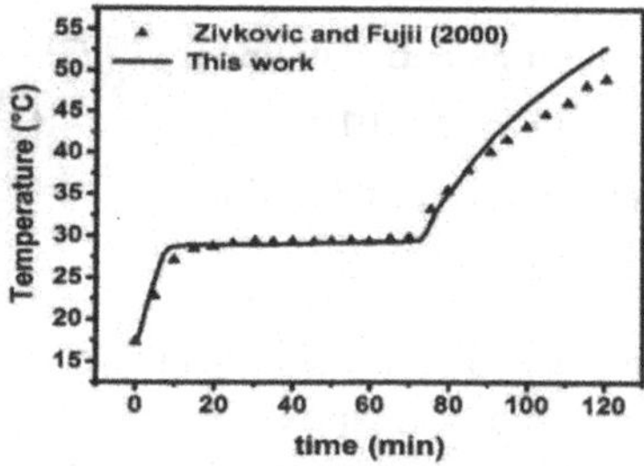

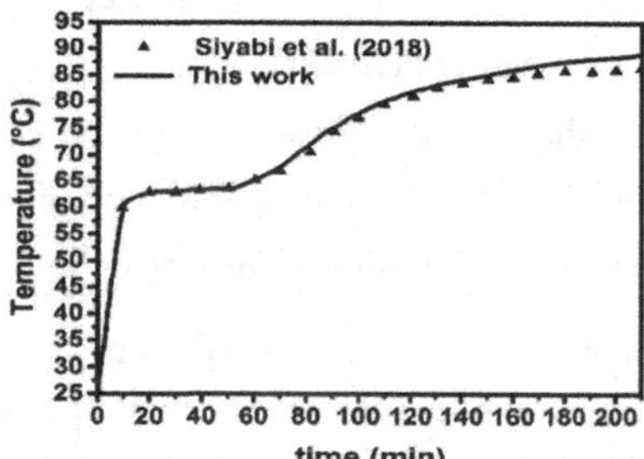

Figure 1: *Validation of present work with (Zivkovic and Fuji, 2000) and (Siyabi et al., 2018) with a maximum error of 2.81% which is within the permission limits.*

3. Results and Discussions

A polished stainless-steel container (Lane 2018), characterised by an outer radius of 16.7 cm, height of 9.1 cm and thickness of 10 mm is utilised as the vessel for housing distinct types of PCM: NaOH, LiBr and NaCl by Kumar *et al.* (2021 and 2022). Although the containers volume remains constant, the mass and thermophysical properties of the various PCMs contained within it exhibit differences. A radially varying constant inward heat flux is applied to the lower surface, while a uniform surface emissivity of 0.1 is attributed to all surfaces for radiation heat transfer.

Initial and boundary conditions serve as essential principles in the realm of scientific and mathematical modelling. When the temporal parameter, denoted as "*t*," holds a value equal to or less than zero ($t <= 0$), we enter the domain of initial conditions. These conditions encapsulate the state of a system at its outset. In a similar, when our attention is directed towards the lower surface of the system, specifically in the inward direction, we encounter boundary conditions. These conditions prescribe how the system interacts with its surroundings and external factors. For instance, when *t* is greater than zero ($t > 0$), the system exhibits a defined temperature of ti = 295 K. Additionally, during this phase, a specific heat flux, represented as "q," maintains a constant value of (q_0). Considering the bottom surface, the inward orientation signifies a particular manner of interaction between the system and its environment. This interaction may encompass phenomena such as energy transfer, material exchange or other relevant processes occurring at this interface by Bashir and Giovannelli (2023). By articulating these initial and boundary conditions with precision, we establish a solid framework for understanding and modelling the behaviour of the system. This clarity empowers us to construct accurate simulations, mathematical models and predictive analyses that faithfully represent real-world situations, capturing the nuances that transpire during this distinct phase of the system's progression. Heat flux (q_0) at the bottom surface is shown in Table 2.

Table 2: Variation in heat flux along the radial direction derived from ray optics simulations conducted on a 1 kW solar parabolic dish collector system, which is operational for six hours daily, imparting solar energy onto the receiver surface (Kumar *et al., 2021*).

R (m)	q_0 (kW/m²)	R (m)	q_0 (kW/m²)
0.003	78.707	0.097	55.741
0.015	75.969	0.105	52.699
0.026	73.384	0.115	49.657
0.039	70.798	0.126	47.072
0.048	68.365	0.138	44.030
0.057	65.931	0.147	39.923
0.067	63.346	0.154	35.969
0.076	61.520	0.160	27.756
0.087	58.631	0.166	18.022

As time increases, gradually the bottom, middle and top portion of PCM temperature increases. Temperature increase from 0th hour to 6th hours from 295 K to 1160 K at the bottom and 890 K at the top. Temperature contours of LiBr PCM at 6th hours in the bottom and middle surface are shown in Figure 2. The temperature progression along the diameter, both within the PCM and the container, is depicted in Figure 3. Additionally, Figure 4 illustrates the temperature distribution along the diameter of the LiBr at the 6th hour, specifically at the bottom, middle and top positions. The highest temperature is observed at the center of the bottom surface. This temperature disparity is attributed to the fluctuating heat flux, resulting in elevated temperatures at the central region of the bottom surface.

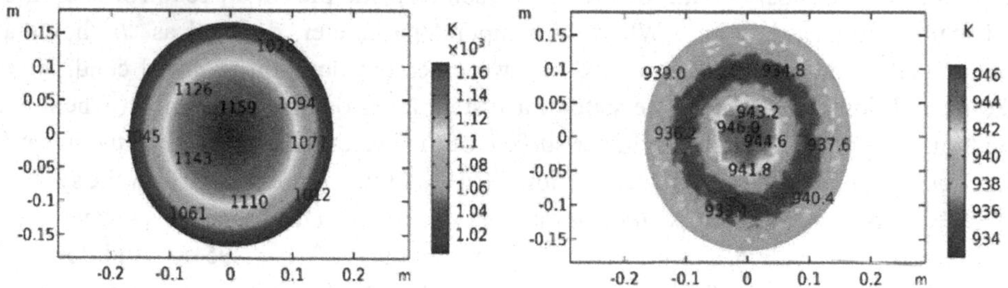

Figure 2: Temperature contour of LiBr at 6th hours on a different surface that is bottom and middle surfaces (left to right).

A comprehensive comparison among various PCMs is illustrated in Figures 3–6. The insights garnered from these figures reveal noteworthy distinctions. Figure 5 distinctly portrays that, at the 6th hour, the total enthalpy for NaOH amounts to approximately 1500 kJ/kg, LiBr records around 800 kJ/kg and NaCl registers approximately 600 kJ/kg. Delving into Figure 4, a clear temporal trend emerges: NaOH attains complete melting by the 3rd hour, LiBr displays 90% melting by the 6th hour, while NaCl undergoes a minimal melting extent, remaining below 0.15% at the same juncture. Turning attention to the aggregate energy storage, it is discernible that NaOH accumulates around 16 MJ, LiBr accumulates about 14 MJ and NaCl stores a mere 6 MJ. Notably, the energy stored in NaOH surpasses that of LiBr by over 12.5% and surpasses NaCl by a substantial 62.5%. Interestingly, despite NaCl exhibiting a higher bulk temperature compared to LiBr and NaOH, the total energy stored by NaOH remains inferior to both LiBr and NaCl.

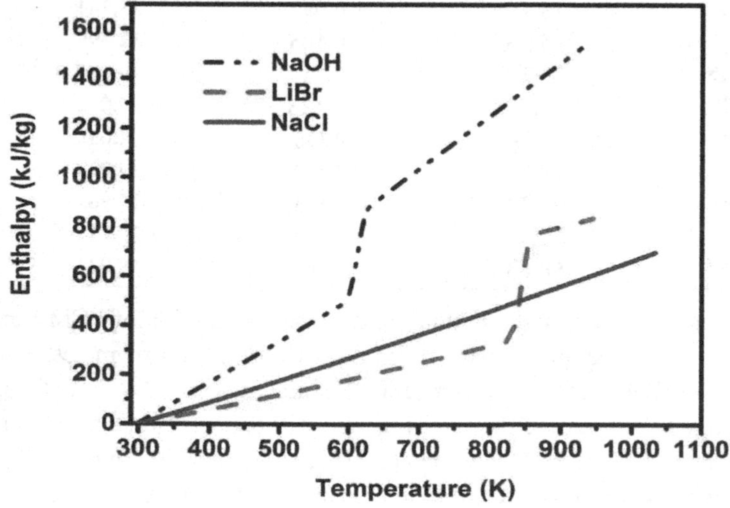

Figure 3: Comparison of enthalpy versus temperature of different PCM.

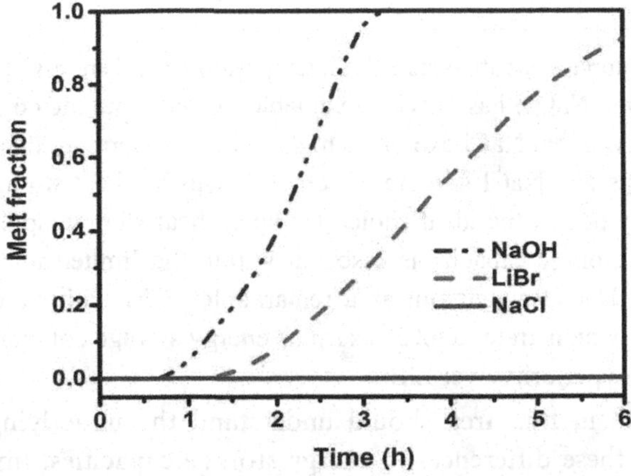

Figure 4: Comparison of melt fraction versus time of different PCM.

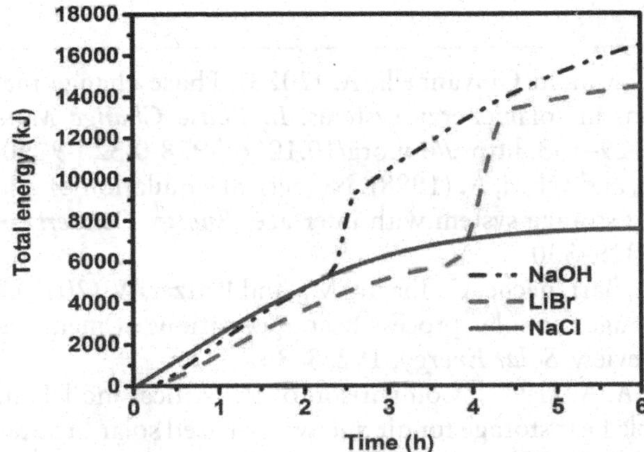

Figure 5: Comparison of stored energy versus time of different PCM.

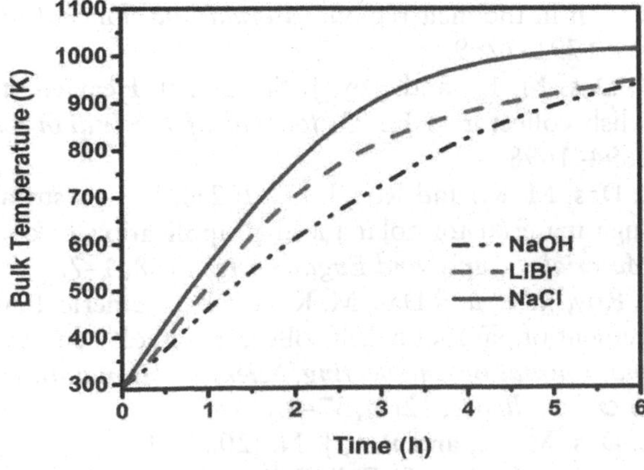

Figure 6: Comparison of bulk temperature versus time of different PCM.

4. Conclusion

In conclusion, the comparative analysis of bulk temperatures and energy storage capacities across NaCl, LiBr and NaOH has provided valuable insights into the complex dynamics of PCMs. It is observed that NaCl exhibits a higher bulk temperature than both LiBr and NaOH. This suggests that NaCl is more effective in sensible heat storage applications. Howcver, NaCl may not be the ideal choice for latent heat storage applications, where maximising energy storage capacity is essential within thc limited temperature range. On the other hand, NaOH stands out as a remarkable PCM with exceptional energy storage potential. It demonstrates a 62.5% surplus energy storage compared to NaCl and a substantial 12.5% superiority over LiBr.

Future research in this area should understand the underlying mechanisms that contribute to these differences in energy storage capacities. Investigating the thermal conductivity, melting and solidification kinetics, and long-term stability of these PCMs could provide further insights.

References

[1] Bashir, M. A., and Giovannelli, A. (2023). Phase change materials (PCMs) applications in solar energy systems. *In Phase Change Materials for Heat Transfer*, 129–153. https://doi.org/10.1016/B978-0-323-91905-0.00004-6.

[2] Costa, M., and Oliva, A. (1998). Numerical simulation of a latent heat thermal energy storage system with interface. *Energy Conversion and Management*, 39, 319–330.

[3] Crespo, A., Barreneche, C., Ibarra, M., and Platzer, W. (2019). Latent thermal energy storage for solar process heat applications at medium-high temperatures – a review. *Solar Energy*, 192, 3–34.

[4] Ghoneim, A. A. (1989). Comparison of theoretical models of phase-change and sensible heat storage for air and water-based solar heating systems. *Solar Energy*, 42, 209–220.

[5] Ismail, K. A. R., and da Silva, M. D. G. E. (2003). Numerical solution of the phase change problem around a horizontal cylinder in the presence of natural convection in the melt region. *International Journal of Heat and Mass Transfer*, 46, 1791–1799.

[6] Kumar, B., Das, M. K., and Roy, J. N. (2020). Heat energy storage using parabolic dish collector in LiNO. *Journal of the Indian Chemical Society*, 97(10a), 1694–1698.

[7] Kumar, B., Das, M. K., and Roy, J. N. (2021). Energy storage using NaOH phase change material for solar thermal application at constant heat flux. *IOP CS: Materials Science and Engineering*, 1168, 1–7.

[8] Kumar, B., Roy, J. N., and Das, M. K. (2022). Numerical analysis of flat receiver placement of parabolic dish collector for solar energy application. *An International Journal of Engineering, Sciences, Humanities & Management, Innovation & Excellence*, 12(1), 37–42.

[9] Kumar, B., Das, M. K., and Roy, J. N. (2022). Design and storage of solar thermal energy production. In D. B. Pal and J. M. Jha (Eds.), *Sustainable and*

Clean Energy Production Technologies. Clean Energy Production Technologies. Springer, Singapore. https://doi.org/10.1007/978-981-16-9135-5_10.

[10] Lane, G. A. (2018). Solar heat storage: latent heat materials volume II. *CRC Press Taylor & Francis Group*. https://doi.org/10.1201/9781351076746.

[11] Li, D., Ding, Y., Wang, P., Wang, S., Yao, H., Wang, J., and Huang, Y. (2019). Integrating two-stage phase change material thermal storage for cascaded waste heat recovery of diesel-engine-powered distributed generation systems: a case study. *Energies*, 12, 1–20.

[12] Sarı, A., and Kaygusuz, K. (2002). Thermal and heat transfer characteristics in a latent heat storage system using lauric acid. *Energy Conversion and Management*, 43, 2493–2507.

[13] Silva, P. D., Gonçalves, L. C., and Pires, L. (2002). Transient behaviour of a latent-heat thermal-energy store: Numerical and experimental studies. *Applied Energy*, 73, 83–98.

[14] Velraj, R., Seeniraj, R. V., Hafner, B., Faber, C., and Schwarzer, K. (1999). Heat transfer enhancement in a latent heat storage system. *Solar Energy*, 65, 171–180.

[15] Zivkovic, B., and Fujii, I. (2001). An Analysis of isothermal phase change of phase change material within rectangular and cylindrical containers. *Solar Energy*, 70, 51–61.

Chapter 44

SOC Prediction of Lithium Ion Battery Based on Machine Learning Algorithm

D. S. R. S. L. Avanthika[1], *B. Arundhati*[2], *Madisa V. G. Varaprasad*[3], *and M. Aruna Kumari*[4]

[1]Students, Vignan's Institute of Information Technology, Visakhapatnam, Andhra Pradesh

[2]Professor, Vignan's Institute of Engineering for Women, Visakhapatnam, Andhra Pradesh

[3]Associate Professor, Vignan's Institute of Information Technology, Visakhapatnam, Andhra Pradesh

[4]Assistant Professor, Vignan's Institute of Information Technology, Visakhapatnam, Andhra Pradesh

Abstract: Electric vehicles (EVs) are four times energy efficient than internal combustion engines (ICE). EV batteries frequently have capacity limitations. The battery lifetime determines how long an EV can be used. The major components used to find the SoC of a battery are its voltage and capacity. The relation between voltage and SoC of a battery is non-linear and also battery degrades over time which effects the capacity of a battery. Yet, due to these nonlinearities of the battery's property, it is challenging to describe the link between the voltage and the current at various temperatures and state of charge (SoC). In this study, a Neural Network (NN)-based approach for estimation of a lithium-ion battery's SoC is introduced. The algorithm takes in variables like cell temperature, current and voltage of a battery. The suggested approach can precisely estimate the SoC by modelling battery activity with a NN. A dataset with a variety of battery operating circumstances is used to train the NN. The algorithm's effectiveness in precisely estimating SoC is shown by the findings, which are based on real-time experimental data.

Keywords: Electric Vehicles (EVs), Internal Combustion Engines (ICE), lithium-ion battery, State of Charge (SoC), Neural Network (NN).

1. Introduction

Many applications require accurate measurement of battery SoC in order to give users an indication of available runtime (Alvarez-Anton *et al.*, 2013). There are numerous ways to determine the SoC of a battery, including Coulomb counting, open circuit voltage (OCV), model-based and machine learning methods (Heeyun

DOI: 10.1201/9781003527442-44

and Suk Won, 2021). Among these, the machine learning method was employed in this study as they are data driven and represents an abstract mapping from an input space consisting of a selected set of predictions to an output space, which represent the largest output (Alexander *et al.*, 2019). The SoC of a battery is influenced by current, voltage and temperature, all of which are important aspects to consider here. Battery management system (BMS) in EVs helps to monitor, control and protect the Li-ion batteries from extreme conditions of misuse (Thiruvonasundari and Deepa, 2021). The experimental dataset that was taken into consideration for this study includes current, voltage, cell temperature and SoC data at temperatures ranging from 10°C, 20°C and 30°C.

The recent approach for estimating SoC was given by Jiang *et al.* "State of Charge Estimation for Lithium-Ion Batteries Using Attention-Based Neural Networks" in 2022 – The authors proposed a NN model which considers the important time steps and battery features for estimating SoC which gives an accurate and robust estimation. The only drawback from this study was that optimisation technique was not implemented for minimising the loss functions and improving the accuracy of the model. Thus, from the above discussion in this study SoC was being estimated using NN with the help of Adam Optimizer for improving the robustness of the model and making it to suit for several research works.

SoC refers to the amount of charge, measured in Ampere-Hours, remaining within the battery at a particular moment of time. In the case of SOC estimation, one of the simplest methods is based on open-circuit voltage (OCV) and Coulomb counting (Carlos *et al.*, 2020). The expression below based on Coulomb counting method depicts the interrelationship among these parameters and their influence on the SoC.

SoC=SoC_initial + (∫(I(t)*Δt)/C_nom)...(1)

where

- SoC_initial is the initial SoC.
- I(t) is the battery current at time t.
- Δt represents the time increment.
- C_nom is the nominal capacity of the battery.

By integrating the current over time and representing the accumulated charge going in and out of the battery, the Coulomb counting method is employed in the calculation above to estimate the SoC.

1.1. Neural Network-Based Algorithm for Machine Learning

In this study, the NN model was trained using 80% of the available data, and the remaining 20% was reserved for testing. The input to the network consisted of voltage, current and cell temperature values recorded at different external temperatures, specifically 10°C, 20°C and 30°C. The goal was to predict the SoC as the output. After training the network with the provided data, its performance was evaluated by comparing the predicted SoC values against the actual values from the testing dataset. This assessment allows for observing how accurately the network can predict the SoC based on the given input features. The NN was developed using Python within a Jupyter Notebook environment in Anaconda Navigator. Objective is to forecast a continuous numerical value based on the input attributes, the model architecture is as follows:

```
model = Sequential([
    Dense(32, activation='relu', input_shape=(3,)),
    Dense(16, activation='relu'),
    Dense(1, activation='linear')
])
```

The code represents the development of a sequential model in a Keras or TensorFlow-based machine learning framework. At bare, a NN consists of at least three computational layers: input layer, hidden layer and output layer (Dickshon et al., 2020). One thing that has to be noted is that neither the loss function nor the optimiser is displayed in the code above. Normally, these components are incorporated independently while building the model before training. The below code will explain about the loss function and optimiser used.

```
model.compile(optimizer=Adam(learning_rate=0.001), loss='mse')
```

The neural network model is built using the mean squared error (MSE) loss function and the Adam optimiser with a learning rate of 0.001. The learning rate option controls the speed at which the optimiser changes the model's weights during training. Typically, slower but more exact convergence results from a lower learning rate. The output of the processing element is calculated by passing the value obtained as a result of the sum function through a linear or nonlinear differentiable transfer function (Fulya and Naim, 2022).

2. Dataset

The EVs' rapid growth can potentially lead power grids to face new challenges due to load profile changes (Morteza *et al.*, 2021). The dataset was collected for a Li-ion battery which has 96s31p, 2976 cells and 50.4kWh. The mass of the vehicle was 1612 kg with a wheel radius of 0.34. Using historical data of charging load and user behaviour, ML algorithms can be utilised to train and learn the trends and patterns from the data (Sakib *et al.*, 2021). The sample data from the size of the dataset of around 16,500 rows under various external climatic temperature conditions are shown in the information supplied in the tables below [1].

Table 2: Sample data at 10°C.

Cell Temperature(°C)	SOC (%)	Current (A)	Voltage (V)
20.24017049	0.744214	–24.0856	11.87315

20.29981467	0.944925	-23.557	11.91544
20.42027915	1.369023	-22.5326	11.99739
20.49097147	1.630309	-21.9495	12.04404
20.54023266	1.818186	-21.5501	12.076
20.59227025	2.022101	-21.134	12.10928

3. Results and Discussions

The outcomes, which are shown in Figure 1, will show the SoC plotted over the data of current, voltage and cell temperature at various temperatures. Additionally, the projected effects of each parameter on the SoC are depicted in Figures 2 and 3, respectively. The results in Figure 1(a), (b) and (c) show the actual and predicted values of SoC for the proposed algorithm. As the actual and predicted values were on the same line the proposed algorithm was better suited for the given conditions. The battery's SOC declines with rising temperatures. As a result, at 30 °C, the battery was over discharged, as illustrated in Figures 2(c) and 3(c). Low temperatures increase the battery's internal resistance, which slows down the chemical reactions. The ideal temperature for Li-ion to operate is between 15°C and 35°C (Carlos et al., 2019). As can be seen in Figures 2(a) and 3(a), the SoC range at 10°C is therefore relatively small. So, the temperature range for using the battery in EV application is 20°C as shown in Figure 2(b) and Figure 3(b). In Figures 2 and 3, the SoC was estimated by considering the voltage and current parameters separately to know the status of the battery with individual effect of each parameter.

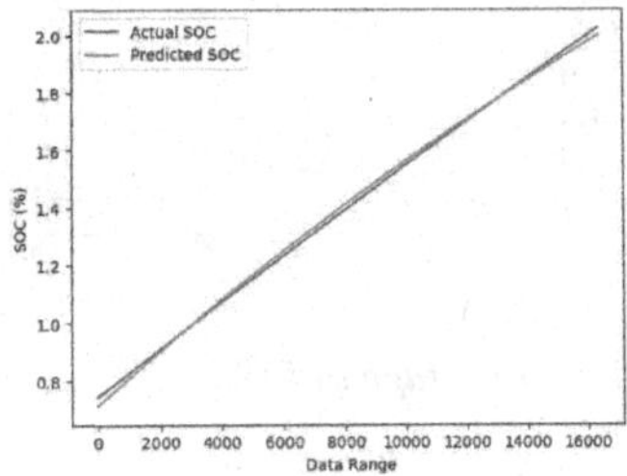

Figure 1(a): At 10°C.

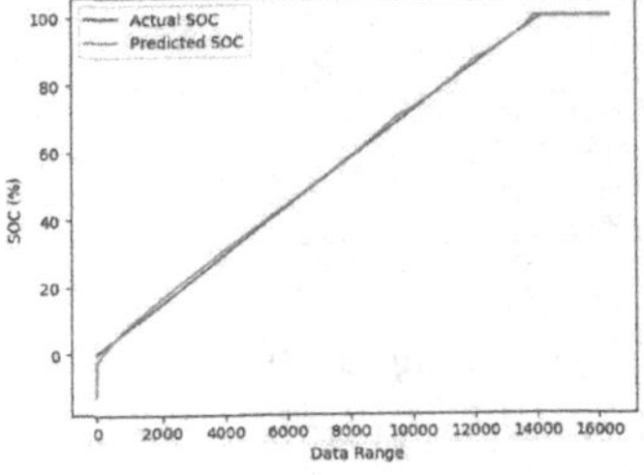

Figure 1(b): At 20°C.

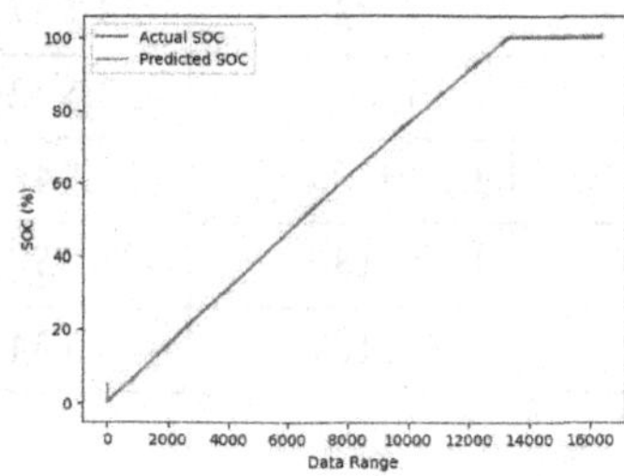

Figure 1(c): *At 30°C.*

Figure 1: *SoC versus data (voltage, current and temperature)*

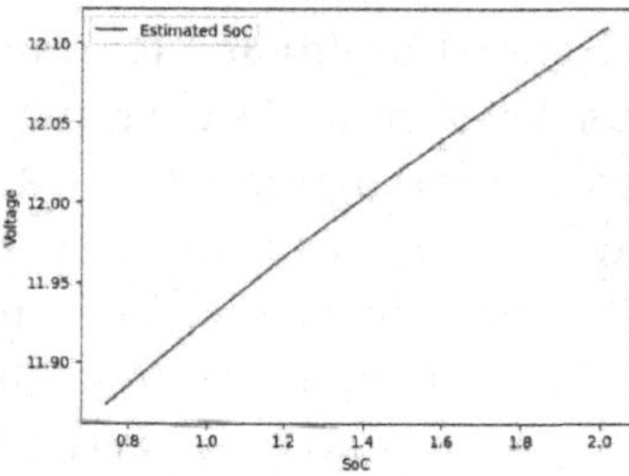

Figure 2(a): *Prediction of SoC over voltage at 10°C.*

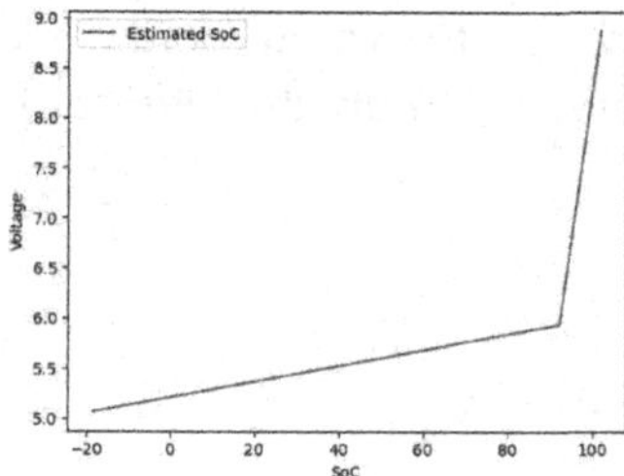

Figure 2(b): *Prediction of SoC over voltage at 20°C.*

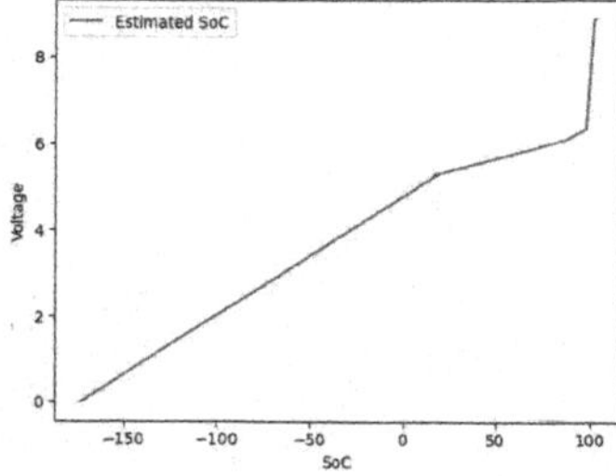

Figure 2(c): *Prediction of SoC over voltage at 30°C.*

Figure 2: *Prediction of SoC over voltage at various temperatures.*

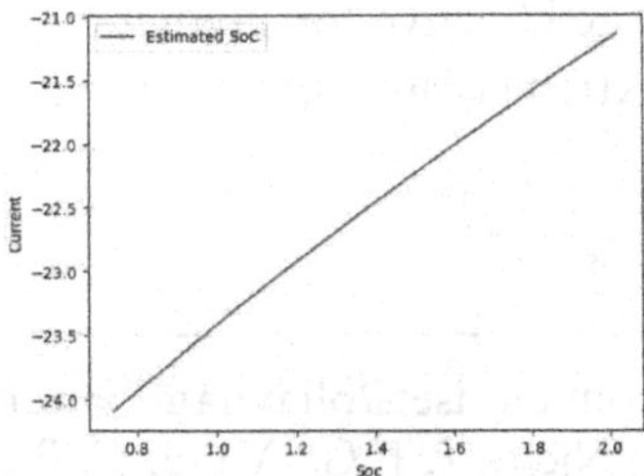

Figure 3(a): Prediction of SoC over current at 10°C.

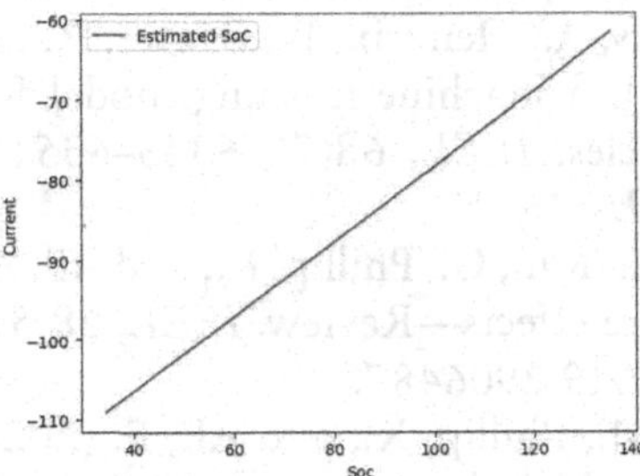

Figure 3(b): Prediction of SoC over current at 20°C.

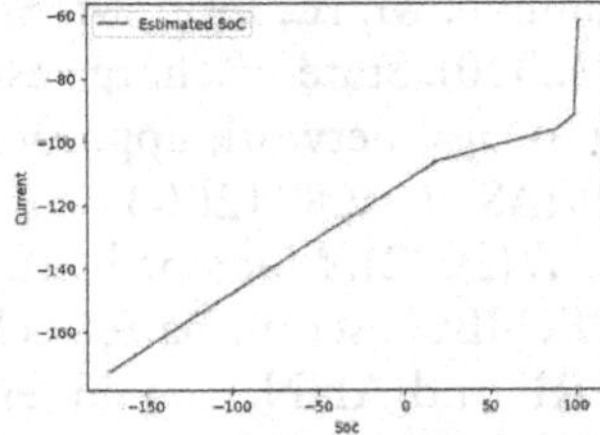

Figure 3(c): Prediction of SoC over current at 30°C.
Figure 3: Prediction of SoC over current at various temperatures.

4. Conclusion

Temperature, voltage and current all have an impact on a battery's SOC which were shown in the above results. The results above demonstrate the effects of temperature, voltage and current on a battery's SoC. The SoC's impact will have an effect on battery life. Continuous SoC monitoring can stop this from happening. It entails keeping track of numerous aspects that have an impact on SoC. Voltage, current, cell temperature and outside temperature were considered in this study. Future research will also consider different battery charging rates and chemical compositions, both of which have a significant impact on a battery's SoC and the proposed algorithm can be used for estimating SoC over these factors also. In order to determine the SoC degradation rate and make future adjustments to the battery's design or usage depending on the results, this algorithm predicts a battery's SoC from relatively little real-time data. The conclusion from this study was that the proposed approach was more effective for predicting the SoC for a Li-ion battery because the MSE value is extremely low and the R-squared value is approximately equal to 1.

References

[1] https://www.kaggle.com/datasets/bhavnam/battery (Dataset).

[2] Alvarez-Anton, J. C., Nieto, P. J. G., Viejo, C. B., and Vilan, J. A. V. (2013). Support vector machines used to estimate the battery state of charge. *IEEE*, 28(12), 5919–5926. https://doi.org/10.1109/TPEL.2013.2243918.

[3] Alexander, S., Byerly, A., Hendrix, B., Bagwe, R. M., dos Santos, E. C., and Miled, Z. B. (2019). A machine learning model for average fuel consumption in heavy vehicles. *IEEE*, 68(7), 6343–6351. https://doi.org/10.1109/TVT.2019.2916299.

[4] Carlos, V., Oliver, G., Ran, G., Phillip, K., and Ali, E. (2019). xEV Li-Ion battery low-temperature effects—Review. *IEEE*, 68(5), 4560–4572. https://doi.org/10.1109/TVT.2019.2906487.

[5] Carlos, V., Pawel, M., Phillip, K., and Ali, E. (2020). Machine learning applied to electrified vehicle battery state of charge and state of health estimation: state-of-the-art. *IEEE*, 8, 2169–3536. https://doi.org/10.1109/ACCESS.2020.2980961.

[6] Dickshon, N. T. H., Hannan, M. A., Lipu, M. S. H., Sahari, K. S. M., Ker, P. J., and Muttaqi, K. M. (2020). State-of-charge estimation of Li-Ion battery in electric vehicles: a deep neural network approach. *IEEE*, *56*(*5*), *5565–5574*. https://doi.org/10.1109/IAS.2019.8912003.

[7] Fulya, A., and Naim, S. T. (2022). Machine learning-based estimation of output current ripple in PFC-IBC used in battery charger of electrical vehicles: a comparison of LR, RF and ANN techniques. *IEEE*, 10, 50078–50086. https://doi.org/10.1109/ACCESS.2022.3174100.

[8] Heeyun, L., and Suk Won, C. (2021). Energy management strategy of fuel cell electric vehicles using model-based reinforcement learning with data-driven model update. *IEEE*, 9, 59244–59254. https://doi.org/10.1109/ACCESS.2021.3072903.

[9] Morteza, D., Amirhossein, M., and Abdollah, K. F. (2021). Reinforcement learning-based load forecasting of electric vehicle charging station using Q-learning technique. *IEEE Transactions on Industrial Informatics*, 17(6), 4229–4237. https://doi.org/10.1109/TII.2020.2990397.

[10] Sakib, S., Al-Ali, A. R., Ahmed, H. O., Salam, D., and Mais, N. (2021). Prediction of EV charging behavior using machine learning. *IEEE*, 8. https://doi.org/10. 1109/ACCESS.2020.3023388.

[11] Thiruvonasundari, D., and Deepa, K. (2021). Machine learning-based optimal cell balancing mechanism for electric vehicle battery management system. *IEEE*, 9, 132846–132861. https://ieeexplore.ieee.org/stamp/stamp.jsp?arnumber=9547315.

Chapter 45

On the Design and Fabrication of a Low-Cost PDMS-based Lenses: Soft Lithography Approach

Ranjitsinha Gidde[1], *Amrjit Kene*[2], *and Avinash Parkha*[3]

[1]Professor, SVERI's College of Engineering Pandharpur, Maharashtra

[2]Associate Professor, SVERI's College of Engineering Pandharpur, Maharashtra

[3]Assistant Professor, SVERI's College of Engineering Pandharpur, Maharashtra

Abstract: Current methods for manufacturing low-cost lenses, such as parallel mould stamping and high-temperature reflow, require complicated engineering controls and costly equipment to produce high-quality lenses. In the proposed study, the steel moulds were designed and lenses were fabricated using replica technique. The lenses were fabricated in in-house facility. By using these lenses, a regular smartphone camera turns into an economical digital dermascope, enabling visualisation of microscopic structures on the skin, such as sweat pores.

Keywords: PDMS lenses, soft lithography, mobile microscopy, micro-fabrication, APT.

1. Introduction

The development of low-cost miniature microscopes has revolutionised imaging and enabled the creation of a new generation of mobile microscope devices that are now widely used in primary telemedicine (Bellina and Missoni, 2009; Menzies *et al.*, 2009; Mudanyali *et al.*, 2010) and global healthcare (Breslauer *et al.*, 2009); (Zhu *et al.*, 2013). As the mobile phone market continues to experience unprecedented growth, innovative low-cost miniature microscopes have emerged, taking advantage of the high-quality digital cameras in many mobile devices. This development has opened exciting possibilities for mobile microscopy, enabling users to easily capture and analyse high-resolution images of samples, all from the convenience of their smartphones (Smith *et al.*, 2011; Skandarajah *et al.*, 2014; Tseng *et al.*, 2010). Several groups have recently created compact, customised devices that leverage the advanced imaging, connectivity and processing features of smartphones to enable various applications such as microscopic imaging, holographic imaging, label-free spectroscopy and image-based quantification of diagnostic tests (Chowdhury and Chau, 2012; Martinez *et al.*, 2008; Ren *et al.*, 2010). Liquid droplets have the

DOI: 10.1201/9781003527442-45

remarkable ability to magnify small objects. By achieving a delicate balance between interfacial energies (i.e., between the liquid, air, and solid surface) and gravity, a hanging droplet can be formed. This unique phenomenon has the potential to be used as a simple and effective magnifying tool, providing a low-cost solution for magnifying small objects with the added benefit of being easily accessible to anyone with access to water and a smooth surface (Duffy *et al.*, 1998; Tadmor *et al.*, 2009). This paper presents three methods for producing low-cost, high-performance lenses using the PDMS-based hanging droplet, micro-pipetting and soft lithography replica technique.

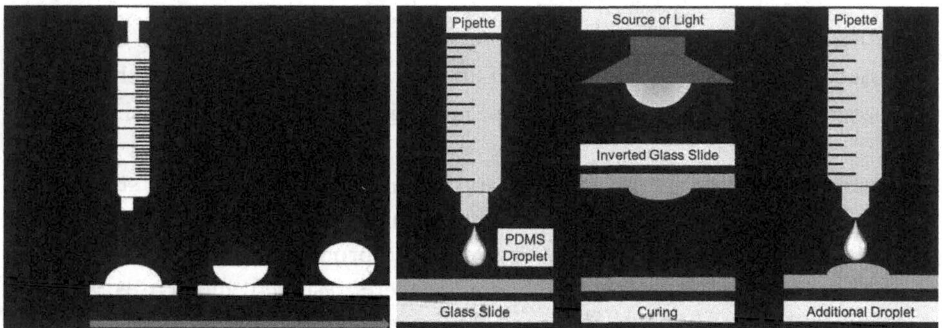

Figure 1(a): Dropping PDMS drops on glass slides. *Figure 2(b): Fabrication approach for lens with primary and additional droplets as per required focal length.*

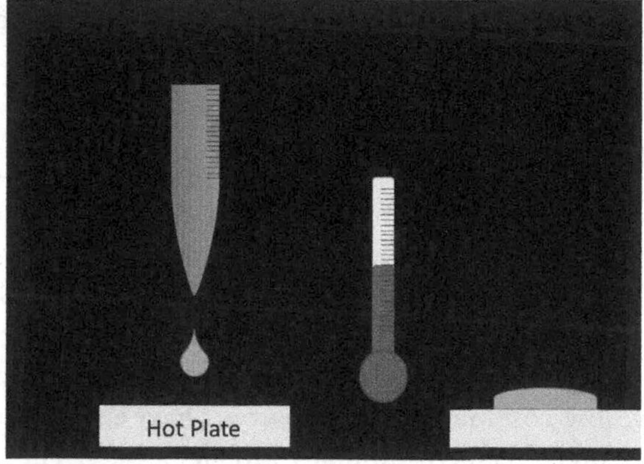

Figure 2: Schematics of fabrication of lenses using micropipetting.

2. Problem Statement

Polymer materials offer substantial commercial and practical advantages, encompassing affordability, excellent optical characteristics, flexibility in shape and high resilience. Due to these advantages cum properties, polymers lenses are produced at mass level and that too at low cost. Polydimethylsiloxane (PDMS) is one of the suited materials for the lenses and more compatible with the glass surface. The main objective is to study magnification achieved through PDMS lenses. In the proposed study, the PDMS-based lenses were fabricated using steel material-based concave moulds of different sizes. The lenses of three different focal lengths were fabricated and tested for magnification.

3. Design and Fabrication Approaches

3.1. Hanging Droplet

An approach for manufacturing lenses with excellent performance at a reduced cost involves the repetitive addition and curing of a suspended PDMS droplet at fixed intervals on a curved PDMS substrate. The even gravitational force applied across the entirety of the inverted PDMS layer aids in creating a precisely shaped refractive lens that can adjust its parabolic properties. The methodology for fabricating a PDMS hanging droplet involves several steps, including the preparation of the PDMS mixture, moulding the PDMS into the desired shape and the droplet to a glass slide. The schematics of the hanging droplet approach are shown in Figure 1.

3.2. Micro-pipetting

In this approach of fabrication, a fixed volume of PDMS is taken through a micropipette and dispensed on the glass slide. The glass slide is then kept at 20-mm height on a hot plate with 200°C temperature. The schematic illustration is shown in Figure 2 (Duffy *et al.*, 1998). In this approach, the infrared pyrometer can be used to measure the temperature for achieving accuracy as per requirement. The slides are cleaned using ethanol to remove any impurities and dust particles, before dispensing the PDMS drop. The curing time is monitored using a digital stopwatch.

3.3. PDMS Replica Moulding

A popular technique for prototyping is the use of soft lithography (a non-optical transfer technique; Duffy *et al.*, 1998). PDMS has been used successfully as the elastomeric material. A photoresist master mould can be created by standard lithographic techniques or any other technique and a liquid PDMS base solution with a curing agent is poured over the master. Due to the minimal surface tension of the PDMS solution, it easily flows and conforms to the shape of the master. Afterwards, the PDMS undergoes curing and is detached, resulting in a negative replica of the master pattern. The soft lithography process flow is shown in Figure 3. A brief idea of soft lithography technique is depicted in Figure 4.

4. Fabrication of Steel Mould and PDMS Lenses

This section is devoted to details of all the steps used such as mixing, degassing, dropping, invert, cure, etc. It includes fabrication of moulds, characterisation of moulds for the dimensions and soft lithography technique. The material chosen for mould preparation is EN31, a steel alloy known for its durability and suitability for various engineering applications. This material has undergone a meticulous manufacturing process that involves five distinct stages: milling, grinding, engraving, hardening and polishing. Figure 5 shows the schematic drawings of various moulds to be fabricated. For the preparation of the PDMS lens, the moulds have been fabricated using milling as the primary machining operation. The rectangular cubes with dimensions 20 mm × 20 mm × 10 mm have been milled using a carbide cutter. Subsequently, the milled material is subjected to grinding to impart a refined surface finish to the material, further enhancing the dimensional accuracy and smoothness of the mould surface. The engraving operation has been used to etch the desired-sized craters onto the surface for further processing of the prepared. This step is crucial in cases where the mould requires specific patterns or details to be transferred onto the final product. The three moulds with different crater depths have been machined that is 0.5 mm, 1.0 mm and 1.5 mm. The prepared mould has been hardened to reinforce its strength and durability, subjecting the material to controlled heating and rapid cooling, resulting in improved hardness and resistance to wear. Finally, the mould material was mirror polished using alumina polishing technique (APT). This final step involves smoothing the surface to an exceptionally high degree, granting the mould a flawless and reflective finish. Polishing enhances the aesthetic appeal and serves practical purposes, facilitating the easy release of moulded objects and minimising friction during manufacturing. The final fabricated moulds for three different levels of focal lengths are shown in Figure 6. The confirmation of dimensional accuracy for the fabricated moulds entailed a thorough examination carried out with specialised in-house Rapid-I Vision microscope, as illustrated in Figure 7. This verification procedure played a pivotal role in establishing the dimensions of the moulds aligns precisely with the intended specifications. This alignment helps to fabricate the moulds with designed dimensions. The fabrication of PDMS lenses was carried out using PDMS replica moulding in the advanced manufacturing laboratory (AML) at the Institute (In-house Facility) using a soft lithography process as depicted in Figure 3. Sylgard 184 silicone elastomer base and curing agent have been mixed as per the specified weight proportion. The vacuum desiccator has been employed for degasification of PDMS mixture. Further, this PDMS mixture has been poured into the fabricated mould. The moulds are then cured by being kept inside the chemical wet bench for 24 hours. Once curing done, the prepared PDMS lenses are peeled off from the mould. The detailed fabrication approach followed and equipment used in the proposed study and lens preparation procedure is depicted in Figure 8.

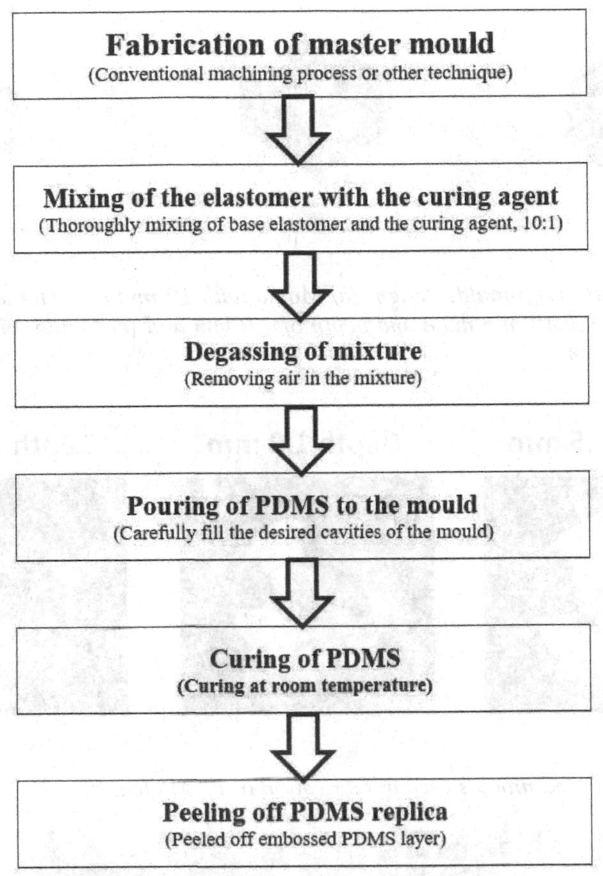

Figure 3: Soft lithography for PDMS replica.

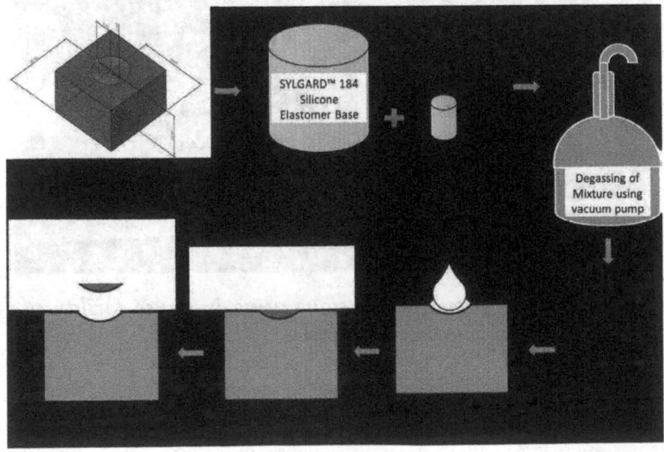

Figure 4: Detailed procedure used for fabrication of PDMS lenses using soft lithography.

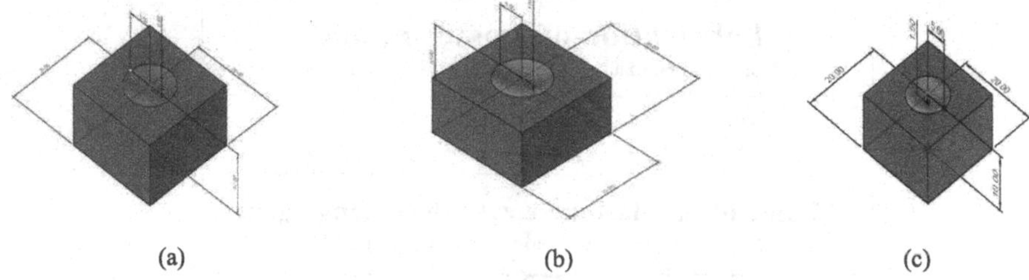

Figure 5: Schematics of moulds designs (a) Mould with 10-mm diameter and depth of 0.5 mm, (b) Mould with 10-mm diam and depth of 1.0 mm and (c) Mould with 10-mm diam and depth of 1.5 mm.

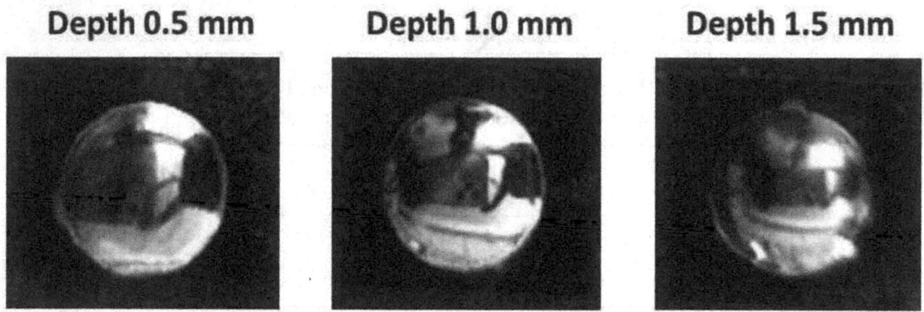

Figure 6: EN31- based moulds used in fabrication of PDMS lenses.

Figure 7: Characterisation of mould dimensions using in-house facility of Rapid-I vision microscope in advanced manufacturing laboratory.

5. Results and Discussion

With our in-house facility, the lenses can be fabricated at mass level. However, with more precise control over the change in the focal length by changing crater depth that will change the focal length. The precise control over the drop size will help to increase the repeatability in precise dimensions and desired magnification. The PDMS lenses have been characterised for magnifications using mobile microscopy. The fabricated lenses were fixed on the surface of the smartphone camera. The three lenses with different focal lengths have been tested for magnification. The images were taken by keeping the smartphone at 20-mm height from the text documents. The results of the magnification are shown in Figure 9. It can be seen that the increase in the focal length that is 0.5 mm, 1.0 mm and 1.5 mm will increase the magnification. A comparison for magnification was carried out with the help of fabricated PDMS lenses capturing photographs of the text by attaching it to the smartphone camera. The fabricated PDMS lenses have shown increased magnification and better resolution with a 1.5-mm depth lens. The proposed fabrication approach can be the best alternative for hanging droplet and/or micro pipetting-based lens fabrication approaches. An addition of a low-cost PDMS-based lens to a smartphone camera which in turn transforms the smartphone into a digital microscope platform. The proposed method of lens fabrication dealt with the obtained irrelevant focal lengths. The PDMS material needs to be poured carefully to achieve the exact focal length of the lens.

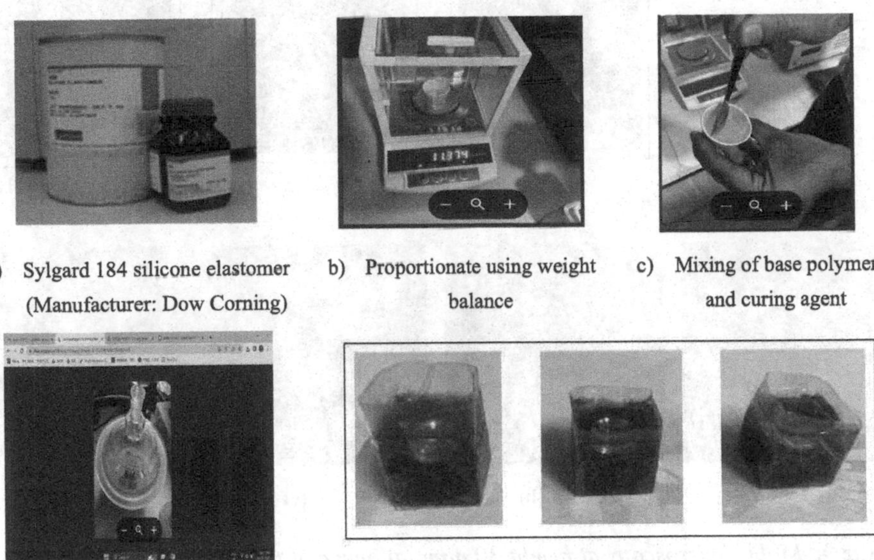

a) Sylgard 184 silicone elastomer (Manufacturer: Dow Corning)
b) Proportionate using weight balance
c) Mixing of base polymer and curing agent
d) Simple desiccator
e) PDMS poured into the moulds made up of EN31

Figure 8: Photographs of steps involved in soft lithography.

6. Conclusions

The research demonstrates the effectiveness of the manufactured lenses for a range of applications that demand precise magnification at a micro-level, even within constrained spaces such as medical devices, micro-biology, fibre optics, microelectronics, miniature optical systems, security and surveillance, material sciences and scientific research. The optical properties are assessed through measurements of focal length and contrast resolution. It is observed that modifying the curing temperature and droplet volume during fabrication allows for precise control over the shape of the lenses, which in turn affects their optical properties. Through the repetition of t [illegible]; possible to enhance the curvature of the lens, thereby leading to a reduction in the focal length of the PDMS lenses (achieving focal lengths of approximately ~2 mm). The lenses with the shortest focal lengths are adept at collimating light emitted from an unmasked LED, and they demonstrate the capability to capture images of minute structures down to approximately 4 μm in size, providing a magnification of 16x. The study has proposed a low-cost lens fabrication method that achieves high-performance lenses at negligible cost. The total cost of fabrication per lens is approximately Rs. 10 per lens. The proposed fabrication approach can be the best candidate to fabricate a low-cost PDMS-based lens which can be added to a smartphone camera which in turn transforms the smartphone into a digital microscope platform.

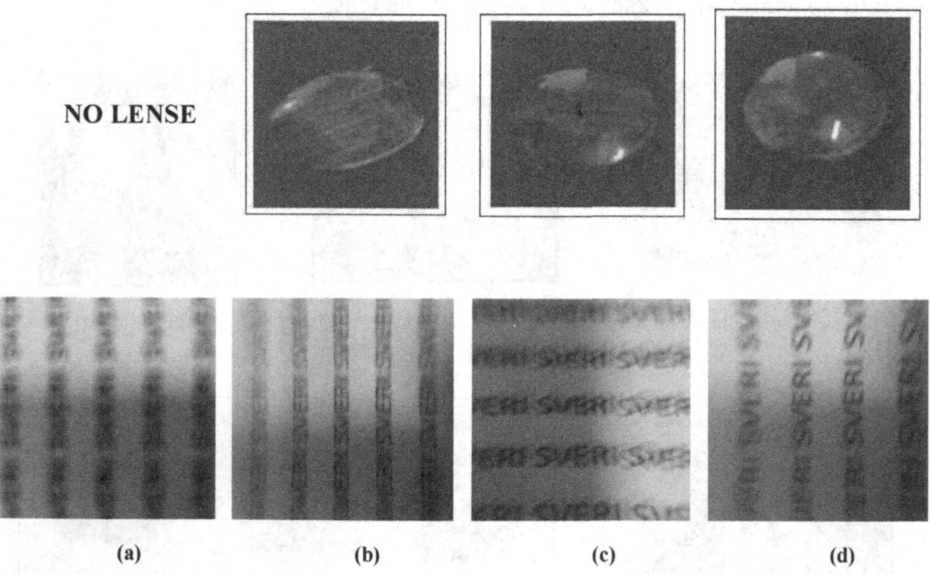

Figure 9: *Mobile microscopy at height 20 mm: (a) image of the text without lens (b) image of the text with lens having focal length of 1.0 mm, (c) image of the text with lens having focal length of 1.5 mm and (d) image of the text with lens having focal length of 2.0 mm.*

Acknowledgment

The authors would like to thank the Punyashlok Ahilyadevi Holkar Solapur University, Solapur for the Financial Support under Grant for Seed Money for Researchers.

References

[1] Bellina, L., and Missoni, E. (2009). Mobile cell-phones (M-phones) in telemicroscopy: increasing connectivity of isolated laboratories. *Diagnostic Pathology*, 4, 19.

[2] Breslauer, D. N., Maamari, R. N., Switz, N. A., Lam, W. A., and Fletcher, D. A. (2009). Mobile phone based clinical microscopy for global health applications. *PLOS ONE*, 4(7), e6320.

[3] Chowdhury, F. A., and Chau, K. J. (2012). Variable focus microscopy using a suspended water droplet. *Journal of Optics*, 14(5), 055501.

[4] Duffy, D. C., McDonald, J. C., Schueller, O. J., and Whitesides, G. M. (1998). Rapid prototyping of microfluidic systems in polydimethylsiloxane. *Analytical Chemistry*, 70(23), 4974–4984.

[5] Martinez, A. W., Phillips, S. T., Carrilho, E., Thomas III, S. W., Sindi, H., and Whitesides, G. M. (2008). Simple telemedicine for developing regions: camera phones and paper-based microfluidic devices for real-time, off-site diagnosis. *Analytical Chemistry*, 80(10), 3699–3707.

[6] Menzies, S. W., Emery, J., Staples, M., Davies, S., McAvoy, B., Fletcher, J., Shahid, K. R., Reid, G., Avramidis, M., Ward, A. M., Burton, R. C., and Elwood, J. M. (2009). Impact of dermoscopy and short-term sequential digital dermoscopy imaging for the management of pigmented lesions in primary care: a sequential intervention trial. *British Journal of Dermatology*, 161(6), 1270–1277.

[7] Mudanyali, O., Tseng, D., Oh, C., Isikman, S. O., Sencan, I., Bishara, W., Oztoprak, C., Seo, S., Khademhosseini, B., and Ozcan, A. (2010). Compact, light-weight and cost-effective microscope based on lensless incoherent holography for telemedicine applications. *Lab on a Chip*, 10(11), 1417–1428.

[8] Ren, H., Xu, S., and Wu, S. T. (2010). Effects of Gravity on the Shape of Liquid Droplets. Optics Communications, 283(17), 3255–3258.

[9] Smith, Z. J., Chu, K., Espenson, A. R., Rahimzadeh, M., Gryshuk, A., Molinaro, M., Dwyre, D. M., Lane, S., Matthews, D., and Wachsmann-Hogiu, S. (2011). Cell-phone-based platform for biomedical device development and education applications. *PLOS ONE*, 6(3), e17150.

[10] Skandarajah, A., Reber, C. D., Switz, N. A., and Fletcher, D. A. (2014). Quantitative imaging with a mobile phone microscope. *PLOS ONE*, 9(5), e96906.

[11] Tseng, D., Mudanyali, O., Oztoprak, C., Isikman, S. O., Sencan, I., Yaglidere, O., and Ozcan, A. (2010). Lensfree microscopy on a cellphone. *Lab on a Chip*, 10(14), 1787–1792.

[12] Tadmor, R., Bahadur, P., Leh, A., N'Guessan, H. E., Jaini, R., and Dang, L. (2009). Measurement of lateral adhesion forces at the interface between a liquid drop and a substrate. *Physical Review Letters*, 103(26), 266101.

[13] Zhu, H., Isikman, S. O., Mudanyali, O., Greenbaum, A., and Ozcan, A. (2013). Optical imaging techniques for point-of-care diagnostics. *Lab on a Chip*, 13(1), 51–67.

Chapter 46

Simulation of U-cup Hydraulic Seal Used in a Hydraulic System Using Ansys

Prasanna Mahankar[1], Anshuman Sinha[2], Aman Singh[2], and Ashwinkumar Dhoble[3]

[1]Assistant Professor, MKSSS'S Cummins College of Engineering for Women, Nagpur, Maharashtra

[2]Student, Visvesvaraya National Institute of Technology, Nagpur, Maharashtra

[3]Associate Professor, Visvesvaraya National Institute of Technology, Nagpur, Maharashtra

Abstract: Seals are mechanical devices that are used in various industrial applications wherein pressure zoning (separation of pressure zones) is desired. In applications wherein a reciprocating or rotating rod moving inside the housing, the seal fits in the housing and rubs against the rod to provide a sealing action between the two sides of the rod. Thus, the seals play a very major role in the pump system devices utilised in the industries. New materials and shapes of seals are being frequently introduced to cater to different applications and their analysis by experimental means becomes a very time-consuming and lengthy task. The simulation method is the most optimum way to recreate a process and somewhat predict the nature and behaviour of the process. With some limited assumptions coming into the picture, one can find results with good accuracy and relevance. The article elaborates on an attempt to figure out the behaviour of a U-cup seal under operation using the simulation method. The system consists of a reciprocating rod purely reciprocates and rubs against the seal thus involving wear behaviour. The article gives a brief overview of the steps involved, the parameters considered and the results obtained after various attempts of simulation.

Keywords: U-cup Seal, simulation of seal.

1. Introduction

Mechanical seals are devices that make the working of the pump systems possible, by preventing fluid leaks and keeping contaminants out. Seals are used in automobiles, concrete pumping equipment, various hydraulic and pneumatic pumps, aerospace, food processing, industrial plant equipment, etc. The mechanical seals basically separate the high-pressure zone from the comparatively lower pressure zone in a

DOI: 10.1201/9781003527442-46

case where there is either rotating or reciprocating motion in the central piston. Seals come in different geometries depending on the application. A U-cup seal, also known as a U-ring seal, is a type of hydraulic seal that is utilised to prevent leakage in hydraulic and pneumatic applications. The U-cup seal has a U-shaped cross-section that creates a tight sealing effect with the shaft of the hydraulic or pneumatic cylinder. It is shown in Figure 1 (Patel *et al.*, 2019; Yang *et al.*, 2011). Physically testing the seal requires dedicated setup, manpower, energy (in setting the assembly into place) and a huge amount of time. Thus, the author has taken up the issue to solve this lengthy problem and derive a simulation solution to ease the problem. The problem statement is to simulate a similar seal testing operation using simulation software and determine the seal behaviour in order to save time, energy and material. The commercial software ANSYS is used to carry out the simulation work. The seal under consideration is a U-cup seal made out of polyurethane material. The main aim of the article is to provide a simulation solution to the experimental approach that was used to test the seals. So, duplicating the entire experiment accurately with the involvement of many parameters and models was the main task. The experimental set up is shown in Figure 1. The seal Figure 1(c) is housed in glands. Two glands were fastened to each other with a bolt and a reciprocating shaft of diameter 70 mm was inserted into it. The reciprocating motion of the shaft is controlled by a pneumatic system. The seal is pressurised with hydraulic oil at different pressures. The trails are run till the seal fails. The failure of the seal is identified by observing the leakage and decrease in pressure value.

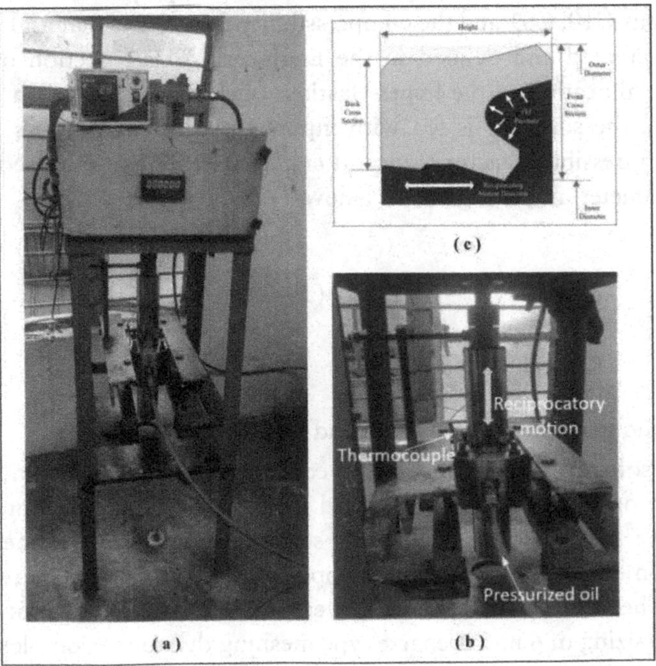

Figure 1: Experimental set up and U-Cup seal.

2. Simulation Model

2.1. Rod and Seal

The system under analysis consists of an assembly of a seal, a rod and an housing. The seal is fixed in the groove of the housing. The housing and the seal assembly is supposed to be stationary and the rod will reciprocate inside the seal and create a sealed contact with the inner surface of the seal. The modelling of the system is done using FUSION 360 software. The seal dimensions used are inner diameter: 69.4 mm, outer diameter 86.14 mm and thickness 12 mm. The system is modelled and then imported in the geometry module of the ANSYS software. The rod will be reciprocating within the seal. The front edge of the rod is provided with a chamfer of 1 mm so that there is no sharp edge involved in rubbing which may cause destructive wear. The rod is designed as a cylinder of diameter 70 mm and length 140 mm.

2.2. Material Modelling and Boundary Conditions

Polyurethane is a hyper-elastic material which is not predefined in the ANSYS material library. Hence, a proper material model needs to be selected to define polyurethane. The options available for hyper-elastic materials are Mooney–Rivlin, Yeoh, Ogden. According to Crudu *et al.* (2012), it is best defined using the Mooney–Rivlin model under the hyper-elastic category. The two-parameter Mooney–Rivlin material model is selected for the proper modelling of polyurethane. It involves the use of constant C10, C01 and the compressibility factor. The material polyurethane is therefore created and defined in the Engineering Data section of the project module. The allocation of the hyper-elastic model is done based on the literature survey. Then, the software is fed with inputs for three parameters namely C10, C01 and compressibility factor (Szurgott *et al.*, 2019). Equations used for Mooney Rivlin 2 parameter model is given as below

$$W = C01\,(I2 - 3) + C10\,(I - 3) + \frac{1}{D1}\,(J - 1)^2 \tag{1}$$

$$\mu = 2\,(C01 + C10)$$

where W is the strain energy potential and μ is the shear modulus.

The ANSYS software uses mathematical equations in the background to calculate the deformation values at the nodes and depict the respective contour for a set of nodes. The meshing process involves selecting appropriate element types and sizes based on the geometry, physical properties and accuracy requirements of the simulation. The elements in the meshing were triangular and the size of each element is 6 mm. The sizing of 6 mm is coarse type meshing due to the complex structure of the seal (He *et al.*, 2018). The fixed support condition is basically used to maintain the seal in a static state and allow the rubbing of the seal against the rod, when the rod exerts a force against the seal during its course of displacement. For the system it has taken a pressure boundary condition to replicate the oil pressure on the valley of the seal (refer to the Figure 1). The oil pressure we have taken is around 0.1 MPa.

3. Contacts and Wear Model

In ANSYS, "Contacts" means the interfaces between two or more geometries or surfaces or bodies that interact with each other in a simulation. Contact modelling can be a complex process that requires careful consideration of the physical behaviour of the system. For our case we have considered frictional contact between the inner faces of the seal and the outer surfaces of the rod including a small chamfer on the same rod. The value of the frictional coefficient is taken to be 0.1. It is shown in Figure 2.

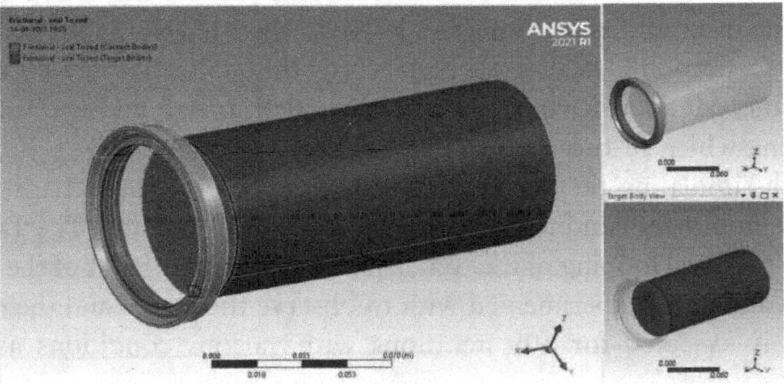

Figure 2: Defining contact surfaces.

The rubbing between the inner surface of the seal and the rod surface causes wear in the softer material that is the inner surface of the seal. In ANSYS, use of the Archard wear model can be incorporated to study the effects of wearing in the seal body. The Archard wear model is an empirical wear model used for predicting wear in cases involving sliding contact between two surfaces. The model assumes that wear occurs as a result of the removal of small particles from the surfaces in contact, due to sliding friction. The equation for the Archard wear model is given by:

$$W = \frac{K}{H} P^m Vrel^n \quad (2)$$

where $\dot{w}$ is the wear volume, *K* is the wear coefficient, *H* is the hardness of the material, *P* is the contact pressure and *Vrel* is the relative sliding velocity. The constants m and n are the pressure exponent and velocity exponent, respectively. The wear coefficient *K* is a material property that depends on various factors like surface roughness, the type of contact and the sliding velocity. The seal in operation faces wear and tear due to thermal and rubbing effects. This cannot be alone simulated by providing the contact information. Appropriate wear model needs to be defined and implemented to find out the wear in the seal after a particular span of operation. The Archard wear model as is implemented to inculcate the wear effects in the simulation using APDL code. The model is implemented by manually giving the parameters to the inbuilt Archard model defined in the ANSYS package. Hence, an APDL code is written under the frictional contact module.

4. Solution and Result Analysis

After hitting the solve button, ANSYS prepares the mathematical model consisting of equations that describe the behaviour of our model under the given loading and boundary conditions. The solving of these equations involves solving for the displacements, stresses and strains. The number of steps to be defined in "Analysis Settings" is a trade-off between the time consumption for the convergence of the solution. The most important thing to update in the Analysis Settings is to turn on the "Large Deflection" because the material of the seal is a hyper-elastic material (polyurethane, Mooney–Rivlin 2 parameter model). It will undergo unpredictable deformations which in turn makes this a "Non-Linear Analysis". The workflow started with simulating the displacement of the rod for a single iteration. All the appropriate boundary conditions mentioned were implemented. The Tabular Data option was finalised looping of the rod displacement, and the rest of the parameters were then varied and experimented with to observe the errors and the output. The simulation is executed for 500 iterations with various conditions as explained below.

Attempt 1: 500 iterations without pressure and wear

The simulation was carried out for 500 iterations, neglecting the pressure consideration and wear. This was the basic base case and executed to confirm if the setup till date was close to correctness. The solution converged for the applied boundary conditions. The deformation characteristics showed a regular trend over the stroke range.

Attempt 2: 500 iterations with pressure and no wear

The base case module was duplicated and a pressure of 100 bar was given initially. It was found that the pressure values were not suiting the case. The pressure was reduced to 1 bar and then it was found that the solution converged. The discrepancies in the pressure values in comparison to the experiment may be due to the fact that the pressure applied was from one side only, while the outer side of the seal was assumed to be a fixed support.

Attempt 3: 500 iterations with pressure and wear consideration

After the successful addition of the pressure boundary condition, the module with 500 iterations and pressure boundary condition (0.1 MPa) was then duplicated and the wear model was added through APDL coding. The two exponents' m and n used in the formula are taken to be 1 and 0.8. There is no specific base to the assumption. It is just done to check the correctness of the code and the effects of the parameters over the wear volume. The hardness value for polyurethane is taken to be 70 A shore, which in terms of N/m2 is 3.45 X 106. The typical values of wear constant k for polyurethane materials in sliding contacts with metals are in the range of 10^{-9}–10^{-7} mm^3/Nm. k is taken to be 10–18 m^3/Nm. It was found that the deformation values remained constant at all the steps. It would be probably due to less number of steps or maybe due to contradicting conditions.

Attempt 4: 500 iterations with wear consideration and no pressure

The last attempt was to check the wear model without pressure boundary condition. The module of the previous attempt was duplicated. The only change that was done was suppressing the pressure boundary condition. It was thought that the pressure over the seal along with the wear involvement would have over constrained the seal. The simulation was run and the data obtained was plotted.

Table 1: Data obtained for various attempts.

	Without pressure and wear	**With pressure and no wear**	**With pressure and wear consideration**	**With wear consideration and no pressure**
Time [s]	**Average Deformation [mm]**			
1	7.02E-15			6.85E-15
100	7.10E-15	9.92E-05	5.63E-02	7.15E-15
200	6.85E-15	9.92E-05	5.63E-02	7.01E-15
300	6.96E-15	9.92E-05	5.63E-02	7.00E-15
400	6.91E-15	9.92E-05	5.63E-02	7.11E-15
500	6.93E-15	9.92E-05	5.63E-02	7.08E-15

The output from the ANSYS software is obtained in terms of dataset. The simulation was carried out considering various parameters and conditions together or one at a time. The values for deformation were the major data that can provide a subtle conclusion. From all the attempts that were made, many results provided a good support to the correctness of the model. The base and the most crucial stage of the simulation is done successfully. All the parameters and factors that were supposed to be included have been explored and included with appropriate considerations. The results that emerged after considering the various cases showcase the effects of individual boundary conditions.

The first attempt showcased a considerable deformation variation. On adding pressure to the first attempt module, the deformation values increased drastically. This might be due to the fact that the pressure given wasn't accurately represented as in the real case. In the real case, there is involvement of the housing support and fluid pressure. This might have made a difference in the simulation case. The data and conditions from the above attempts can be utilised and modified in a way that the conditions fit well and provide a solution considering all the required conditions.

5. Conclusion

The problem statement was discussed and appropriate literature survey was done via the papers and articles, already existing in this domain. Various information regarding the selection of the material model, appropriate software, etc. was successfully obtained during this course of action. The modelling of the system was done using FUSION 360 software and then it was imported in ANSYS. The simulation was carried out for the appropriate boundary conditions as per the requirement of the problem. The application of displacement to the rod was tried to be done using various methods, but some or the other limitations of the available options landed us to the "Tabular Data" option. For simulations with 500 iterations, almost 10–15 hrs were required for each attempt. For a simulation of 500 steps, considerable change in the seal geometry was not observed. The seal used in the experimental analysis worked for about 1 lakh cycles and then showed some cracks and permanent deformation signs. However, ANSYS has a limitation of giving only 10,000 displacement entries using the "Tabular Data" option. Therefore, the software limitation is one such conclusion that needs to be noted for further work in this domain. The iteration time for 10,000 cycles was predicted to be very large. Considering the fact that 500 iterations took almost 10 hrs, 10000 cycles would take almost 100 hrs.

The future work in this can be extending the same setup to some other software with no displacement data limitations and a faster computational speed.

References

[1] Crudu, M., Fatu, A., Cananau, S., Hajjam, M., Pascu, A., and Cristescu, C. (2012). A numerical and experimental friction analysis of reciprocating hydraulic 'U' rod seals. *Proceedings of the Institution of Mechanical Engineers, Part J: Journal of Engineering Tribology*, 226(9), 785–794.

[2] He, Y., Bayly, A. E., and Hassanpour, A. (2018). Coupling CFD-DEM with dynamic meshing: A new approach for fluid-structure interaction in particle-fluid flows. *Powder Technology*, 325, 620–631.

[3] Patel, H., Salehi, S., Ahmed, R., and Teodoriu, C. (2019). Review of elastomer seal assemblies in oil & gas wells: Performance evaluation, failure mechanisms, and gaps in industry standards. *J Pet Sci Eng.*, 179, 1046–1062.

[4] Szurgott, P., and Jarzębski, Ł. (2019). Selection of a hyper-elastic material model-a case study for a polyurethane component. *Latin American Journal of Solids and Structures*, 16.

[5] Taravella, B. M., and Rogers, C. T. (2017). A computational fluid dynamics analysis of an ideal anguilliform swimming motion. *Marine Technology Society Journal*, 51(6), 21–32.

[6] Wang, B. Q., Peng, X. D., and Meng, X. K. (2019). A thermo-elastohydrodynamic lubrication model for hydraulic rod O-ring seals under mixed lubrication conditions. *Tribology International*, 129, 442–458.

[7] Xin, L., Gaoliang, P., and Zhe, L. (2014). Prediction of seal wear with thermal–structural coupled finite element method. *Finite Elements in Analysis and Design*, 83, 10–21.

[8] Yang, B., and Salant, R. F. (2011). Elastohydrodynamic lubrication simulation of O-ring and U-cup hydraulic seals. *Proceedings of the Institution of Mechanical Engineers, Part J: Journal of Engineering Tribology*, 225(7), 603–610.

Chapter 47

Supply Chain Management in Life Insurance companies in India

Mahesh R. Shukla[1], *Bhavana W. Khapekar*[2], *Supriya Gupta Bani*[3], *and Shailesh N. Khekale*[4]

[1]Associate Professor, Department of Mechanical Engineering, Cummins College of Engineering for Women Nagpur, India

[2]Assistant Professor, Lady Amritbai Daga College & Smt. Ratnidevi Purohit College India

[3]Assistant Professor, Department of Data Science (CSE), Shri Ramdeobaba College of Engineering and Management (RCOEM), Nagpur, India

[4]Assistant Professor, Cummins College of Engineering for Women Nagpur, India

Abstract: This study intends to reduce the operational expenses of insurance company proposal forms and policy papers by using SCM in the operations department of life insurance firms. The complete process of how the forms are delivered from suppliers to consumers, going through several departments and their flow from one point to another for various underwriting criteria is set out. Various expenses related with the distribution network of policy papers have been thoroughly examined. The time it takes for a proposal form to become a part of a customer's policy document is also examined. The emphasis is on reforming processes that are no longer necessary yet still exist. Based on data obtained mostly through questionnaires and personal interviews, the study develops a framework for assessing supply chain performance. According to the report, despite high costs paid in document processing at all levels, customer difficulties, failures in channel expansion owing to operational issues, loss of market share, loss of customer base and loss of business, enterprises have not identified any other means of operation. The current study urges insurance businesses to implement more cost-effective business practices and provides context for why SCM is vital to them. A few critical ideas have been offered in this article on how operational delays may be eliminated while operating costs are reduced, as well as how customer retention and channel growth can be done properly and methodically.

Keywords: Supply chain management, life insurance, performance measures, operations management, process control.

DOI: 10.1201/9781003527442-47

1. Introduction

Supply Chain Management (SCM), which was established in the early 1980s (Oliver and Webber, 1982), is now extensively used in both the industrial and service industries, resulting in enhanced logistics and efficient material handling, which leads to higher productivity at lower costs. SCM's primary task is to connect important company activities and procedures with cross-enterprise business operations (Barroso *et al.*, 2011). In other terms, SCM is a concept "whose major goal is to integrate and manage the procurement, flow, and control of materials across different functions and many layers of suppliers using a whole systems approach" (Monczka *et al.*, 1998). As a result, we discover that "the entire notion of SCM is truly foreseen in integration" (Pagell, 2004). A large deal of study has been conducted on supply chain integration (Gimenez and Ventura, 2005), which is regarded to be of tremendous strategic and operational value (Zailani and Rajgopal, 2005). Numerous studies based on primary data and empirical investigations have been conducted to conclude that SCM integration and implementation (Chen *et al.*, 2004) will result in improved performance (Agami *et al.*, 2012) Even fewer are the research studies conducted on the deployment of SCM in the Operations Department of the Insurance industry (Ashill *et al.*, 2009). Various research investigations on service sectors have found that fundamental quality control processes fail owing to poor operating systems. Industry pressures have fueled innovations aimed at reducing costs in crucial tasks. As has been observed in many sectors, if a sector is opened up to private actors, the incumbents always suffer difficulties. With their solid financial backing, excellent technology assistance, and comprehensive study and accurate and right knowledge, the new entrants found it much simpler to expand and succeed. Interestingly, this did not apply to the Indian life insurance business, where LIC continues to dominate even 21 years after the sector was opened up to private competitors. Private players were also needed to expand swiftly in order to compete, and they were particularly aggressive in this regard. They did not understand the large upfront costs involved with their growth, and their breakeven time, which was supposed to be 10 years, has now increased to 30 years. This has put a strain on the private sector. In addition, they are also dealing with diminishing volumes as policy violations increase year after year. As a result, it is obvious that the insurance business should seek for a flexible delivery company capable of efficient operations. According to the IRDA 2019 annual reports, there are 29 life insurance firms operating in India, contributing 2.3% of the country's overall economy, amounting to 7,87,072 crores of rupees. The fundamental causes of waste in insurance management techniques are needless delays in policy docket issuing, duplicate issues, poor operating systems, excessive costs and a mismatch of customer demands. This necessitates the proper administration of a wide variety of procedures with various measures, ranging from the acquisition of the appropriate insurance forms through the issuing of policy dockets to clients. The current policy documentation methodology is excessively complicated, and policy issuing takes much too long. Figure 1 displays the current routing of insurance forms used by insurance firms. Forms are typically obtained from a single printer stationed at a certain location throughout the nation. They are subsequently allocated to the zonal offices, which are then routed to the state head office and ultimately to the branch offices. The forms are distributed by the branch offices to the rural areas of that region. Forms are made available to sales managers in the branch office, who then distribute them to agents, who then deliver them to consumers. Typically, the forms vary based on the type of product (ULIP/Traditional, for example). Customers complete the forms with the assistance of agents and return them to the branch office together with the appropriate papers. As a result, the branch office may get completed proposal forms from consumers either directly or indirectly via the same path but in the other way.

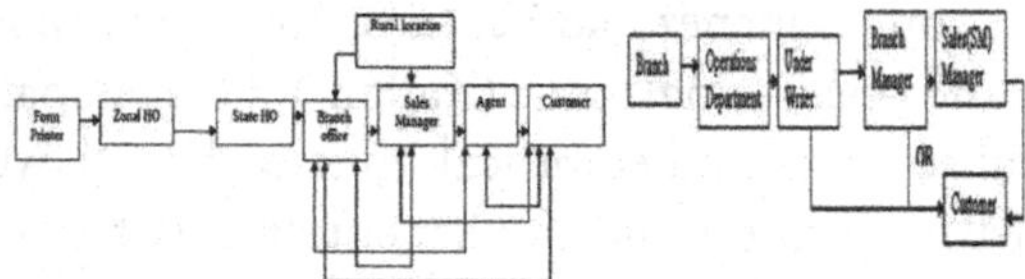

Figure 1: *Routing of proposal forms & policy dockets.*
Figure 2: *Process of issuance of policy dockets to the customer.*

When the completed forms arrive at the branch office, the policy documents are thoroughly screened by the operations department, examined by the underwriter and then returned to the consumer, which is a lengthy and time-consuming procedure. Figure 2 displays the method used by insurance firms to issue policy documents to clients. The underwriting staff distributes policy paperwork directly to clients in urban areas but to the branch manager in rural areas. The branch manager then gives the policy documents to the sales manager, who distributes them to the consumer either personally or through agents. In the event of a discrepancy from either the client or the branch operations team, a new method, represented in Figure 3, is used to route forms from the branch to the underwriter and back to the consumer.

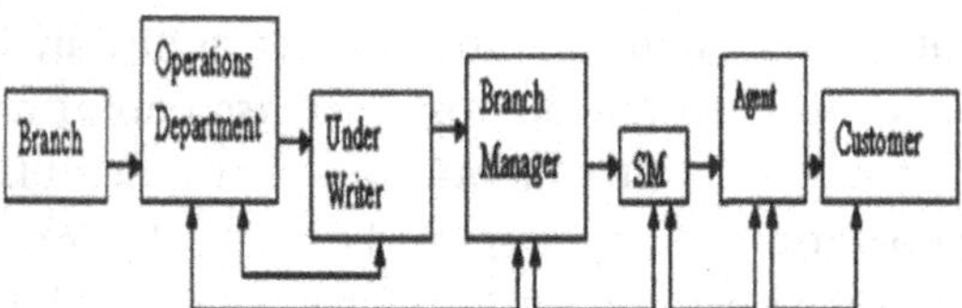

Figure 3: *Process adopted by insurance companies in case of refusal by the underwriting department.*

The whole procedure represented in Figure 2 is repeated, and the client is addressed once again if any clarification about the given details is necessary. This frequently results in the client cancelling the coverage owing to (i) the long time it takes the firm to complete his application or (ii) his refusal to meet the conditions stated by the company. Businesses must help streamline and counter insurance plans and advances that have resulted to suit the specific needs of policyholders as a first step forward into reducing organisational latency, with such a special focus on reducing operational costs, which account for nearly 7% of the total of the company's different costs lead to improper mobility of regulatory papers (Kumar and Thomas, 2016). In terms of policy form production and distribution, they are printed in large quantities without any market study and sent to branch offices via state head offices. At no point is a product need analysis performed, and each transaction incurs significant financial and human costs. In reality, there is no feedback system in place to learn about the requirements of certain types of forms, their availability or their use. Furthermore, no one in the firm is held accountable for these documents. As a result, it is critical for the sector to spend not just in development and distribution but also in processing and customer service. As a result, this paper is an attempt to establish a process-oriented framework in the operations department of life insurance firms in order to minimise operations expenses and issuance time, hence enhancing customer satisfaction and influencing company growth. The study also discusses trends, difficulties and potential solutions for logistics management in SCM that will be implemented in the insurance sector using principles from operations research (OR) disciplines applied to specific domains. The study finishes with ideas for prospective applications of the suggested framework in the insurance business.

2. Context

In terms of SCM adoption in the service sector, it has been implemented in the health, IT and financial sectors, but it has not been used in the insurance business nor has any study been conducted in this respect (Shukla and Shrivastava, 2019). The authors gained firsthand knowledge of the operational systems of the firms' operations departments while working with numerous life insurance companies in various roles (Hammer and Champy, 1993). During their time with the firms, they thought that the intricate subtleties involved in form processing should be eliminated or simplified, which would undoubtedly aid in decreasing operational delays and costs, as well as boosting customer happiness and company success. The following are some of the reasons why the insurance sector should use supply chain in the operations department: Shukla and Shrivastava (2017).

1. Insurance underwriting forms are critical to the firm since they serve as a contract between the company and the consumer. However, it entails a lot of actions that are needless and consuming since they require regular tracking, involve duplication, and become too expensive for a firm without bringing value to the organisation in terms of client retention and organisational expansion goal..
2. The expense of processing documents at multiple levels and in different places is enormous. Customers have troubles, personnel are de-motivated, channel development fails owing to operational issues and customers are lost.

3. Customers are frequently dissatisfied due to the lengthy and complex process of issuing policy documents. Furthermore, if the policy registration fails for any reason, customers are inconvenienced because they must repeat the process.
4. If customers are dissatisfied, they will cancel their policies, resulting in a significant loss for the company. Despite the fact that insurance companies have been dealing with these issues for a long time, they have yet to find a solution (Ba and Johansson, 2008).
5. There have been no empirical studies to date on the implementation of supply chain in the operations departments of life insurance companies. If insurance companies use a supply chain system diligently, the problems mentioned above may be eliminated or at least minimised.

2.1. *Objectives of Study*

The current researchers believe that it is necessary to investigate various areas such as the availability of proposal forms in respective branches, its tracking for issuance from underwriting to respective customers, and the factors that cause delays in the issuance of policies to customers. At the same time, the researchers believe that cost savings can be achieved by making changes to the routing of forms in the operations department (Shukla and Shrivastava, 2019). The objectives of pursuing the present study are to show the advantages of implementing SCM in the life insurance operations department such as:

a. Minimising operational delays
b. Minimising the routing of proposal forms
c. Reducing unproductive time in tracking for issuance of policies
d. Reducing operations cost (Craighead *et al.*, 2009)
e. Improving customer satisfaction and their retention rate
f. Increasing service quality (Quayle, 2003)
g. Increasing company's profit
h. Increasing and improving overall performance of the company.

The other objective is to propose a conceptual framework for SCM with propositions of further research addressing the role of SCM in operations department in insurance sectors.

2.2. *Hypothesis Development*

The SCM practises employed in this study can be broadly classified into four types: (a) Efforts to improve operational efficiency by reducing operation delays, (b) Lowering operational costs through the implementation of quality practises, (c) Efforts to improve customer satisfaction and service and (d) Increasing the company's performance (Volker *et al.*, 2013). The hypotheses which will be tested are as follows:

H1 There is corelation between minimising operation delays and growth of company.

H2 There is corelation between reducing operations cost and growth of company.

H3 There is corelation between implementing SCM and improving customer satisfaction.

H4 There is corelation between implementing SCM and increasing the performance of life insurance companies.

3. Research Methodology

This study employs both qualitative and quantitative research methods. The survey method was chosen to collect primary data. The factors identified by factor analysis of branch managers', operations managers', sales managers' and agents' perceptions were considered dependent variables, while position, experience and age were considered independent variables. Two open-ended questions were asked to gather information about issues related to SCM implementation to improve customer service and company growth. The personal responses of all groups to these open-ended questions were used as qualitative data (Chopra *et al.*, 2009). A semi-structured questionnaire was developed from the revised instrument based on Brownell (18) and Ng and Pine (19). The population comprised of Branch managers, Operation managers, Sales managers and agents associated with eight leading private life insurance companies from different states of India which were in existence for more than 10 years. The sample consisted of 16 Branch Managers, 16 Operations Managers, 200 Sales Managers and 200 Agents from eight leading Life insurance companies: ICICI Prudential Life Insurance, Aviva Life insurance, Tata AIA Life Insurance, Kotak Mahindra Old Mutual, Max Life, SBI Life, Exide Life Insurance and Birla Sun Life. Questions were send through survey monkey.com. Out of a sample of 432, only 256 responded. The questionnaire consisted of eight questions (two questions for each objective). Primary data was gathered through the distribution of questionnaires to managers and individual interviews with them. Secondary data on the establishment and operation of affiliated firms in important industries such as retail, healthcare, telecommunications and airlines was gathered from the available literature (journals, research papers, annual reports and so on).

Survey Questionnaire: The survey-questionnaire contained four questions. The five items for Question 1 on factors responsible for operational delays were: form non-availability, too many check centres for submitted forms, denial of access to sales team, form duplication and company norms constraining. Participants were asked to rate their experiences on a five-point Likert scale. (5 = very important, 4 = important, 3 = doesn't matter, 2 = not important and 1= least important).

Question 2 focused on the factors that contribute to lower operating costs. Respondents were asked to rate the importance of factors that could help reduce operations costs. The five items chosen were: spot issuance, the same type of forms for all products, the availability of forms in one location, technological use and clearly defined company norms. Respondents were once again asked to rate themselves on a five-point scale.

Question 3 was about increasing customer satisfaction. The five items chosen for this were: too many requirements that were not previously disclosed, internal customer dissatisfaction, timely receipt of policy documents, customer switching to another company and policies that were not transparent to customers.

Question 4 focused on the factors responsible for increasing and improving the company's performance. The five items chosen were: rapid issuance of policy documents, customer base development, customer trust building, profit/revenue margin improvement and business expansion and growth.

3.1. Research Design

The record keeping provided trustworthiness and reliability by chronicling specifics of the methods and decisions made at all stages of the research (Craighead and Meredith, 2009). The researchers added value by purposefully grouping themselves across interviews and analyses to gain a better understanding of the participants' perspectives. The audio tapes, which recorded the participants' actual statements, added to the reality. The questionnaires used to interview the 16 branch managers and 16 operations managers are listed below.

1. What factors contribute to insurance industry operational delays?
2. What steps can be taken to cut operating costs?
3. What steps can be taken to enhance customer satisfaction?
4. What efforts should be made to boost and improve the performance of the company?

Following meeting direction, the meetings were interpreted from sound tapes to electronic organisation. Willig's four-phase phenomenological investigation was used to lead the information examination. The meetings were read a few times as an underlying experience with the meeting scripts. A slew of unfocused notes was produced. The second phase of the investigation included identifying and marking the emerging topics, both subordinate and central subjects. These subjects noticed a few repeating designs cutting across the data. The pith of what was found in the text was used to code recognised topics. Topic rundown tables were created (Romsdal, 2008) Ultimately, four significant subjects emerged from the data. The Table 1 below lists the top topics and sub-topics that emerged from the stories.

Table 1: Themes and sub-themes.

Master Themes	Sub-themes
I Reducing Operation delays	i. non-availability of forms ii. too many check centres for the submitted forms iii. denial of access to sales team iv. duplicating of forms v. the constraining company norms

II. Reducing operations cost	**i. spot issuance ii. same type of forms for all products iii. availability of forms at one place iv. technological use v. clearly defined norms of the company**
III. Improving Customer satisfaction	**i. too many requirements not earlier disclosed ii. dissatisfaction of internal customers iii. not receiving policy documents in time iv. switching of customers to other company v. no transparency of rules**
IV. Increasing the performance of life insurance companies	**i. quick & fast issuance of policy documents ii. developing customer base iii. improving profit/revenue margin iv. building trust with the customers v. business expansion and growth**

The classifications were educated by the review's motivation, the analysts' information and the implications were made express by the actual members.

4. Data Analysis

The current study sought to investigate the perceptions of branch managers, operations managers, sales managers and agents regarding significant factors that can reduce insurance company operation delays and costs. The goal was also to determine whether implementing SCM would increase customer satisfaction, thereby contributing to the company's growth and development. The findings in this section are based on a preliminary examination of the interview transcripts. Interview questions guide the presentation of findings.

H1: The first hypothesis stated that there is corelation between minimising operation delays and growth of company.

The first open-ended question, about the factors that cause operational delays in the insurance sector, was answered by 211 out of 217 participants. The majority (72%) believe that direct transaction costs increase branch and company revenue while decreasing profits. Processing costs add no value to the company or the customer. Underwriting requirements lengthen the turnaround time, causing problems with policy issuance. All related functions are tracked in terms of both time and cost. Customers lose faith in the company as a result of the policy document delay. In the long run, ethical concerns arise. Long-term overlap between new and existing business appears. The number of policies that are available for free review is growing. Everything has an effect on the company's growth. Thus the first hypothesis is accepted.

H2: There is corelation between reducing operations cost and growth of company.

The second question from the questionnaire was considered to determine the significant relationship between lowering operations costs and company growth. According to the responses gathered, there are too many process stations and manpower involvement, which leads to complications and loss of effective man hours because the time required for tracking each discrepancy is too high and involves almost everyone involved with that job. Minimising this may assist the company in making the best use of its available resources and manpower. Also, the logistics of the insurance forms from the printer from which it was sourced to the final policy document that reaches the customer are too high, which can be reduced by either spot issuance or using technology, where all forms should be scanned and sent to the underwriter. Managers at all levels agreed that the timely and proper availability of the forms and their issuance saved them a lot of time tracking the status and allowed them to focus more on business, which led to the company's growth. Thus, the 2nd hypothesis may be accepted.

H3: There is corelation between implementing SCM and improving customer satisfaction.

This open-ended question was answered by 202 of the 217 respondents. There were 15 missing responses. The majority of participants (58%) believed that customers were generally dissatisfied with the company's services due to late issuance of their policies or other requirements of either documents or information to be provided at various times by the underwriting team. This time, the entire team in charge of the case is involved. The absence of even one member of the team causes delays in issuance, which leads to employee and customer frustration. The current study's overall findings support the third hypothesis. In other words, implementing SCM would almost certainly increase customer satisfaction.

H4: According to the fourth hypothesis, there is a link between implementing SCM and improving the performance of life insurance companies. 211 out of 217 participants responded to the open-ended question about steps to be taken to increase and improve the company's performance. Most participants agreed that quick, faster and timely policy issuance was critical to increasing and improving company performance because employees and channel partners were highly motivated. Customers stated that because of the reduced processing time and requirements, they began to develop trust in the company and its operations. Branch managers stated that it assists the company in accepting and confronting business challenges from competitors, as well as creating goodwill in the markets. Thus, the overall findings indicate that the fourth hypothesis, stating that there is a significant relationship between implementing SCM and increasing the performance of life insurance companies, is proven.

5. Findings

According to the survey results, the basic input cost is significantly higher due to the existing complex operation system. The direct transaction cost increases branch and company revenue, lowering profits. Figure 7 depicts the location of form sourcing. In our study, location refers to whether the forms are to be sourced at the regional (state) level, urban (in cities) or rural (remote) level. Form sourness primarily refers to various types of forecasts.

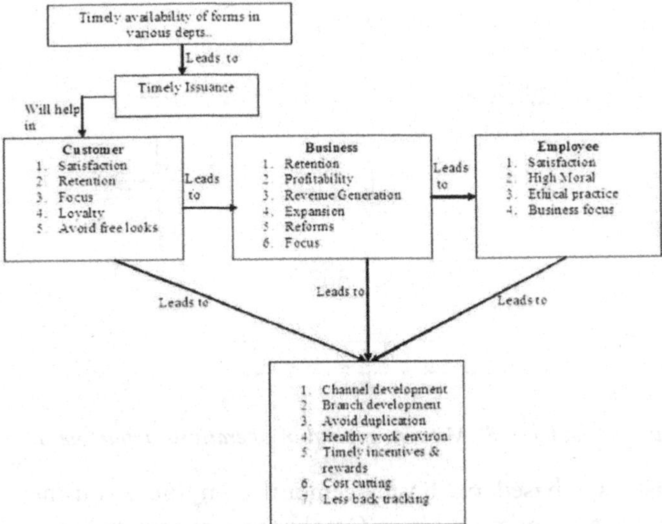

Figure 4: Research findings in illustrated form.

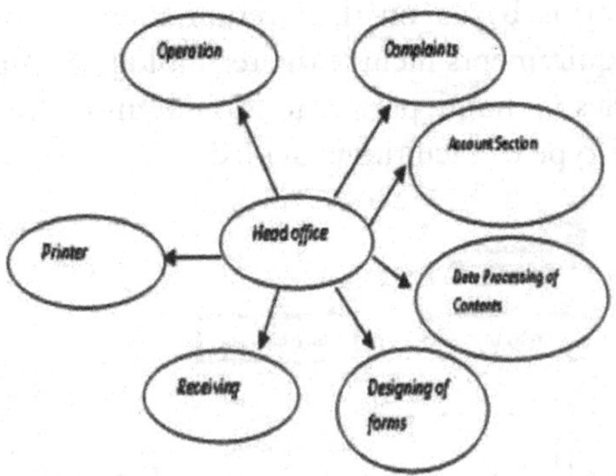

Figure 5: Operational chart.

The cost of processing adds no value to the company or the customer. Underwriting requirements increase turnaround time, causing policy issuance issues. Parallel tracking requires both time and money from all related functionaries. Customers lose faith in the company as a result of the delay in receiving policy documents. Ethical issues emerge, and the rate of policy review increases. Long-term duplication of new and existing businesses is possible. Expansion plans have been pushed back due to business losses and shrinking profit margins. Finally, the customer base decreases and quantum buisness comes down (Cvetic, 2009). The findings of the research can be represented by the following diagram. Figure 5 depicts the operational chart for the process to be followed in any organisation in the sector discussed in this paper.

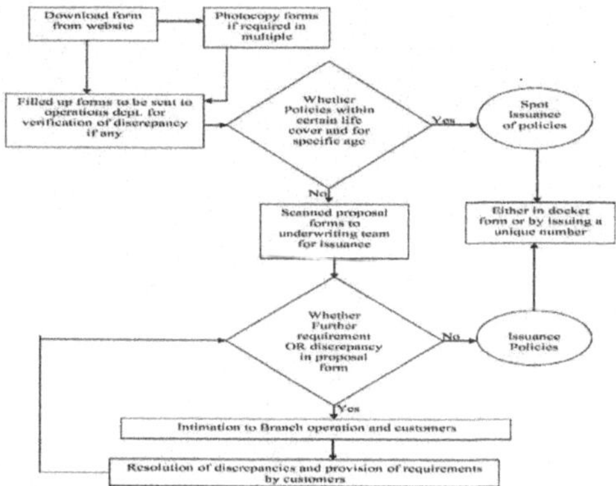

Figure 6: Proposed model of SCM to be applied in operations department.

These forecasts are based on business months (in India, business months for insurance products are primarily from October to January), the team size operating in a specific area, the company's past records regarding the requirements and the company's expansion plan, if any, in terms of team size or branches. These forecasts will assist businesses in determining the number of forms required. Companies can purchase those forms based on their requirements. Aside from forecasting, factors that lead to requirements include the technological support available in the specific area, inventories on hand, permission to download required quantities and information about the type of facilities provided.

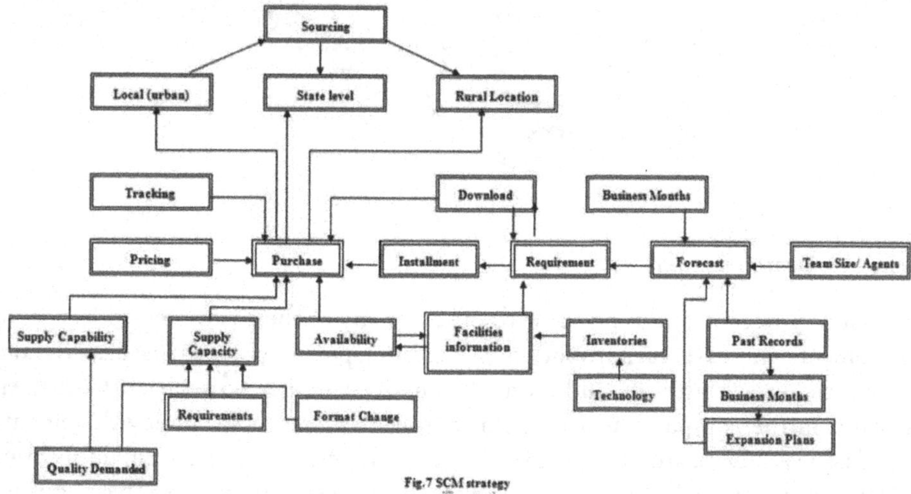

Figure 7: SCM strategy.

Once the requirements have been thoroughly understood, it is critical to determine whether orders should be placed in bulk or in installments. There are numerous factors that influence the purchase of forms. The company must track the supplier who has the ability and capacity to provide the forms in the required quality, quantity and price at the required locations at the required times. To achieve strategic fit with the competitive strategy of a business organisation there should be a balance between need & supply, demand & want which the chain aims at (Figure 5). Future Research Directions: The findings of this study will be useful for other examination researchers who need to effectively research and implement this innovation in other areas of assistance. If carried out correctly, the review will help in cost reduction as unnecessary cycles will be eliminated from the framework. This study will lead to the essentials and non-essentials in study of sourcing of forms. Figure 8 reveals that time, speed, storage, handling and ethics become non-essentials and types of business channels, types of forms required, inventory management, quality of forms, business location and condition of forms become the service essentials (Supply Chain Council, 2006).

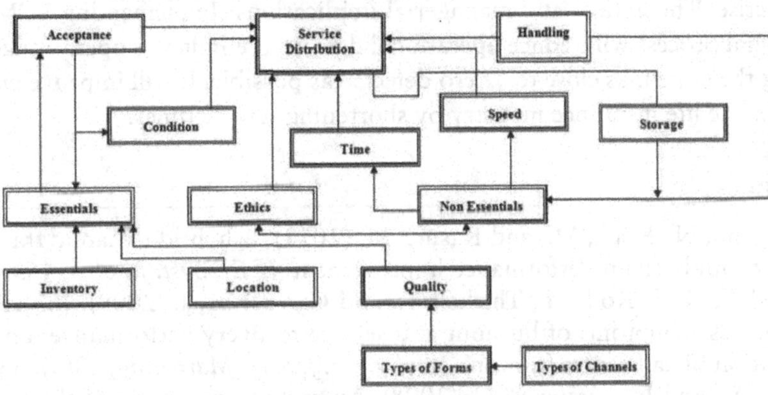

Figure 8: Insurance forms Distribution supply chain Management The Essentials and Non-Essentials.

6. Suggestions

SCM application implementing the above-mentioned model in the operations department of insurance companies will undoubtedly reduce operation delays, reduce operational costs and improve customer satisfaction, thereby influencing insurance company growth and expansion. Underwriting and non-underwriting forms can be combined. Forms can be downloaded directly and printouts for required quantities can be obtained in the respective branches (Gimenez and Ventura, 2009). Policies within a certain range of insured amount can be issued directly by the branch manager or the branch operations team. Customers should not be given paper policy documents but instead a unique ID. The agents, customers and sales managers all agreed that medical underwriting and requirements were major impediments to the fast issuance of policies and should be communicated to the underwriting team as soon as the customers' medical exams are completed. After being called into the branch office, remote location policy documents can be handed over to the agents. All forms that require underwriting can be scanned and sent to the underwriting team for review before being returned to the branch for final printouts and issuance (Figure 6).

7. Conclusion

This paper explains how implementing SCM in the insurance industry can reduce operation delays, lower operating costs and increase customer satisfaction, all of which affect the growth and expansion of insurance companies. The supply chain has had a significant impact on industry, but the academic community is still learning about this powerful strategy. As a result, it will be incumbent on the academic community to provide well-founded theories to explain the phenomenon of supply chain. In other words, supply chain lacks a theoretical foundation, and it is our responsibility as academics to bridge the gap between supply chain theory and practise. Theoretical and managerial implications: Implementing SCM in the operational process will reduce operational defects. It eliminates operational issues, bringing the system as close to "Zero defects" as possible. It will improve customer service in the life insurance industry by shortening service times.

References

[1] Agami, N., Saleh, M., and Rasmy, M. (2012). A hybrid dynamic framework for supply chain Performance improvement. *IEEE Syst. J.*, 6(3), 469–478.

[2] Ashill, N. F., Rod. M., Thirkell. P., and Carruthers, J. (2009). Job resourcefulness, symptoms of burnout and service recovery performance: an examination of call centre frontline employees. *J. Serv. Marketing*, 23(5), 338–350.

[3] Ba, S., and Johansson, W.C. (2008). An exploratory study of the impact of eservice process on online customer satisfaction. *Prod. Oper. Manage.*, 17(1), 107–119. Barroso, A. P. et al. (2011). The resilience paradigm in supply chain management: A case study. *Proceedings of 2011, IEEE,* IEEM, 928–932.

[4] Chen, H., Themistocleous, M., and Chiu, K. H. (2004) Approaches to supply chain Integration followed by SMEs: an exploratory case study. *In the Proceedings of the tenth Americas conference on information systems*, New York, 2610–2620.

[5] Chopra, S., Menddl, P., and Kalara, D. V. (2009). Supply chain management, strategy, planning and operation. *Prentice hall* 3rd edition.

[6] Craighead, C. W.,. Ketchen Jr. D. J,. Dunn K.S, and. Hult G.T.M. (2011). Addressing Common Method Variance: Guidelines for Survey Research on Information Technology, Operations, and Supply Chain Management. *IEEE Transactions on Engineering Management*, 58(3), 578–588.

[7] Craighead, C. W., and Meredith, J. (2008). Operations management research: Evolution and alternative future paths. *Int. J. Oper. Prod. Manage.*, 28(8), 710–726.

[8] Craighead, C. W., Hult, G. T. M., and Ketchen, D. J. (2009). The effects of innovation-cost strategy, knowledge, and action in the supply chain on performance. *J. Oper. Manage.*, 27(5), 405–421.

[9] Cvetic, B. (2012). A conceptual model for supply chain performance management and improvement. *Advances in Business-Related Scientific Research Journal (ABSRJ)*, 3(1), 62–75.

[10] Gimenez, C., and Ventura, E. (2005). Logistics-production, logistics-marketing and external integration – their impact on performance. *International Journal of Operations and Production Management*, 25(1), 20–38.

[11] Hammer, M., and Champy, J. (1993). Reengineering the corporation: manifest for business revolution (1st ed.). New York, NY: Harper Business. *Management Review*, 28(2), 65–71. IRDA 2019, Annual Report.

[12] Klassen, R. D., and Menor, L. J. (2007). The process management triangle: an empirical investigation of process trade-offs. *Journal of Operations Management,* 25(5),1015–1034. Kumar, S., and Thomas, B. (2006). Overcoming scale disadvantages in life insurance operations. *Infosys viewpoint.*Lance, C. E., Dawson, D., Birkelbach, D., and Hoffman, B. J. (2010). Method effects, measurement error, and substantive conclusions. *OrganizationalRes. Methods*, 13(3), 435–455.

[13] Liu, Y., and Li, H. (2009). Research on DCOR-based integrated information resource management platform of supply chain system. *International Forum on Information Technology and Applications, IEEE*, 513–516.

[14] Mentzer, J. T., DeWitt, W., Keebler, J., Min, S., Nix, N., Smith, C., et al. (2001). Defining supply chain management. *Journal of Business Logistics,* 22(2), 1–25.

[15] Monczka, R., Trend, R., and Handfield, R. (1998). Purchasing and supply Chain Management. Cincinnati, OH: South-Western College Publishing. Nyere, J. (2007). The design-chain operations reference-Model.www.supply-chain.org, 2007.

[16] Oliver, R. K., and Webber, M. D. (1982). Supply chain management: logistics catches up with strategy. *Logistics: The Strategic Issues*, Pitman, London, 63–75.Pagell, M. (2004). Understanding the factors that enable and inhibit the integration of operations, purchasing and logistics. *Journal of Operations Management*, 22(5), 459–487.

[17] Supply Chain Council, The Design Chain Operations Reference Model (DCOR) Version 1.0 www.supply-chain.org. 2006.

[18] Tan, K. C. (2001). A framework of supply chain management literature. *European Journal of Purchasing & Supply Chain Management*, 7, 39–48.

[19] Quayle, M. (2003). A study of supply chain management practice in UK industrial SMEs. *Supply chain management: An International Journal*, 8(I), 79–86.

[20] Romsdal, A. (2008). Action research in supply chain management; investigating the appropriateness from an organisation theory perspective. Norwegian University of Science and Techn., 39, 68–73.Shukla, M. R., and Shrivastava (2017). Supply chain management process: process innovation and evaluation of a process oriented framework in operations department of Lie Insurance Company. *1st International conference on Best Practices in Supply Chain Management*, SOA University, Bhubaneshwar.

[21] Shukla, M. R., and Shrivastava. (2019). Supply chain management process, a life insurance perspective. *3rd International conference on Best Practic-*

es in Supply Chain Management, SOA University, Bhubaneshwar.Tomlin, B. (2006). On the value of mitigation and contingency strategies for managing supply chain disruption risks. *Kenan-Flagler Business School*, Chapel Hill, North Carolina, 52(5), 639–657.

[22] Volker, M. G., Constantin, B., and Martin, C. S. (2013). Antecedents of proactive supply chain risk management – a contingency theory perspective. *International Journal of Production Research*, 51(10), 2842–2867.

[23] Zailani, S., and Rajagopal, P. (2005). Supply chain integration and performance: US versus East Asian companies. *Supply Chain Management: An International Journal*, 10(5), 379–393.

Chapter 48

Design and Development of Manually Operated Lanzan Seed Decorticator for Livelihood of Tribal Community in Eastern Vidarbha region in India

Yogesh Dandekar[1], *Prasanna Mahankar*[1], *and Dr. Shailesh Khekale*[1]

[1]Assistant professor, MKSSS's Cummins College of Engineering for Women, Nagpur

Abstract: Lanzan, the tree species, is commercially very useful and found abundantly in the eastern Vidarbha region. Although the chironji nuts have been used extensively by decorticating the kernel, the lack of a simple, hand-operated machine forces the tribal community to decorticate these seeds manually, resulting in the migration of tribal people into other businesses. A manually operated seed decorticating machine has been developed to reduce manual efforts and improve the productivity of decorticating process as compared to the manual one. The results were encouraging and it has been found that some more improvements in the designed decorticators will again increase the efficiency and help tribal people return to their traditional business for livelihood. In the design, adjustment by the operated in the roller gap is allowed and gap 5 mm, 6 mm, 7 mm and 8 mm is achieved by the screw and spring arrangement. It was found after several trials that 7-mm roller gap gives maximum decortication of complete seeds as compared to 5 mm and 6-mm gap, where the amount of crushed seeds is more. Higher roller gap of 8 mm resulted in loss of un-decorticated kernels in considerable amount. This product has been awarded a Design Patent by the Government of India Patent Office, Kolkata with Design number 301100.

Keywords: Chironji, decorticator, migration of tribal, socially and economically beneficial.

1. Introduction

Lanzan (chironji) is a tree species that is commercially very useful and is found abundantly in the eastern Vidarbha region. The queen of species is used in sweet meals as a substitute for almond kernels, and its kernel oil is mostly used in cosmetics manufacturing as a substitute for olive oil and almond oil. (Hemvathy *et al.*, 1987). The average annual seed collection is 300–1200 quintals in Vidarbha.

DOI: 10.1201/9781003527442-48

On average, 40–50 kg of fresh fruit is produced per tree, which yields 8–10 kg on drying, resulting in 1–1.5 kg of finished product per tree. Chironji is an almost evergreen, moderately sized tree with a straight, cylindrical trunk, up to a height of 10–15 m. The fruits of chironji mature in 4–5 months and are harvested manually in the season (months of April and May; Pandey, 1985). The green-colored skins of harvested chironji fruits turn black on storage, which has to be removed before shelling to remove the skin; fruits are usually soaked overnight in plain water and rubbed between palms or with a jute bag. The cleaned nuts are then dried in the sunlight and stored for further processing. At present, many machines have been incorporated for decorticating operations (Kumar *et al.*, 2012). Although the chironji nuts and kernels have been used extensively, the lack of a simple hand-operated machine forces the tribal community to decorticate these seeds manually. This results in the migration of tribal people to other business sources or they simply sell collected seeds to traders at a very low price. The entire business is taken over by wealthy businessmen who do the entire process with the help of motorized machines and sell them to the third market at more profit. It is difficult for the tribal people to afford those big machines. To assist them economically and socially, so that they can be self-employed and get good output in less time, there is a need to develop this machine. It is a simple machine with a pair of indented and extruded rollers with a tightening screw and a handle to rotate the rollers. Furthermore, the seed grader and kernel strainer are also attached to it.

2. Materials and Methodology

The number of points was considered during the design. This includes the cost of construction, being simple in operation, being light in weight, being safe to use, etc. Also considered in the design was the ease of replacement of parts in case of damage or failure. Various designs were thought for and studied before the decision. This required several brainstorming sessions, suggestions from experts, field and market surveys, etc. The set of two rollers as mentioned in the Figure 1 are fabricated on a lathe machine. The mechanism to rotate these rollers is designed and installed with a hopper to feed the un-decorticated seeds.

Figure 1: Roller mechanism.

It is made up of (1) a pair of rollers, (2) two sets of gears, (3) handle to rotate the gears and (4) a screw and spring arrangement to adjust the pitch of the gears according to the variations in the size of the seeds. The screw given in the machine is placed on the top of the plate (5) which slides inside the slot (6) provided. As the screw is adjusted it provides pressure to the plate as a result of which gear is moved towards the second gear and the pitch gets reduced. This action is supported by the spring attached between two rollers. When the screw is released, the spring regains its size and thus the roller comes out of the other, increasing pitch this mechanism is useful in decorticating various sizes of chironji seeds (5–9 mm; Omoruyi *et al.*, 2015) The design of rollers was a crucial part of the proper functioning of the machine. The plain cylindrical rollers were inefficient to decorticate seeds; hence the idea of cup-profiled rollers was adopted. As shown in Figure 2 the cup-shaped rollers can perfectly hold the seeds from its suture line and can easily decorticate it.

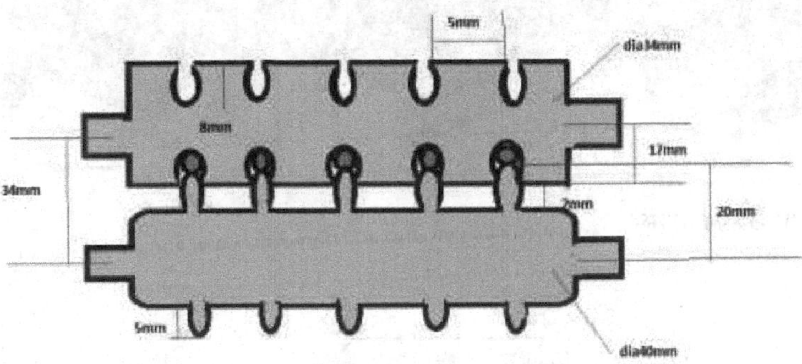

Figure 2: Grooved rollers.

The grooves are made of 8-mm deep and extruded grooves are of 5 mm. The extruded grooves travel 3 mm inside the 8-mm deep groove as a result of which 5 mm of working depth can be obtained to accommodate the various size of seeds which varies from 5 to 8 mm. The cup-shaped grooves hold seeds and grip them correctly so that they get decorticated easily. When the rollers are rotated with the help of gears, extruded portion of the rollers travels inside the deeply grooved portion of another roller breaking the seed shell neatly into two parts, thus wholesome seeds can be obtained. The percentage of crushed or partially broken seeds is less as compared to square-shaped grooved rollers.

3. Analysis

The finite element analysis was done by considering 50 N crushing force and torque of 1,300 Nmm. The following analysis is shown: nevertheless, this machine could apply to the tribal community at a very affordable price and can play a vital role in regaining the community's trust in this business. Fig 3 shows the ANSYS outcome of the CAD model

3.1. Chain Value Analysis of Chironji Seeds in Maharashtra Market

According to the chain value analysis, seeds from the tribal community are not sold directly to the retailer or third market as a result of which they suffer a loss. If the decorticating machine is provided, they can sell processed and decorticated seeds directly to the consumer gaining more profit thus assisting them socially and economically. Loss to the community per season per person as described in chain value analysis is expected to be Rs. 1,488 which is the number to be precisely affecting daily livelihood and emphasises the need for such a hand-operated decorticator. Figure 5 indicates clearly that how business has been accumulated to wealthy traders and tribal community is not benefited because non-availability of the simple hand-operated decorticating machine.

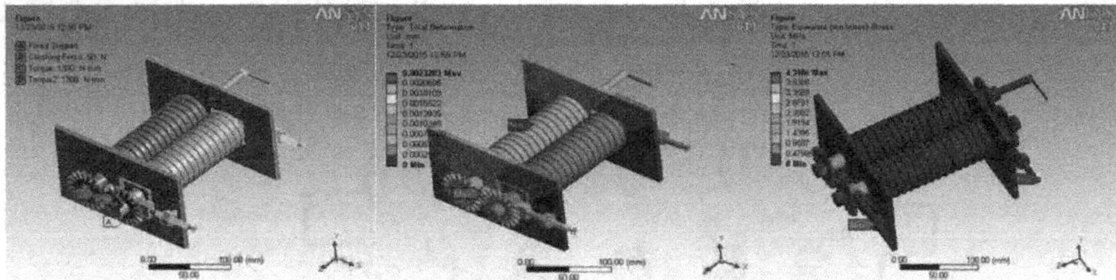

***Figure 3:** Ansys analysis.*

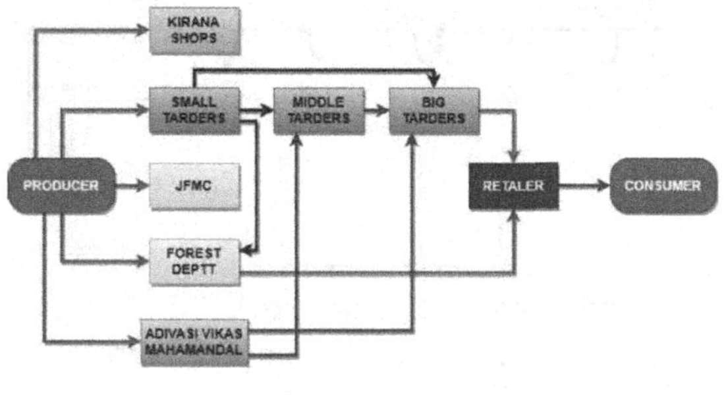

***Figure 4:** Value chain for Chironji (Source: Report on Mapping Livelihoods Value Chains in Maharashtra under project UMED, the Maharashtra State Rural Livelihoods Mission)*

After several trials on the machine, it is found that 65% of the seeds are getting proper decortication and hence further study is required to increase the efficiency of the product. Many times, seeds due to oil content within are getting stuck to the rollers and need frequent cleaning of the device during the operation. This manually operated chironji seed decorticator can be beneficial to tribal people in gaining more seeds and profit which can assist them socially and economically encouraging them to get their traditional business back. It is clear from Table 1 that the total average income per household can be increased from Rs. 125/kg to Rs. 937.50/kg and can be obtained if a proper machine is provided to the community.

Table 1: Income per household per season.

Details	Raw Chironji
Production/tree/season	5.5 Kg
Collection/HH/Day	1.5 Kg
Total Production/HH/Season	8 Kg
Quantity for domestic use	0.5 Kg
Total quantity sold/season	7.5 Kg
Avg rate/season	Rs. 125/kg
Total avg income/HH/Season	Rs. 937.50

4. Testing

The machine was tested for several runs of decortication operation. Every run carried out for 1 Kg of seeds and results were recorded. Average results are tabled as follows. It is likely to have more crushed seeds with roller gaps and likely to have un-decorticated seeds by keeping the roller gap higher. As per the results shown in Table 4, It was found that keeping roller gap 7 mm gave better results. Figure 6 explains the output results for nine different trial runs carried out with varying the roller gap from 5 to 8 mm. Table 2 and Table 3 below explains this in detail.

Table 2: Input parameters of decorticating/per run of 1 Kg seeds.

Sr No.	Input Parameters	Quantity
1	Weight of seeds introduced	1 kg
2	Total time required to decorticate	48 minutes
3	No. of times machine remained idle	4 times in one run
4	Idle time of machine	1.5 minutes per times/run
5	Actual time required	42 minutes

Table 3: Output parameters of decorticating/per run of 1 Kg seeds kernel with roller gap 7mm.

S.No.	Output Parameters	Quantity
1	Weight of kernel	1000 gms
2	Weight of shell	345 gms
3	Weight of fully decorticated seed	482 gms
4	Weight of partially cracked seed	75 gms
5	Weight of crushed seeds	125 gms
6	Weight of unbroken seeds	50 gms
7	Approximate efficiency of the machine	48.2 %

Table 4: Result of decortication keeping roller gap 7 mm.

Run No	Shell (GMS)	Decorticated Seeds (GMS)	Partially Cracked Seeds (GMS)	Crushed Seeds (GMS)	Unbroken Seeds (GMS)
2	377	450	70	94	49
3	336	454	71	97	53
4	364	436	64	98	47
5	358	470	69	92	47
6	366	444	69	94	43
7	326	461	76	93	58
8	347	469	70	97	53
9	367	482	69	94	49
10	369	471	72	93	49

5. Conclusion

The cracking rollers were found to be able to detach the shell from the seeds. However, the rollers could rupture the shell into smaller and irregular pieces. Using a mechanised system to achieve shell-free seed recovery directly from whole fruit can save production time and labour wages. Moreover, the design of the system is known to be highly productive, low-maintenance and easy to operate towards sustainability. With the study of chain value analysis, the design is modified and tries to increase the efficiency of the product. The machine, after several tests, is found to be 40.5% efficient, and more research on the topic is required to make it more efficient.

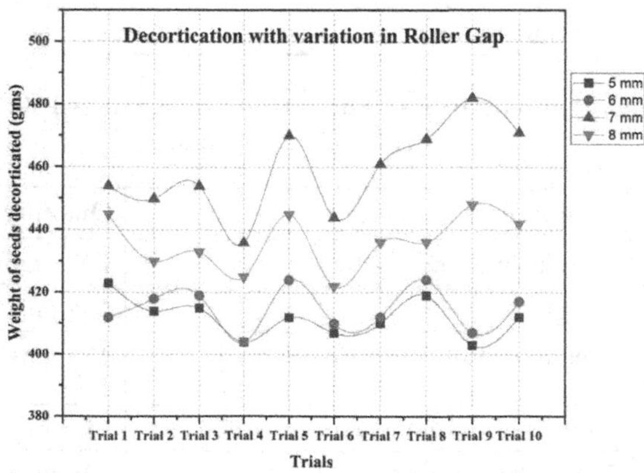

Figure 5: Decorticated seeds with variation in roller gap.

References

[1] Ghafari, A., Chegini, G. R., Khazaei, J., and Vahdati. (2011). Design, construction and performance evaluation of the walnut cracking machine. *Int. J. Nuts Relat. Sci.*, 2(1), 11–16.

[2] Hemavathy, J., and Prabhakar, J. V. (1988). Lipid composition of Chironji (Buchanania lanzan) kernel lipid classes and fatty acid analysis. *J. Food Compos. Anal.*, 370, 366–370.

[3] Jimoh, M. O. (2013). Design of an effective automated machine for quality palm kernel production. *IOSR J. Mech. Civ. Eng.*, 6(1), 89–97.

[4] Kumar, J., Vengaiah, P. C., Srivastav, P. P., and Bhowmick, P. K. (2012). Chironji nut (Buchanania lanzan) processing, present practices and scope. *Indian J. Tradit. Knowl.*, 11(1), 202–204.

[5] Lim, B. Y., Shamsudin, R., Hang, B. T., Baharudin, T., and Yunus, R. (2014). The performance of a Jatropha fruit shelling machine and the future improvement. *Univers. J. Appl. Sci.*, 2(7), 233–236.

[6] Maghsoudi, H., Khoshtaghaza, M. H., Minaei, S., and Zaki Dizaji, H. (2012). Fracture resistance of unsplit pistachio (Pistacia vera L.) nuts against splitting force, under compressive loading. *J. Agric. Sci. Technol.*, 14(2), 99–310.

[7] Omoruyi, A., and Ugwu, K. C. (2015). Optimization and performance evaluation of palm nut cracking machine. *Int. J. Sci. Res.*, 4(7), 646–653.

[8] Pandey, G. P. (1985). Effects of gaseous hydrogen fluoride on leaves of Terminalia tomentosa and Buchanania lanzan trees. *Environ. Pollution. Ser. An Ecol. Biol.*, 37(4), 323–334.

[9] Pradhan, R. C., Naik, S. N., Bhatnagar, N., and Vijay, V. K. (2010). Design, development and testing of hand-operated decorticator for Jatropha fruit. *Appl. Energy.*, 87(3), 762–768.

Chapter 49

Unveiling the Dynamics and Optimisation of Go-Kart Performance: An In-depth Analysis and Evaluation

Akshay Anjikar[1], *Manish Moroliya*[1], *Shailendra Daf*[1], *and Shailesh Khekale*[2]

[1]Assistant Professor, Department of Mechanical Engineering, Priyadarshini Bhagwati College of Engineering, Nagpur

[2]Assistant Professor, Department of Mechanical Engineering, Cummins College of Engineering for Women, Nagpur

Abstract: Designing a go-kart entails a multifaceted and intricate process, involving factors like safety, functionality, performance and maneuverability. This research project aims to develop a go-kart that excels in these aspects while adhering to competition rules, environmental considerations and cost constraints. The design group focuses on creating an aesthetically pleasing and ergonomic layout, ensuring a rigid and torsion-free frame, and optimising the powertrain for superior performance. The braking group emphasises a reliable and efficient braking system, while the steering group strives for a responsive and accurate steering system. The engine and transmission groups meticulously select and optimise the powertrain to reduce emissions and noise without compromising performance. Computer-aided design (CAD) software and finite element analysis (FEA) are employed to simulate performance, optimise the design and identify areas for improvement. Prototyping and testing are crucial to evaluate the design's effectiveness and identify any deficiencies. Continuous testing, evaluation and feedback from experts and industry professionals ensure a high-quality design that meets industry standards.

Keywords: Go-Kart analysis, powertrain optimisation, aerodynamic analysis, power loss minimisation, Computational Fluid Dynamics (CFD), crashworthiness, performance metrics, comparative analysis, data collection, Finite Element Analysis (FEA).

1. Introduction

Go-karts are popular recreational vehicles known for their compact size, speed and thrilling driving experience. Designing a go-kart involves a multifaceted and intricate process, encompassing various engineering aspects such as safety,

DOI: 10.1201/9781003527442-49

functionality, performance and maneuverability. This research paper aims to analyse and optimise go-karts, taking into account the findings from previous studies and utilising advanced analysis techniques.

The project focuses on the analysis of go-karts to identify areas for improvement and develop innovative solutions that enhance performance, safety and comfort. To ensure a comprehensive analysis, extensive research has been conducted on existing go-kart designs and their performance characteristics. By studying previous works and considering their strengths and weaknesses, this research aims to build upon existing knowledge and contribute to the development of superior go-kart designs.

Mehta *et al.* (2011) suggested that environmental factors, such as noise pollution and emissions, are crucial considerations in the design process. The engine and transmission groups are dedicated to selecting and optimising the powertrain components to minimise emissions and noise while maintaining high-performance standards. Balancing the trade-off between power and environmental impact is an essential aspect of go-kart analysis.

Tuck (2007) highlighted that the cost-effectiveness is another vital aspect considered during the design process. The project operates within a specified budget, necessitating careful evaluation of material and component costs. By striking the right balance between performance and affordability, the goal is to create go-karts that are competitive in the market without compromising on quality.

Dabhade (2009) elaborated that continuous testing and evaluation play a pivotal role in ensuring that each subsystem meets the desired objectives. Utilising computer simulations, data analysis and feedback from experts and industry professionals, the design group iteratively refines and adjusts the design to achieve optimal performance and safety standards. Adhering to industry regulations and standards is paramount to ensure that the final design meets the highest quality benchmarks.

Patil *et al.* (2014) briefed about the designing a go-kart is a complex and challenging task that demands meticulous attention to detail. The project encompasses a multidisciplinary approach, with dedicated groups responsible for different subsystems, including design, braking, steering, engine and transmission. Each group collaborates to integrate their expertise and insights, resulting in a holistic analysis and optimisation process.

By exploring various aspects of go-kart analysis, this research aims to contribute to the development of advanced go-kart designs that surpass current standards. The findings from this study will aid enthusiasts, manufacturers and industry professionals in enhancing go-kart performance, safety and overall enjoyment (Malianup *et al.*, 2014)

2. Chassis

Murray *et al.* (2004) suggested that the chassis is a fundamental component of a go-kart, providing structural integrity and support for various subsystems. The design and analysis of the chassis play a crucial role in determining the vehicle's overall performance, stability, and safety. This section will discuss the chassis design parameters and specifications.

Gianluca *et al.* (2008) briefed that to ensure strength and durability, the chassis is constructed using seamless tube AISI 4130, a high-strength alloy known for its excellent mechanical properties. The material selection is influenced by its ability to withstand the dynamic forces experienced during go-kart operation.

The wheelbase, which is the distance between the front and rear axles, is an essential parameter affecting the go-kart's handling characteristics. In this analysis, the wheelbase is set at 1020 mm (40.15 inches), ensuring a balanced weight distribution and stability during cornering.

The overall length of the vehicle is determined as 1625.6 mm (64 inches), providing adequate space for the driver and accommodating the necessary components without compromising the go-kart's maneuverability.

3. Methodology

The methodology employed in this research paper on go-kart analysis encompasses a comprehensive approach, integrating insights from previous studies and utilising advanced software tools for design and analysis.

3.1. Literature Review

Conduct a thorough review of existing literature and case studies to gain knowledge and insights into go-kart design and analysis. This step involves studying references that provide valuable information on hybrid go-karts by Mehta *et al.* (2011). Chassis design, super-karts, static analysis of go-kart chassis, compression ratio effects, previous go-kart projects, land speed record go-karts, stress analysis of automotive chassis, strength of materials and design of machine elements by Tuck (2007). This literature review serves as a foundation for the research, providing a comprehensive understanding of the subject matter.

3.2. Problem Identification

Identify practical problems encountered during dynamic events, such as the INTERNATIONAL GO KART CHAMPIONSHIP. This step involves studying the challenges faced by go-karts during high-speed maneuvers, cornering, braking and acceleration. By identifying these problems, the research aims to address them through the design and analysis process, enhancing the performance and safety of the go-kart.

3.3. Design

Utilise SOLIDWORKS 12.0 software to design the go-kart frame and other components. This step involves considering the rulebook guidelines, safety criteria for motorsports events and design principles established in the literature review.

The design process considers factors such as chassis geometry, material selection, component integration and ergonomic considerations.

3.4. *Analysis*

Perform a comprehensive analysis of the go-kart design using ANSYS WORKBENCH 14.5 and SOLIDWORKS SIMULATION software. These software tools enable simulations and analysis of various aspects, including structural integrity, stress distribution, deformation, vibration and performance optimisation. The analysis aims to validate the design, identify potential areas of improvement and optimise the go-kart's performance, safety, and durability.

4. Material Selection

Patil *et al.* (2011) explained that the material selection process for the go-kart frame design was a crucial aspect in ensuring the chassis met the technical requirements of the competition while maintaining optimal performance. Drawing references and the material chosen for the frame was AISI-4130 due to its excellent properties and suitability for the application.

AISI-4130 was selected for its favorable characteristics, including good weldability, softness and the ability to be strengthened. These properties make it an ideal choice for constructing the go- kart's frame, ensuring the necessary strength and durability.

To absorb energy during high-impact situations, the material needed to possess exceptional structural properties. AISI-4130 was chosen for its high strength-to-weight ratio, enabling the frame to withstand dynamic forces while minimising weight. For the frame design, a 1-inch diameter tube with a wall thickness of 2 mm was employed, ensuring the material dimensions met the required properties. The material properties of AISI-4130 were evaluated, and various parameters were found to be suitable for the application. These properties include the ultimate tensile strength, yield tensile strength, yield bulk modulus, shear modulus, modulus of elasticity, Poisson's ratio and percentage of elongation. The material's chemical composition was also considered, confirming its appropriateness for the chassis design.

5. Different View of the Go-Kart

In this Section, we present four distinct views of the go-kart, each captured from a different angle to provide a comprehensive understanding of its design and features. Figure 1 offers a frontal perspective, showcasing the go-kart's front-end details, including the steering mechanism and front wheel positioning. Figure 2 presents a profile view, allowing for a thorough examination of the go-kart's chassis and body design. Figure 3 highlights the rear aspect, emphasising the rear-end components, such as the engine, exhaust system and rear suspension. Finally, Figure 4 provides an overhead view, offering insights into the layout and arrangement of top-mounted features like the driver's seat and engine compartment.

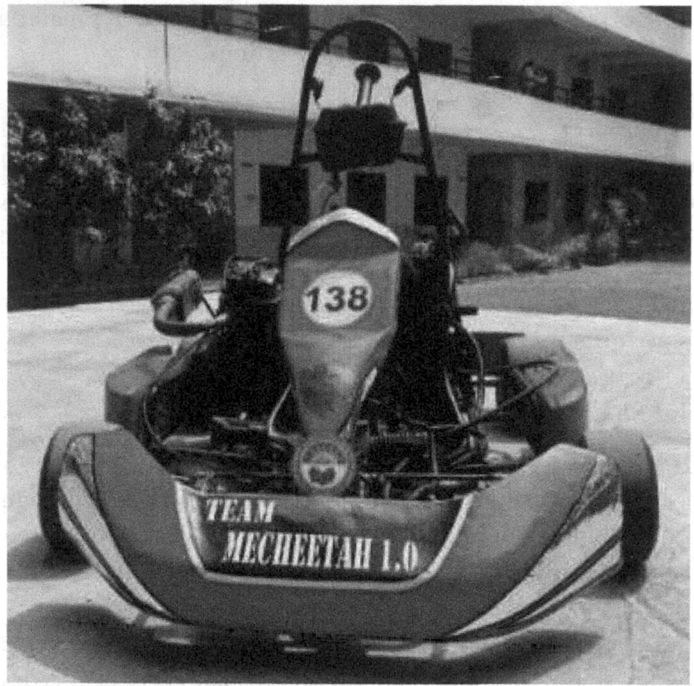

Figure 1: Front view of the go kart.

Figure 1 provides a frontal view of the go-kart, offering a clear insight into its design and features as seen from the front. This perspective highlights the layout of the steering mechanism, front wheel positioning and the overall visual appeal of the go-kart's front end.

Figure 2: Side view of the go kart.

Figure 2 displays a side view of the go-kart, giving a profile perspective of the vehicle. This view allows for an in-depth examination of the go-kart's chassis, body design and the arrangement of its components from a lateral angle.

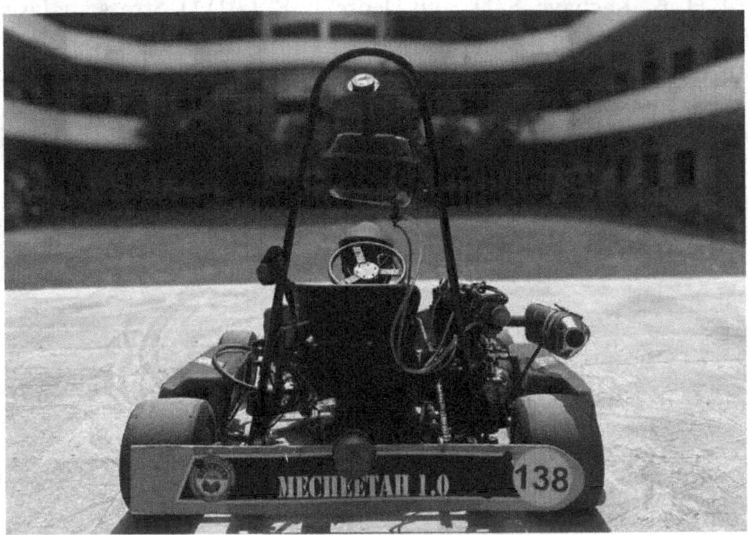

Figure 3: Back view of the go kart.

Figure 3 showcases the back view of the go-kart, emphasising its rear-end components and distinctive design elements. This angle might reveal details about the engine, exhaust system and rear suspension configuration.

6. Conclusion

In summary, this research paper concentrated on analysing a racing go-kart chassis using FEA. It aimed to enhance performance by addressing maximum deflection issues. The study successfully identified deflection and compared theoretical and simulated results, noting some disparities due to simplifications and real-world variations. It underscores the need to blend theoretical and practical knowledge for accurate results. The study has limitations, like assumptions, suggesting future research for refined models and experimental validation. Overall, this research contributes to advancing racing go-kart design through static and FEA, benefiting motorsports engineering.

References

[1] Dabhade, H. (2009). Design and fabrication of super–kart HD-250. 2(5), 209–212.

[2] Gianluca, M., and Marco, U. (2008). Design of a land speed record Go-kart. 222–225.

[3] Malianup, D., and Yadav, S. D. (2014). The effect of compression ratio on performance of 4-stroke spark ignition engine. 2(7), 405–410.

[4] Mehta, A. V., Padhiar, N., and Mendapara, J. (2011). Design and analysis of hybrid go-kart. 2, 277–288.
[5] Murray, J., and Lentine, T. (2004). Met 214 Go kart Project. 112–118.
[6] Patil, H. B., Kachave, S. D., and Deore, E. R. (2011). Stress analysis of automotive chassis with various thickness. 6, 86–92.
[7] Patil, N. R., Kulkarni, R., Mane, B. R., and Malve, S. H. (2014). Static analysis of go-kart chassis frame by analytical and solidworks simulation. 3, 99–104.
[8] Tuck, L. W. (2007). Design and fabrication of ump go-kart chassis. 102–108.

Chapter 50

Ergonomic Design of Visual Inspection Workstation Using Digital Human Modelling

Archimans Ray[1] *and Prasad V. Kane*[2]

[1]PG student, VNIT Nagpur, Nagpur

[2]Assistant Professor, VNIT Nagpur

Abstract: The empirical study of man-machine interaction is known as ergonomics. Ergonomics' main goal is to fit machine and man together in order to increase job efficiency, minimise stress and exhaustion at work. Ergonomics is extremely important in areas where manual activities directly impact an employee's overall health. This paper proposes a new design of a visual inspection workstation based on ergonomic evaluation in a pump production plant. For analysis, a questionnaire method is applied followed by an anthropometry check & measurement. Image analysis and rapid upper limb assessment (RULA) tools from Ergofellow software are used for evaluation and checking the effectiveness of the existing design and modification proposed design to improve it. The digital modelling of the manikin along with the modifications in the workstations is carried out. The RULA Score analysis is reduced from 5 to 2 in the modified workstation. The Lehman's test also shown significant energy reduction with the modified workstation. These modifications will have a good impact on the reduction of work-related musculoskeletal disorders (WMSDs) & would also help in maintaining healthy postures while working at the production line.

Keywords: Ergonomic assessment, WMSDs, digital human modelling, RULA, workstation design.

1. Introduction

Ergonomics is concerned with making the workplace as efficient safe and comfortable as possible. It is employed to improve employee efficiency, reduces stresses and fatigue by ergonomic workplace design to maintain a balance between worker characteristics and task demands (Bridger, 2008). Ergonomically designed workstations enhance the worker's productivity; provide safety, physical and mental well-being, and job satisfaction (Delleman and Dul, 2002). In developing countries, now industries are focusing on productivity improvement through micro and macro means (Kumar *et al.*, 2002). Ergonomics is also an area of improvement and hence it has gained importance in the last decades

DOI: 10.1201/9781003527442-50

(Makhbul *et al.*, 2022). Most industries are taking an interest in re-designing their workplace to improve the performance and efficiency of the workers (Kushwaha *et al.*, 2016). An ergonomically deficient workplace causes low productivity, poor quality of work, emotional stress and work-related musculoskeletal disorders in the workforce (Yeow and Sen, 2003). The review revealed that the workstation design was based on anthropometry and ergonomic considerations which became important input to carry this work. (Das and Sengupta, 1996; Kumari, 2018; Sauter *et al.*, 1991) This work is carried out in one of the automobile ancillary manufacturing plants in Rajasthan. They are involved in the manufacturing of fuel pumps. In the pump production plant, there are almost 446 workstations in the production line with 500–550 machine operators and workers. There are 13 different sections in total with 46 visual inspection (VI) stations. Due to the eight-hour shift and unfavourable working postures, the workers were facing musculoskeletal hazards. This work is focused on the ergonomic analysis and re-design of VI workstation. It was observed that the assembly line provided with the old design of the workstation and chair. It was found that the workers used to follow sitting and standing posture due to incorrect dimensions of the workstation. This causes more stress in their body during work. For these reasons, an evaluation of the VI workstations was carried out and modifications are suggested using the digital modeling of the workstation.

2. Methodology

The methodology followed in this work is depicted in the flow chart as shown in Figure 1.0. For the ergonomic evaluation of the workstation, proper data related to the issues faced by workers related to their WMSDs, the dimensions of the workstation and the human interaction involved with the workstation have been collected (Soares and Rebelo, 2016). Initially, the survey was carried out based on the questionnaire method and the related data was collected from the workers of the workshop. Based on the responses to the questionnaire, analysis of the severity of WMSDs and fatigue in body muscles is identified. The Lehman test was performed to understand the energy consumption of a worker in an eight-hour shift using the Ergofellow software. Ergonomic analysis tool rapid upper limb assessment (RULA) was applied, based on the image analysis of the body postures. The image analysis was done to identify the various angles of the joints of the body to analyse posture (Vyavahare and Kallurkar, 2012). The outputs of image analysis are applied as input to RULA which provides the ergonomic score indicating the ergonomic compatibility of the workstation in the current situation. It was observed that the sitting/standing posture needs to be changed to the sitting posture by modifying the workstation as per reach range rules. In this work, the existing and modified workstation is modelled in a digitised form with a human manikin in this virtual world. The results of the Lehmann Test and RULA are presented and discussed.

3. Ergonomic Analysis and Results

3.1. MSD Questionnaire Findings

The MSD questionnaire was designed to identify the occurrence of issues related to muscular due to workstation postural stresses. The questionnaire responses are plotted in Figure 3 where it is found that a significant number of workers reported the musculoskeletal pain they have been suffering from last 6 months. The cumulative plot has been shown in one time frame in Figure 3, while Figure 2 shows the occurrence of MSDs in different body parts in the last 7 days. The values indicated on the bar chart indicate the occurrence of the percentage of the operator who reported the particular MSD issue. From the above plots, it can be observed that the upper part of the body that is the shoulder and hands are the most affected body elements. Lehman test was applied using the Ergofellow software which gives the metabolism rate while a worker was working for a long period of time. In this test, the posture of working is needed for giving input in the Ergofellow software. Figure 4 shows that for 7 hours of manual labour-intensive working, metabolism consumed 630 kcal. The input assigned to this analysis for the existing workstation where the posture of operators was in standing position for a major portion of the working hours.

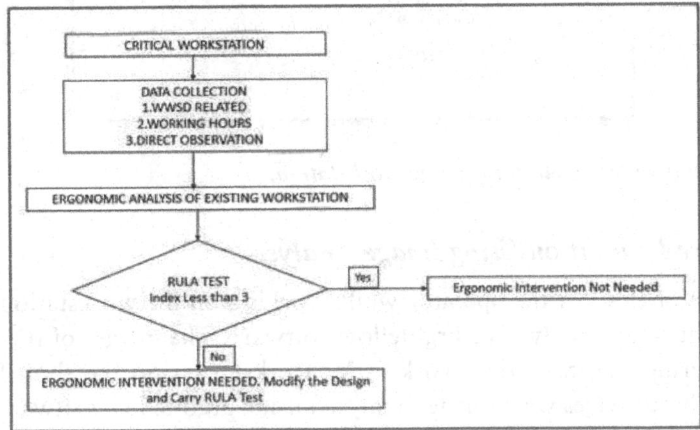

Figure 1: Work methodology.

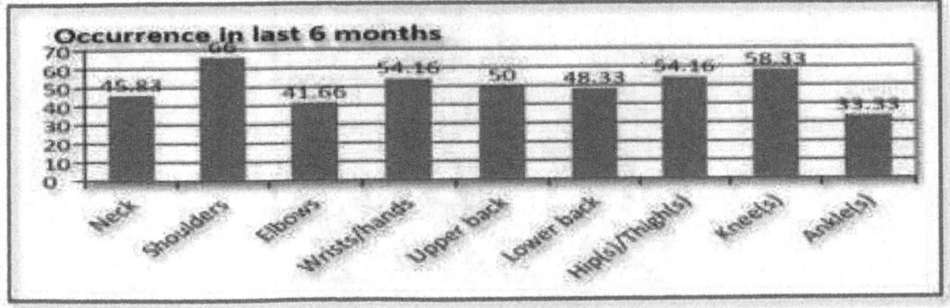

Figure 2: MSD occurrence in 7 days.

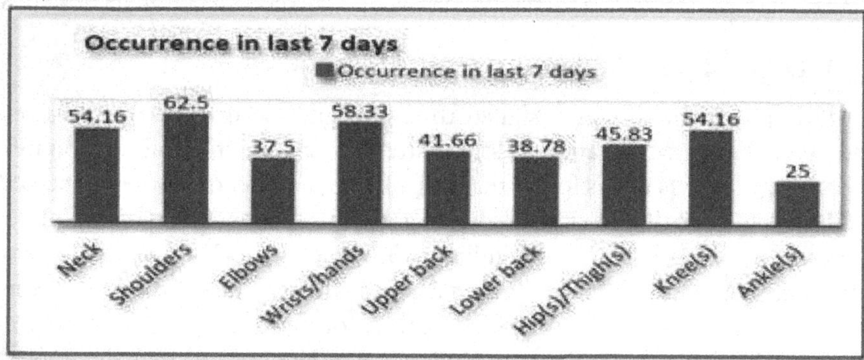

Figure 3: MSD occurrence in 6 months.

Figure 3 indicates the Lehman worksheet with the calculation behind the metabolism estimation.

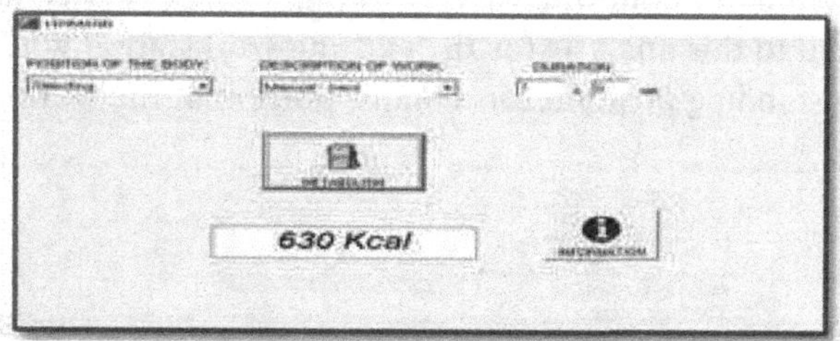

Figure 4: Lehman test result for existing workstation.

3.2. Posture Evaluation Using Image Analysis

The posture analysis of the operator while working on the workstation is carried out using the image analysis of Ergofellow software. The images of the operator's postures during the inspection work at VI workstation on the shop floor were captured. These images were further analysed using ergo-fellow software as shown in Figures 5 and 6. From this image analysis, the angles at the joints of arm, wrist,

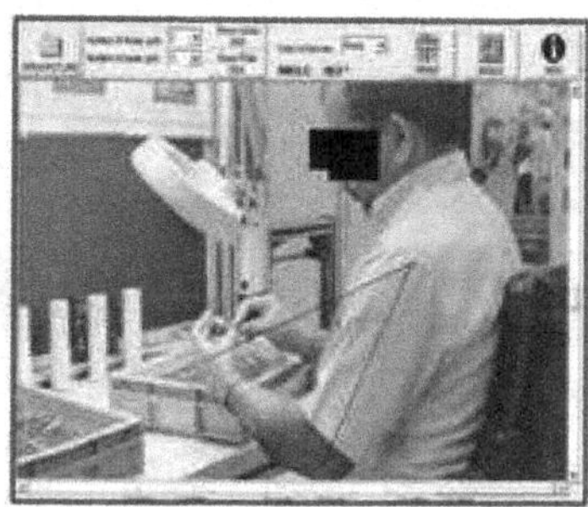

Figure 5: Image analysis of upper arm.

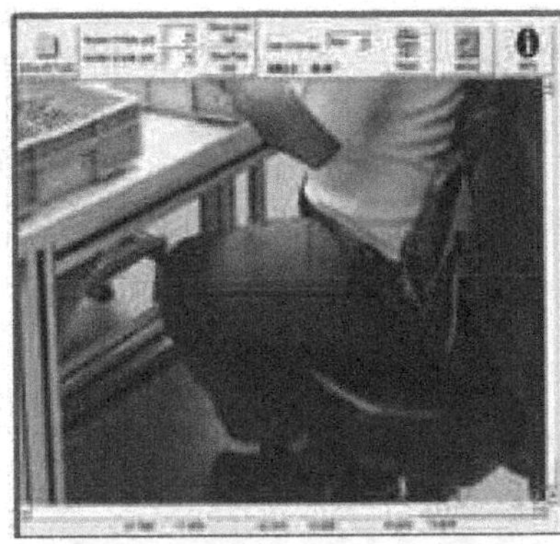

Figure 6: Image analysis of legs.

neck, trunk and leg were obtained. Figures 5 and 6 the module of the image analysis are used to obtain the angles of the upper arm and leg. The angles obtained from this analysis are mentioned in the Table 1.

Table 1: Final results of image analysis-joint angles of body posture.

Body Posture	Upper Arm	Lower Arm	Trunk	Wrist	Neck	Legs
Values (Degrees)	14°<X<32°	23±3°	6±2°	14°	18±2°	35°<X<62°

3.3. RULA Score-Evaluation

The image analysis revealed the working angles at the limb joints which are used as input in RULA for the operator working at the VI workstation and the methodology followed is similar to the methodology reported by other researchers (Ansari and Sheikh, 2014). Figure 7 shows the plugging in of upper and lower arm angles in the software module and Figure 8 shows the plugging in of the angles at the wrist and other joints.

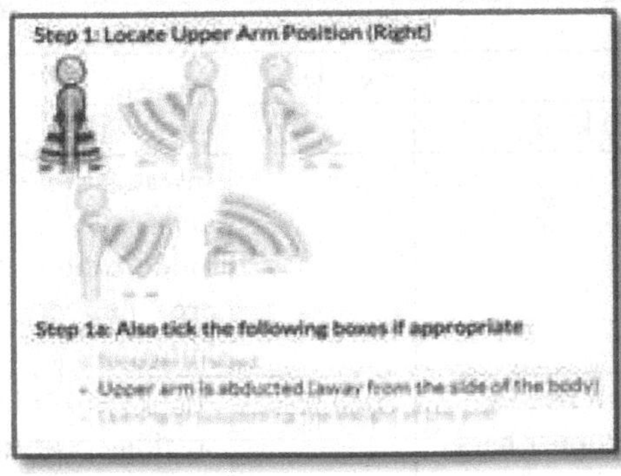

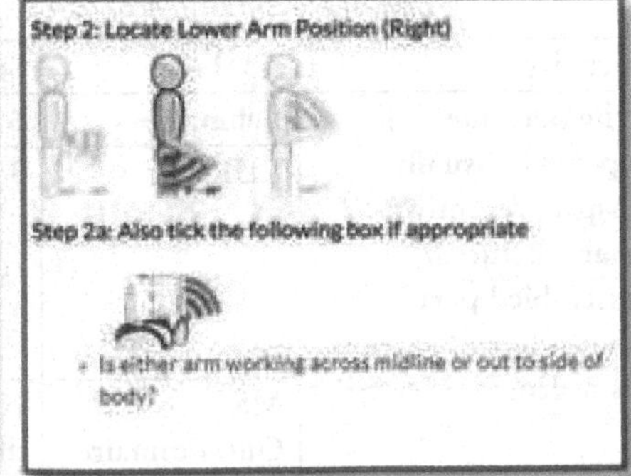

Figure 7: RULA test of upper & lower arm.

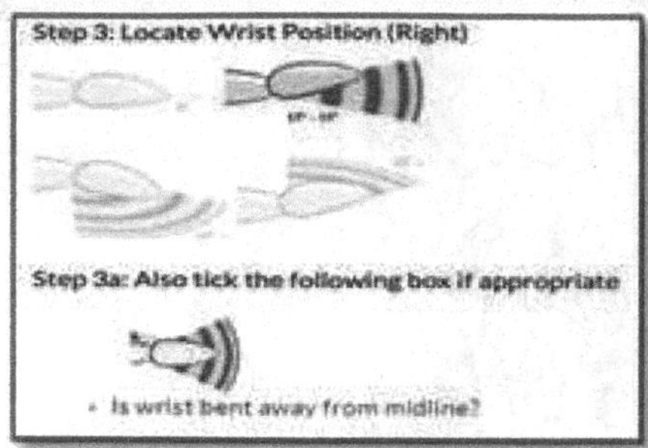

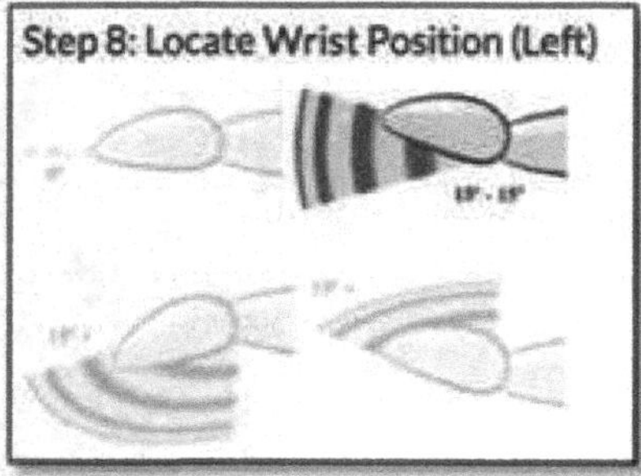

Figure 8: RULA test of wrist position test.

The RULA test has been carried out in two sections. One is for the right side and the other one is for the left side of the body. As shown in Figure 9, the average score of 20 workstations is estimated to be 5. This score indicates that the ergonomic score of the workstation-operator interaction is not acceptable and the operator is in danger of having MSD over the period of time and changes are needed in the existing workstation. The overall results of the ergonomic evaluation of the existing workstation are presented in Table 2.

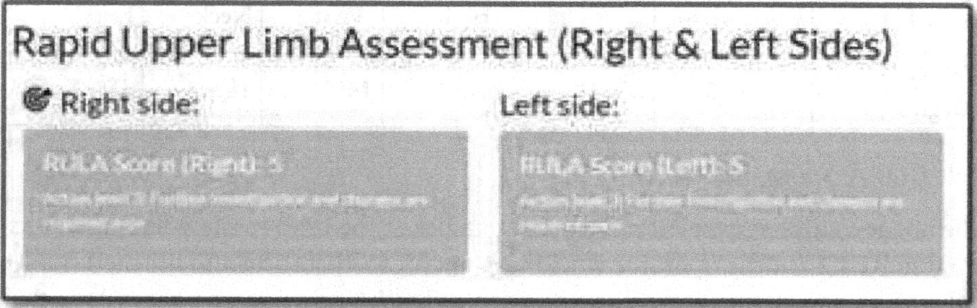

Figure 9: Final RULA score.

Table 2: Overall result of ergonomic assessment

Activity	Tool	Score	Risk level	Action
The machine operator visually inspect recently manufactured/ assembled part sincerely and mark it as accepted/rejected	Lehman	630 kcal	---	---
	RULA	5	High	Investigation and modification are required immediately
	MSD Questionnaire	High risk in the upper back, wrist, and neck	High	Improvement in working posture

4. Ergonomic Changes and Modification

4.1. Initial Design & Reach Range Problem

The existing design of the workstation was studied and modeled. It was found that in the design of that workstation, there was a reach range problem according to the reach range criteria of 50 percentile Indian manikins (Soares and Rebelo, 2016; Vyavahare and Kallurkar, 2012). Several tools, actuators, bins and buttons are falling out of reach range. These important tools are needed to relocate to bring them inside reach range.

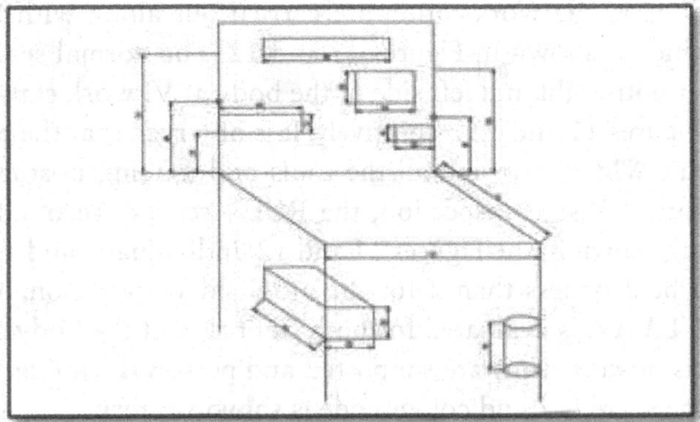

Figure 10: Existing workstation CAD model.

The layout of the existing workstation for VI was drawn in the CAD software which is shown in Figure 9. It was observed that due to the reach range problem, the chair provided in the workstation is not effectively used. The operator prefers to be in a standing posture as important tools to be used are outside reach range. It is inferred that the gauge toolbox and magnifying glasses toolbox should be brought within reach range. It was also felt that there is a need to relocate the rejection tray position and the water bottle holder position location so that provides more open space in legroom. The rejection tray and bottle holder's position must be kept inside the reach range. Thus posture of the operator could be changed from standing to sitting by modifying the workstation with the consideration of anthropometry.

4.2. Final Workstation Design and Changes

Figure 10 indicates the layout of the workstation drawn in CAD software. In the workstation, there are seven important tools in the VI workstation that are relocated during modification. Magnifying glass, rejection tray, gauge toolbox 1 & 2, book holder tray, instruction monitor and water bottle holder are these tools and actuators are taken into consideration and their locations in the VI workstation are rearranged and relocated. During modification, efforts are made to relocate in positions such that it would be inside the reach range for 50 percentile manikins.

4.3. Digital Human Modelling & RULA Test

Apart from the simple Ergofellow software analysis, the digital modelling of the workstation is carried out along with the digital manikin carried out (Das and Sengupta). Many researchers applied the digital modelling technique and reported that it is the most preferred tool along with Ergofellow software for ergonomic assessment and redesign of workstations. (Alipour *et al.*, 2021; Anghel *et al.*, 2014; Hussain *et al.*, 2019). Based on the anthropometry of the 50% manikin for Indian populations, the dimensions of the workstations are carried out and these changes are done for designing new workstations. A digital human model as well as a digital workstation are modelled with the consideration of anthropometric data. The modifications in the workstation are carried out along with the RULA test applied to it which is shown in Figures 11 and 12. The normal seating posture of the operator for both right and left side of the body at VI work station is modelled and shown in Figures 11 and 12 respectively. It is observed that the posture is static and intermittent. While accessing all the tools and making postures accordingly during operation of Visual Inspection, the RULA scores are obtained from this digital modelling shown in the Figures 11 and 12. Individually and as a total RULA score found to be 2 or less than 2 for the modified workstation. The ergonomic score of the RULA test is evaluated for both right side of the body and left side of the body. In this posture arms are supported and person is leaning. The posture is acceptable as the score is 2 and colour code is showing green.

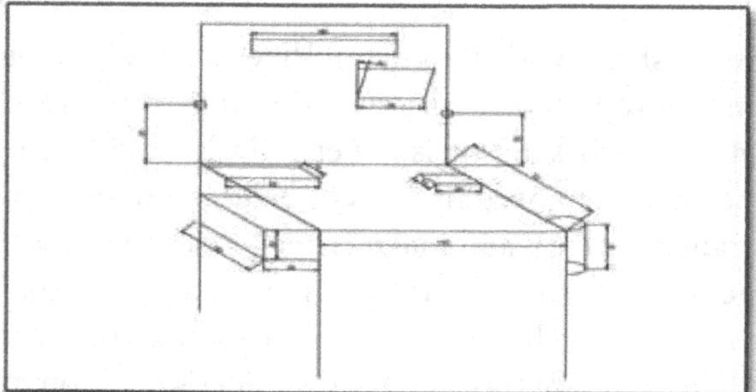

***Figure 11:** Modified sketch of workstation.*

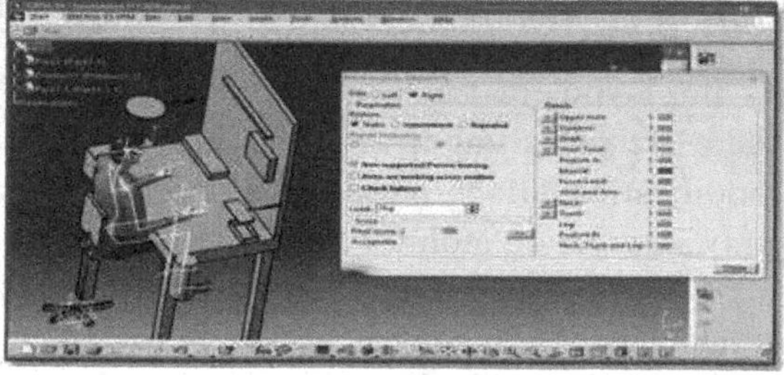

***Figure 12:** Ergonomic assessment in digital model (R).*

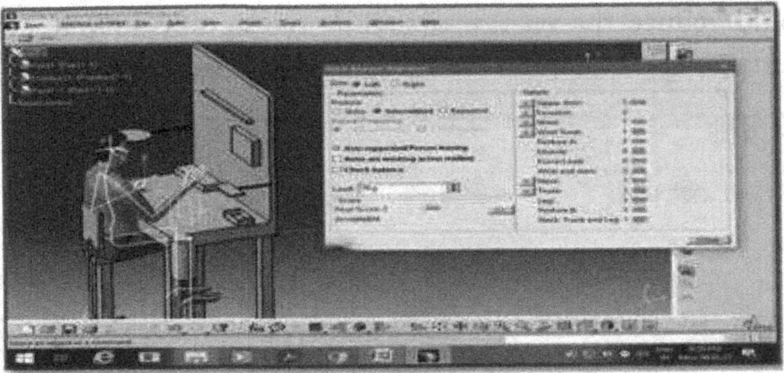

Figure 13: Ergonomic assessment in digital model (L).

4.4. Reach Envelope Analysis

The digital modelling software provides the module for the reach envelop analysis which has been applied by many researcher for analysing the modified workstation (Kushwaha and Kane, 2016). In this work, for the modified design of the workstation, the same analysis is carried out. The envelop analysis helped to evaluate the position of gauges, tools on it and dimensions of the chair based on anthropometry. A reach envelope is defined such that it indicates all of the possible positions that the manikin can reach using only the arm and forearm. Figure 14 shows all the bins, tools and actuators are within the reach zone.

4.5. Lehman Test

In the existing workstation, the operator of this VI workstation works in a standing position continuously for seven hours where all the tasks of work are manual and energy intensive. The Lehman test for this posture and for the duration of one shift is evaluated using the Ergofellow software and found to be 630 Kcal which was discussed in section 3.1. After the design modification of the workstation, Lehman test is again applied and the energy consumption is found to be 490 Kcal as shown in Figure 14.

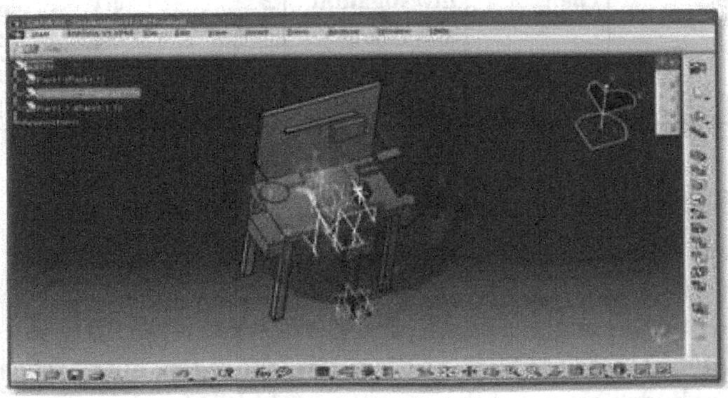

Figure 14: Reach envelope analysis of digital human model.

The change of posture of the worker from the standing position to the seating position would reduce the energy consumption significantly for the shift of 7 working hours. Figure 15 depicts the result obtained from the Lehman test model.

Figure 15: Lehman test result for the modified workstation.

4.6. Effectiveness of Modification

The effectiveness of the modifications of the design of the workstation is evaluated by applying ergonomic assessment tests and the results obtained are summarised in Table 3. The overall change in RULA score is found to be significantly reduced to value of 2. For the existing workstation, the RULA score was found to be 5 and after the re-design of the workstation based on the anthropometry and reach envelop analysis, the RULA score is reduced to 2 for both standing and sitting posture. Therefore, the proposed modifications would increase the productivity of the workers.

Table 3: Comparison of the effectiveness of the workstation design modification.

Tools	Before Modification			After Modification		
	Score	Risk level		Score	Risk level	
Lehmann	630 kcal	---	---	490 kcal	---	140 kcal
RULA	5	High	Investigation and changes required immediately	2	Very Low	Changes may be required in far future

5. Conclusion

Ergonomic audit was carried out for the VI workstation and inferred that it needs ergonomic intervention. The prevalence of work-related musculoskeletal disorders is understood by analysing the questionnaire. The MSD reported in the survey of the operators provided the motivation and input for the redesigning of the workstation. The analysis of body posture, anthropometry and MSD for the existing workstation is carried out. The existing workstation is assessed with the help of image analysis, RULA test and Lehman test in Ergofellow software. The existing workstation is analysed and modelled to know the necessity and importance of redesigning. Based on the anthropometry data, the work reach envelope analysis is carried out for the tools, actuators and bins used in the VI workstation by the worker. This analysis helped to understand the effectiveness of changes for near-reach range and far-reach range of the operator. It also indicates the effectiveness of the usage of tools in both the sitting and standing posture of workers. The RULA score was calculated using image analysis and energy consumption is obtained using the Lehman Test. The RULA score obtained in both sitting and standing posture after modification is reduced to 2 and the energy consumption is reduced. Hence the provision of an ergonomic workplace is a proactive step to improve the quality of work life on the assembly line of the production shop.

References

[1] Alipour, P., Daneshmandi, H., Fararuei, M., and Zamanian, Z. (2021).

[2] Ergonomic design of manual assembly workstation using digital human modeling. *Annals of Global Health*, 87(1).

[3] Anghel, D. C., Belu, N., and Rachieru, N. (2014). How to redesign ergonomic workstations, using neural networks and the Rula method in Catia V5. *Advanced Materials Research*, 1036, 995–1000.

[4] Ansari, N. A., and Sheikh, M. J. (2014). Evaluation of work Posture by RULA and REBA: a case study. *IOSR Journal of Mechanical and Civil Engineering*, 11(4), 18–23.

[5] Bridger, R. (2008). *Introduction to Ergonomics*: CRC Press.

[6] Das, B., and Sengupta, A. K. (1995). Computer-aided human modelling programs for workstation design. *Ergonomics*, 38(9), 1958–1972.

[7] Das, B., and Sengupta, A. K. (1996). Industrial workstation design: a systematic ergonomics approach. *Applied ergonomics*, 27(3), 157–163.

[8] Delleman, Nico J, & Dul, Jan. (2002). Sewing machine operation: workstation adjustment, working posture, and workers' perceptions. *International Journal of Industrial Ergonomics*, 30(6), 341–353.

[9] Hussain, M. M., Qutubuddin, S., Raja, K. P., and Kesava, R. Ch. (2019). Digital human modeling in ergonomic risk assessment of working postures using RULA. *Paper presented at the Proceedings of the International Conference on Industrial Engineering and Operations Management*, Bangkok, Thailand.

[10] Kumar, R., Banga, H. K., Kumar, R. S., Sehijpal, S., Scutaru, M. L., and Pruncu, C. I. (2021). Ergonomic evaluation of workstation design using taguchi experimental approach: A case of an automotive industry. *International Journal on Interactive Design and Manufacturing* (IJIDeM), 15, 481–498.

[11] Kumari, A. (2018). An ergonomic approach for modifying the workstation design of food processing enterprises. *Journal of Ergonomics*, 8(5), 1–5.

[12] Kushwaha, D. K., and Kane, P. V. (2016). Ergonomic assessment and workstation design of shipping crane cabin in steel industry. *International Journal of Industrial Ergonomics*, 52, 29–39.

[13] Makhbul, Z. K. M., Shukor, Md. S., and Muhamed, A. A. (2022).
[14] Ergonomics workstation environment toward organisational competitiveness. *International Journal of Public Health*, 11(1), 157–169.
[15] Sauter, S. L., Schleifer, L. M., and Knutson, S. J. (1991). Work posture,
[16] workstation design, and musculoskeletal discomfort in a VDT data entry task. *Human Factors*, 33(2), 151–167.
[17] Soares, M., and Rebelo, F. (2016). Ergonomics in design: Methods and techniques.
[18] Vyavahare, R. T., and Kallurkar, S. (2012). Anthropometric and strength data of Indian agricultural workers for equipment design: a review.
[19] Yeow, P. H., and Sen, R. N. (2003). Quality, productivity, occupational health and safety and costeffectiveness of ergonomic improvements in the test workstations of an electronic factory. *International Journal of Industrial Ergonomics*, 32(3), 147–163.

Chapter 51

Application of FMECA and FTA for the Engines of Diesel Locomotive

Biswajit Ghosh[1], Swapnil Gundewar[2], and P.V. Kane[3]

[1]Visvesvaraya National Institute of Technology, Nagpur, India
[2]Faculty of Engineering and Technology Sawangi, Maharashtra
[3]Assistant Professor, Visvesvaraya National Institute of Technology, Nagpur, India

Abstract: To be proactive and have continuous improvement of processes is at the core of all the public and private enterprises in this century for survive and growth. This work is intended to improve the existing maintenance process by implementing the tools of reliability centered maintenance (RCM) techniques, failure modes, effects and criticality analysis (FMECA) and fault tree analysis (FTA). These qualitative RCM techniques are implemented in the maintenance practices of engines at diesel locomotive shed to give additional insight into the conventional maintenance methods. The existing maintenance practices are based on the conventional periodic and preventive maintenance. This paper elaborates on identifying major risks using the standard FMECA technique to evaluate the risk priority number (RPN). The FMECA is applied by having a brainstorming session and it reports the occurrence, severity and detection possibility of failure. This analysis is further appended by FTA which provides the root causes of the failure using the deductive approach. The outcome of the work is to aid the maintenance crew to decide the priority of their maintenance actions thereby eliminating the possible failures along with the root cause analysis.

Keywords: Reliability-centered maintenance, FMECA, FTA, maintenance planning.

1. Introduction

The public sector and private sector industries have become conscious of the continuous improvement of all their processes (Kumar *et al.*, 2023; Liu *et al.*, 2013). Maintenance is one of the most important and critical functions in the public sector transport industry such as road, rail and air transportation (Cheliyan and Bhattacharyya, 2018). The railway transport sector has the most important maintenance activity for their loco engines, bogies, tracks and electrical systems. This paper elaborates on the efforts made to implement the RCM-based technique

DOI: 10.1201/9781003527442-51

at the diesel locomotive shed located in central India where the regular maintenance, repair and periodic overhaul of diesel locomotive engines is carried out. The existing system implements a preventive periodic maintenance schedule as recommended by the by research design and standards organisation. The existing maintenance process has the potential to improve by implementing the RCM technique. Implementing the tools such as FMECA and FTA. Efforts are made to improve the existing maintenance policy by providing insight into the decision of the prioritisation of maintenance activity. The root cause analysis implemented gives the maintenance department the idea to focus on the causes of the critical failures of the functions of the system and subsystems (Yazdi *et al.*, 2023). The FMECA is a tool to identify the risk priority number (RPN). This number is based on the occurrence of the failure, the ability to detect the failure and the consequences that is severity of the failure. These parameters are obtained by brainstorming with the engineers and workers. These parameters are rated on the scale of 1–10 which is discussed in the next section. The reliability centered maintenance (RCM) breaks the monotonous practice and introduces techniques that will be much faster and more efficient. For this very purpose in this work FMECA and FTA methods are used which are very specific and will help to know the proper causes and effects of the problems and the level of risks associated with them (Cristea and Constantinescu, 2017).

The FMECA applied for risk estimation by identifying the causes of failure and examining the possible causes of failure in the functionality of the system. It is applied to alleviate the failure by taking advantage of the early identification through brainstorming and anticipating the failure. It found its genesis in the 1940s when the U.S. military applied it to their systems and later found to be popular in industries. It was further widely applied in the aerospace and automotive industries. Narayanagounder and Gurusami (2009) reported the effectiveness of the application of the RPN to prioritise the maintenance and carry the corrective maintenance actions. In case of two RPN number is the same for two or three failure modes, the authors presented the application of analysis of variance (ANOVA) for deciding the priority of the maintenance action. Liu *et al.* (2013) and Xie (2013) discussed the limitations of the FMECA and FTA due to the subjectivity involved in it. Ahmed and Mahmoud (2014) reported the successful application of the FMEA in the construction projects. They have also presented how the FMECA can be implemented along with pairwise comparison and Markov chain. Josiah and Keraita (2018) reported the implementation of the RCM technique where they identified the failure modes of the systems of the corn milling plant. They reported that by applying the early detection method can reduce the possibility of failure thereby eliminating the possibility of plant shutdown. This improves the overall equipment effectiveness (OEE).

FTA is a root cause identification technique based on a top-down approach which is basically a deductive failure analysis. In this analysis, the top event indicates the failure state of the system using the Boolean logic to combine a series of intermediate events and the lower level events. FTA find applications for the detailed failure analysis and its elimination (Ansori, 2023; Cristea and Constantinescu, 2017; Kabir, 2017; Liu *et al.*, 2015; Yazdi, 2023). Liu and Yang (2015) presented the application of the FTA method along with a quantitative analysis technique to investigate the causes of high-speed railway accidents. Darwish *et al.* (2017) proposed a hybrid Bayesian-based technique to model with which the system designers could improve the quality of services that are related to active assisted living systems. Kabir (2017) reported the model-based dependability analysis (MBDA). Cheliyan and Bhattacharya presented a probabilistic failure analysis to study the leakage of the oil and gas in a sub-sea production system. They presented an application of FTA and fuzzy FTA. The fuzzy FTA application was intended to obtain the failure probabilities of the intermediate events and the basic events (Cheliyan and Bhattacharyya, 2018). The authors reported that this evaluation assisted in identifying the nature of the dependence of the top events on the basic and intermediate events. This analysis could identify the weakest links which is the cause of leakage in the sub-sea production system. Sihombing and Torbol (2018) presented a new algorithm by using parallel computing termed it a general purpose graphic processor unit (GPGPU) which attempts to solve a fault tree.

2. FMEAC and FTA Analysis

The methodology adopted in this work is shown by flow charts shown in Figure 1. For this work the data of past failures was collected for the duration of the last 12 years. This data was categorised based on failures that occurred in different engines. Out of 46 engines, 15 such engines were selected for detailed analysis which witnessed the frequent failures. In the methodology, the first step is to study the existing maintenance practices being performed and study the various components and systems. The second step is to collect the past failure and maintenance data. The data obtained is further segregated on the basis of which type of failure. Based on this data, the failures are divided on the basis of the system or subsystem affected by that particular failure.

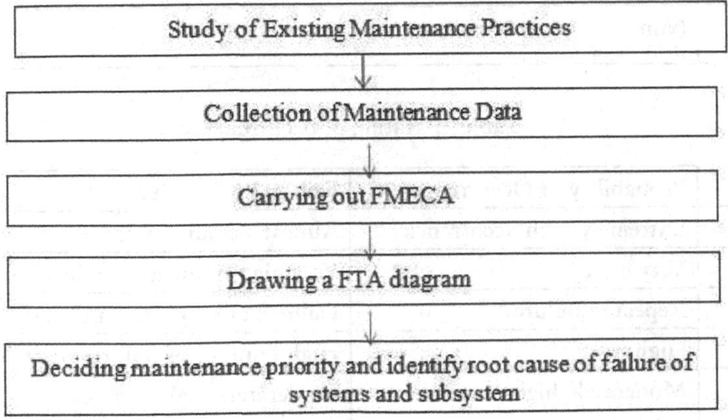

Figure 1: *Flow chart of methodology.*

The failure modes & effects analysis is carried out to identify the RPN associated with each of the components. Each system is further divided into subsystems, components and elements for identification of probable failure. The severity, occurrence and detection ratings of various probable failure modes of each system are determined. The severity, occurrence and detection of failure are obtained on a scale of 1–10 by having brainstorming sessions with the maintenance engineers, and operators of the maintenance division. Tables 1, 2 and 3 show the ratings given for the severity, occurrence and detectability, respectively.

Table 1: Rating of severity.

Ranking	Effect	Severity of Effect
10	Hazardous	Failure has hazardous consequences and it occurs without warning. The operations are halted and/or noncompliance with government regulations or standards.
9	Serious	Hazardous consequences and/or noncompliance with government regulations or standards.
8	Extreme	Primary function affected and the product is not operating.
7	Major	Partial functioning but product performance is severely affected.
6	Significant	Degraded Product Performance. The functions of comfort or convenience functions may not operate
5	Moderate	Product performance is moderately affected. Repair is needed due to the failure.
4	Low	Product performance or system performance is slightly affected. The product does not require repair.
3	Minor	The system performance has a minor effect.
2	Very minor	Product performance has a effected to a very minor effect.
1	None	No effect

Table 2: Rating of occurrence.

Ranking	Probability of Occurrence	Failure Probability
10	Extremely high occurrence	Almost certain failure
9	Very high	Very high number of failures occurrences
8	Repeated failures	Failures are repeated and easy to detect
7	High	High number of failures occurrence
6	Moderately high	Moderately high number of failures
5	Moderate	Moderate number of failures

4	Relatively low	Few occurrences
3	Low	Very few occurrences
2	Remote	Rare number of failures occurrences
1	Nearly impossible	Failure unlikely. History shows no such failure

Table 3: Ranking of detection.

Ranking	Detection	Detection Rating
10	Absolute uncertainty	Not able to detect a failure.
9	Very remote	Very remote chance to detect failure.
8	Remote	Remote chance to detect failure.
7	Very low	Very low chance to detect failure.
6	Low	Low chance to detect failure
5	Moderate	Moderate chance to detect the failure.
4	Moderately High	Moderately high chance to detect the failure .
3	High	High to detect the failure.
2	Very high	Very high chance to detect the failure.
1	Almost certain	It is certain to detect the failure at an early stage.

Air brake system is one of the most important system of the diesel locomotive. Table 4 shows the FMECA of air-brake system of an engine. Rating of severity is an expert elicitation.

Table 4: FMECA of air brake system.

Air compressor V-belt broken	Under braked	Fatigue	4	3	3	36	V belt to be changed
MR tank air pressure pipe uncouple and MR pressure	Reduce d flow	Pipe broken	6	5	6	180	Auxiliary reservoir delivery pipe to be changed
Throttle not respondin g	Reduce d flow	Pipe broken	5	6	4	120	Broken pipe to be replaced
Thread of pipe broken	Reduce d flow	Overuse	5	3	8	120	New Threading to be applied
Heavy smoke coming from air compressor	Can cause damage to compress or	Lack of lubrication	4	6	5	120	Proper lubrication done

Vacuum not getting developed	Under braked	Hole in vacuum pipe	7	4	3	84	Small hole in vacuum pipe to be repaired filled
Potential Failure Mode	Potential Failure Effects	Causes	S	O	D	RPN	Action
Auxiliary reservoir pipe broken	Harpedal	Delivery pipe broken	6	4	5	120	Auxiliary reservoir delivery pipe to be changed
Unusual sound coming from compressor	Rubbing	Friction	3	4	4	48	Loco to be checked
Loco not moving both side	Struck in braking position	Struck in non-braking position	3	4	4	48	Governor rubber diaphragm to be changed
Air compressor V-belt broken	Under braked	Fatigue	4	3	3	36	V belt to be changed
MR tank air pressure pipe uncouple and MR pressure	Reduce d flow	Pipe broken	6	5	6	180	Auxiliary reservoir delivery pipe to be changed
Throttle not respondin g	Reduce d flow	Pipe broken	5	6	4	120	Broken pipe to be replaced
Thread of pipe broken	Reduce d flow	Overuse	5	3	8	120	New Threading to be applied
Heavy smoke coming from air compressor	Can cause damage to compress or	Lack of lubrication	4	6	5	120	Proper lubrication done
Vacuum not getting developed	Under braked	Hole in vacuum pipe	7	4	3	84	Small hole in vacuum pipe to be repaired filled

The failure mode of auxiliary reservoir pipe broken is found to have the high RPN because if the reservoir pipe is broken then air will not fill in the reservoir that will not allow the train to move as train runs when the pressure is maintained, release of pressure causes braking action. Therefore, it is quite severe, however it does not occur very frequently and it can be detected by detecting from which coach or bogie the air has leaked. The brake pedal will become hard. Due to fatigue, the V-belt of air compressor may get broken due to which the braking action will not be proper. However, it can be rectified by changing the belt so the risk associated is less. Sometimes when the pipe breaks the throttle does not respond properly, which causes reduction in flow of air. The severity is moderate and occurrence is frequent but can be detected with some effort. It can be rectified by replacing the broken pipe. Lubrication in running equipment is very necessary, lack of which can cause catch fire in extreme case and sometimes heavy smoke comes from air-compressor which found to be little frequent; however, the severity and detectability is moderate. The FTA of the braking system is also carried out and the fault tree diagram is shown in Figure 2.

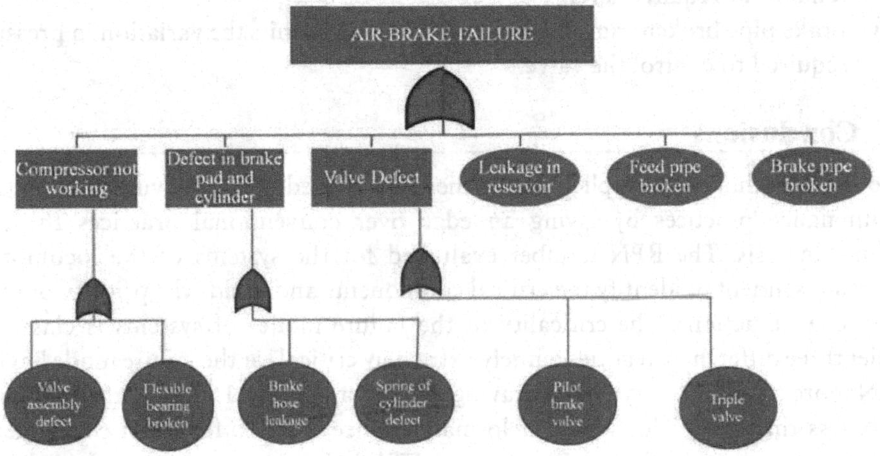

***Figure 2:** Fault tree analysis of air-brake system.*

Figure 2 shows the FTA of air-brake System. The following are the major causes of failure of air-brake system and if one of them occur then the whole system will fail and therefore there is "OR" relation with system. The analysis of the various subsystems involved in FTA is as follows

i. Compressor not working: If there is any failure in compressor, then it will not compress air for the system. The air-brake system requires compressed air for maintaining the train in running position, for braking action it releases compressed air, and hence compressor plays a vital role in air-brake system. There are mainly two reasons for failure of compressor that is defects in valve assembly and due to breakage of flexible bearing. Any of these two failures cause failure of compressor.
ii. Damaged brake pad and cylinder: Brake pads are steel backing plates with friction material bound to the surface that faces the disc brake rotor. A brake cylinder is located in the top of each wheel above the shoes. It exerts force onto the shoes to bring them in contact with drum so as to stop the train. So,

both of these components are very critical for the functioning for air-brake system. The reasons for damaged brake pad and cylinder are leakage in brake drum and defect in the spring of cylinder.

iii. Valve defect: The valve controls the flow of air from reservoir to the brake. The main reasons for valve defects are breakage of pilot valve (controls high-pressure or low-pressure feed and triple valve (controls the automatic air brake by regulating by three openings the intake, exhaust, and the equalisation of compressed air in the brake piston).

iv. Leakage in reservoir: There are two types of reservoirs in air-brake system, main reservoir and auxiliary reservoir. If there is air leakage from the reservoirs then brakes will not be applied.

v. Feed pipe broken: It feeds air to the brake pipe and ensures brake pipe pressure remains at required level.

vi. Brake pipe broken: runs the length of train, transmits the variation in pressure required to control the valve.

3. Conclusions

The RCM technique is applied in the diesel loco shed for improving the existing maintenance practices by giving an edge over conventional practices through failure analysis. The RPN number evaluated for the systems of the locomotive helps department to identify the critical components and decide the priority of their maintenance actions. The criticality of the failure modes of systems is classified under three different categories namely extremely critical for the failure mode having RPN more than 150, very critical having RPN from 100 to 150 and critical having RPN less than 100. This would help maintenance team to focus on components or failure modes with high RPN number. FTA is implemented and the air-brake system is presented in this paper. This FTA criticality evaluates the root causes of failure of any system which depends on its sub-systems and their sub-systems in a hierarchy. FTA analysis would give insight to the maintenance team, how the top event of fault tree diagram is linked with different subsystem. The detailed analysis would help maintenance team to put various sensors, pressure gauges, redundant components at appropriate locations to reduce RPN number. Thus, FMECA and FTA are powerful tools to improve maintenance performance based on risk assessed. Application of these reliability centred maintenance technique is a proactive step to improve system performance.

References

[1] Ahmed, and Mahmoud, A. M. M. (2015). Composite FMEA for risk assessment in the construction projects based on the integration of the conventional FMEA with the method of pairwise comparison and Markov chain.

[2] Ansori, I., Dwitya, H. W., and Mutharuddin, M. (2023). Enhancing brake system evaluation in periodic testing of goods transport vehicles through FTA-FMEA risk analysis. *Automotive Experiences*, 6(2), 320–335.

[3] Cheliyan, A. S., and Bhattacharyya, (2018). Fuzzy fault tree analysis of oil and gas leakage in subsea production systems. *Journal of Ocean Engineering and Science*, 3(1), 38–48.

[4] Cristea, G., and Constantinescu, D. M. (2017) A comparative critical study between FMEA and FTA risk analysis methods. *IOP Conference Series*, 252, 012046. https://doi.org/10.1088/1757-899x/252/1/012046.

[5] Darwish, D. L. F., Molham, and Almouahed, S. (2017). The integration of expert defined importance factors to enrich Bayesian fault tree analysis. Ideas. Repec.Org.

[6] Kabir, S. (2017). An overview of fault tree analysis and its application in model based dependability analysis. *Expert Systems With Applications*, 77, 114–135.

[7] Kumar, S., Bhatkulkar, T., and Kane, P. V. (2023). FMEA and FTA of coal handling system of power plant. *Materials Today: Proceedings*, 90, 197–200. https://doi.org/10.1016/j.matpr.2023.06.012.

[8] Liu, H. C., Ling, L., and Nan, L. (2013). Risk evaluation approaches in failure mode and effects analysis: a literature review. *Expert Systems With Applications*, 40(2), 828–838.

[9] Liu, P., Lixing, Y., Ziyou, G., Shukai, L., and Yuan, G. (2015). Fault tree analysis combined with quantitative analysis for high-speed railway accidents. *Safety Science*, 79, 344–357.

[10] Narayanagounder, S. (2023). A new approach for prioritization of failure modes in design FMEA using ANOVA. https://publications.waset.org/6947/a

[11] new-approach-for-prioritization-of-failure-modes-in-design-fmea-using-anova.

[12] Sihombing, F., and Marco, T. (2018). Parallel fault tree analysis for accurate reliability of complex systems. *Structural Safety*, 72, 41–53.

[13] Xie, L. (2013). A new method for failure modes and effects analysis and its application in a Hydrokinetic Turbine System. Scholars' Mine, n.d. https://scholarsmine.mst.edu/masters_theses/7124/.

[14] Yazdi, M., Javad, M., He, L., Hong, Z. H., Zarei, E., Reza, G. P., and Sidum, A. (2023). Fault tree analysis improvements: a bibliometric analysis and literature review. *Quality and Reliability Engineering International*, 39(5), 1639–1659.

Chapter 52

Analysis of Location of Emergency Department in Multi-speciality Hospitals

Shailesh Narrayanrao Khekale[1], Ramesh D. Askhedkar[2], Rajesh Parikh[3], Yogesh Dandekar[4], Sushi Lanjewar[4], Aditya Kawadaskar[4], Mahesh Shukla[5], Manish Moroliya[6], and Akshay Anjikar[6]

[1]Assistant Professor, Mechanical Engineering Department, Cummins College of Engineering for Women, Nagpur

[2]Principal (rtd.), K.D.K. College of Engineering, Nagpur

[3]Professor, K.D.K. College of Engineering, Nagpur

[4]Assistant Professor, Cummins College of Engineering for Women, Nagpur

[5]Associate Professor, Cummins College of Engineering for Women, Nagpur

[6]Assistant Professor, Priyadarshini Bhagwati College of Engineering, Nagpur

Abstract: Health care system plays very important role in the general welfare of society. The emergency department (ED) is a core clinical unit of a hospital. The location of the ED in the hospital is very crucial to provide faster treatment to the critical patients. Proper location of ED in the hospital helps to minimise the time required to reach the patient to the ED. After analysing the location of the ED in different hospitals in Central India, some drawbacks were found out. Time study and observations were used to measure the problems. Corrective measures to remove these drawbacks were also worked out, these measures were recommended to the hospital administration and some of them were implemented which helped the patient as well as caretakers and reduced patient shifting time from ED to other important departments when every second is crucial in saving life of the patient in critical condition.

Keywords: Emergency Department (ED), layout, patient, patient shifting.

1. Introduction

Health care system plays very important role in the general welfare of society. An emergency department (ED) is a facility, which specialises in the critical emergency care of ill health patients who report without a prior appointment. The performance of ED affects the performance of the hospital to a larger extend. Steptoe *et al.* (2011) defined ED by many perspectives like academic, hospital and patient.

DOI: 10.1201/9781003527442-52

Objectives of ED are to provide prompt treatment (which should be started immediately within golden time) to patients in critical condition and the time period required to stabilise the patient. Also the length of stay of the patient in the ED should be minimised so that the patient can be shifted speedily to the ICCU/ICU for achieving further improvement. EDs is intended to provide rapid access to essential care for acutely ill patients (Schull *et al.*, 2003). As per National Health Profile of India 2009, trauma emergencies are listed as the 3rd leading cause of death in India. Total societal cost of injury in India to be approximately 3% of India's GDP as per stated by recent calculations by the Planning Commission of India. (Report of the Working Group on Emergency Care in India, 2013)

The location of the ED in the hospital is very crucial to provide faster treatment to the critical patients. Proper location of ED in the hospital helps to minimise the time required to reach the patient to the ED. The proper layout of the ED minimises unnecessary movements of doctors and other health care personnel and helps in providing medical help to patients in the shortest possible time. Serious patients and accidental cases should get treatment in minimum time. Therefore, the location of ED should be near to the entry gate of the hospital and easily accessible.

2. Literature Review

In health care, the implications of improper location decisions affect largely well beyond cost and patient service considerations. If too few facilities are utilised and if they are not located properly, affect the mortality (death) rate and morbidity (disease) can result (Daskin *et al.*, 2010). Facilities layout represents most significant opportunities for expanding and improving the performance of any system including healthcare. Khadem *et al.* (2009) evaluated revised layout of the ED of public hospitals with 75% reduction in average waiting time of patients and 10% per month increase in the capacity of the department. Khadem *et al.* (2009) showed the development and application of a high-level methodology for simulating and optimising hospital designs by considering the elements like hospital location and typology, patient needs and goals, processes, staffing, hospital resources and building design. How fast the patients will be placed in beds and be seen by a casualty medical officer or doctor depends on good location and physical layout of the ED and it is also necessary to accelerate patient flow which depend on how promptly the patient can be registered and get triaged (Hall *et al.*, 2007). It is found that, very small work has been done to associate the hospital layout and design with hospital operational efficiency and effectiveness (Bacon, 2012). Andrea *et al.* (2022) analysed the ED in various contexts like functional layout, structural, technical features, design features and amenities. Aygün *et. al.* (2021) studied ED architecture, it's space requirements, location and general design factors of EDs.

3. Methodology

The methods used in the study included an understanding of the system by observations of different EDs in multi-speciality hospitals located in central India. Various multi-speciality hospitals with ED facilities were visited and studied. Interviews with patients, their caretakers, healthcare professionals including casualty medical officer in ED, intensivists, specialists, hospital administrators, nursing staff and ward boys were conducted in these hospitals. Interviews were based on the norms as per Indian Public Health Standards (IPHS).

IPHS for the location of ED are about clearly identification of ED from all entrances, location of ED on the ground floor for the ease access, separate entry for the ED other than main gate, easy ambulance approach and adequate space for free passage of vehicles, covered area for arriving patients, signboards to ED should be displayed at all the entry gates of the hospital with directions arrows and covered parking should be available for appropriate number of ambulances. On the basis of application of this methodology in various multi-speciality hospitals in Nagpur, Maharashtra, various problems faced by hospitals as well as by patients were identified. Out of these hospitals, six prominent hospitals were concentrated for problems for location of ED. Time study was conducted to understand the actual time consumed for shifting of patient to ED due to its improper location. Layouts of these hospitals were studied with location of ED. Modified location was recommended to hospital authorities. Simulation methodology was used to elaborate the problem of long travelling distance of the patients to reach the ED from entrance gate of hospitals. After implementation of recommendations about new location of ED, time study was again conducted and with the help of simulation methodology, improvement in time saving was achieved.

3.1. Observation and Time Study

It was observed that the location of ED in multi-specialty hospitals violated the norms prescribed by IPHS. Location of ED in some hospitals was away from the diagnostic centre and ICCU/ICU ward. It was not identified properly by signboard and did not have separate entry gate. Parking space for ambulance in front of ED for shifting the patient was not proper and sufficient. Patient shifting area from ED to other departments and from ambulance to ED was not aerially covered. The patients had to search ED and lose substantial time to reach ED. Patients caretakers had to move extra distance for registration because registration counter is away from the ED.

The location and layout of the ED were studied in various multi-speciality hospitals in Central India. Out of these hospitals, six prominent hospitals in Nagpur were concentrated for study and critically analysed. Figure 1 shows layout of ground floor with ED in one of the selected multi-speciality hospital.

Figure 1: Layout of ground floor with ED in multi-speciality hospital.

That selected multi-speciality hospital is spread over 1.5-acre area. It has three blocks, Block I, II and III. Block 1 consists of CCU-III, physiotherapy and pharmacy at the ground floor, CCU-II at the first floor, CCU-I at the second floor, operation theatre at third floor and deluxe ward at the fourth floor. Block II has four floors. Block II consists of reception, account section, diagnostic centre and semi-private wards (SPW) at the ground floor, general ward and operation theatre at the first floor, male and female general wards at the second floor, administration office and ICCU–IV at the third floor. Block III has three floors. Block III consists of ED and PRA office at the ground floor, paediatric ward and gynaecology at the first and second floors, respectively.

The ED's layout is shown in the Figure 2. The ED consists of patient admission room, minor operation theatre (OT), emergency critical care unit, personal relational assistant (PRA) office, area for stretcher, wheel chair, nursing station, casualty medical officer's (CMO) office, store room, doctor's resting room and space for patient caretaker. ED has total seven beds, three in admission room, and three in CCU and one in minor OT.

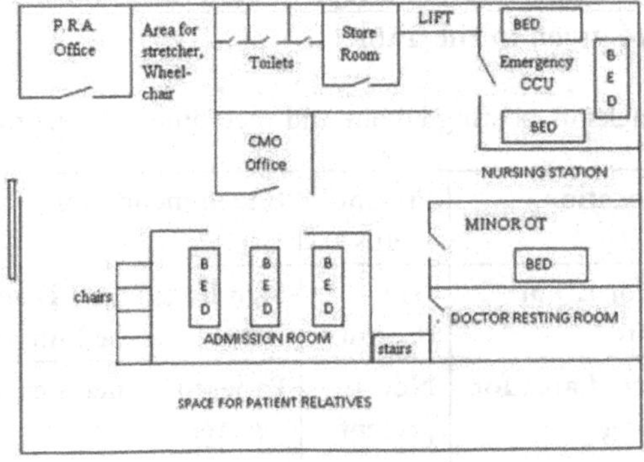

Figure 2: Layout of the ED.

Patients had to be shifted from ED to another departments including CCU- I to CCU- IV, diagnostic centre, male/female general ward, SPW, deep ward and spatial ward. The distance in between ED and another departments and time required (calculated by time study) for these patients shifting activities are given in the table 1. Time study was carried out with 180 samples. Standard deviations of respective elements were calculated. Registration counter is also away from the ED.

Table 1: Distance during shifting of patients between ED and other departments (Walking distance and average time require)

ED	Destination	Distance in meter	Time required average(second)
ED	CCU- III (Old building, ground floor)	35	70
ED	CCU-I, CCU-II (Old building, I and II floor)	50	130 and 170
ED	CCU-IV (Old building, III floor)	57	120
ED	Diagnostic Centre (New building, Ground floor)	40	90
ED	Male/Female general ward(New Building, II floor)	40	120
ED	Semi-private ward	33	70
ED	Deep ward (Old building, Backward area, ground floor)	57	120
ED	Spatial ward (Old building, III floor)	41	120

From the Table 1, it was observed that the distances of ICU and Wards form ED were large and shifting of patient took 1–2 minutes. For the sake of saving the life of the critically ill health patient, every second plays important role.

4. Recommendations

After analysing the location of the ED, some draw backs were found out. Corrective measures to remove these draw backs were also worked out, these measures were recommended to the hospital administration. These drawbacks and recommendations are given in the Table 2.

Table 2: Drawbacks in existing layout and recommendations for improvements.

Sr. No.	Norms for location of ED	Present Status	Recommendations
1	Sign Board for ED at Entrance Gate	Not present	Sign Board for ED should be located on left side of the Entrance Gate
2	Aerially covered area for patient shifting	Not present	Patient shifting area should be aerially covered.
3	Separate Entry gate for ED	Not present	Separate gate should be at leftside of present gate

4	**Parking for Ambulance and private vehicles for patient caretaker**	**Not present**	**Due to less available space, parking area cannot be enlarged.**
5	**Place of ED is near the Diagnostic centre**	**Not present**	**ED should be shifted to position of Semi-private ward (SPW) at ground floor of Block II**

As per recommendations, ED was shifted to place of SPW at ground floor of block II and SPW to place of ED at ground floor of Block III which is shown in Figure 2. Due to recommended location of ED near the diagnostic centre and ICCU or wards, the patient's shifting distance and time are reduced. New gate is located at leftside of present one, will help to provide separate entry for critical patients and separate space for patients coming from ambulance or private vehicle for ED and other visitors should enter in the hospital only by old one. Sign board is at left of the new gate, helps the patient to find the location of ED very easily. During rainy season patient shifting from ambulance to ED is difficult, therefore patient shifting area is aerially covered. Parking for ambulance and private vehicles for patient caretaker is made available. The proposed changes including new location of ED, separate entry gate for emergency patients with position of signboard are shown in the Figure 3. Figure 2 shows modified layout of ground floor with ED in multi-speciality hospital.

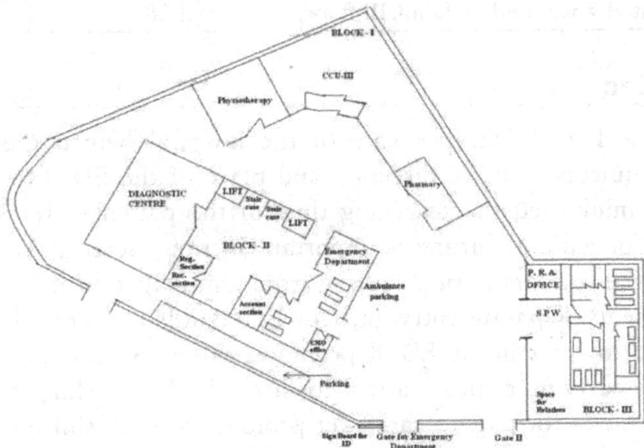

Figure 3: Modified location of ED in layout of hospital.

Modified layout decreased the patient shifting distance and respective time from ED to different departments and from these departments to ED. Reduced distances are given in the Table 2.

5. Implementations and Result

All recommendations in Table 3 were implemented. Time study was again carried to find the reduced time after applying modified layout shown in Figure 2. For time study 80 samples were collected with standard deviations. Modified layout reduced the distance of all important units and facilities from the ED. Shifting distance of patients between ED and other departments (Walking distance) were reduced and shown in Table 3. Improved time specifically for critical patients requiring CCU admission was shown in Table 3 which was satisfactory result of this project.

Table 3: Reduced distance and average time during shifting of patients between ED and other departments (Walking distance).

ED	Destination	Distance in Meter	Time Required (second)
ED	CCU- III (Old building, ground floor)	18	20
ED	CCU-I, CCU-II,(Old building, I and II floor)	24	45 and 90
ED	CCU-IV (Old building, III floor)	28	60
ED	Diagnostic Centre (New building, Ground floor)	4	30
ED	Male/ Female general ward(New Building, II floor)	10	100
ED	Semi-private ward	33	70
ED	Deep ward (Old building, Backward area, ground floor)	27	100
ED	Spatial ward (Old building, III floor)	28	120

6. Discussion

Sign board for ED at entrance gate of the hospital helped the patient and caretakers to quickly find the direction and place of the ED. The path showed (layout) in it minimised the searching time of the patient caretakers. Aerially covered area for patient shifting is important in rainy season. Aerially covered area protects the patients during shifting from ambulance to ED or from ED to other departments. Separate entry gate for ED avoids any arrival problems for the ambulance to enter in the ED. It promotes safe entry for the ambulance or private vehicle carrying critical patient towards the ED. Parking for ambulance and private vehicles for patient caretaker promote smooth shifting of patient in the ED of the hospital. It also helps in transferring the patient from ED to other departments, transferring the patients to other hospitals for further health service or discharging the patients. Modified location of ED near the diagnostic centre decreased the patient shifting distance and time as well as decreased the risk of handling critical patients. Modified location of ED reduced the patient shifting distance and time in between ED and other departments. Specifically patient shifting time is very important when every second is important for saving the life of the critical health patients.

7. Conclusion

Observations of different EDs in multi-speciality hospitals located in Central India as well as time study carried and simulation helped to find out the problems in those hospitals. Corrective measures were recommended to the administrations of hospitals, many of these were implemented. Modified location of ED in the hospital helped the patients and caretakers to identify the ED by reducing their efforts and time of finding and searching the ED. It reduced the distance between ED and other departments for patient shifting purpose. Reduction in distance decreased the patient shifting time when every second is important for ill health patients. Distance between ED and registration counter was also reduced and helped the patient caretakers.

References

[1] Bacon, M. (2012). Occupancy analytics—a new science for energy efficient hospital design. Health and Care Infrastructure and Innovation, Centre Conference.

[2] Brambilla, A., Mangili, S., Das, M., Lal, S., and Capolongo, S. (2002). Analysis of functional layout in emergency departments (ED). Shedding light on the free standing emergency department (FSED). *Model Applied Sciences*, 12, 1–19.

[3] Daskin, M. S., and Dean, L. K. (2010). Operations Research and Health Care. 43–76. https://link.springer.com/chapter/10.1007/1-4020-8066-2_3

[4] Hayward, R., Brocklehurst, D., and Sharma, S. B. (2014). Mapping patient journeys and optimising hospital designs for people flow: agent-based modelling techniques. http://www.vtdv.be/l/library/download/urn:uuid:d7eb0613-8d0b-46fc-b65c-14a274590df0/18a

[5] Hazal, A., and Ercin, C. (2021). Evaluation of hospital's emergency departments according to user requirements. *European Journal of Sustainable Development.*, 10(1), 103–122.

[6] Khadem, M., Bashir, H., and Azri, Y. F. (2009). Evaluating the layout of the emergency department of a public hospital using computer simulation modelling: a case study. *IEEE conference.*

[7] Randolph, H. W. (2006). Patient flow: reducing delay in healthcare delivery, Los Angeles, CA: Springer Science+Business Media LLC.

[8] Report of the Working Group on Emergency Care in India. (2013). http://healthmarketinnovations.org/sites/default/files/Centralized%20Ambulance%20Trauma%20Services%20Supporting%20Document%202.pdf

[9] Schull, M. J., Lazier, K., Vermeulen, M., Mawhinney, S., and Morrison, L. J. (2003). Emergency department contributors to ambulance diversion: a quantitative analysis. *Annals of Emergency Medicine*, 41(4), 467–476.

[10] Steptoe, A. P., Corel, B., Sullivan, A. F., and Camargo, C. A. (2011). Characterizing emergency departments to improve understanding of emergency care systems. *International Journal of Emergency Medicine*, 4(42), 1–8.

Chapter 53

Implementation of DMAIC Methodology in Belt Industry for Minimisation of Fabric Wastage

Shailesh Narayanrao Khekale[1], Sushil Lanjewar[1], Yogesh Dandekar[1], Aditya Kawadaskar[1], Vikram Dandekar[1], Manish Moroliya[2], Mahesh Shukla[3], and Shailendra Daf[2]

[1]Assistant Professor, Cummins College of Engineering for Women, Nagpur

[2]Assistant Professor, Priyadarshini Bhagwati College of Engineering, Nagpur

[3]Associate Professor, Cummins College of Engineering for Women Nagpur, Maharashtra

Abstract: Belt manufacturing industry faces operational wastages during manufacturing processes due to several reasons. These can be minimised by applying six sigma DMAIC methodologies by identifying and eliminating them. DMAIC (Define, Measure, Analyse, Improve and Control) was applied to decrease fabric wastages in belt manufacturing. Processes associated to fabric wastages were recognised to articulate the problem, main issues and pain areas. Data was collected about concerned processes to determine the actual execution and process capability as well as root causes were recognized. Technical solutions were determined to achieve the decided improvement. To maintain this improvement, various tools were properly applied for tracing the concerned processes. Fabric wastages were reduced in belt production in terms of reduction in DPMO (Defects per Million Opportunities) from 49,5443 to 17,240. Sigma level was improved from 1.5 to 2.7.

Keywords: Six sigma, belt industry, fabric wastages.

1. Introduction

Production with minimum wastage is goal for every manufacturing industry. To be in the competition, price and performance of the products are the major concerned areas for the industries. Profit level is to be affected by price of the products. Price of the products used to be increased by the operational waste. To maintain the profit level, industries try to minimise the operational wastages. About 100% final products cannot be achieved by any process. There are always some undesirables' bi-products. Final cost of the product is going to be increased by these undesirables bi-products and they are not going to increase any worth to the final product. These

DOI: 10.1201/9781003527442-53

undesirables bi-products are named as operational wastages. DMAIC methodology of six sigma is to be used to solve operational wastage problem. Six-sigma is statistical process enhancement method which deals with the finding the defects as well as eradicating their reasons in every operation. This methodology is useful to focus the goal which is most important to costumers. DMAIC methodology was used in belt manufacturing industry as a industrial case study. This belt manufacturing industry is major one in Maharashtra. To reduce operational wastages different tools of six sigma were used including DMAIC.

2. Literature Review

Customer delight is to be achieved by reducing number of defects to a level of 3.4 defects per million opportunities in products, processes and service by applying six sigma methodology (Rathore et al., 2021). Sigma is a measure of "variation about the average" in to a level of 3.4 defects per million opportunities in products, processes. Six sigma improvement drives is the latest and most efficient technique in quality engineering and management spectrum. To analyse the root causes of business problems, six sigma is used as full technique which is high-performance and data-driven approach (Rout et al., 2004). DMAIC and DFSS are two tools in six sigma approach. DMAIC methodology is a complete technique to resolve problem by translating real-world problem in to statistical problem. It also helps to find the statistical solution. These solutions are used to be transformed into practical solutions and execution of these practical solutions is properly conducted in the organisations (Mittal et al., 2023).

3. Problem Definition

Rubber compound, biased fabric and chord are the leading raw materials in the belt manufacturing industry. Excess utilisation of raw material was not considered seriously due to its reusability. At the same time there was drastic increase in consumption of other raw material like biased fabric and cord. Manufacturing cost of the belts was greater than expected due to high cost of fabric and cord. Higher amount of wastages increased the cost of the belt. Cord wastages, fabric wastages and in-process wastages were major wastages in the belt manufacturing process. Cord and fabric wastages were found out in the course of drum building process. In-process wastages were found out in belt cutting operation. These wastages decreased the profit of margin. Biased fabric wastage was selected for the study.

4. Methodology

To minimise the fabric wastage, DMAIC of six sigma methodology was decided after discussion with management of the plant. DMAIC is technology applied for continuous improvement which is also closed-loop process that eliminates unproductive steps (Mittal et al., 2023). To solve the problem of wastages, information about all processes and their primary data was collected with due permission of the authority. Define, measure, analyze, improvement and control are the steps of DMAIC methodology which were implemented to reduce the fabric wastage. DMAIC is innovative measurements and realistic technology for continuous enhancement with the purpose of eliminating the unproductive steps (Khekale et al., 2010). DMAIC methodology is to be implemented in five steps. It was recognised by Motorola. Define step consists of identification and definition of the problem. Main processes are identified and then data collection has to carry for calculating the performance in the measure step. Root causes of the identified problem are acknowledged in analysis step. In control step, solutions for problem are identified and implementations of the solutions are carried out. Improvement is retained in this step (Lynn Vining et al., 2005).

4.1. Define Step

In this step objective and scope of the study was determined. Data regarding the related processes was collected. Determination of end customers and deliverables to customers were found out. In this step, tools used are project charter, project plan and process flow map (Desai et al., 2008).

Complete work was completed as Define: 02 Month; Measure: 02 Month; Analyse: 01 Month; Improve: 02 Month; Control: 02 Month.

SIPOC is a structured method for recording crucial process information at high level. Identification of the key elements of a project is achieved by SIPOC before beginning of the work. This tool describes "S" means the Suppliers of the process, "I" means Inputs to the process, "P" means the Process which are going to improve, "O" means the Outputs of the process and "C" means the Customers that receive the process outputs (Antony et al., 2002). Suppliers, inputs/requirements, key process steps, outputs/requirements, customers and critical-to-quality elements of a business process are to be defined and documented in SIPOC diagram. It develops a high-level understanding of the process, the process steps (sub processes) and their relationship (Khekale et al., 2010). Table 1 elaborates the SIPOC of belt manufacturing.

Table 1: SIPOC of belt manufacturing.

Supplier	Input	Process	Output	Customer
Planning	Planning sheet	Belt Manufacturing	Belt of define specification	End Customer
Purchase	Cord		Production Yield Reports	B.S.R.
Stores	Bias fabric		Test Report	-
HR	Man power		Fabric wastage report	-
R & D	specification		In process wastage report	-
Engineering	Spares for machine		-	-

After studying and analysing the processes in belt manufacturing unit, drum building and Bias fabric cutting were found as pain areas where fabric wastage was occurred. Critical factors responsible for wastages were found out after finding the pain areas. Critical to quality (CTQ) of the bias fabric wastage and critical factors are mentioned in Table 2.

Table 2: CTQ specification table.

Critical to quality	Definition of operation	Driver	Definition of defect
Fabric waste	(Weight of fabric waste/ weight of total fabric issued) × 100	Drum Building and Bias cutting	Weigh. of wastage fabric

4.2. *Measure Step*

In this step, performance of operations in pain areas was determined and data of operations was collected. The data collection for fabric wastage was carried out for the period August to November which is mentioned in Table 3.

Table 3: Month-wise fabric wastage.

Month	May	June	July	August	September	October	November
Fabric Wastage (%) kg	7.8	8.1	7.5	7.8	9.1	9.5	9.8

4.3. *Normality Test*

Normality test was carried for the data collected in terms of fabric wastage as per data collection plan. Minitab-14 software (Anderson Normality Test) was used for determination of the normality of fabric wastage data and is represented in Figure 1. As value of p for the data was greater than 0.05, data for fabric wastage was normal.

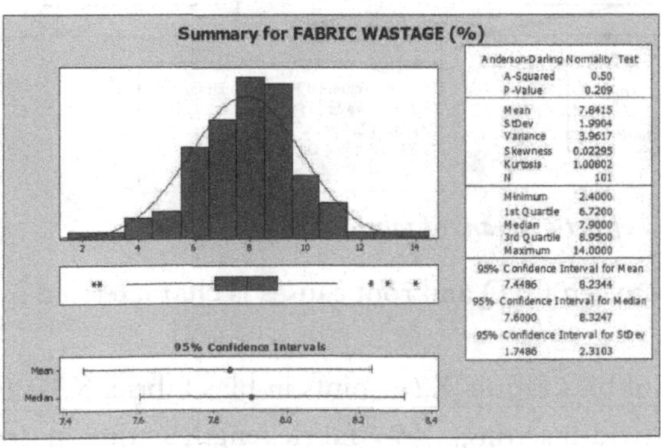

Figure 1: Normality test of fabric wastage.

4.4. Process Capability Test

Process capability test (Figure 2) was applied to determine the performance of the drum building process for the fabric wastage after the normality test. DPMO is calculated by process capability test as 49,5443. Sigma level was found as 1.37 as per DPMO value.

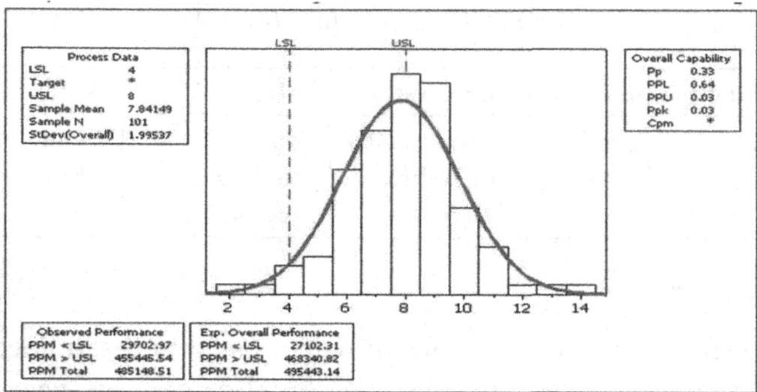

Figure 2: Process capability test of fabric wastage.

4.5. Analyse Step

Third step in the DMAIC improvement cycle is analyse step which consists of cause and effect diagram to find out the probable causes. This step described the probable causes which have the more influence on the fabric wastages.

4.6. Cause and Effect Diagram

A cause-and-effect diagram (Figure 3) for fabric wastage described effect of man, material, machine, environment and method on fabric wastage.

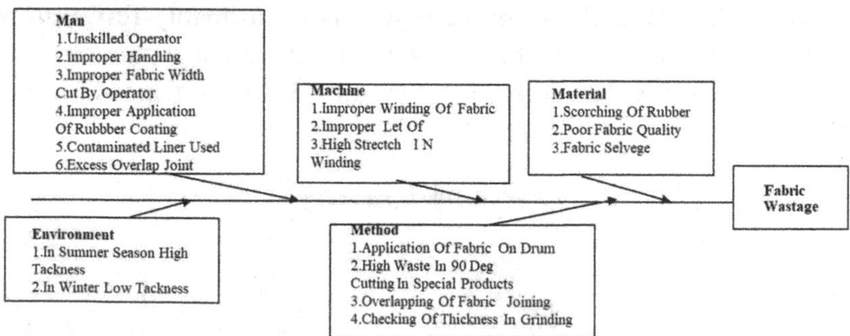

Figure 3: Cause and effect diagram of fabric wastage.

The correlation between CTQ and root causes is characterised by

Y = F(x)

X1 = rubberizing of bias fabric, X2 = joints in bias fabric, X3 = folding,

X4 = side cutting (width setting), X5 = fabric length setting in drum building.

Following Pareto chart describes the influence of different causes of the wastages. Fabric length setting in drum building was the causes which play major role in fabric wastage.

Pareto chart (Figure 4) (Y-axis = Wastages) (X-axis = Causes) demonstrates details about the causes and their effect on wastages

Following Pareto chart (Y-axis = Fabric wastage) demonstrates cause. It shows their impact on the fabric wastages. Rubberising of bias fabric, side cutting, joints in bias fabric, folding and fabric length setting in drum building played important roles in fabric wastages. For prioritisation of these root causes, relative data was collected which was evaluated by statistical techniques. After evaluating, fabric length setting in drum building and side cutting were found out as the major dominant causes for high amount of fabric wastage.

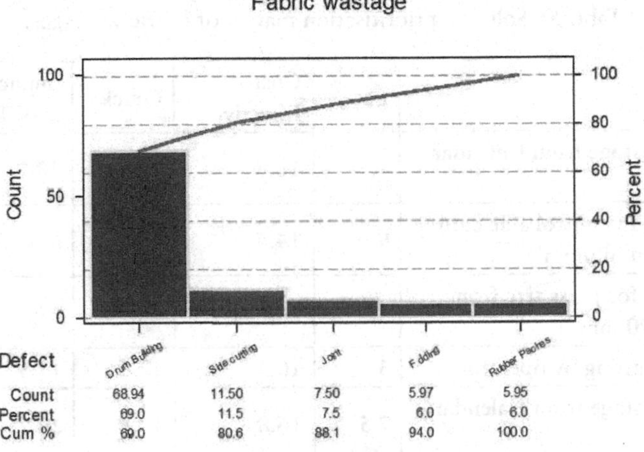

Figure 4: *The Pareto chart presenting various causes for fabric wastages.*

4.7. *Improve Step*

It is the fourth step in DMAIC and was used to acknowledge the measures. Implementation of these measures to solve the problem was conducted.

Table 4: Represents the suggested solutions to the fabric wastages.

Critical to Quality	Cause Authenticated	Suggested Solutions
Fabric wastage	Fabric length setting in Drum Building	Training for the operators during measuring the length of fabric according to specification should be conducted. Monitoring of operators (shift wise machine wise).
	Rubberizing of bais fabric	Proper care that is cooling of rollers must be taken during rubberising process of fabric in friction calendar machine.

	Joints and folding in Bais fabric	Use of proper tension during rolling of fabric.
	Side cutting (width setting Bias cutting table)	Reduce 640 mm width of fabric to 600 mm for manual and 610 mm for servicer. For that use two square rod welded to the bias cutting table maintaining width of 600 mm in between them

Categorisations of the solutions were carried according to different criteria by consulting the company's manager and engineers.

To carry out the brainstorming session, solution prioritization matrix was used. it is shown in Table 5 as per weight-age criterion. Implementation of these solutions was rigorously carried out.

Table 5: Solution prioritisation matrix of fabric wastages.

Solution	Easy	Cost Effective	Quick	Impact on CTQ	Total
Control of wastage from Calendar Machine	7.5	16.8	12.6	20.7	57.6
Reduce width to control side cutting in poetize from 600mm	9	14.7	8.4	6.9	39
Reduce width for press size from 640 mm to 590 mm	12	14.7	18.9	23	68.6
Exact fabric cutting by operator	3	10.5	12.6	11.5	37.6
Control of wastage from Calendar Machine	7.5	16.8	12.6	20.7	57.6

Normality test was carried for the data collected in terms of fabric wastage as per data collection plan. Minitab-14 software was used for this purpose. Again Anderson Normality Test (Figure 5) was used to determine the normality test of fabric wastage data. It was found that value of p for the given fabric data was greater than 0.05, the data was normal.

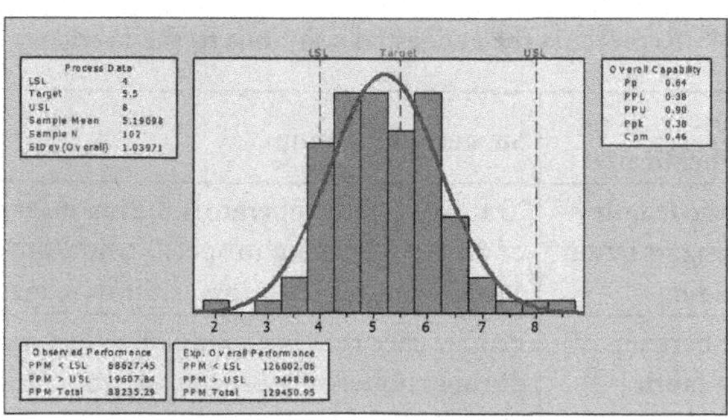

Figure 5: Normality test of fabric wastage.

Process capability test was applied to determine the performance of the drum building process with respect to fabric wastage after the normality test. DPMO was calculated by process capability test as 17240. Sigma level was found as 2.7 as per DPMO value.

4.8. Control Step

The last step of DMAIC methodology is control step. It consists of rigorously implementation of proper solutions. In this step, as proper implementation of solutions is critical factor, stress was given for continuous observation for implementing the suggested processes. For that, inspection of check charts was executed regularly. Monitoring of pain area was carried as per shift, day and month. Training to the staff, implementing various incentives schemes were executed for achieving and sustaining the improvements. (Lynn Vining et al., 2005). Fabric wastage is shown in fabric wastage by graph (Figure 6). Figure 6 shows fabric wastage from the month of May to April. It was found that fabric wastage was reduced from January.

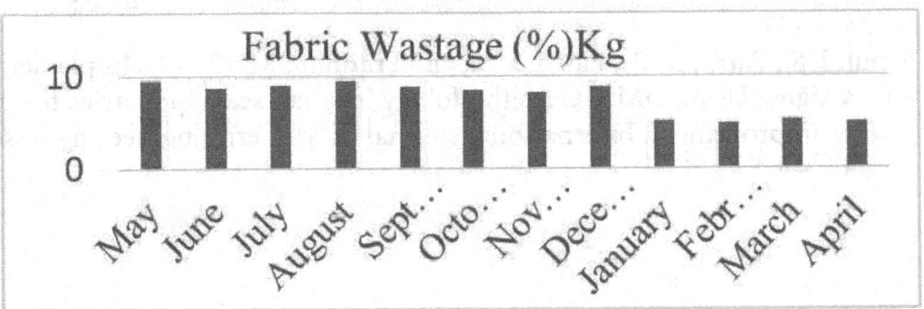

Figure 6: Month wise fabric wastage.

5. Result and Discussion

To optimise the variables of fabric wastage, DMAIC (six sigma)-based methodology was implemented. The improved results obtained about fabric wastage are in the form of DPMO (Previous = 49,5443, Improved = 17,240), sigma level (Previous = 1.5, Improved = 2.7). Due to six sigma DMAIC methodology it was found that organisation successfully attained breakthrough in decreasing fabric wastage. It was already proved that six sigma was found to be the greatest motivational methodology for every person in the organisation and it consists of continuous improvement. Due to involvement in this case study, all employees at workplace had successfully established the important statistical thinking. In this case study, many benefits of proper implications was found to be enormous. For getting this breakthrough in belt manufacturing processes, further research is possible in achieving maximum production with least possible wastages. Thus, six sigma is continues upgrading process comprising all production operations carried out in the work place.

References

[1] Antony, J., and Banuelas, R. (2002). Key ingredients for the effective implementation of six Sigma program. Measuring Business Excellence, 6(4), 20–27.

[2] Desai, T. N., and Shrivastava, R. L. (2008). Six sigma – a new direction to quality and productivity management. Proceedings of the World Congress on Engineering and Computer Science, San Francisco, USA.

[3] Khekale, S. N., Chatpalliwar, A. S., and Thakur, N. V. (2010). Minimisation of cord wastages in belt industry using DMAIC. International Journal of Engineering Science and Technology, 2(8), 3687–3694.

[4] Lynn Vining, et al. (2006). Quality Management Coordinator, Clinical Outcomes and Resource Management.

[5] Mittal, A., Gupta, P., Kumar, V., Owad, A., Mahlawat, S., and Singh., S. (2023). The performance improvement analysis using Six Sigma DMAIC methodology: a case study on Indian manufacturing company. Heliyon, (9).

[6] Rathore, R., and Patidar, P. (2021). A Review of Six Sigma DMAIC Methodology, Implementation and Future Research in the Manufacturing Sector. International Research Journal of Engineering and Technology (IRJET), 8(1), 1335–1339.

[7] Rout, I. S., Patra, D. R., Patro, S. S., and Pradhan, M. (2014). Implementation of six sigma Using DMAIC Methodology in small scale industries for performance improvement. International Journal of Modern Engineering Research, (4), 44–49.

Chapter 54

Design and Analysis of Manually Operated Lifting Crane for Small-Scale Industries

Vikram Dandekar[1], Shailesh Narayanrao Khekale[1], Yogesh Dandekar[1], and Sanjay Aloni[1]

[1]Assistant Professor, Department of Mechanical Engineering, Cummins College of Engineer for Women, Hingna, Nagpur

Abstract: The transportation of loads, workers tool, is one of the most important activities to be performed. The project aims at designing and fabrication of such devices like portable and manually operated lifting crane for our college workshop as well as small-scale industries that cannot afford other types of high-cost cranes. The crane is expected to work with minimal technical challenges and sustain weight of about 1.5 ton (1,500 Kg). This portable crane is designed which can lift a heavy load with a maximum capacity of 1 ton for which the stress on each section of beam and column has been calculated by analysing the weight of 1 ton to be lifted. The result shows that the calculated bending stress for each section of structure for 1 ton capacity of load is within its permissible limit. It shows that it will be very useful to handle load nearby 1 ton capacity. This can be used for industrial as well as domestic purpose.

The advantage of this crane is that it is powered manually by operating pulley block chain and hence no electrical energy is required. The complete design has been made in CAD using Cero Parametric 2.0. The analysis of the structure (crane) is also done on ANSYS.

Keywords: ANSYS, beam, bending stress, CAD, crane, lifting.

1. Introduction

The transportation of loads, workers tool, is one of the most important activities to be performed for which various operations are carried out at domestic levels, construction sites, workshops and industries at a greater height.

Human muscle power has no limit to the weight or height that can be lifted. Heavy lifting necessitates frequent bending and twisting, which can lead to low back disorder.

The project aims at designing and fabrication of such devices, portable and manually operated lifting crane for our college workshop as well as small-scale

DOI: 10.1201/9781003527442-54

industries which that afford other types of crane that are high in cost. The crane is expected to work with minimal technical challenges and sustain weight of about 1.5 ton (1,500 Kg). Also it should carry load efficiently. The crane will lift the weight with the help of chain pulley block.

The manually operated lifting crane is being used for the up-lifting of the weight that a human being cannot lift or even if it can lift it is not possible without the help of any other person. Similarly the problem we faced was when in our college workshop the dismantling of the engine block was going on. But the main problem that we face was when we opened the head of the engine block. It was very heavy and it was not at all possible to lift the engine head. But our college did have the solution for the up-lifting of the engine head that is tripod. But the disadvantage was that it was very tall and did not have any kind base support system so that it can get fixed and lift the weight up. So we got the idea of manually operated lifting crane which would not only help our college workshop but also small-scale industries that can't able to afford much expensive machinery for lifting up the weight.

2. Literature Review

Mabrou (2019) designed light duty gantry crane and also developed the crane control circuit involves modelling dynamic crane behavior, selecting an appropriate control strategy and validating the circuit using a gantry crane prototype for use in workshop. Another paper by Ugale (2020) designed and carried structural analysis of mechanical forklift with the help of ANSYS software and it can lift goods up to 500 Kg. The analysis on ANSYS has demonstrated that the design is safe for some accepted parameters. Researcher given the analysis of the goods lift mechanism, specifically the rope guide, was conducted to ensure its durability against bending and sagging failure. These shows that lift has capacity for the task involves lifting a weight of 500 kg at a height of 6 m with minimal effort. Pan (2018) represented the idea of forklift which is operated by battery and also discussed about the various components and parts used in it. The development of mechanical forklift assured that the machine is designed to save time for operators or workers by reducing manual lift with a maximum load capacity of 200 Kg at a height of 1250 mm. The paper of Chukwulozie *et al.* (2017) designed the passenger lift by using scissor assembly with the help of hydraulic jack and sustainable gear drive which optimised ultimately drive configuration for high load conditions. Richard *et al.* (2001) explored crane-related injuries, safety devices and procedures, offering recommendations for improved prevention and future research on crane safety. Other paper of Spruogis *et al.* (2012) showed a mathematical model which is proposed to evaluate the hydraulic damper's impact on the crane's dynamic features during lifting, reducing vibration damping period.

2.1. Problem Identification

In mechanical workshop, there is no any convenient device for lifting heavy material, the material should be handle by human being. So there is a issue of human safety. In this way, we decide to make a lift which should have maximum capacity of 1 ton, that will be helpful to workshop attendant to operate and handle the heavy load.

3. Methodology

The design consist of square bar on which I-section bar is mounted on box section bar there is a plate to which chain pulley block is attached to I-section. On square box there is a gear box attached through which chain is being managed for the movement of the plate horizontally. The problem faced in this design is it is not easy to manage which requires continues lubrication and it is stationary not movable. To overcome these problems and to move the crane, wheels can be provided to bottom.

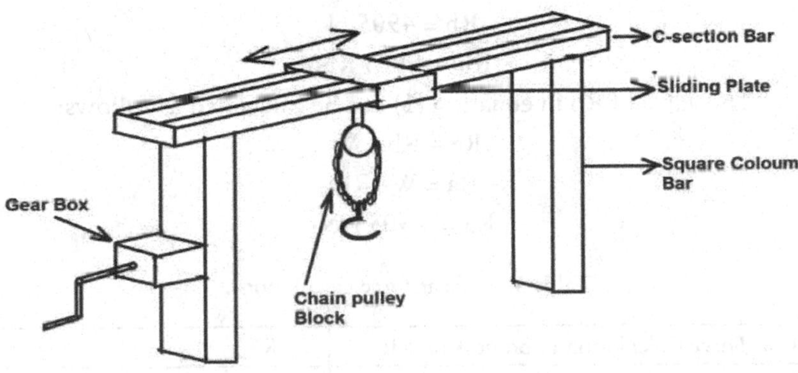

Figure 1: Design plan.

3.1. Design Calculations

Calculations for Bending Moment (BMD) and Shear Force Diagram (SFD)

Table 5.1: Dimensions of the lift structure.

Dimensions of Rectangular Bar	**Length of the bar**	**6 ft (1.828 m).**
	Height of the structure	**6 ft.**
	Weight	**1000 kgs (1 ton)**
Dimension of the hollow rectangular bar	**Length of each side of the rectangular hollow pipe**	**56 mm**
	Thickness of the hollow rectangular pipe	**5 mm**
Dimension of I beam	**length of the I beam** **Thickness of the I beam**	**91.43 mm** **5 mm**

Calculations $\Sigma Fx = 0$; $\Sigma Fy = 0$

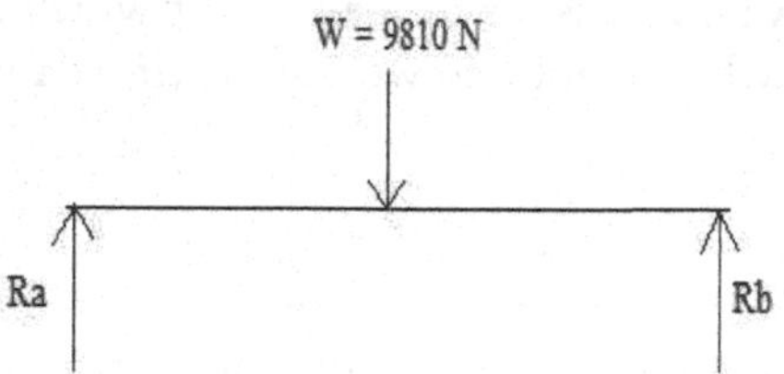

Reaction load

$$Ra + Rb = W \text{ ....(i)}$$

Taking moment about point A

$$\Sigma Ma = 0$$

$$Rb = W/2$$

$$= 9810/2$$

$$Rb = 4905 \text{ N}$$

$$Rb = 4.905 \text{ KN}$$

The value of Rb in equation (1) can be calculated as follows:

$$Ra + Rb = W$$

$$Ra = W - Rb$$

$$Ra = 4.905 \text{ KN.}$$

Table 5.2: Shear force calculations.

Shear Force Calculation at point A on left	**= 0 KN.**
Shear Force Calculation at point A on right	**= 4.9 KN.**
Shear Force Calculation at point C on left	**= 4.9 KN.**
Shear Force Calculation at point C on right	**= 4.9 – 9.81 = –4.91 KN.**
Shear Force Calculation at point B on left	**= –4.91 KN.**
Shear Force Calculation at point B on right	**= 0 KN.**

Table 5.3: Bending moment calculations.

Moment at point A (Ma)	$= 4.9 \times 1.828 - 9.81 \times 0.914$	$= 0 \text{ KN}^{-m}$
Moment at point B (Mb)	$= 0 \text{ KN}^{-m}$	$= 0 \text{ KN}^{-m}$
Moment at point C (Mc)	$= 4.905 \times 0.914$	$= 4.483 \text{ KN}^{-m}$

Shear Force Diagram (SFD)

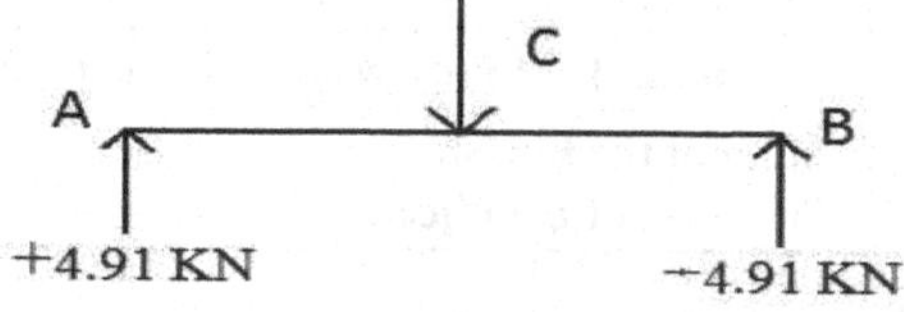

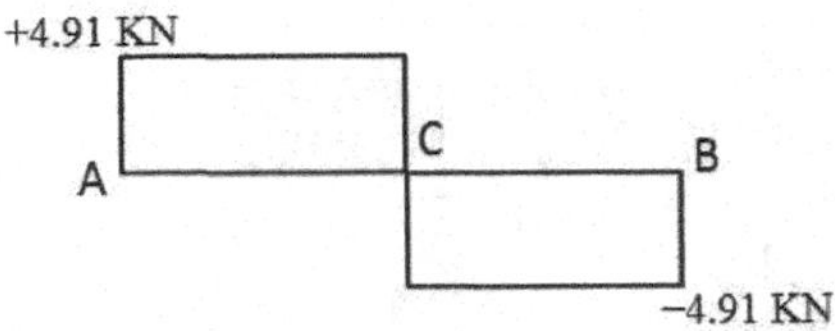

Shear force diagram (SFD)

Bending Moment Diagram (BMD)

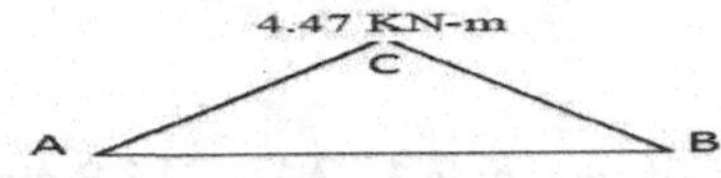

Bending moment Diagram (BMD)

Calculations of the stress acting on the box structure

Calculating stress on the hollow rectangular pipe

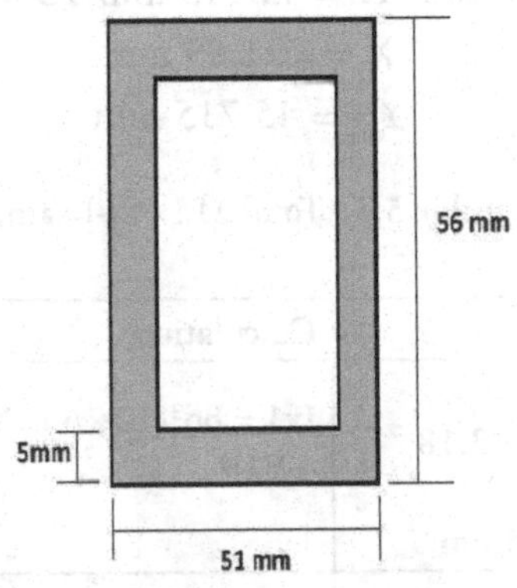

Table 5.4: Moment of inertia for box section.

Parameter of Box Section	Calculation	Result
M.I of outer rectangle: Length of the rectangle(b) = 51 mm Width of the rectangle(d) = 56 mm	Ixx1 = bd3/12 = 51× (56)3/12	= 746.36 × 103 mm
M.I of inner rectangle: Length of the rectangle(b) = 41 mm Width of the rectangle(d) = 46 mm	Ixx2 = bd3/12 = 41×(46)3/12	= 332.56 × 103 mm
Total M.I of the figure	IXX = Ixx1+ Ixx2	= 413.8 × 103 mm

Radius of Gyration, Y = 56/2 = 28 mm

Bending stress in the whole beam

= 0.303×10^3 N/mm2

Calculating stress on I section

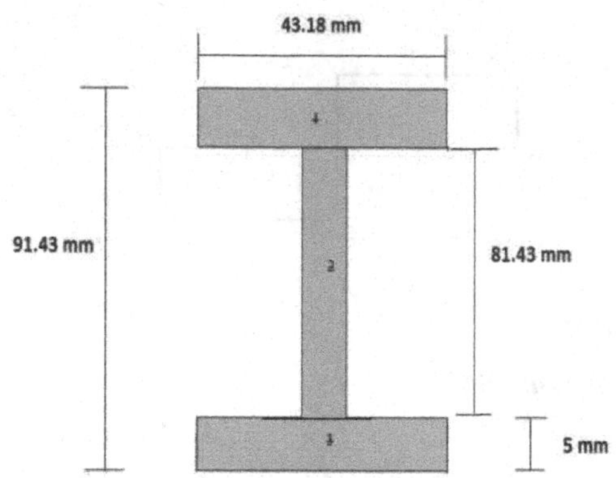

Area of rectangle 1 - A1 = 215.9 mm^2

Area of rectangle 2 - A2 = 407.15 mm^2

Area of rectangle 3 - A3 = 215.9 mm^2

X1 = 21.59 mm X2 = 21.59 mm X3 = 21.59 mm

Y1 = 88.93 mm Y2 = 45.715 mm Y3 = 88.93 mm

X = = 21.59 mm

Y = = 45.715 mm

Table 5.5: Total M.I of I-beam.

Parameter of Box Section	Calculation	Result
M.I of Rectangle 1: Length of the rectangle(b) = 43.18 mm Width of the rectangle(d) = 5mm	$Ix1 = bd^3/12+ A1\ (Y - Y1)^2$	= 403650.071 mm^2
M.I of Rectangle 2: Length of the rectangle(b) = 5 mm Width of the rectangle(d) = 81.43 mm	$Ix2 = bd^3/12+ A2\ (Y - Y2)^2$	= 224979.0418 mm^2
M.I of Rectangle 3: Length of the rectangle (b) = 43.18 mm Width of the rectangle(d) = 5mm	$Ix3 = bd^3/12+ A3\ (Y - Y3)^2$	= 403650.071 mm^2
Total M.I of the figure	IXX = IX1+IX2+IX3	= 1032279.184 mm^2

Bending Stress (σ) = I/M xY = 0.386 N/mm^2

3.2. CAD/CAE Software for Design

PRO/E -For 3D component design

PRO/E Assembly- For assembling components

ANSYS Workbench 14.0 – For analysis.

3.3. Description in Creo

Square pipe We have used mild steel (M.S) square pipe which is hollow. This is used because it is responsible for the stability of the crane. The basic dimension of the bar is 72 × 72 × 4.0 mm.

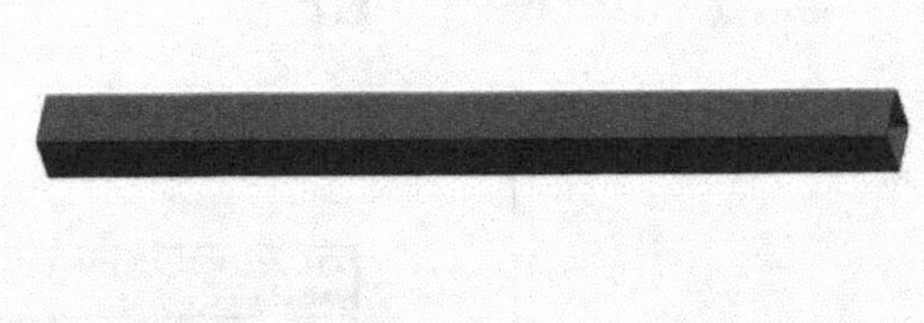

Figure 2: Square pipe in CREO.

I beamThe material used for I-beam is M.S which is of 150 mm in length with thickness about 5 mm.

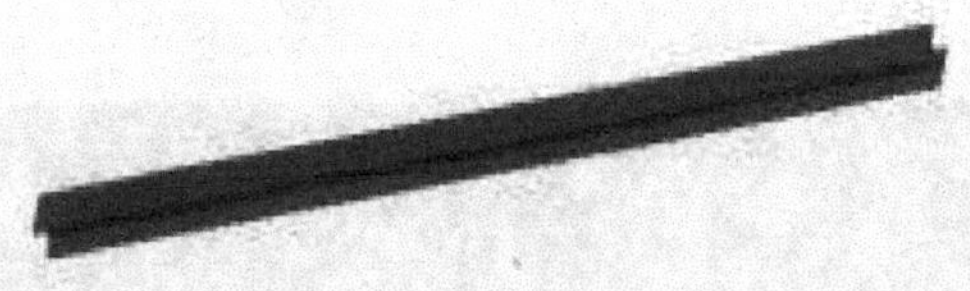

Figure 3: I-beam in CREO.

AssemblyThis is the model of the crane in CAD which is tested on ANSYS workbench software.

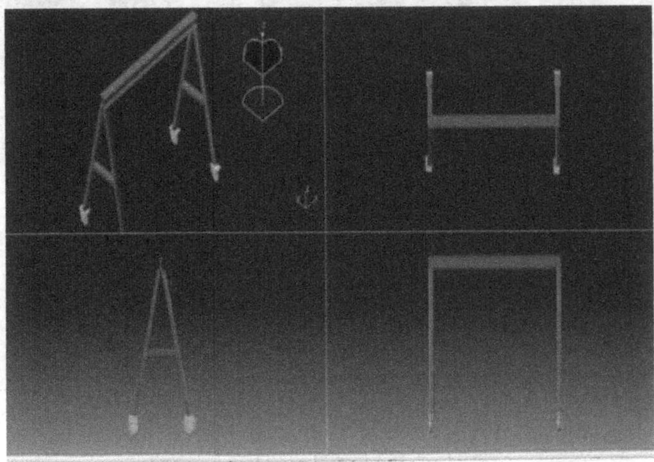

Figure 4: Full assembly of crane in CAD sheet.

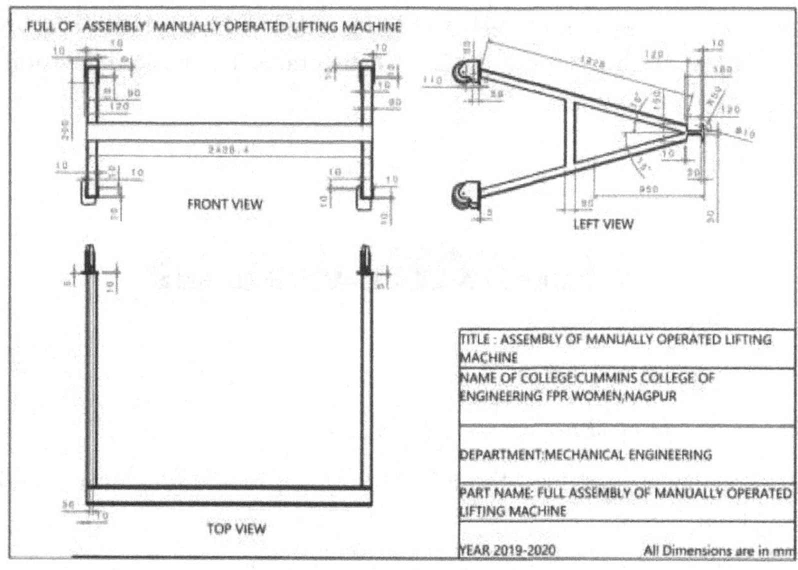

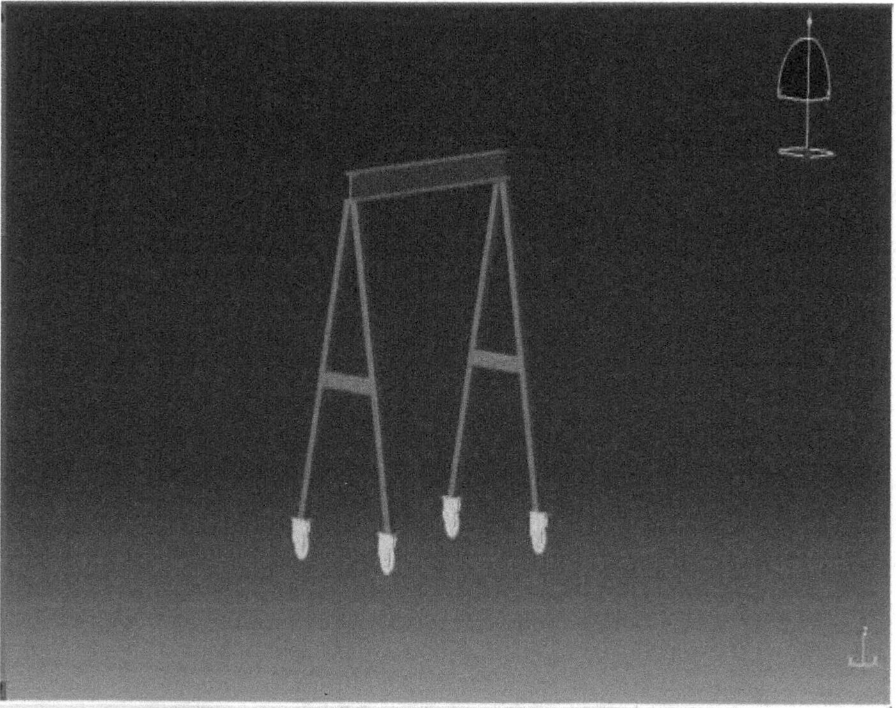

Figure 5: Full assembly in CATIA.

ANSYS workbench report I-beam section

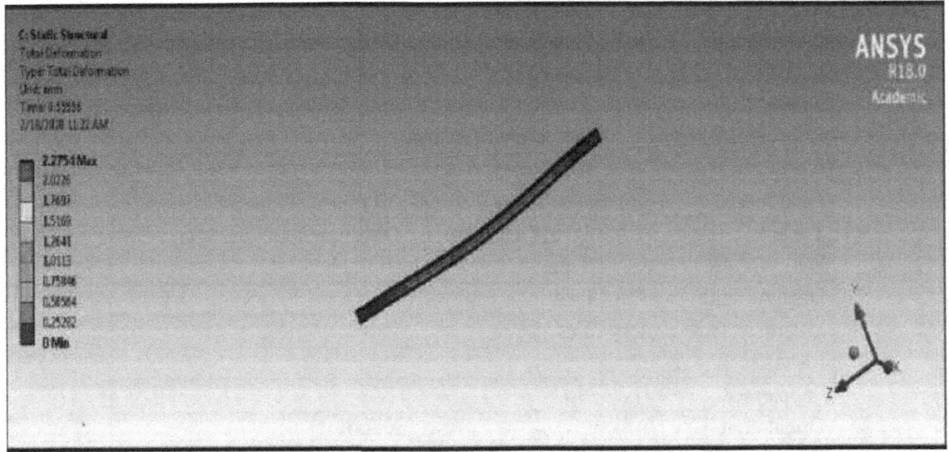

Figure 6: ***Deformation of length 6 feet with a load of 2.5 ton of I-beam.***

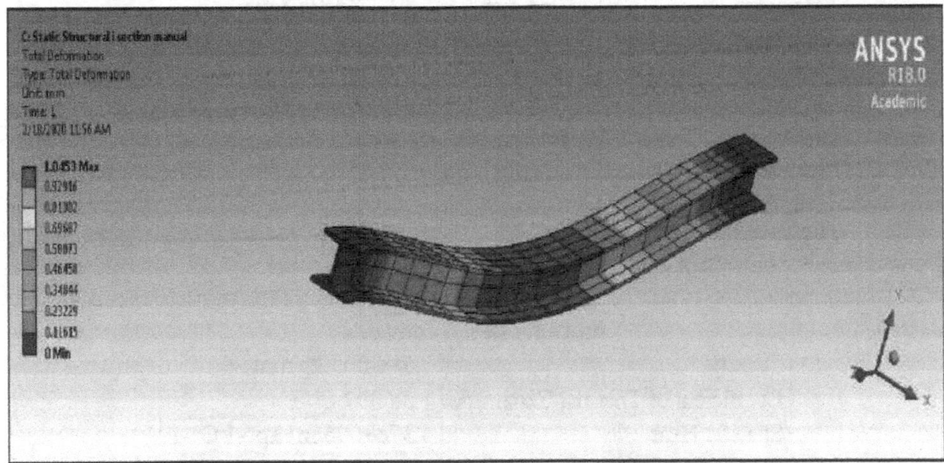

Figure 7: *Deformation of length 6 feet with a load of 1.5 ton of I-beam.*

Table 5.1
Results

Object Name	Total Deformation	Directional Deformation
State	Solved	
Scope		
Scoping Method	Geometry Selection	
Geometry	All Bodies	
Definition		
Type	Total Deformation	Directional Deformation
By	Time	
Display Time	Last	
Calculate Time History	Yes	
Identifier		
Suppressed	No	
Orientation		X Axis
Coordinate System		Global Coordinate System
Results		
Minimum	0. mm	-1.006e-003 mm
Maximum	1.0456 mm	3.2765e-003 mm
Minimum Occurs On	SYS-2\MSBR	
Maximum Occurs On	SYS-2\MSBR	
Information		
Time	1. s	
Load Step	1	
Sub step	1	
Iteration Number	1	

Table 5.2
structural steel > constants

Density	7.58e-007 kg mm^-3
Coefficient of Thermal Expansion	1.3e-007 C^-1
Specific Heat	4.37e+006 mJ kg^-1 C^-1
Thermal Conductivity	6.50e-002 W mm^-1 C^-1
Resistivity	1.7e-003 ohm mm

Table 5.3
structural steel > alternating stress mean stress

Alternating Stress MPa	Cycles	Mean Stress MPa
3900	10	0
2857	30	0
1806	50	0
1414	100	0
1070	200	0
445	2000	0
265	10000	0
250	20000	0
158	1.e+005	0
104	2.e+005	0
86.2	1.e+006	0

Table 5.4: Strain life parameter.

Strength Coeff. Mpa	Strength Exponent	Ductility Coeff.	Ductility Expo	Cyclic strength Mpa	Cyclic strength
925	–0.107	0.217	–0.57	1000	0.2

Table 5.5: Isotropic elasticity.

Temp. C	Young Modu.	Poison's Ratio	Bulk Modu.	Shear Modu.
22	$2x10^5$	0.3	$1.6767x10^5$	76950

Table 5.6: Isotropic permeability.

Relative Permeability
10000

4. Result

From the analysis of ANSYS software the bending stresses acting on the steel structure is less than its permissible limit.

1. Bending stress on I section is **0.386 N/mm^2**
2. Bending stress on box section is **0.303 N/mm^2**

Figure 8: Actual photograph of lifting crane.

Figure 9: Fabricated crane in loaded condition.

5. Conclusion

ANSYS results shows that designed model of manually operated lifting crane is safe for the maximum load on the top side.

We have successfully fabricated A-Section structure portable manually operated crane which gives nearly 6.8 ft without the use of electricity.

This can be used in any workshop and small-scale industry where there is no electricity because it is operated with non-conventional energy.

We have performed testing of lift with 1000 kgs load (1 ton).

Additional modifications and analysis can be conducted to optimise the design and identify other crucial parameters related to manually operating a lifting crane.

References

[1] Chukwulozie, O. P., Obika, E. N., Andrew, A. O., and Ebieladoh, S. J. (2017). Steel work design and analysis of a mobile floor crane. *British Journal of Applied Science & Technology*, 13(5), 1–9.

[2] Mabrouk, M., and Sherif, M. M. A. (2019). Design and implementation of a light duty gantry crane. *International Journal of Engineering Research & Technology*, 3(12), 2278–0181.

[3] Pan, L., Du, Q., and Chunshan, H. (2018). Design research on hydraulic system of working device of a forklift. *5th International Conference on Advanced Design and Manufacturing Engineering*. Richard, L. Neitzel, N. S. S, and Kyle, K. R. (2001). Review of crane safety in the construction industry. *Applied Occupational and Environmental Hygiene*, 16(12), 1106–1117.

[4] Spruogis, B., Jakstas, A., Turla, V., Iljin, I., and Sesok, N. (2012). Research of dynamics of lifting equipment. *Journal of Vibroengineering*, 14(1), 1392–8716.

[5] Ugale, S. (2020). Design and structural analysis of mechanical forklift using ANSYS Software. *International Journal of Research in Advent Technology*, 2(5), 234–237.

Chapter 55

Design and Simulation of Robotic Arm Joint for Sanitisation

Sushil Lanjewar[1], Sandip Khedkar[2], Shailesh khekale[1], Aditya Kawadaskar[1], and Mahesh Shukla[3]

[1]Assistant Professor, Cummins College of Engineering for Women, Nagpur

[2]Assistant Professor, Yeshwantrao Chavan College of Engineering, Nagpur

[3]Associate Professor, Cummins College of Engineering for Women, Nagpur

Abstract: The integration of robotics, particularly the design and simulation of a robotic arm for sanitisation and health monitoring, is indeed a promising approach to combat the COVID-19 pandemic and enhance safety measures. The COVID-19 pandemic has become a major challenge for the healthcare system. It is essential to follow precautionary measures to avoid the spread of the virus. Emerging technologies such as robotics can prove to be a boon to sustaining the pandemic. The focus of the paper is the design and simulation of the robotic arm joint for sanitisation influences to automatically sanitizing the workspace without actual human intervention. The robotic arm is a 3R robotic arm joined with mechanical joints. It is situated on a trolley carrying a sanitiser tank and electronic components. Kinematics of the arm is analysed for its reach and movement. Raspberry Pi is used with sensors for its automatic movement. Each component is designed and analysed on CAD software. Calculations of each component are performed. Motion simulation and stress analysis are performed on the design to test its sustainability. Further, the features of the robot can be extended by changing the end effector, which is a nozzle, with the sensors for examining COVID-19 health conditions of a person without any human contact.

Keywords: 3R robotic arm, degree of freedom, IoT, CAD.

1. Introduction

"Robotics" is a field that aims to design machines that can help and assist humans. It integrates various fields such as mechanical engineering, computer science, mathematics, etc. The word "Robotics" was coined and first used by a Russian-born American writer Issac Asimov in 1942. He was also the one to propose three Laws of Robotics. The Origin of the word "Robot" is in the Czech word "Robota" which means either a slave or a mechanical device that would help its master.

DOI: 10.1201/9781003527442-55

We have been surrounded by robots for many years now. These Robots or machines have made our lives simpler and easier. In 1969, General Motors installed the first robot for spot welding automobiles. Since then, there has been numerous research and work put into the field of Robotics until today, where David Hanson created a social humanoid robot called "Sophia".

The world was facing the COVID-19 pandemic. It is very essential to follow certain rules such as social distancing, contactless working to slow down the rate of spread of the disease. It is very important to sanitise various workspaces such as hospitals, commercial buildings, schools and colleges periodically to prevent people from getting infected. There is a high risk to the person who will perform sanitisation. This automatic disinfectant system will perform sanitisation without any actual human interference preventing the worker from direct contact with infected areas. Absolutely, enabling a worker to control the robot remotely from a safer location provides a significant advantage in terms of safety and efficiency. This setup not only minimises the worker's direct exposure to potentially contaminated areas but also allows for precise and controlled sanitisation processes.

The smart robot will be able to perform various activities such as taking temperature and other such medical factors of a person and then generating a database with the unique ID of each person to keep a check whether the medical parameters are within the prescribed range.

2. Literature Review

The study of Beasley Ryan A (2012) found the impact of robots in multiple medical domains. Medical robotics is a young and relatively unexplored field made possible by technical improvements. The advancements in medical robotics have the potential to revolutionise healthcare by improving efficiency, accuracy and accessibility to medical services. Kumar (2012) explored the study of the physical structure and dimensions of a robot, including the arrangement of its links, joints and end-effector. He also studies Robot kinematics that is the study of motion and position of robots without considering the forces or torques involved. It involves analysing the relationships between joint angles, link lengths and the position/orientation of the end-effector. Forward kinematics calculates the end-effector position based on joint angles, while inverse kinematics determines the joint angles needed to achieve a desired end-effector position. Study by Muhammad (2013) successfully designed and fabricated a multipurpose articulated robotic arm having 3-DOF capable to perform various industrial tasks. By utilising the inverse kinematic equations the designed robotic arm has performed the all designated tasks assigned in a selected work envelop. Another study of Puran-Singh et al. (2013) was found that end-effectors are a device at the end of a robotic arm, designed to interact with the environment. The study of Lanjewar (2016) discussed omnidirectional technology in vehicles typically refers to a design or technology that allows a vehicle to move in any direction without requiring traditional maneuvers like turning or reversing. This technology often involves the use of specialised wheels or propulsion systems that provide exceptional maneuverability.

Gandhare (2017) provided detailed insights into the design, control and performance of AGVs utilising omnidirectional wheels for material handling applications, along with experimental results and potential real-world applications. An overview of prominent research and commercially available robotic devices for assistive and rehabilitation purposes. The main considerations for the design of rehabilitation robotic devices are discussed by Sirlantzis (2019). Bhandari (2021) gives different approach to developing the hardware and software for an accelerometer-controlled robotic arm.

3. Methodology

3.1. Construction

The *3R* (3 links and 3 joints and 3 *DOF*) robot consists of a trolley and a robotic arm. [2] The trolley frame consists of four vertical and 12 horizontal SAE1020 steel bars (L-sections). The steel (SAE1020) frame trolley houses a sanitiser tank (Acrylic tank; 12 liters capacity) and a wooden (ply) box for electrical and electronics equipment. The sanitiser tank and wooden box are supported on steel (SAE 1020) plates and can slide along the slides (SAE 1020) for refilling and repairing purposes. High torque D.C. motor-powered castor wheels (rubber and Cast Iron) along with the wheels (100-mm diameter and 50-mm width), made of Nylon and P.U. rubber, provide smooth movement of the robot according to the signals received via a *Bluetooth chip* installed. [3] The robot arm consists of 3 links (OD of 35 mm and ID of 25 mm each) of SAE1020 steel; joined by 3 joints (one twisting joint and two knuckle joints). The first link is 300-mm long, the second link is 150-mm long and the third is 75-mm long. Three stepper motors are mechanically connected at the three joints to index and power the movements of links. The first link is connected to the second link by a knuckle joint (180° rotation) thus allowing *pitching* of the second link. The end-effector that is a brass, *mist* nozzle (3-mm OD, 1-mm ID and 35-mm length) is mechanically connected to the third link. The second link is connected to the third link by a *knuckle joint* (SAE1020) which can rotate through 180° thus allowing *yawing* of the third link. The first link of the robot arm is connected to the trolley through a *twisting joint* which can rotate through 360° in the clockwise and counterclockwise direction. The sanitiser through the tank is pumped through a PVC pipe to the nozzle for sanitisation using a motor pump. A camera lens and a light are also provided near the nozzle for monitoring. A temperature (DHT11) sensor and ultrasonic sensor are provided for temperature sensing of patients and obstacle avoidance respectively using a Raspberry Pi, breadboard, LCD screen, motors, motor drive and battery. Python codes are provided for both purposes.

3.2. Working

According to the signals received from the controller, the commands are given to the motor and electronic circuit, and using two motors the robot is directed employing castor wheels. The three stepper motors placed; one at twisting joint and two at knuckle joints; rotate the joints and thus the links; to reach the surfaces by a combination of *twisting, pitching and yawing* movements which are pre-calculated by parametric equations obtained by *D-H Parameter Matrices.* [4] According to the requirements; the settings are done to pump the sanitiser from the sanitiser tank through a PVC pipe to the nozzle for sanitisation. Temperature sensing is done using the temperature sensor and Raspberry Pi by replacing the end effector (nozzle) with the temperature sensor. An ultrasonic sensor detects and helps in avoiding obstacles in the robot's path.

Table 1: Robot specifications.

Description	Value
Gross weight	50 kg
Degree of freedom	3
Floor surface area occupied	0.1 m^2
Maximum height of the robot	1.07 m
Maximum reach of the arm and the end effector	0.255 m
Storage volume occupied	0.1068 m^3
Sanitiser Tank Capacity	12.68 l
Mass rate of flow of sanitiser through the nozzle	3.7 g/s
Time between refilling of the sanitizer tank	50 minutes

(*Source: Author's compilation)*

3.3. Computer Aided Design (CAD) Modelling and Simulation

CAD software is used to create initial conceptual designs based on requirements and ideas. CAD tools to develop detailed, precise models of the product, considering materials, dimensions, tolerances and other specifications. The twisting joint makes twisting motion among the output and input link. During this process, the output link axis will be vertical to the rotational axis. The output link rotates in relation to the input link. The material of twisting joint used is mild steel. Twisting joint will allow 360° rotation of the immediate link. The CAD model of the twisting joint was done on CATIA V5R18 as shown in Figure 1.

Motion simulations can run to study the motion and behaviour of mechanical parts and mechanisms, ensuring proper functionality and safety. The motion simulation of robot parts and joints was carried out on CREO Parametric 3.0. The sliding motion of the control box and sanitiser tank followed by the front and back movement of wheels, twisting, yawning and pitching motion of links and joints was carried out.

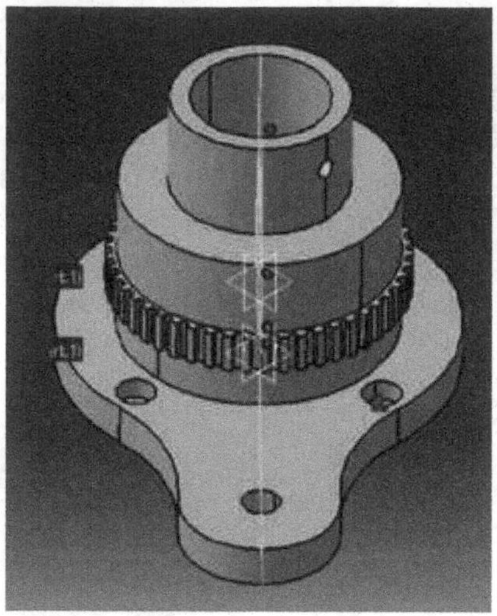
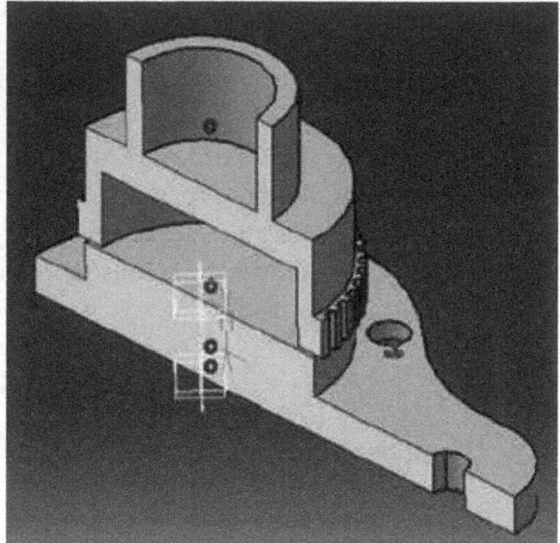

Figure 1: CAD model of robot twisting joint and section view.

4. Calculations

4.1. Calculation for Forward Kinematics

Robot kinematics deals with the study of the geometry of the motion of kinematic chains having a multi-degree of freedom. It provides the basic relationship between the robot's joint coordinates and its spatial layout with reference to a fixed coordinate system. This analysis can be done using the Denavit–Hartenberg parameters (D–H Parameters). The four D–H parameters for robot link are shown in Table 2.

Table 2: DH parameter for robotic arm.

i	α	a	d	θ
1	0	0	0	θ_1
2	0	L_1	0	θ_2
3	0	L_2	0	θ_3

The D–H method is a tool for kinematic analysis of 3R robotic arm. The 3R robotic arm consists of three degrees of freedom with three revolute or rotational joints.

In forward kinematics, the origin is fixed at the bottom-most joint (first joint), as well as the axes are fixed. The rotation and length of link 1 are defined as and respectively. A series of matrices are used to describe the transformation from the first joint to the last joint and thus by considering all these transformation matrices we get, the total transformation matrix which provides us with coordinates of the end effector.

Total Transformation Matrix =

$$\begin{bmatrix} C_{12}C_3 - S_{12}S_3 & -C_{12}S_3 - S_{12}S_3 & 0 & L_1C_1 + L_2C_{12} + L_3C_{123} \\ S_{12}C_3 - C_{12}S_3 & -S_{12}S_3 + C_3C_{12} & 0 & L_1S_1 + L_2S_{12} + L_3S_{123} \\ 0 & 0 & 1 & 0 \\ 0 & 0 & 0 & 1 \end{bmatrix}$$

Let, X and Y be the coordinates of the end effector and are as follows:

$$X = L_1 \cos \theta_1 + L_2 \cos (\theta_1 + \quad) + \text{L3}\cos (\theta_1 + \theta_2 + \theta_3)$$

$$Y = L_1 \sin \theta_1 + L_2 \sin (\theta_1 + \theta_2) + L_3 \sin (\theta_1 + \theta_2 + \theta_3)$$

Case 1- Consider $L_1 = 300\text{mm}$, $L_2 = 150$ mm and $L_3 = 75\text{mm}$

$$\theta_1 = 90°, \theta_2 = 0°, \theta_3 = 0°$$

$$X = 300 \cos 90° + 150 \cos(90°+0°) + 75 \cos(90°+0°+0°)$$

$$= 0$$

$$Y = 300 \sin 90° + 150 \sin(90°+0°) + 75 \sin(90°+0°+0°)$$

$$= 525$$

$$= (0, 525)$$

Case 2- consider $L_1 = 300$ mm, $L_2 = 150$ mm and $L_3 = 75\text{mm}$

$$\theta_1 = 90°, \theta_2 = 45°, \theta_3 = 0°$$

$$X = 300 \cos 90° + 150 \cos (90°+45°) + 75 \cos (90°+45°+0°)$$

$$= -159.075$$

$$Y = 300 \sin 90° + 150 \sin(90°+45°) + 75 \sin(90°+45°+0°)$$

$$= 459.075$$

$$= (-159.075, 459.075)$$

Case 3- consider $L_1 = 300\text{mm}$, $L_2 = 150$ mm and $L_3 = 75\text{mm}$

$$\theta_1 = 90°, \theta_2 = 90°, \theta_3 = 0°$$

$$X = 300 \cos 90° + 150 \cos (90°+90°) + 75 \cos (90°+90°+0°)$$

$$= -225$$

$$Y = 300 \sin 90° + 150 \sin (90°+90°) + 75 \sin (90°+90°+0°)$$

$$= 300$$

$$= (-225, 300)$$

Similarly, inverse kinematics is an analytical process to obtain the joint and link parameters when the coordinates of the end effectors are known.

Inverse kinematic is an analytical process to obtain the joint and link parameter when the coordinates of the end effector are known. The coordinates of the end effector are known as task space parameter.

$$x = L_1C_1 + L_2C_{12} + L_3C_{123}$$

$$y = L_1S_1 + L_2S_{12} + L_3S_{123}$$

$$\varnothing = \theta_1 + \theta_2 + \theta_3$$

The values of the θ_1 and θ_2 can be easily calculated by using the normal algebraic equation in order to calculate θ_3. Thus

Let, $X = L_1C_1 + L_2C_{12}$

$$Y = L_1S_1 + L_2S_{12}$$

Squaring and adding X and Y, we get,

$$X^2 + Y^2 = L_1^{\,2} + L_2^{\,2} + 2L_1L_2C_2$$

$$\theta_2 = \cos^{-1}\frac{X^2 + Y^2 - L_1^{\,2} - L_2^{\,2}}{2L_1L_2}$$

Thus, we got two values, one +ve and none –ve for θ_2, respectively.

Again consider,

$X = L_1C_1 + L_2\left(C_1C_2 - S_1S_2\right) = K_1C_1 - K_2S_1$ (For easy understanding using K_1 and K_2)

$X = \cos(A + \theta_1)$ equation 1

Similarly,

$Y = \sin(A + \theta_1)$ equation 2

Dividing equation 2 by equation 1, we get.

$$\tan\left(A + \theta_1\right) = \frac{Y}{X}$$

$$\theta_1 = \tan^{-1}\frac{Y}{X} - \tan^{-1}\frac{K_1}{K_2}$$

Thus we get two values for θ_1.

Now using the basic equation of task space parameter we can get the values for θ_3.

$$\theta_3 = \varnothing - \theta_1 - \theta_2$$

After solving these algebraic equations we get the rotations of each links. For the same given position and orientation, we get two sets values and the most correct value can be further found by using work space method.

5. Result

FEA is a common simulation technique used to analyse stress, strain, thermal effects and other physical properties of components under various conditions. Ansys 16.2 software was used to carry out the stress–strain and total deformation analysis on the twisting joint and the values are well within the permissible design limits. Figure 2 shows the total deformation of the twisting joint.

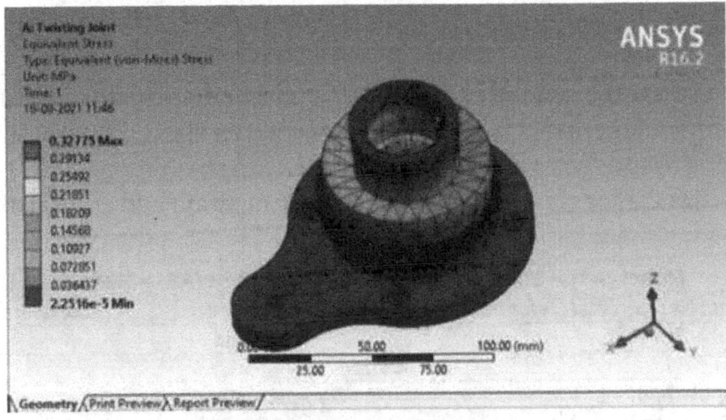

Figure 2: Twisting joint: Total deformation equivalent stress.

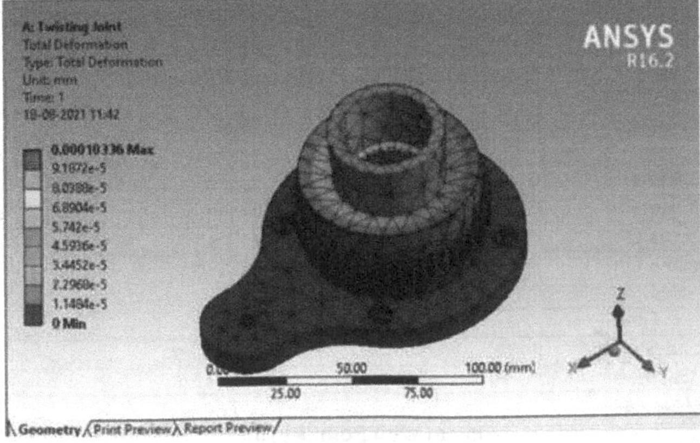

Figure 3: Twisting joint (equivalent stress).

The equivalent stresses are shown in the Figure 3 for the twisting joint as per the Ansys analysis. Also the equivalent strain is analysed and the result is shown in the Figure 4.

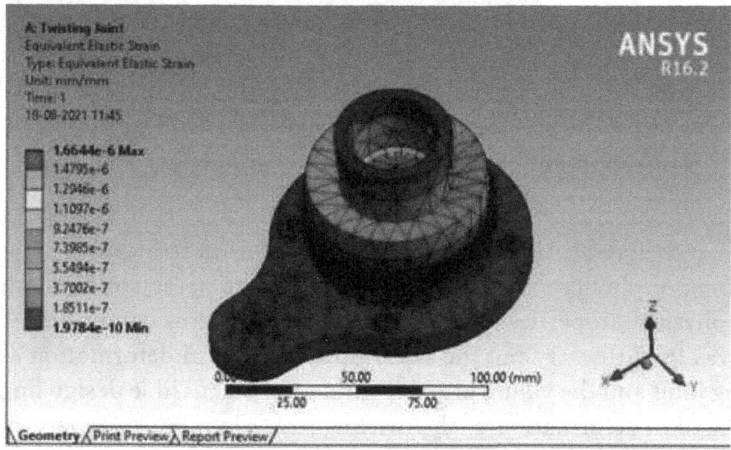

Figure 4: Twisting joint (equivalent stress).

The Ansys software generates the report and results for the twisting joint total deformation equivalent stress and equivalent strain is shown below in Table 3.

Table 3: Results for total deformation, equivalent strain and equivalent stress.

Model (A4) > Static Structural (A5) > Solution (A6) > Results

Object Name	*Total Deformation*	*Equivalent Elastic Strain*	*Equivalent Stress*
State	Solved		
Scope			
Scoping Method	Geometry Selection		
Geometry	All Bodies		
Definition			
Type	Total Deformation	Equivalent Elastic Strain	Equivalent (von-Mises) Stress
By	Time		
Display Time	Last		
Calculate Time History	Yes		
Identifier			
Suppressed	No		
Results			
Minimum	0. mm	1.9784e-010 mm/mm	2.2516e-005 MPa
Maximum	1.0336e-004 mm	1.6644e-006 mm/mm	0.32775 MPa
Minimum Occurs On	PartBody		
Maximum Occurs On	PartBody		
Information			
Time	1. s		
Load Step	1		
Substep	1		
Iteration Number	1		
Integration Point Results			
Display Option		Averaged	
Average Across Bodies		No	

6. Conclusion and Discussion

The design and simulation of a robotic arm for sanitisation of workspaces proves as an efficient model for sustaining and overcoming the COVID-19 pandemic and helps an individual to carry out hygiene while following prescribed norms, to overcome the spread of disease. The features provided by the robotic arm allow people to adopt modern ways of sustaining the pandemic using advancements such as robotics in the medical field.

It is concluded that the use of a robotic arm for sanitisation meets the requirements and acts as a COVID-19 warrior to overcome the pandemic situation.

References

[1] Beasley, R. A. (2012). Medical robots: current systems and research directions. *Journal of Robotics*, 1–15.

[2] Bhandari, P. K., and Jain N. (2021). Design and structural analysis of pick & place robotic arm. *IJIRT*, 8(4), 83–90.

[3] Gandhare, H. N., Pitale, J. R., Joshi, P. N., Tiwari S. R., and Lanjewar S. (2017). Design fabrication and simulation of omnidirectional wheel operated automated guided vehicle. *International Journal on Research & Modern Trends in Engineering & Management*, 2(1), 1–7.

[4] Kumar, V. (2012). Introduction to robot geometry and kinematics, book chapter.Lanjewar, S., and Sargaonkar, P. (2016). Fabrication and simulation of mecanum wheel for automation. *International Journal of Innovations in Engineering and Science*, 1(1), 1–6.

[5] Muhammad, N. H. (2013). Design fabrication and controlling of multipurpose 3-DOF robotic arm. *IOP Conference Series Materials Science and Engineering*.

[6] Puran-Singh, Kumar, A., and Vashisht, M. (2013). Design of robotic arm with gripper and end effector: for spot welding. *Universal Journal of Mechanical Engineering*, 1(3), 92–97.

[7] Sirlantzis, K. (2019). Robotics, Handbook of Electronic Assistive Technology. 311–345, Science Direct.

D. Emerging Trends in Basic & Applied Sciences

Chapter 56

Municipal Solid Waste Dumpsites' Effects on the Environment and Public Health

Zeba Khatoon[1], *Raghvendra Singh*[1], *and Bhojraj N. Kale*[1]

[1]Department of Civil Engineering, Ujjain Engineering College Ujjain, MP

Abstract: World's most severe industrial disaster, the leaking of MIC gas from a Union Carbide factory on December 3, 1984 was witnessed in Bhopal, the capital of Madhya Pradesh, which is the biggest state in India. It is crucial to evaluate the groundwater quality in Bhanapur, Bhopal, given the global interest in the overall condition of ground water, which is being harmed by human activity, overexploitation, excessive pumping as well as permeation of waste products, sewage from manufacturing plants and due to geological alterations. The ground water sections in Bhanpur, Bhopal were examined physically, chemically and bacteriologically, and the results are presented in this paper. According to the study, the majority of the parameters' concentrations fall less than the maximum allowable limits for TDS, alkalinity, magnesium, chloride, sulphate and fluoride. There have been reports of bacterial contamination near Bhanpur, which serves as a municipally owned waste disposal site, with some locations even exceeding acceptable levels and being unfit for drinking. The sample was collected, preserved and pre-treated in accordance with the BIS technique, which is the accepted international standard for sample collection.

Keywords: Air, contamination, environment, health, water.

1. Introduction

Life on our planet depends on water in many ways. The growth and constitution of cells, the transport of nutrients to cells, along with body metabolism all depend on water. Contaminants in the water alter the mechanism's tendency, leading to both long- and short-term illnesses. Water becomes biologically impure when living things like algae, bacteria, protozoa and viruses are present. Numerous microbes in water can cause a variety of problems. Impurities can be classified into several categories, including organic, inorganic, biological and as well radiological contaminants (Sonel *et al.*, 2010). It is possible to quantify organic pollution using the chemical characteristic of water. Organic materials may contain impurities that have the potential to seriously harm human health, including cancer, hormonal imbalances

DOI: 10.1201/9781003527442-56

and neurological disorders. Waste from industries as well as soils or objects that the water goes through may contain radioactively contaminated material. Regular research yields certain technologies and processes for removing contaminants from water (Sonel *et al.*, 2010; Telang & Chaturvedi, 2008).

Many locations, which includes Bhopal, was listed as being indigenous to contamination as a result of the removal of untreated solid waste at the Bhanpur, Bhopal disposal location. Groundwater supplies have been contaminated by pollutant releasing and improper water recharge. Additionally, groundwater as well as surface water have been contaminated by leachates from farms, industrial waste, as well as primarily solid waste from municipalities. We have chosen ground water bodies in Bhanpur, Bhopal because of the location of the municipal solid waste disposal site, in order to investigate the severity of pollution in ground water. The utilisation of ground water along with proximity to the population sources were factors in choosing these locations (Sonel *et al.*, 2010; Telang & Chaturvedi, 2008). The adverse changes to the natural environment, which are entirely the result of human activity and have direct or indirect effects on alterations to the energy distribution, radiation levels, along with chemical and physical makeup of organisms, are causing constant alterations in the worldwide climate. The most common forms of pollution as well as pollutants released, came across in waste from homes and businesses waters, as well as the possible impact on water quality. These modifications can impact man either directly or via his sources of water, farming and other living things. Chemicals are among the leading causes of water contamination because they are released when water traverses' geological materials. Man-made chemicals can also be hazardous (Kataria *et al.*, 2011).

The total quantity of waste generated every day in all households has increased due to the expanding global population and increasing need for food as well as additional necessities. Ultimately, this garbage is disposed of in municipal dumps, and because of inadequate supervision and ineffectiveness, these sites become potential danger to the environment and to health for residents who live close to them (Adam *et al.*, 2015). According to Garrod and Willis (1998) and Palmiotto *et al.* (2014), operating a landfill is typically accompanied by unpleasant odours, loud, disrupting noise from excavators, litter, dust, an abundance of rodents, unplanned landfill fires, bio aerosol emissions and volatile organic pollutants that can contaminate surface and groundwater (Garrod & Willis, 1998; Palmiotto *et al.*, 2014). As a massive quantity of carbon dioxide and CH4 are produced during the process of decomposition of organic wastes in solid waste landfills, they play a significant role in the world's anthropogenic greenhouse gas (GHG) emissions (Kumar *et al.*, 2004). Humans who continuously breathe in CH4 may experience loss of communication, nausea, vomiting or possibly death (Okeke & Armour 2000). Numerous diseases have been linked to the illegal dumping of MSW. Residents of areas with poor waste management are more likely to contract respiratory infections, diarrhoea and malaria (Kafando *et al.*, 2013).

According to prior studies, people who live nearest to landfill sites are more likely to experience medical conditions like asthma, cuts, diarrhoea, stomach pain, recurrent flu, cholera, malaria, cough, skin irritation, and tuberculosis. Chronic chemical exposure and inhalation of hazardous gases and particles of landfills are now the primary contributors of health issues (Adeola, 2000; Bridges *et al.*, 2000; Sankoh *et al.*, 2013). People can be exposed to diseases along with other contaminants when they bathe in, irrigate their food with or drink water that has been contaminated by solid wastes. People who live close to landfills have also been found to frequently experience respiratory symptoms, skin irritation, nose and vision issues, gastrointestinal problems, exhaustion, migraines mental health issues and allergies (Abul, 2013; Alam *et al.*, 2013). This study aims to assess the effects of solid waste dumpsites on both the environment and human health.

2. Materials and Methods

2.1. Study Location

The study is carried out in the Madhya Pradesh's capital city, Bhopal. The Bhanpur dumpsite is chosen as the location for the current study's solid waste disposal. The Bhanpur dumpsite, an abandoned landfill with a total footprint of 57.80 acres, is situated at 230°17'47.59°N, 770°25'11.54°E.

2.2. Data Collection and Analysis

An equivalent cross-sectional technique is used. Randomly chosen families respondents are given a semi-structured questionnaire. The three sections of the questionnaire are: comprises of socio-demographic data, environmental data and local residents' health issues details.

To assess the validity and completeness of the responses, each questionnaire is carefully examined. Participants in the survey are the heads of family or adults (especially females) from the neighbourhood who lived close to the dump. For this study, 100 Bhanpur residents are chosen as respondents in total.

3. Findings of Study and Analysis

3.1. Respondent Socio-Demographic Information

Table 1 lists the social and demographic details of the respondents. According to the findings of this study, the majority of the respondents to this study are between the ages of 31–50. According to Shomoye and Kabir (2016), the significant number of women in the research is likely a result of their accessibility, greater understanding and awareness of the health issues that influence their kids and other household members. Most of the respondents were self-employed (businessmen) and had lived in or close to both study areas for more than five years. The distribution of respondents' educational levels reveals that most of them had no formal education and that only a small number had a tertiary degree. Uneducated people who have fled to the dumpsite and its surroundings in order to survive call it home Omoniyi (2014).

Table 1: Social and demographic details of the respondents.

Details	Values (N = 100)	Details	Values (N = 100)
Gender		Drinking water Source	
Male	40%	Tube Wells/Bore well	14%
Female	60%	Govt. Water tank/ pipeline	86%
Age		Occupation	
21–30 years	22%	Daily wage worker	22%
31–40 years	30%	Businessman/Shopkeeper	54%
41–50 years	40%	Govt. Service	18%
51 and above	8%	Private Service	6%
Literacy Level		Family Size	
No Schooling	40%	2–3 members	28%
Primary	10%	4–6 members	52%
High School	32%	6–8 members	16%
Tertiary education	18%	Above 9 members	4%
Stay duration near Bhanapur			
1–5 years	4%	5–10 years	16%
11 years and above	80%		

(Source: Author's compilation)

3.2. Respondents' Opinions Regarding Environmental Issues in the Area

According to the research's findings, it is bad for a person's health to live close to a landfill. A large percentage of respondents objected that rats as well flies have a dominant influence in the surroundings regarding the prevalence of rats as well fleas from dumpsites. A large proportion of respondents believed that one of the most serious problems related to living nearest to the dumpsite were the uncomfortable odour and various health issues. Most of the family's respondents voiced complaints about the odour coming from the dumpsite when discussing the impact on the standard of the environment surrounding the site. This refers to the breakdown of organic solid wastes that were stored for an extended period of time before they became an agent of environmental contamination and health risks for the local population. According to Ohwo (2011) garbage left unattended over a long time poses a serious risk and yields a revolting odour that can pose a serious health risk to those who live nearby. Additionally, unrestrained waste burning in open landfills could lead to polluted air and a rise in greenhouse gas emissions, both of which have been linked to climate change. Burning waste is typically a poor choice for waste management from an environmental standpoint because it releases a dangerous mixture of pollutants into the environment that can cause cancer.

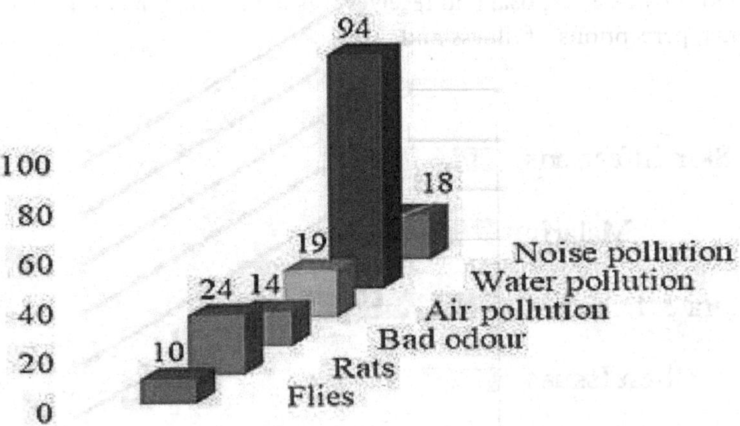

Figure 1: Respondents perceptions on environmental problems.

It can be dangerous to dispose of solid waste improperly, which frequently manifests itself by contaminating surface and ground water by way of leachate. About 90% of those surveyed stated that leachate from dump sites had contaminated the groundwater. Studies have revealed that groundwater contamination from landfills is unavoidable because leachate seeps into the ground by means of membrane cracks and does so because of the high bacterial content. The previous section of this study examined groundwater through laboratory analysis and discovered contamination from both study areas; however, the municipality supplied the area's source of drinking water. When asked about noise pollution, 16% of respondents said it was caused by the continual motion of vehicles transporting solid waste of the metropolitan area to the landfill; this could be because the landfill is closer to the city. Respondents' perceptions on the environmental problem is represented in figure 1.

3.3. Respondents' Perceptions of Illness and Disease

The study sought to determine the health effects that the landfills were having on the nearby residents. The number of respondents with typhoid, chest-related illnesses, cholera and diarrhoea, malaria, and skin infections, near the Bhanpur dumpsite is recorded as 52%, 9%, 23%, 14% and 7% respectively. Similar findings were reported by other studies, which demonstrated that the locals living close to the waste dumpsites frequently contracted respiratory infections, typhoid fever, skin infections and diarrhoea (Addo & Adei, 2015).

Additionally, according to the WHO (2000) people who live close to waste dumps are more likely to experience respiratory infections, severe intestinal infections, dermatological conditions, blood and retinal cancer, and cholera threats (Addo & Adei, 2015). Together, this study shed light on the possible negative effects of poor waste management on health. According to this study findings, Bhanpur dumpsite reported health issues largely. This is as a result of Bhanpur dumpsite being closed for waste disposal and receives waste on a daily basis. Figure 2 depicts respondents' perceptions of illness and disease.

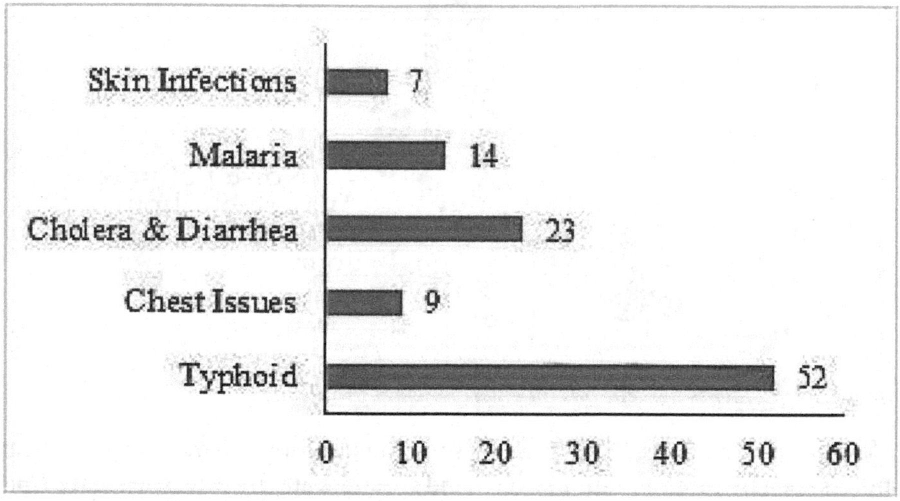

Figure 2: Respondents perception of on illness/ disease.

4. Conclusion

This study looked at how people close to the Bhanpur dumpsite in Bhopal, Madhya Pradesh, affected the natural environment and people's health. The outcomes of the data analysis showed that people of Bhanpur suffered from related diseases as a result of the dumpsite's proximity to their homes. The survey revealed that those who live close to the dumpsite are the ones who are most impacted by it. As a result, they suffered from typhoid, diseases of the skin, malaria, cholera and diarrhoea. This is due to the fact that the dumpsite is active and has numerous effects on local residents' health related to the environment. The study's respondents' overall health status and the pollution from the dumpsite are related. Additionally, it has been noticed that due to both ground water pollution along with obnoxious and disease-carrying odour, the severity of air and water pollution is direr during the rainy season. The study comes to the conclusion that the dumpsite needs to be properly maintained and managed in order to reduce its negative environmental effects. It is important to the Bhopal Municipal Corporation that these residents are relocated in a secure and hygienic environment for the improvement of the health status of those who live less than 50 metres from the dumpsite.

References

[1] Abul, S. (2010). Environmental and health impact of solid waste disposal at Mangwaneni dumpsite in Manzini: swaziland. *J Sustainable Dev Afr.* 12(7), 64–78.

[2] Adam, B. A. A., El-gader, A. B., and Abdelrhman, I. E. A. (2015). Health and environmental impacts due to final disposal of solid waste in Zalingy Town - Central Darfur State – Sudan. *International Journal of Research - Granthaalayah*, 4(11), 92–100.

[3] Addo, I. B., Adei, D., and Acheampong, E. O. (2015). Solid waste management and its health implications on the dwellers of Kumasi metropolis, Ghana. *Curr Res J Social Sci.*, 7(3), 81–93.

[4] Adeola, F. O. (2000). Endangered community, enduring people: toxic contamination, health, and adaptive responses in a local context. *Environ. Behav*, 32, 209–249.

[5] Alam, P., and Ahmade, K. (2013). Impact of solid waste on health and the environment. *Int J Sustainable Dev Green Econ.*, 2, 1.

[6] Bridges, O., Bridges, J. W., and Potter, J. F. (2000). A generic comparison of the airborne risks to human health from landfill and incinerator disposal of municipal solid waste. *Environmentalist.*, 20, 325–334.

[7] Garrod, G., and Willis, K. (1998). Estimating lost amenity due to landfill waste disposal. *Resour. Conserv. Recycl.*, 22, 83–95.

[8] Kafando, P., Segda, B., Nzihou, J., and Koulidiati, J. (2013). Environmental impacts of waste management deficiencies and health issues: a case study in the City of Kaya, Burkina Faso. *J Environ Prot.*, 4, 1080–1087. doi:10.4236/jep.2013.410124.

[9] Kataria, H. C., Gupta, M., Kumar, M., Kushwaha, S., Kashyap, S., Trivedi, S., Bhadoriya, R., and Bandewar, K. (2011). Study of physico-chemical parameters of drinking water of Bhopal city with Reference to Health Impacts. *Curr World Environ*, 6(1), 95–99 https://doi.org/10.12944/CWE.6.1.13.

[10] Kumar, S., Gaikwad, S. A., Shekdar, A. V., Kshirsagar, P. S., and Singh, R. N. (2004). Estimation method for national methane emission from solid waste landfills. *Atmos. Environ.*, 38, 3481–3487. https://dx.doi/10.1016/j.atmosenv..02.057.

[11] Ohwo, O. (2011). Spatial analysis of the quality of borehole water supply in Warri-Effurun Metropolis, Delta State, Nigeria. Ikogho. *A Multi-Disciplinary Journal.*, 9, 91–103.

[12] Okeke, C. U., and Armour, A. (2000). Post-landfill siting perceptions of nearby residents: a case study of Halton landfill. *Appl. Geogr.*, 20, 137–154.

[13] Omoniyi, O., and Olusosun. (2014). From waste to wealth. New Telegraph Newspaper, http://new telegraphonline.com/olusosun-dumpsite-from- waste-to-wealth (accessed 12 September 2015).

[14] Palmiotto, M., Fattore, E., Paiano, V., Celeste, G., Colombo, A., and Davoli, E. (2014). Influence of a municipal solid waste landfill in the surrounding environment: toxicological risk and odour nuisance effects. *Environ. Int.*, 68, 16–24.

[15] Sankoh, F. P., Yan, X., and Tran, Q. (2013). Environmental and health impact of solid waste disposal in developing cities: A case study of Granville brook dumpsite, Freetown, Sierra Leone. *J Environ. Prot.*, 4, 665–670.

[16] Shomoye, F. B., and Kabir, R. (2016). Health effects of solid waste disposal at a dumpsite on the surrounding human settlemens. *Journal of Public Health in Developing Countries*., 2(3), 268–275.
[17] Sonel, M., Mourya, M., Dwivedi, S., and Tiwari, D. R. (2010). Physico-chemical and bacteriological studies of ground water layers in Bhanpur Bhopal (M.P.). *Curr World Environ.*, 5(2), 379–382. https://doi.org/10.12944/CWE.5.2.27.
[18] Telang, S., and Chaturvedi, A. (2008). Status of ground water quality in Bhopal city. *Curr World Environ.*, 3(2), 313–316. https:// doi.org/10.12944/CWE.3.2.19.

Chapter 57

Electrochemical Properties of Polyaniline-Doped Nickel Oxide Nanocomposites

Arti M. Chaudhari[1] *and Bharti S. Anerao*[2]

[1]Department of Physics, Yeshwant Mahavidyalaya, Wardha, India

[2]Department of Physics, Dada Ramchand Bakhru Sindhu Mahavidyalaya, Nagpur, India

Abstract: Polyaniline-based inorganic nanocomposites have been attracting more attention since attachment of inorganic nanoparticles with conducting polymer turns out higher capacitances. We have synthesised nanocomposites that consist the combination of nickel oxide and polyaniline (PANI) and studied their electrochemical performance. The morphological characterisations and structural properties were carried out by using transmission electron microscopy (TEM) and X-ray diffraction (XRD). For nanocomposites electrode characterisation, an electrochemical impedance spectroscopy was conducted.

Keywords: Electrochemical impedance spectroscopy, pani/nickel oxide nanocomposites, x-ray diffraction.

1. Introduction

Polymer inorganic nanocomposites possess benefit of both polymer and inorganic nanoparticles (Wang and Shiff, 2007). Nanocomposites consist of filler in nanoform size particles which enhancing the properties of the nanocomposites (Gowrishankar *et al.*, 2006). The conducting polymer "PANI" have been chosen due to their high electrical conductivity and stability. PANI is useful in a variety of applications such as solar cell (Alet *et al.*, 2006), rechargeable batteries (Liu *et al.*, 2022), electrochemical sensors, electrochromic displays, capacitors and active corrosion protector (Pan *et al.*, 2022). Therefore, there has been increasing demand of the researchers to prepare PANI/ nanocomposites (Kamatchi *et al.*, 2018). The literature review on conducting polymer nanocomposites shows that PANI has been successfully utilised for the preparation of nanocomposites (Sun *et al.*, 2016) and to study further electric and magnetic properties (Ping *et al.*,2007). We focus on the effect of nickel oxide nanoparticles on the electrochemical properties of the PANI/ nanocomposites (Chunming *et al.*, 2009).

DOI: 10.1201/9781003527442-57

2. Synthesis of NiO and PANI/NiO Nanocomposites

NiO nanoparticles were synthesised by the simple method of chemical precipitation in which nickel carbonate hexahydrate ($NiCo_3.6H_2O$) used as precursor and starch was used as a capping agent (Li *et al.*, 2007). *In situ* chemical oxidative polymerisation of aniline using ammonium persulphate as an oxidant in the H_2SO_4 solution was utilised for the synthesis of PANI/NiO nanocomposites (Kalyane & Khadke, 2017). The electrochemical properties were studied by using a synthesised material as a working electrode as shown in reference (Babu *et al.*, 2015).

3. Results and Discussion

3.1. Transmission Electron Microscopy

Figure 1(A) shows that NiO nanoparticles were formed in spherical shape with size 25 nm. Figure 1(B) of PANI/NiO nanocomposite indicating the coating of PANI on the NiO nanoparticles. This result showed strong influence of NiO nanoparticles on prepared PANI (Qi *et al.*, 2009). This is due to the agglomerating tendency of nanoparticles, as well as the formation of clusters or agglomerates which decrease the interfacial interaction between the PANI matrix and NiO nanoparticles, resulting in the decreased specific capacitance values (Zhong-ai *et al.*, 2006).

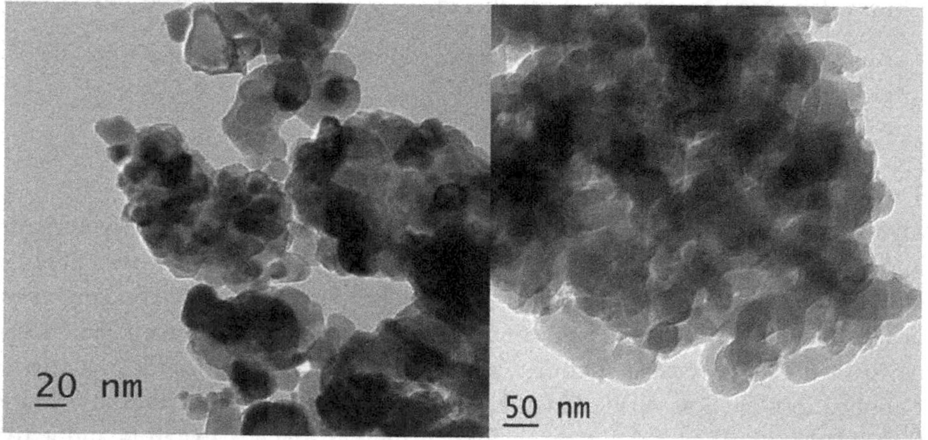

Figure 1: (A) TEM image of NiO. (B) TEM image of PANI/NiO nanocomposites.

3.2. Electrochemical Impedance Spectroscopy

Nyquist plots of synthesised NiO and PANI/NiO electrode were shown in Figure 2 in order to study electrochemical behavior of NiO and PANI/NiO electrode. In the nyquist plot of NiO, vertical line indicates psedocapacitive performance of the NiO electrode and diffusion behaviour of the electrolyte. Semicircle in high-frequency region indicating Faradic charge transfer resistance (Rct) which reflects surface properties of the electrode material. Furthermore, at high frequencies, PANI/

NiO nanocomposite electrode materials have an almost semi-circle appearance illustrating the lower charge transfer resistance (Rct) but it is higher than that of NiO electrode (Bhise *et al.*, 2019).

From the nyquist plot of the NiO electrode, low-charge transfer resistance showed high value of specific capacitance as compared to PANI/NiO electrode due to smaller grain size (23.54 nm) and higher surface area (Singh *et al.*, 2013). These remarkable and ideal properties of the NiO electrode material make it super electrode material as compared to the PANI/NiO electrode. Hence NiO electrode could be used for important application of supercapacitor for energy storage applications (Anerao *et al.*, 2020).

Figure 3 shows the XRD spectra of pure NiO and PANI/NiO nanocomposites as shown in reference (Kondawar *et al.*, 2012). XRD spectra of NiO showed (111), (200), (220), (311) and (222) hkl reflections at 2θ =37.25°°, 43.29°, 62.90°, 76.35° and 79.41°. During the reflux process, grain size reduction occurs, which results in broadened diffraction peaks.

3.3. X-ray Diffraction

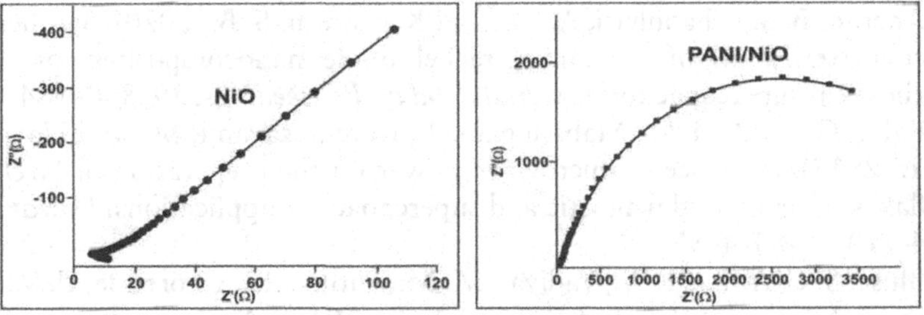

Figure 2: Nyquist plots of NiO and PANI/NiO electrode.

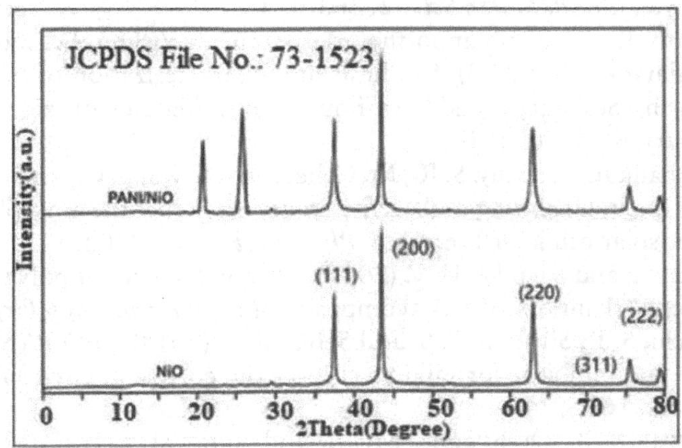

Figure 3: XRD spectra of NiO and PANI/NiO nanocomposites.

The average crystallite size of the NiO was found to be 23.54 nm using Debye Scherrer formula (Gobinath *et al.*, 2023). XRD pattern of PANI/NiO nanocomposites shows that strong interfacial interactions occur between polyaniline and nickel oxide nanoparticles.

4. Conclusion

PANI-nickel oxide nanocomposites were prepared by the method of chemical oxidative polymerisation with the use of ammonium persulphate as an oxidant in the H_2SO_4 solution. TEM picture of PANI-nickel oxide nanocomposites indicates NiO nanoparticles coated inside polyaniline with size 50 nm. The electrochemical impedance spectra of NiO electrode acquire enhanced pseudocapacitive nature because of smaller crystallite size with a spherical morphology, which indicates remarkable properties for an application in supercapacitor field.

References

[1] Alet, P. J. Palacin, S. Roca, P. I. C., Kalache, B., Firon, M., and Bettignies, R. (2006). Hybrid solar cells based on thin film silicon and P3HT. *Eur. Phys. J. Appl. Phys.* 36(3), 231–234.

[2] Anerao, B. S., Chaudhari, A. M., and Kondawar, S. B. (2020). Synthesis and characterization of polyaniline-nickel oxide nanocomposites for electrochemical supercapacitor. *Materials Today: Proceedings*, 29, 880–884.

[3] Babu, G. A., Ravi, G., Mahalingam, T., Kumaresavanji, M., and Hayakawa, Y. (2015). Influence of microwave power on the preparation of NiO nanoflakes for enhanced magnetic and supercapacitor applications. *Dalton Trans.* 44(10), 4485–4497.

[4] Bhise, S. C, Awale, D. V., Vadiyar, M. M., Patil, S. K., Ghorpade, U. V., Kokare, B. N., Kim, J. H., and Kolekar, S. S. (2019). A mesoporous nickel oxide nanosheet as an electrode material for supercapacitor application using the 1-(2,3 –dihydroxy propyl)-3- methylimidazolium hydroxide ionic liquid electrolyte. *J. Bull. Mater. Sci.* 42, 263, 1–10.

[5] Gobinath, E., Dhatchinamoorthy, M., Saran, P., Vishnu, D., Indumathy, R., and Kalaiarasi, G. (2023). Synthesis and characterization of NiO nanoparticles using Sesbania grandiflora flower to evaluate cytotoxicity. *Results in Chemistry.* 6(1011043),1–7.

[6] Gowrishankar, V., Scully, S. R., McGehee, M. D., Wang, Q., and Branz, H. M. (2006). Exciton splitting and carrier transport across the amorphous silicon/polymer solar cell interface. *Appl. Phys. Lett.* 89, 252102, 1–3.

[7] Kalyane, S., and Khadke, U. V. (2017). AC conductivity of polyaniline/mixed metal oxide (Pani/$NiCoFe_2O_3$) composites. *Int J Pure and App Phy.* 13, 201–9.

[8] Kamatchi, S. P., Sivakumar, S. and Selvaraj, S. (2018). NiO-PANI composite as potential inhibitor for mild steel in acidic corrosion environment. *J. Int. Chem. Sci.* 16(2), 268.

[9] Kondawar, S. B., Deshpande, M. D., and Agrawal, S. P. (2012). Transport properties of conductive polyaniline nanocomposites based on carbon nanotubes. *Int J Compos Mater.* 2(3), 32–36.

[10] Li, L. C., Jiang, J., and Xu, F. (2007). Synthesis and ferrimagnetic properties of novel Sm-substituted Li-Ni ferrite polyaniline nanocomposite. *Materials Letters*, 61(4), 1091–1096.

[11] Liu, Y., Chen, Y., Zhang, X., Lin, C., Zhang, H., Miao, X., Lin, J., Chen, S., and Zhang, Y. (2022). A high voltage aqueous rechargeable zinc polyaniline hybrid battery achieved by decoupling alkali acid electrolyte. *Chem. Engg. J.* 444, 136478.

[12] Pan, W., Dong, J., Gui, T., Liu, R., Liu, X., and Luo, J. (2022). Fabrication of dual anti-corrosive polyaniline microcapsules via Pickering emulsion for active corrosion protection of steel. 18(14), 2829–2841.

[13] Ping, X., Xijiang, H., Jingjing, J., Xiaohong, W., Xuandong, L., and Aihua, W. (2007). Synthesis and characterization of novel coralloid polyaniline/BaFe12O19 nanocomposites. *J. Phys. Chem.* C. 113(34),12603–12608.

[14] Chunming, Y., Haiyin, L., Dongbai, X., and Zhengyan, C. (2009). Hollow polyaniline/Fe_3O_4 microsphere composites: preparation, characterization, and applications in microwave absorption. *Reactive and Functional Polymers*. 69(2), 137–144.

[15] Qi, Y., Zhang, J., Qiu, S., Sun, L., Xu, F., Zhu, M., Ouyang, L., and Sun, D. (2009). Thermal stability, decomposition and glass transition behavior of PANI/NiO composites. *J. of Therm. Anal. and Calorim.* 98(2), 533–537.

[16] Singh, A. K., Sarkar, D., Khan, G. G., and Mandal, K. (2013). Unique hydrogenated Ni/NiO core/shell 1D nano- heterostructures with superior electrochemical performance as supercapacitors. *J. of Mater. Chem. A.* 1(41), 12759.

[17] Sun, B., He, X., Leng, X., and He, J. (2016). Flower-like polyaniline–NiO structures: a high specific capacity supercapacitor electrode material with remarkable cycling stability. *J. RSC Adv.* 6, 43959–43963.

[18] Wang, W., and Shiff, E. A. (2007). Polyaniline on crystalline silicon heterojuction solar cells. *App. Phys. Lett.* 91(13),133504-1–133504-3.

[19] Zhong-ai, H., Zhao, X. H., Kong, C., Yang, Y., Shang, X., Ren, L., and Wang, Y. (2006). The preparation and characterization of quadrate $NiFe_2O_4$/polyaniline. *J. Materials Science: Materials in Electronics*. 17(11), 859–863.

Chapter 58

Study of E-Waste Management Practices in Maharashtra State (India)

O.M. Ashtankar[1], Apeksha Agarwal[1], Sonali Bhor[1], Kirti Bothe[1], Akash Jadhav[2], and K. Sonimindia[2]

[1]Assistant Professor, BCACS Pune, Savitribai Phule Pune University, Pune, India
[2]Research Scholar, BCACS Pune, Savitribai Phule Pune University, Pune, India

Abstract: Rapidly increasing population, increase in GDP, purchasing power and technological advancement have resulted into electronic waste generation in huge quantity in the country, which is a matter of great concern for various countries across the world. Insufficient e-waste management infrastructure in India has resulted in approximately 95% of e-waste in India being illegally recycled by "kabadiwalas" or "raddiwalas" who are informal waste pickers resulting in uncontrolled mounting of e-waste across India. India stands in top five among e-waste producing countries across the world. Due to issues like awareness, law enforcement in India the problem of increasing e-waste is increasing. After careful literature review, research gap was identified in terms of public awareness as well as enforcement of e-waste management rules. This research work employees' descriptive study wherein data was collected by snowball sampling method through social media (online) platform. Interview of six local government bodies' officers of the concerned department and online survey of 385 persons in Maharashtra state was conducted for this research study. The specific objectives of this research are ascertaining e-waste management practices, e-waste awareness level of public and loopholes in e-waste policy implementation policy in India. The outcome of this research work will be useful in designing appropriate policies for management of e-waste by people and various local government bodies. It can also be useful for designing appropriate e-waste awareness campaigns.

Keywords: E-waste, recycle, legislature, kabadiwalas, raddiwalas, GDP, MT.

1. Introduction

Yoheeswaran (2013) defined E-waste as "waste electrical and electronic equipment, whole or in part or rejects from their manufacturing and repair process, which are intended to be discarded". In this techno-savvy world, uses of electronic devices have increased at an enormous pace in last two decades. Around

DOI: 10.1201/9781003527442-58

the globe, annually 50 million MT of estimate electronic waste is generated and among that India contributes around estimated e-waste of eight MT. In this fourth industrialisation revolution, e-waste in this big amount accounts for the upcoming challenge to human kind, environment, climate, etc. Issue become more worst due to the fact that majority of the developing and poor countries either lack basic infrastructure or are not aware about proper way of disposal or recycling of e-waste. Garg *et al.* (2019) said in India too e-waste is not disposed by the standard way. For many persons involved in e-waste activities like kabadiwalas, their understanding about e-waste management is earning livelihood that is collection of scrap without segregation of recyclable or throwable electronic devices. This kabadiwalas when sell or process the collected scrap they often prefer the cost-efficient way to manage which results in improper way of e-waste management. They prefer methods like open burning, use of hazardous chemical and many more irrelevant ways. But knowingly or unknowingly they are producing the crude of e-waste though some are recycled but maximum is either thrown with the ordinary garbage or are landfilled, which ultimately results in the generation of toxic material, soil erosion, etc.

1.2. E-Waste Generation Scenario in India

Indian Govt. Ministry of Environment report mentioned that only 32.9% e-waste is being recycled, whereas 67% (nearly 10,74,024 tonnes) remained unprocessed and lying idle in the environment posing greater challenge to the environment.

E-waste Metals	Health impact on organs
Arsenic	Skin Dieses, Lungs Cancer
Lead	Kidney, Reproductive/nervous system
Barium	Heart Muscles
Chromium	Liver, Kidney, Lungs
Mercury	Immune & Nervous system, liver, Kidney, brain
Cadmium	Bones, spine
CFC	Skin Cancer, Genetic Damage

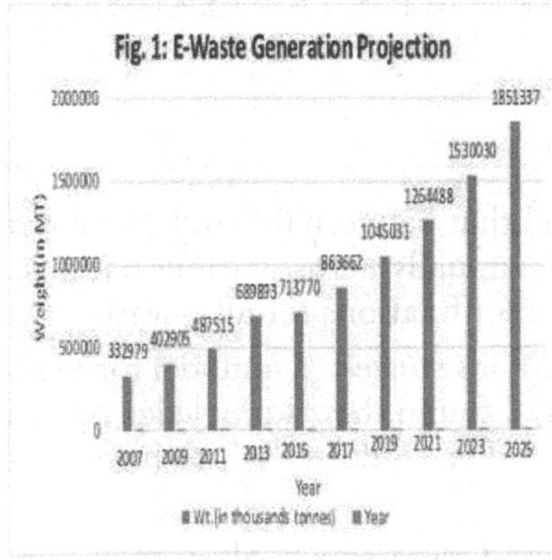

Fig. 1: E-Waste Generation Projection

1.3. *Challenges in Electronic Waste Management for Policy Design*

Following are the challenges being faced while formulating policies for e-waste management.

1) ***Lack of accurate information about e-waste generation across the country:*** For designing effective system of processing, collection and transportation, accurate information about quantity, composition and flows is necessary. In India this task is assigned to State Pollution Control Boards (SPCB). Sales and import data of electronic gadgets in the country can be useful to measure of e-waste, but this data is not available with SPCB
2) ***Processing by informal sector:*** In India majority of the e-waste is being processed by informal sector due to their efficiency in sourcing e-waste as compared to formal sector. Lack of awareness among citizens and cost of returning to formal e-waste collection centres are the main causes for reduced sourcing by formal sector.
3) ***Conservative Culture:*** As against "Use & throw" culture in Europe and America, India has a conservative culture. Hence, people do not like to throw away any item at free of cost. They use it to the maximum possible extent. Many at times it creates friction in the market as it becomes very difficult for formal sectors to source e-waste from the market.

1.4. Negative Effects of E-waste on Human Health E-waste contain various toxic substances like lead, lithium, mercury, barium, cadmium, polybrominated flame retardants, etc., which are very harmful for human health. These chemicals can damage heart, kidney, brain, liver and skeleton system of human body.

1.5. *Negative Effects of E-waste on Soil, Water & Air*

When e-waste is informally disposed-off, dismantled, burnt and melt harmful gas released in the environment resulting into air pollution and danger to respiratory health of living beings. Chronic dieses, allergies, cancer and asthma like diseases are caused due to air pollution caused due to e-waste. If e-waste is dumped in landfills or on ground without proper treatment heavy metals and flame retardants seep directly into the soil and ground water. Thus contamination of heavy metals in soil, water and crops endangers the life of various microorganisms, animals and other species.

2. Literature Review

- **Fadaei (2022)** found that between 0.8 and 30 kilogramme of electronic waste are produced annually across various nations. Landfills (safe or unsafe) are used by 73% of nations for all e-waste.
- **Sharma *et al.* (2023)** has suggested a model for explaining barriers to circular economy and found that "knowledge barrier" is one of the prominent barrier in oil and gas industry which hampers the implementation of circular economy.
- **Sant'Anna (2022)** used regression analysis to optimise the waste proportion

in manufacturing toner and its process. This model can be adopted in similar industries for e-waste control.

- **Charkraborty *et al.*** **(2022)** discussed about the metals extraction processes from e-waste. Research also mentioned about e-waste, metal and environmental pollution caused due to waste and extraction and challenges in extraction.
- **Sharma *et al*** **(2023)** discussed about how green logistics impacted circular economy adoption in the context of industry 4.0. Researcher found that the usage of digital technologies and circular economy has significant impact on reduction of e-waste.
- **Vijayan *et al.*** **(2022)** said supportive rules and regulations are necessary to promote recycling activities and alter behaviour.
- **Ismail (2021)** conducted study in the context of ASEAN countries and evaluated the potential economic value of e-waste generation. e-waste from mobile phone and its potential economic value was also the part of this study.
- **Lakshmi *et al.*'s** **(2021)** study highlighted the soil and water pollution due to e-waste. Also it mentioned about recycling it in sustainable and eco-friendly approach as also metals extraction revenue.
- **He, *et al.*** **(2021)** discussed about the pollution problems of e-waste due to its fast generating and colossal volume. Classification of e-waste and ways to manage based on that classification.
- **Neha and Adhana (2019)** conducted their study with help of secondary data. In research paper they highlighted the amount of e-waste and at which position India is placed in production and recycling.
- **Almulhim, A. I. (2022)** studied about awareness level of household and found that majority of them are unaware of the methods and benefits of segregation of e-waste. Lack of potential recycling options is also one of the causes of poor e-waste management. The researcher has advocated for educational interactions and efforts for raising public awareness to tackle the issue of e-waste
- **Yoheeswaran (2013)** in his research paper he raised scenario of legislation, waste recycling for the formal and informal sector in nation.

3. Research Methodology

Descriptive study was conducted for this research work wherein data was collected from 385 citizens in Maharashtra by snowball sampling method by online mode. The objectives for this research are as follows:

1. Ascertaining the e-waste management scenario & challenges in India.
2. Ascertaining the relationship between educational level of people and e-waste management awareness levels
3. To ascertain general awareness and level of awareness about e-waste management issues among the people of Maharashtra state

3.1. Hypothesis

$H_0$1: No correlation exist between education level and e-waste management awareness of people in Maharashtra state.

$H_0$2: People in Maharashtra State are unaware about e-waste management issue and its hazardous effects.

***3.2. Type of Research*:** Survey. Objectives of the study are related with ascertaining awareness level of people about e-waste management practices; hence, survey method is appropriate for this research work.

***3.3. Research Approach*:** Online/Digital mode. Since the focus of study is about the people who are mostly using electronic items like computers, mobile phone, etc. Online platform was thought appropriate for data collection.

Sample Units: People using electronic gadgets in Maharashtra State

Sample Size = 385

It is calculated based on the following formula.

Sample Size $n = \{Z^2 * P * Q * N\}/\{[e^2 (N-1)] + [z^2 * P * Q]\}$

$= \{(1.96)^2 * 0.5 * 0.5 * 112{,}374{,}333\}/\{[(0.05)^2 (112{,}374{,}333-1)] + [(1.96)^2 0.5 * 0.5\}$

$= 384.15$

where e = margin of error. Assuming it as 5%, n= Sample Size

N= Population size = 112,374,333(Maharashtra Population (Source: Census 2011)

Z= Normal standard variate...related to degree of confidence (for CL = 95%, Z = 1.96)

P= Estimated proportion of an attribute that is present in a population – the data regarding number of people using electronic gadgets across the country is not available but the information available on www.statista.com mentioned about 50% penetration rate of internet users in India. Hence, it is assumed that P = 0.5, Q = 1–P

4. Data Analysis

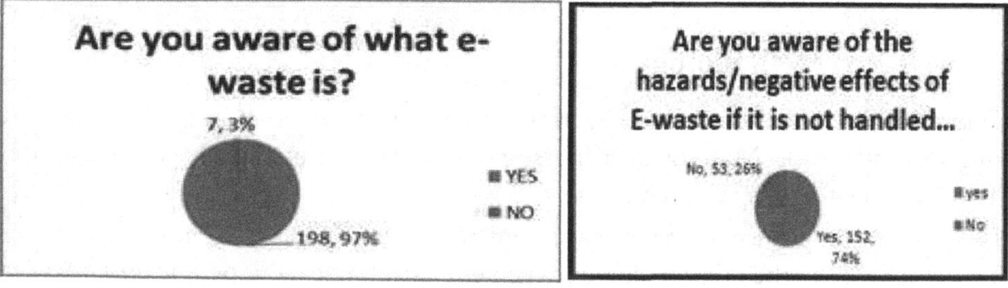

Figure 2: Awareness about e-waste and its hazardous effects.

Figure 2 above exhibits that the awareness of people in Maharashtra is 97%, and 74% people are having the information about its hazardous nature.

Table 2: Awareness of people about e-waste and related issues.

S.N.	Question	Mean
1	**I do have a comprehensive idea what electronic waste**	3.14878
2	**Aware of available options for disposal of e-wastes**	3.270732
3	**I am aware about e-waste but not sure for items falling in that category**	3.182927
4	**I am aware about e-waste but unaware actual application to my life.**	3.139024
5	**I am aware about Govt. regulation for e-waste management**	3.036585
6	**Aware of health and environmental ill impact due to e-waste**	3.113659
7	**I use to buy electronic gadgets even if older one is still working**	3.030244
TOTAL		3.131707

Table 1 above exhibits that the mean score of awareness of people in Maharashtra State about e-waste, management of it, government regulation, health hazards, etc. is 3.1317 on a five-point rating scale which shows that people are aware about the concept of e-waste and its effect on environment. But the level of awareness is low and it seems they are neutral in terms of their attitude towards e-waste management that is indifferent on the issue of e-waste management. These results indicate that extensive and impactful promotional campaign for e-waste awareness is needed in Maharashtra.

Therefore, Null hypothesis number 1 is rejected and it is concluded that the majority of the people in Maharashtra state are aware about the e-waste issues and their level of awareness is low.

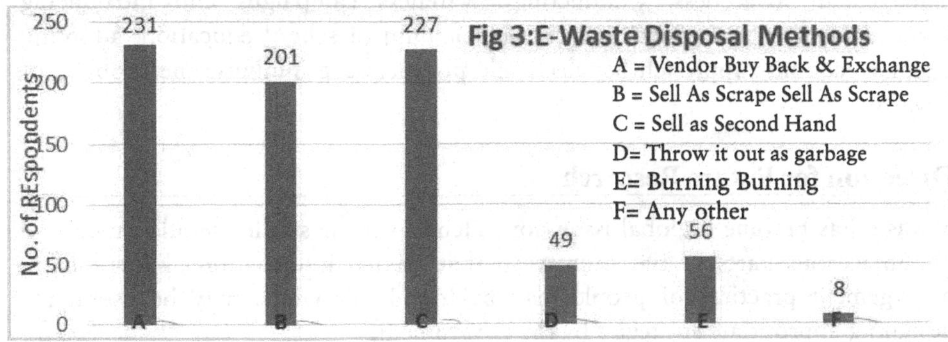

Figure 3: E-waste disposal methods.

Figure 3 above exhibits following details:

When it comes to disposing off the e-waste, around 60% either exchange the electronics for new one, 52% sell it as a scrap, 58% sell it as a second-hand item, surprisingly 35% and 14% of respondents throw and burn the e-waste, respectively.

Table 3: Education versus Awareness of people about e-waste & related issues.

Education Level	Below 10th Std	SSC	HSC	Degree	P.G.	PG Above
Awareness Level	3.0412	3.1223	3.4232	3.6344	3.8933	3.9254

Pearson correlation Coefficient: $R = \Sigma(X_i - \bar{x})*(Y_i - \bar{y}) / \sqrt{\Sigma(X_i - \bar{x})^2 * \Sigma(Y_i - \bar{y})^2}$

$\bar{x} = 3.5, \bar{y} = 3.5033, \Sigma(X_i - \bar{x})^2 = 17.5\ \Sigma(Y_i - \bar{y})^2 = 0.7077, \Sigma(X_i - \bar{x})*(Y_i - \bar{y}) = 3.46$

$R = 3.46/\sqrt{(17.5*0.7077)} = 0.9832$

Above result mention about Pearson correlation value indicates about significant but positive relationship between two variables that is X and Y, ($R(4) = 0.983$, $p < .001$). Hence, the null hypothesis is rejected. Therefore, it is concluded that there exists strong positive correlation between education and awareness about e-waste management & related issues. Therefore, Null hypothesis number 2 is rejected and it is concluded that there exists strong positive relationship between education level and e-waste management awareness level of people.

5. Conclusion

E-waste or electronic waste management problem is preaching more vigorously as a new major problem for human life existence. This research concludes that although majority of the people are aware about the concept of e-waste and its effect on environment, but the level of awareness is low and it seems they are neutral in terms of their attitude towards e-waste that is indifferent on e-waste management issue. It is found that majority of the people use to dispose of their e-waste by way of exchange and selling it as scrap to vendors. A strong positive correlation is found between the educational qualification of people and e-waste management awareness. Conducting awareness campaigns and introducing E-waste management practices in the curriculum of school education, adopting circular economy approach for electronic products can minimise the problem of e-waste.

Direction for Future Research

E-waste has become a global issue now. Hence, similar studies should be carried out in various parts of the country so that factual information about e-waste management practices of people may be found out which may be useful for designing appropriate strategies by the government.

References

[1] Almulhim, A. I. (2022). Household's awareness and participation in sustainable electronic waste management practices in Saudi Arabia. *Ain Shams Eng. J*. 13, 101729.

[2] Fadaei, A. (2022). E-waste management status worldwide: major challenges and solutions. *Proceedings of the International Academy of Ecology and Environmental Sciences*. 12(4), 281–293.

[3] Charkraborty, S. C., et al. (2022). Metals extraction processes from electronic waste: constraints and opportunities. *Environmental Science and Pollution Research International*. 29(22), 32651–32669.

[4] Neha, G., and Adhana, D. K. (2019). E-waste management in India: A study of current scenario. *International Journal of Management, Technology & Engineering*. IX (I).

[5] Sant'Anna, A. M. O. (2022). Statistical process monitoring for e-waste based on beta regression and particle swarm optimization. *The International Journal of Quality &Reliability Management, Bradford*. 39(7), 1663–1675.

[6] Yoheeswaran, E. (2013). E-waste management in India. *Global Research Analysis*. 2(4).

Chapter 59

A Short Review on Conducting Polymer "Polyaniline"- Conduction Mechanism and Applications

Shubhangi Bompilwar[1] *and Sneha Uttarwar*[2]

[1]Department of Physics, Cummins College of Engineering for Women, Nagpur, Maharashtra, India

[2]Department of Mathematics, Cummins College of Engineering for Women, Nagpur, Maharashtra, India

Abstract: The discovery of conducting polymers has spurred significant advancements in the realm of material science, paving the way for novel prospects in science and technology. The 21st century has borne witness to remarkable progress in technological domains such as microelectronics, sensory, biosensors as well as chemical and biochemical engineering, all of which centre around the utilisation of innovative nanostructured materials. Conducting polymers, a relatively recent material category, have undergone extensive exploration in the past decades. Notably, a new avenue of research and development has emerged, focusing on the nanostructurisation of conducting polymers and their composites with other materials, leading to the creation of intelligent materials tailored for current and future applications. A prominent subset within this area is the investigation of conducting polymer nanocomposites, which has garnered substantial interest. These specialised materials find their principal applications in diverse areas including electric energy storage systems, chemical and biosensors, biodegradable and medically compatible biomaterials, corrosion protection and electronic devices. Amid the various types of conducting polymers like polyaniline, polyacetylene, polypyrrole and PEDOT, polyaniline (PANI) has particularly stood out due to its straightforward synthesis, processability and stability. The focal point of this research article centres on comprehending the conduction mechanisms and exploring applications of the conducting polyaniline and its hybrid compositions across various sectors.

Keywords: Conducting polymer, nanomaterials, polyaniline, nanocomposite, applications.

DOI: 10.1201/9781003527442-59

1. Introduction

In the late 19th century, a significant breakthrough emerged with the revelation that certain types of polymers, while inherently poor conductors, could attain conductivity levels comparable to metals through chemical or electrochemical treatment. This discovery by Shirakawa *et al.* (1977) marked the inception of a fascinating realm in material exploration. These conducting polymers (CPs), unconventional plastic materials capable of conducting electricity, swiftly captured attention following their discovery by MacDiarmid *et al.* (1979). The appeal of CPs stems from their ability to amalgamate attributes like mechanical robustness, lightweight nature, resilience, flexibility and malleability with electrical conductivity akin to metals. The realm of CPs catalysed advances in comprehending the underlying chemistry and physics of π-bonded macromolecules, fostering fresh models to elucidate the mechanisms governing charge transport. Following the discovery of conducting polyacetylene by Shirakawa *et al.* (1977), the quest for additional CPs continued availability of roughly 25 distinct CPs to date. These conducting polymers, both in their pure form and composites, have found utility across diverse domains such as electronic devices, sensors, supercapacitors and biomedical engineering (Hui *et al.*, 2019; Ghorba, 2020). In recent times, the emergence of carbon nanotubes (MWCNT) has spurred the exploration of novel avenues. The hybrid of MWCNT with polyaniline (PANI) has taken centre stage in recent research. Leveraging its more planar configuration, PANI-CNT composites have found extensive application in electronics. They have demonstrated their prowess in photovoltaic cells, organic light emitting diodes, electromagnetic shielding, electrostatic dissipation, antennas and batteries (Sobha *et al.*, 2017).

2. Conduction Mechanism

Conjugated polymers exhibit an intriguing electronic property whereby chemical bonding results in the presence of a single unpaired electron (referred to as a π electron) per carbon atom. This electron delocalisation creates a pathway for efficient charge movement along the polymer chain, akin to a "highway" for charges. The electrical conductivity of these polymers is contingent on both the degree of doping and the alignment of the polymer chains. Consequently, doped conjugated polymers display remarkable conductivity due to two primary factors:

1. Doping introduces additional charge carriers into the polymer's electronic structure. As each repeating unit serves as a potential redox site, these conjugated polymers can be doped either in the n-type (reduced) or p-type (oxidized) direction, resulting in a significant density of charge carriers (Chiang *et al.*, 1978).
2. The attractive interaction between an electron within a given repeating unit and the nuclei in neighbouring units leads to the dispersion of charge carriers along the polymer chain. This delocalisation facilitates the mobility of charge carriers, extending not only along the polymer's length but also across three-dimensional space through inter-chain electron transfers.

Polyaniline, the second most researched conducting polymer after polypyrrole, has garnered attention due to its favourable doping–dedoping characteristics. Its responsiveness to environmental changes, manifested as alterations in resistance, renders it exceptionally versatile for diverse applications. These conducting polymers possess an electronic band configuration, wherein the energy gap differentiating the highest occupied π electron band (valence band) from the lowest unoccupied band (conduction band) dictates their inherent electrical and optical attributes. Through the process of doping, the polymer's electronic band arrangement undergoes modification, generating fresh electronic states within the energy gap, leading to observable colour shifts. Doping induces a shift in optical absorption towards lower energies, and the discernible colour disparity between the doped and undoped states is linked to the energy gap's properties (Bredas, 1986). The polymer chain acquires a charge upon doping, triggering a structural shift in the polymer chain's configuration. This transformation fosters the creation of distinct types of localized perturbations referred to as "solitons" or "polarons" (radical ions) and "bipolarons" (radical dions) (Nowak, 1862).

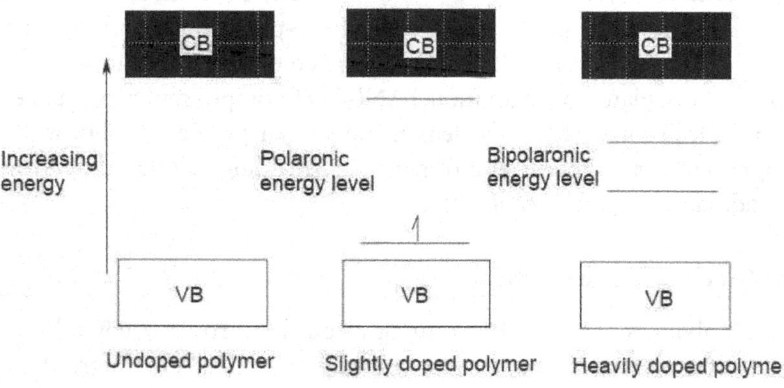

Figure 1: Formation of energy bands in CPs.

PANI, a prominent conducting polymer, holds a pivotal position within the realm of promising conductive polymers. It has garnered significant attention due to its favourable blend of economic viability, adeptness in combining chemical and physical traits, remarkable resilience in various environments and its wide array of practical applications. Over the last three decades, PANI has remained a focal point of intensive research due to its intricate chemistry and multifaceted properties. The inception of PANI traces back to the isolation of its monomer, aniline, from the pyrolytic distillation of indigo. This aniline, initially termed "Crystalline" displayed the ability to form well-defined crystalline salts when treated with sulfuric and phosphoric acids. In 1840, Fritzsche obtained colourless oil from indigo and named it aniline, derived from the Spanish term "Ail," meaning indigo. This substance was then oxidised to form polyaniline. Although the first conclusive report on polyaniline surfaced in 1862 by Letheby, scattered references to its structure emerged periodically throughout the early 20th century, continuing until the 1980s. A pivotal turning point arrived when MacDiarmid reexamined the earlier work of Josefowicz and stumbled upon the realisation that polyacetylene could attain electrical conductivity through protonic doping. Preceding MacDiarmid's findings, similar substantial conductivity had been observed in polypyrrole, an organic semiconductor and derivative of polyacetylene. Moreover, McGinness and collaborators had showcased a high-conductivity "ON" state within a bistable switch composed of melanin—a mixed copolymer of polyaniline, polyacetylene and polypyrrole. Subsequently, the exploration of polyaniline exploded, resulting in an extensive body of literature encompassing its synthesis, characteristics and practical applications.

***2.1. Oxidation States of Polyaniline*:** Following figure shows the structure of polyaniline polymerised from the aniline monomer. Polyaniline can be found in one of three idealised oxidation states (Lu *et al.*,1995).

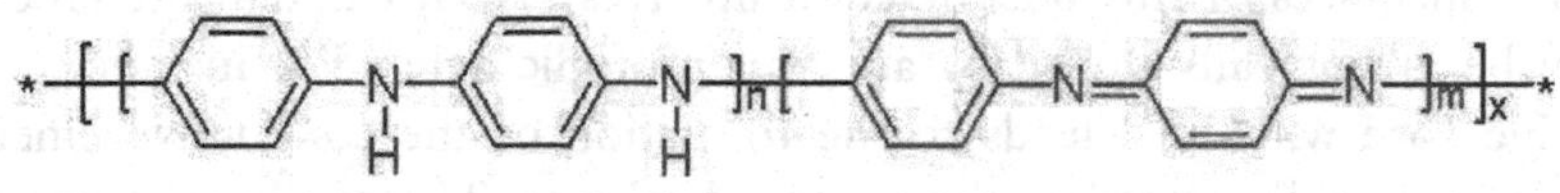

Figure 2: Main polyaniline structures, n + m = 1, x = degree of polymerisation

Leucoemeraldine (LE) – white/clear; Emeraldine (EB, ES) – green or blue; Pernigraniline (PN) – blue/violet.

In above figure, x equals half the degree of polymerization (DP). Leucoemeraldine with n = 1, m = 0 is the fully reduced state. Pernigraniline is the fully oxidised state (n = 0, m = 1) with imine links instead of amine links. The emeraldine (n = m = 0.5) form of polyaniline, often referred to as emeraldine base (EB), is either neutral or doped, with the imine nitrogen protonated by an acid. Emeraldine base is regarded as the most useful form of polyaniline due to its high stability at room temperature and the fact that upon doping the emeraldine salt form of polyaniline is electrically conducting. The only conducting state of PANI is the green coloured emeraldine salt (ES), which is the protonated form of EB. Leucoemeraldine and pernigraniline are poor conductors even when doped with an acid. All these oxidation states of PANI are depicted in the following Figure.

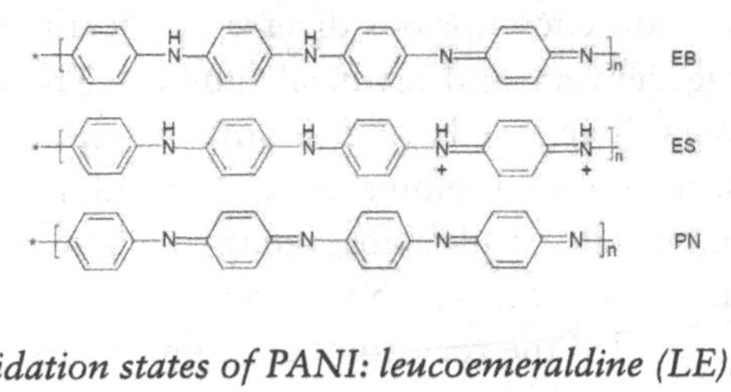

Figure 3: Different oxidation states of PANI: leucoemeraldine (LE).

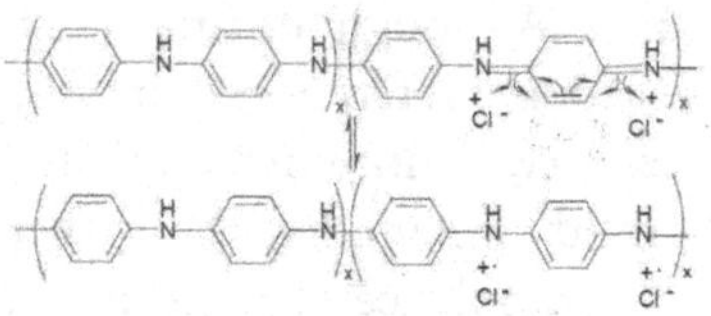

Figure 4: Protonic doping in PANI.

Emeraldine base (EB), emeraldine salt (ES) and pernigraniline (PN).

The fully reduced leucoemeraldine form of polyaniline (PANI) can undergo oxidative doping to transform into the conductive emeraldine state, achieved through either chemical or electrochemical doping methods. The process of gradually oxidising leucoemeraldine base to emeraldine salt and ultimately to pernigraniline through electrochemical means has been documented using cyclic voltammetry. Protonic doping is a process where a conjugated polymer is doped to enhance its conductivity without altering the polymer's electron count. One of the pioneering instances of this phenomenon involves PANI, wherein the emeraldine base can be protonically doped by aqueous protonic acids. For instance, treating emeraldine base with HCl leads to the formation of the conductive emeraldine hydrochloride, as illustrated in the provided scheme. The resulting dictation can adopt a resonance structure comprising two distinct polarons.

3. *Applications of PANI*

The versatility of PANI, which can exist in various forms through acid/base treatment and oxidation/reduction processes, has positioned it as a highly adaptable member within the family of conducting polymers. This adaptability has led to a broad spectrum of applications, with significant focus in electronics and the biomedical field (Pernaut and Reynolds, 2000).

3.1. Biomedical Engineering

In the realm of bioengineering and biomedicine, the convergence of medicine and engineering demands innovative technologies. Neuroscientists seek solutions that address nerve weaknesses and advance neuroscience. Biocompatible conductive scaffolds play a pivotal role in rectifying organ disorders. Notably, PANI's potential has been prominently explored in delivery systems, including novel structures like electro-drug delivery mechanisms.

The biomedical domain encompasses diverse applications such as biosensors, tissue engineering, drug delivery and artificial muscles. These applications centre on interactions between living cells and biomaterials, pivotal for diagnosing and treating diseases. Traditional biomaterials derived from nature were once used for their biocompatibility and biodegradability. However, the discovery of new biocompatible materials like carbons, ceramics, metals, composites and nanocomposites has expanded the repertoire of biomaterials. These materials offer enhanced properties such as improved compatibility, processability, mechanical strength and tunable characteristics through optimized synthesis methods (Li, 2018; Prakash & Pivin, 2015).

Notably, PANI-coated conductive cotton fabrics exhibit potent antibacterial and antifungal properties. The antibacterial function of PANI stems from its ability to donate or accept electrons, leading to electrostatic interaction with bacteria. This interaction disrupts bacterial walls, causing leakage of intracellular fluid and eventual demise (Pandiselvi *et al.*, 2015). Research has shown PANI composite materials, like chitosan-ZnO-PANI, to possess excellent antibacterial efficacy against both gram-positive and gram-negative bacteria compared to pure components. PANI's capability to undergo morphological and dimensional changes upon doping and de-doping, as well as its compatibility with bioactive materials, positions it as an attractive candidate for tissue engineering and artificial neural networks (Athukorala, 2021).

Nanofibrous polymeric scaffolds, mimicking the natural extracellular matrix, have been extensively explored, particularly in cardiac tissue engineering, using PANI and polymers such as PLA or gelatin (Ostrovidov *et al.*, 2017; Wang, 2017). PANI has demonstrated compatibility with skin and has shown negligible irritation and sensitisation effects (Humpolice *et al.*, 2012). Furthermore, the porous nature and high surface area-to-volume ratio of PANI-based nanocomposites make them appealing for targeted drug delivery systems, particularly crucial in diseases like cancer. In essence, the multifaceted properties of PANI, ranging from its diverse forms to its interactions with biological entities, have led to its widespread utilisation in various applications, prominently in the biomedical field.

3.2. Electronics

Research and development in nanomaterials focuses on innovating novel devices essential for meeting the current demands in areas such as microelectronics, sensors and energy storage. One prominent nanomaterial is PANI and its composites, which exhibits versatile applications in electronics due to its controllable conductivity spanning from insulator to metallic states. Conducting Polymers (CPs) can be synthesised in diverse structural forms such as 1D, 2D, quantum dots and amorphous materials.

Gas sensing is critical due to the adverse impact of toxic gases on human health and the environment. Detection of harmful gases like CO_2, CO, SO_2 and NH_3 is pivotal. CPs due to their acid-base chemistry has demonstrated effectiveness as sensor materials compared to metal oxides. PANI, unique in its doping–dedoping mechanism, displays distinct chemical structures and response mechanisms to various gases. Gas sensors translate chemical interactions into electrical signals. PANI and its composites have been explored for sensing gases like NH_3 (Liu *et al.*, 2018; Du *et al.*, 2011; Pawar *et al.*, 2012; Sharma *et al.*, 2013), CO (Misra *et al.*, 2004; Liu *et al.*, 2012), Hydrogen (Chen *et al.*, 2011; Fowler, 2009) and H_2S (Shirsat *et al.*, 2009; HasanSuhail *et al.*, 2019), among others. In the realm of energy storage, traditional materials have limitations in terms of conductivity and electrochemical behaviour. PANI stands out with its high capacitance, adjustable conductivity and variable oxidation states, making it a suitable candidate for energy storage devices like capacitors and supercapacitors. PANI composites with materials like NiO (Gautham *et al.*, 2022; Nandapure *et al.*, 2013), MnO_2, CNTs and Graphene (Chao Yang *et al.*, 2017; Zhang, 2009; Hui, 2019) have effectively contributed to high-performance energy storage applications. Pranob and Pratap (2011) studied Schottky Junction with PANI thin films and aluminium metal and found diode ideality factor much larger than unity.

PANI-CdS nanocomposites have been explored for solar cell applications, acting as active layers in photovoltaic devices. Employing a straightforward screen-printing technique, an organic–inorganic diode based on PANI and CdS has been fabricated, demonstrating rectification effects. The successful fabrication of photovoltaic devices using PANI-CdS nanocomposites and nano frequency-tuned devices (Kondawar *et al.*, 2011) has also been reported. Polystyrene /PANI/Ag composite has been synthesised by environmentally friendly method by Liao (2018) and found its effective anticorrosive properties over pure PANI.

4. Conclusion

Conducting polymers (CPs) with conjugated π bonds have gained significant attention due to their unique electronic and optical properties. These properties enable their use in a wide range of applications, from insulators to conductors. Polyaniline stands out due to its rich acid–base chemistry, which makes it highly versatile for various applications. Its doping and de-doping processes can tune its conductivity and other properties. PANI has found applications in electronics, sensors, biosensors, textiles, transparent electrodes, energy storage devices (supercapacitors) and many more. This versatility stems from its adjustable properties and compatibility with different substrates. While PANI offers remarkable properties, its inherent hydrophobicity limits its use in certain applications, such as scaffolds and biosensors that require good cell adhesion. Surface functionalisation methods are being explored to improve biocompatibility and cell adhesion, broadening its potential applications. PANI nanocomposites with high surface area have been effectively employed in drug delivery systems. The incorporation of nanofillers allows for improved drug loading and controlled release, capitalising on PANI's conductivity and surface properties. A challenge with PANI nanocomposites lies in achieving uniform distribution and dispersion of nanofillers within the PANI matrix. Surface modification methods have been developed to address this issue, ensuring optimal performance and properties of the composite. PANI composites with nanofillers like carbon black, carbon nanotubes, carbon nanofibers, graphene and fullerene have garnered significant research interest. These nanofillers can be strategically incorporated to tune mechanical, optical and conducting properties of the composite material.

References

[1] Athukorala, S. S., Tran, T. S., Balu, R., Truong, V. K., Chapman, J., Dutta, N. K., Choudhury, R. N. (2021). 3D printable electrically conductive hydrogel scaffolds for biomedical applications: a review. *Polymers*, 13(3), 474.

[2] Bompilwar, S. D., Kondawar, S. B., and Tabhane, V. A. (2010). Impedance study of nanostructure cadmium sulfide and zinc sulfide. *Archives of Applied Science Research*, 2 (3), 225–230.

[3] Bredas, J. L. (1986). Handbook of conducting polymers. Ed. T ASkotheim, Marcel Dekken, New York, 849.

[4] Chiang, C. K., Gau, S. C., Fincher, C. R. Jr., Park, Y. W., and MacDiarmid, A. G. (1978). Polyacetylene, (CH)x: n-type and p-type doping and compensation. *Appl. Phys. Lett.*, 33(1), 18–20.

[5] Chiang, C. K., Fincher, C. R. Jr., Park, Y. W., Heeger, A. J., Shirakawa, H., Louis, E. J., Gau, S. C., and Alan, G. M. (1977). Electrical conductivity in doped polyacetylene. *Phys. Rev. Lett.* 39, 1098.

[6] Fattahi, P., Yang, G., Kim, G., and Abidian, M. R. (2014). A review of organic and inorganic biomaterials for neural interfaces. *Adv. Mater.* 26, 1846–1885.

[7] Feast, W. J., Tsibouklis, J., Pouwer, K. L., Groenendaal, L., and Meijer, E. W. (1996). Synthesis, processing and material properties of conjugated polymers. 37(22), 5017–5047.

[8] Fowler, J. D., Virji, S., Kaner, R. B., Weiller, B. H. (2009). Hydrogen detection by polyaniline nanofibers on gold and platinum electrodes. *J. Phys. Chem. C.* 113, 6444–6449.

[9] Gautam, K. P., Acharya, D., Subedi, V., Das, M., Neupane, S., Chhetri, K., and Yadav, A. P. (2022). Nickel oxide-incorporated polyaniline nanocomposites as an efficient electrode material for supercapacitor application. *Organics*, 10(6):86. Ghorbani, F., Zamanian, A., and Aidun (2020). A. conductive electrospun polyurethane-polyaniline scaffolds coated with poly(vinyl alcohol)-GPTMS under oxygen plasma surface modification. *Mater. Today Commun.* 22, 100752.

[10] Hui J., Wei D., Chen J., and Yang Z. (2019). Polyaniline nanotubes/carbon cloth composite electrode by thermal acid doping for high-performance supercapacitors. Polymers. 11:2053.

[11] Humpolicek, P., Kasparkova, V., Saha, P., and Stejskal, J. (2012). Biocompatibility of polyaniline. *Synth.Metals*.162, 722–727.

[12] Kondawar, S. B., Thakare, S. R., Khati, V., and Bompilwar, S. (2009). Nanostructured titania reinforced conducting polymer composites. *International Journal of Modern Physics B*, 23(15), 3297–3304.

[13] Letheby, H. (1862). On the production of a blue substance by the electrolysis of sulphate of aniline. *J chemical Society*, 15, 161–163.

[14] Liao, G. (2018). Green preparation of sulfonated polystyrene/ polyaniline/ silver Composites with enhanced anticorrositive properties. *International Journal of Chemistry*,10(1), 81–86.

[15] Li, Y. (2018). Challenges and issues of using polymers as structural materials in MEMS: a review. *J. Microelectromech. Syst.* 27, 581–598.

[16] Liu, C., Tai, H., Zhang, P., Yuan, Z., Du, X., Xie, G., and Jiang, Y. (2018). A high-performance flexible gas sensor based on self-assembled PANI-CeO_2 nanocomposite thin film for trace-level NH_3 detection at room temperature. *Sensor & Actuator. B: Chem.*, 261, 587–597.

[17] Lu, W. K., Elsenbaumer, R. L., and Wessling, B. (1995). Corrosion protection of mild steel by coatings containing polyaniline. *Synth. Met.* 71(1–3), 2163–2166.

[18] Nandapure, B. I., Kondawar, S. B., Salunkhe, M. Y., and Nandapure, A. I. (2013). Magnetic and transport properties of conducting polyaniline/nickel oxide nanocomposites. *Adv. Mat. Lett.* 4(2), 134–140.

[19] Nigrey, P. J., MacDiarmid, A. G., and Heeger. (1979). Electrochemistry of polyacetylene,(CH)x: electrochemical doping of (CH) x films to the metallic state. *Chem.Commun.* 96, 594–595.

[20] Nowak, S. S., Rughooputh, V., Hotta, S., and Heeger, A. J. (1987). Polaronas and bipolarons on a conducting polymers in solution. *Macromolecules*, 20(5), 965–968.

[21] Ostrovidov, S., Ebrahimi, M., Bae, H., Nguyen, H. K., Salehi, S., Kim, S. B., Kumatani, A., Matsue, T., Shi, X., and Nakajima, K. (2017). Gelatin–polyaniline composite nanofibers enhanced excitation–contraction coupling system maturation in Myotubes. *ACS Appl. Mater. Interfaces.* 9, 42444–42458. Pandiselvi, K., and Thambidurai, S. (2015). Synthesis, characterization and antimicrobial activity of chitosan–zinc oxide/polyaniline composites. *Materials Science in Semiconductor Processing.*, 31, 573–581.

[22] Park, J. H., Park, O. O., Shin, K. H., Jin, C. S., and Kim, J. H. (2001). An electrochemical capacitor based on a Ni (OH) 2/activated carbon composite electrode. *Electrochem.Solid State Lett.* 5(2), H7–H10.

[23] Pawar, S. G., Chougule, M. A., Sen, S., and Patil, V. B. (2012). Development of nanostructured polyaniline–titanium dioxide gas sensors for ammonia recognition. *J. Appl. Polym. Sci.*, 125,1418–1424.

[24] Pernaut, J. M., and Reynolds, J. R. (2000). Use of conducting electroactive polymers for drug delivery and sensing of bioactive molecules. *A Redox Chemistry Approach J. Phys. Chem. B*, 104, 4080.

[25] Prakash, J., Pivin, J. C., and Swart, H. C. (2015). Noble metal nanoparticles embedding into polymeric materials: from fundamentals to applications. *Adv. Colloid Interface Sci.* 226, 187–202.

[26] Suhail, M. H., Gh. Abdullah, O., and Ghada Ayad, K. (2019). Hydrogen sulfide sensors based on PANI/f-SWCNT polymer nanocomposite thin films prepared by electrochemical polymerization. *Journal of Science: Advanced Materials and Devices.*, 4(1), 143–149.

[27] Shirakawa, H., Louis, E. J., MacDiarmid, A. G., Chiang, C. K., and Heeger, A. J. (1977). Synthesis of electrically conducting organic polymers: halogen derivatives of polyacetylene, $(CH)_x$. *J Chem. Soc. Chem. Commun.*, 578–580.

[28] Sobha, A. P., Sreekala, P. S., and Narayanankutty, S. K. (2017). Electrical, thermal, mechanical and electromagnetic interference shielding properties of PANI/FMWCNT/TPU composites. *Prog. Org. Coat.*, 113, 168–174.

[29] Wang, L., Wu, Y., Hu, T., Guo, B., and Ma, P. X. (2017). Electrospun conductive nanofibrous scaffolds for engineering cardiac tissue and 3D bioactuators. *ActaBiomater.*, 59, 68–81.

[30] Yang, C., Zhang, L., Hu, N., Yang, Z., Su, Y., Xu, S., Li, M., Yao, L., Hong, M., and Zhang, Y. (2017). Rational design of sandwiched polyaniline nanotube/layered graphene/polyaniline nanotube papers for high-volumetric supercapacitors. *Chemical Engineering Journal*, 309 (1), F 89–97.

[31] Zhang, K., Zhang, L. L., Zhao, X. S., and Wu, J. (2010). Graphene/polyaniline nanofiber composites as supercapacitor.electrodes. *Chem. Mater.*, 22, 1392–1401.

Chapter 60

Amendments on Amalgamation, Categorisation and Solicitation of Polypyrrole Nanocomposites (PPy NC) and their optimization – An update

Subodh K. Sakhare[1], T. D. Kose[2], and Rajdip Utane[3]

[1]Department of Chemistry, Shri Sadguru Saibaba Science & Commerce College, Ashti, Affiliated to Gondwana University Gadchiroli (MH), India

[2]Dr. Khatri Mahavidyalaya, Tukum Chandrapur, Affiliated to Gondwana University Gadchiroli (MH), India

[3]Department of Chemistry, Sant Gadge Maharaj Mahavidyalaya, Hingna, Affiliated to RTM Nagpur University, Nagpur (MH), India

Abstract: In recent study, nano composites of polypyrrole have been synthesised by novel and green approach towards the synthesis such as microwave and ultra-sonication in polymerisation of pyrrole. The nanocomposites were synthesised by mixing polypyrrole and silver nitrate (AgNO3) in different weight percentages. The development of nanocomposites and uncertainties in the physical substantial composites were investigated by characterising the samples using FTIR, XRD, SEM and EDX examination. The dimensions of the constituent part of nano composites was analysed by XRD. Those synthesized and isolated PPy-NC utilised for pharmacological activity and the effect of different PPy-NC with route of synthesis were compared.

Keywords: polypyrrole, nanocomposites, microwave, ultra-sonication pharmacological activity.

1. Introduction

In the context of a lack of new antibiotic manufacture, the development of antibiotic resistance by microorganisms has been viewed as a global emergency and one of the most important issues for global public health. Consequently, it becomes very vital to explore methods of avoiding the usage of antibiotics. The use of chemical composites at the nanoscale in place of conventional medications offers significant benefits for the inactivation of new resistance mechanisms acquired by various microbes. The use of conducting polymers as antibacterial agents whether alone, in combination or in conjunction with antibiotics has been viewed in this approach as

DOI: 10.1201/9781003527442-60

a viable method aimed at novel antiseptic structures. Now individual, polypyrrole devises remained quoted such as unique of the furthermost noteworthy carbon-based resources in the collected works. It has been efficaciously used in plentiful solicitations. Thanks to its excellent electrical properties, simplicity in synthesis, high stability in ambient conditions and favourable redox properties (Hullmann and Meyer, 2003).

1.1. Green Nanotechnology

Nanotechnology is seen as a crucial 21st-epoch technology and partakes created a lot of eagerness around the ecosphere nonetheless him partakes slackened downcast due to a lack of acquaintance approximately the menaces concomitant with it and a deficiency of dogmata to control evolving concerns. However, canvassers keep working hard to overcome obstacles related to managing, producing, funding, regulatory and technical elements. Green industrial and green chemistry are dual subfields of green expertise that kind use of plant-based materials. Green nanotechnology is one such subfield (Balbus *et al.*, 2007; Zou *et al.*, 2008). Through the use of less material and renewable inputs when available, it decreases the consumption of energy and fuel. Additionally, by safeguarding underdone ingredients, energy and water as well as by miserable conservatoire emanations besides unsafe surplus, nanotechnological things, developments and applications are estimated to critically subsidise to conservation and climatic fortification. The crucial roles of green nanotechnology embrace amplified liveliness effectiveness, worse unwanted and greenhouse smoke emanations and declined intake of non-renewable renewed constituents. Green nanotechnology is a fantastic opportunity to prevent negative impacts from happening in the first place (Benn and Westerhoff, 2008; Besley *et al.*, 2008; Guo *et al.*, 2007).

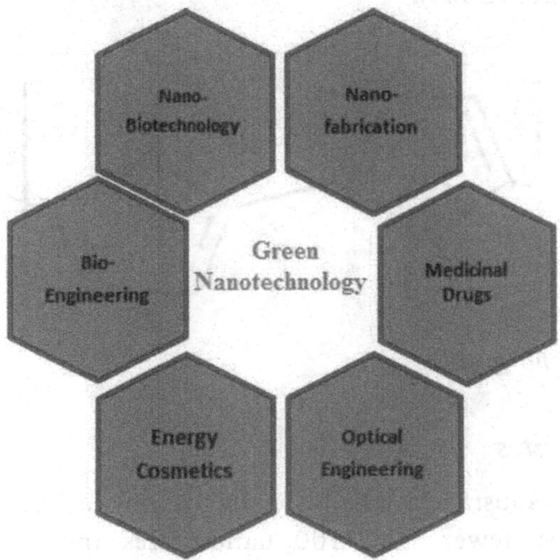

***Figure1:** Global fragments of green nanotechnology.*

Green chemistry and green engineering are the foundations upon which green nanotechnology is built, rather than it emerging from nothing. Green nanotechnology applications could include the use of nanomaterials in clean production processes that synthesise nanoparticles using sunlight or by recycling industrial waste products into nanomaterials, aside from such apparent fields as the development of solar cells, biofuels and fuel cells (Hristozov and Ertel, 2009; Jayapalan *et al.*, 2013; Som *et al.*, 2011). There is some "really" green nanotechnology, such as the full growth of nanoelements in vegetal life but these exertions will certainly not extent the weighbridge obligatory for the manufacture of nanomaterials on an industrial scale. Green nanotechnology requires a thorough process assessment in order to get firm conclusions, much like other industrially produced goods.

1.2. Polypyrrole

An organic polymer called polypyrrole (PPy) is created when pyrrole is subjected to oxidative polymerisation. The solid has the chemical formula $(C_4H_2NH)_n$. It is a polymer with inherent conductivity that is utilised in the disciplines of optics, biology and medicine. Chemical sensors and electrochemical applications are the two main uses of PPy and similar conductive polymers in electronic devices. PPy is an insulator but its oxidised derivatives work well as conductors of electricity. The oxidation conditions and reagents utilised have an impact on the material's conductivity. The range of conductivities is 2–100 S/cm. Larger anions, such tosylates, are linked to higher conductivities. The material must expand during doping in order to accommodate the charge-compensating anions. The physical alterations brought on by this charging and discharging have been compared to synthetic muscle. Ionic diffusion through PPy films exhibits anomalous diffusion patterns and has fractal features on the surface (Hutchison 2008; Lam *et al.*, 2006; Rickerby and Morrison, 2006).

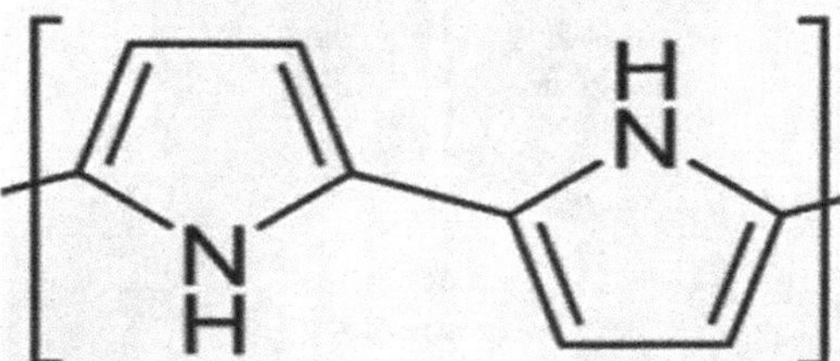

Figure 2: *Structure of polyprrole.*

1.3. Nano Composites

A multiphase solid substance called a nanocomposite has one, two or three dimensions that are fewer than 100 nanometres (nm) or structures with nanoscale reappearance reserves between the various phases that make up the material (Carnovale *et al.*, 2016; Eastlake *et al.*, 2012; Schulte *et al.*, 2013).

The perception overdue nanocomposite is to enterprise and fabrication new materials with noteworthy springiness and perfection in their corporal chattels by consuming erection slabs with distances in the nanometre assortment. In its broadest sense, this description can refer to porous media, colloids, gels and copolymers, although it is more frequently used to describe the solid mixture of a bulk matrix and nano-dimensional phase(s) with dissimilar chattels as a product of basic and substance metamorphoses. The nanocomposite will have ominously altered automatic, electrical, thermal, optical, electrochemical and catalytic assets on or after its apparatuses. For these impacts, size precincts obligate remained suggested.

1.4. PPy Nano Composites

Silver's antimicrobial properties have been widely documented in the literature. The operative exploit of silver contrary to microbes is a extent reliant on headway, as has remained beforehand predictable for Pd-based schemes. Subdivisions with a size in the order of 7 nm penetrate the nuclear content and release Ag+ ions, which is favoured by the high accessible surface area of nanoparticles. (Khan *et al.*, 2014; Manuel *et al.*, 2018)

Based on this circumstance, the level of aggregation of structures that may be easily led from the growth of composite edifices determines the incorporation of silver nanoparticles on the polymeric matrix. Different synthesis methods have been developed to provide appropriate dispersion of silver nanoparticles on various support morphologies, as follows: the character accumulation of methyl orange creates microscale chambers that are investigated as templates for the polymeric growth in the process of creating cylindrical polymeric templates embellished with silver chloride of (Centi *et al.*, 2011 and Hua *et al.*, 2012).

The polymerisation occurs when ammonium persulfate is used as an oxidising agent in the presence of silver nitrate and pyrrole monomers and the polypyrrole chains develop on tubular supports, taking on the form of hollow tubes embellished with silver chloride nanoparticles. Upadhyay *et al.* described the creation of silver nanoparticle-adorned polypyrrole nano amalgams from *in situ* reductions of silver nitrate as a different option. The amount of silver in the composite and the overall antibacterial activity of the finished product were found to be directly correlated by the authors. The general mechanisms underlying this antibacterial system involve the release of Ag+ ions into the cells and the electrostatic contact between the negatively charged bacteria and the conductive layer of polypyrrole. Huh and Kwon (2011) and Munguia and Nizet (2017) disclosed another intriguing characteristic for the silver nanoparticle-decorated nanostructures. In this procedure, pyrrole monomers were used to saturate silica nanoparticles and silver nitrate was investigated as a photosensitiser. Silver nanoparticles were created in silica PPy composites as a result of the polymerisation, which was started by ultraviolet irradiation by Bansal *et al.* (2019).

2. Literature Review

2.1. *Methodology (Polymerization of Pyrrole)*

2.1.1. Microwave Method

Using pyrrole monomer and various oxidants, the chemical oxidative polymerisation method was used to create PPy. In a beaker with 100 ml of distilled water, the chemical polymerisation was conducted by combining various molar ratios of pyrrole monomer, an oxidant and a surfactant.

2.1.2. Ultra-sonication Method

Via pyrrole monomer and various oxidants, PPy was created using a chemical oxidative polymerisation process. Different molar ratios of pyrrole monomer, an oxidant and a surfactant were used to do the chemical polymerisation in a beaker containing 100 ml of pure water.

2.1.3. Polypyrrole Nanocomposites

Polypyrrole nanotubes were dissolved in 25 ml of DD water, and then ultrasonically processed for 20 minutes. AgNO3 was calculatedly added and stirred into the PPy-NT solution for up to two hours. Separately, sodium borohydride solution was made by combining it with 75 ml of DD water. Drop-by-drop, magnetic stirring was used to combine the silver nitrate solution with the sodium borohydride solution for 24 hours at room temperature. Throughout the synthesis, a 1:2 molar ratio of sodium borohydride to silver nitrate was maintained. The filtrate was then dried after being extensively washed with ethanol and DD water. AgNO3 concentrations of 3, 6, 9, 12 and 15 weight percentage with respect to PPy-NTs were changed to create four distinct compositions with variable concentrations of Ag-NP.

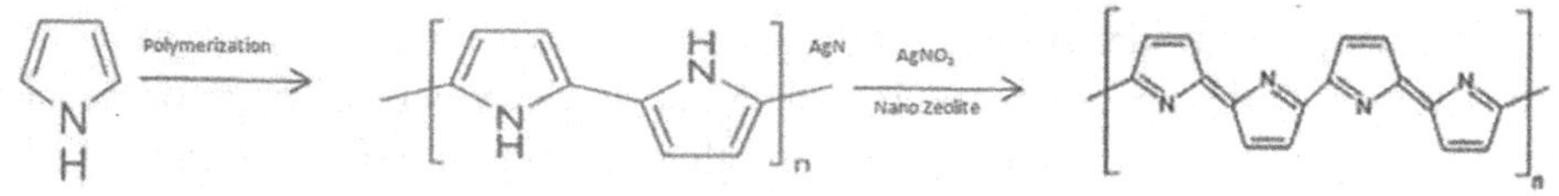

Figure 3: *Syntheis of polypyrrole nanocomposites.*

3. Characterization

3.1. *FTIR*

Fourier transform infrared (FTIR) spectroscopy, electrochemical comportment of the revised conductors was through by cyclic voltammetry (CV) in nanocomposite redox autopsy. As revealed by UV spectrophotometer, the absorption peaks ranged from 280 to 295 nm for all synthesised nanoparticles.

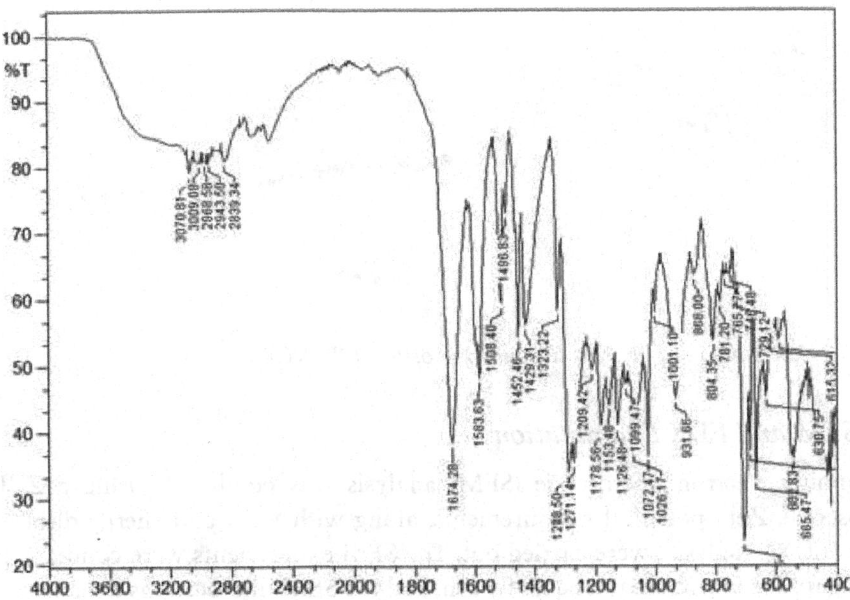

Wavenumber (cm^{-1})

Figure 4: FTIR data of polypyrrole nanocomposites (PPy NC).

3.2. XRD

X-ray diffraction (XRD) spectra of synthetic zeolites were obtained using the Bruker AXS D8 Advance analysis system, with measurements conducted in a temperature range from 1800 °C to 4300 °C. The measuring circle diameters used were 435 mm, 500 mm and 600 mm, with a predefined angle range of 3600. The maximum usable angular range was set between 30 and 1350 degrees. The X-ray source utilised was copper (Cu), with a wavelength of 1.5406 angstroms. The detector employed was Si(Li) PSD.

Samples were scanned within the 2θ range of 10°–50°, where θ represents the angle of diffraction. Various crystalline phases present in the samples were identified with the assistance of JCPDS (Joint Committee on Powder Diffraction Standards) files for inorganic compounds. The quantitative assessment of the crystallinity of the synthesised zeolite was determined by summing the heights of the major peaks observed in the X-ray diffraction patterns.

High-energy X-ray diffraction (HE-XRD) diffractograms provided insight into the crystalline structure of the compound. Crystallite size calculations were performed using the X-ray diffraction pattern of the PPy nanocomposite. The diffraction peaks for chemically synthesised materials were associated with their corresponding lattice/Miller indices in parentheses. However, XRD patterns obtained from plant extracts were devoid of discernible peaks.

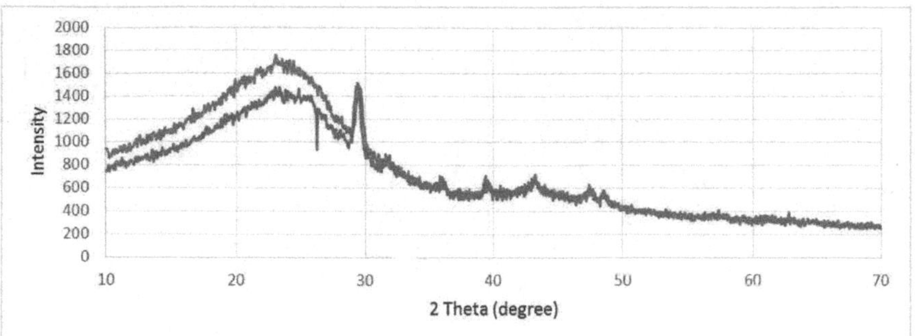

Figure 5: XRD data of polypyrrole nanocomposites (PPy NC).

3.3. SEM and EDX Examination

A scanning electron microscope (SEM) analysis was conducted using a ZEISS microscope. Zeta potential measurements, along with SEM and energy-dispersed X-ray (EDX) analyses, were carried out. The SEM experiments were conducted to investigate the morphology of coal fly ash, and the SEM micrographs depicted coal fly ash at 6.32 KX magnification and zeolite structure at 3.12 KX magnification.

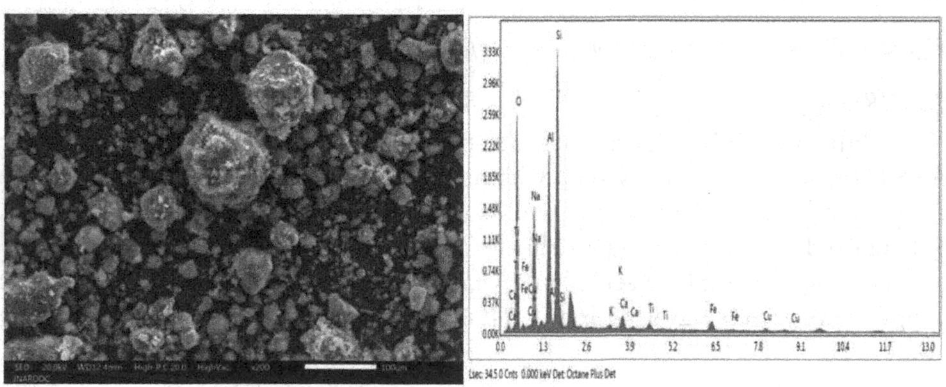

Figure 6: SEM with EDX data of polypyrrole nanocomposites (PPy NC).

4. Pharmacological Activity

4.1. *Antibacterial Activity*

In the context of a lack of new antibiotic manufacture, the development of antibiotic resistance by microorganisms has been viewed as a global emergency and one of the most important issues for global public health. Consequently, it becomes very vital to explore methods of avoiding the usage of antibiotics. The use of chemical composites at the nanoscale in place of conventional medications offers significant advantages for the inactivation of new resistance mechanisms acquired by various microbes.

The use of conducting polymers as antibacterial agents whether alone, in combination or in conjunction with antibiotics has been viewed in this approach as a viable methodology for new antibacterial systems. Particularly, polypyrrole has been cited in the literature as one of the most significant organic compounds, having been successfully investigated in a wide range of applications by utilising its outstanding electrical properties. excellent stability under ambient conditions, simplicity of synthesis and beneficial redox characteristics. Polypyrrole finds widespread applications in various fields, including the fabrication of electrodes for supercapacitors, sensors, corrosion-resistant surfaces, the elimination of heavy metal ions and traces from wastewater, adsorption of dyes, selective adsorption of components, electromagnetic wave absorption and as antibacterial agents. Due to the dearth of traditional antibiotics, the use of antimicrobial compounds is highly encouraged in a variety of sectors, including healthcare, food and textiles. These alternative materials typical environmental friendliness, strong biodegradability and inherent activity against antibiotic-resistant pathogens serve as design cues. These uses include the sterilisation of water and food packaging. The goal is to prevent methicillin-resistant Staphylococcus aureus (MRSA) and create antibacterial textiles using nanocomposites of poly(lactic acid) and chitosan. Polypyrrole is modified with positive charges strategically placed at intervals of three to five monomers along its main chain that results in the intrinsic antibacterial activity of polypyrrole. The resulting polymeric chain, as stated, has a significant antibacterial activity due to this pertinent cationic property.

4.2. Hemolysis Activity

This is an explanation of a process involving the evaluation of hemolysis activity *in vitro* for PPy-NT:Ag-NP nanocomposites using the method described as hemolysis is a phenomenon where red blood cells (RBC) break down or are destroyed, leading to the release of the hemoglobin they contain into the surrounding environment. To conduct the *in vitro* hemolysis activity test of PPy-NT:Ag-NP nanocomposites, the following steps were taken: mammalian blood was drawn and placed in a heparinized tube containing 4% sodium citrate. The tube was then subjected to centrifugation at 3000 rpm for 20 minutes at 4 °C. Afterward, phosphate saline buffer (PBS, pH 7.4) was used to wash the erythrocytes repeatedly. To prepare the hematocrit, 5% of the erythrocytes were gently resuspended in PBS. PPy-NT:Ag-NP nanocomposites were dispersed in PBS at various concentrations, including 1.25 g/ml, 2.5 mg/ml, 5 mg/ml and 10 mg/ml. Triton X-100, a substance known to harm red blood cells, was used as the positive control, while PBS served as the negative control. In multiple microfuge tubes, 100 µl of the dissolved samples at different concentrations were combined with 1900 µl of hematocrit and incubated at 37 °C for 1 hour. Following the incubation, the samples were centrifuged at 3000 rpm for 5 minutes and then placed in an ice bath for 60 seconds. The concentration of free hemoglobin in the supernatants was measured by determining absorbance at 540 nm, which was used as an indicator of hemolysis. The percentage of hemolysis was subsequently calculated based on the obtained data.

5. Conclusions

A promising strategy to combat the proliferation of antibiotic resistance involves the development of alternative antibacterial agents based on conducting polymers. These agents can be enhanced through the addition of fillers and physical techniques like phototherapy and electrical excitation to overcome limitations caused by the gradual attachment of microorganisms to the polymer surface. These approaches have applications in various industries, including medicine, food and textiles. Creating substrates with high porosity, surface area and flexibility to serve as prototypes for wound dressing devices activated by external stimuli is a promising approach for effective treatment of microbial infections.

The interaction of carbon derivatives, metal nanoparticles and polypyrrole within a porous substrate is a significant prototype for antibacterial applications due to the high surface area, conductivity, inherent activity of polypyrrole (for electrostatic interaction) and carbon nanotubes (for physical membrane disruption) in composites. These properties can be optimised through heat treatment and reactive oxygen species (ROS) generation to enhance their effectiveness. For instance, PPy-chitosan composites exhibit a positive zeta potential, promoting electrostatic interactions with oppositely charged species, which benefits the material's polycationic behaviour against both gram-positive and gram-negative bacteria.

The primary focus of this research was the development of environmentally friendly methods for synthesising polypyrrole nanocomposites using microwave and ultrasonication techniques. Characterisation was carried out using FTIR, XRD, SEM and EDX analyses. This study underscores the importance of green nanotechnology, which emphasises eco-friendly materials and processes to reduce the potential risks associated with conventional nanomaterials. By adopting methods that minimise energy consumption, waste generation and environmental impact, this research highlights the potential of green nanotechnology for sustainable advancements.

The study also sheds light on the properties and applications of polypyrrole, an organic polymer known for its exceptional electrical properties, stability and redox behaviour. Polypyrrole has diverse applications in sensors, supercapacitors and antibacterial agents. Its intrinsic antibacterial activity due to cationic behaviour makes it a promising candidate for addressing antibiotic resistance.

Furthermore, the research explores nanocomposites, emphasising their significance in engineering materials with improved properties through the incorporation of nanoparticles. Specifically, the study focuses on polypyrrole nanocomposites, which combine the antimicrobial properties of silver nanoparticles with the conductive and redox properties of polypyrrole. Various synthesis methods are discussed, and the antibacterial activity of these nanocomposites is investigated. The combination of silver nanoparticles and polypyrrole enhances the antibacterial effectiveness of the resulting nanocomposite.

In the methodology section, the study elaborates on the experimental procedures used to synthesise polypyrrole nanocomposites via microwave and ultrasonication methods. The preparation of samples with varying silver nitrate concentrations is detailed, emphasising the importance of controlled synthesis to achieve desired properties. Characterisation techniques, including FTIR, XRD, SEM and EDX analyses, are employed to gain insights into the structural, morphological and compositional properties of the synthesised nanocomposites. These techniques provide a comprehensive understanding of the materials' characteristics, aiding in the optimisation of their properties for specific applications. Additionally, the study explores the pharmacological activities of the synthesised polypyrrole nanocomposites, particularly their antibacterial and hemolysis activities, highlighting their potential as alternatives to traditional antibiotics and their importance in addressing the global challenge of antibiotic resistance.

Acknowledgments

We would like to thank PBRI, Bhopal, for providing the necessary support.

Conflicts of Interest

The authors declare no conflict of interest.

References

[1] Balbus, J. M., Florini, K., Denison, R.A., and Walsh, S. A. (2007). Protecting workers and the environment: an environmental NGO's perspective on nanotechnology. *J. Nanopart. Res.*, 9, 11–22.

[2] Bansal, R., Jain, A., Goyal, M., Singh, T., Sood, H., and Malviya, H. (2019). Antibiotic abuse during endodontic treatment: a contributing factor to antibiotic resistance. *Journal of Family Medicine and Primary Care*, 8(11), 3518–3524.

[3] Benn, T. M., and Westerhoff, P. (2008). Nanoparticle silver released into water from com mercially available sock fabrics. *Environ. Sci. Technol*, 42, 4133–4139.Besley, J.C., Kramer, V.L., and Priest, S. H. (2008). Expert opinion on nanotechnology: risks, benefits, and regulation. *J. Nanopart. Res.*, 10, 549–558.

[4] Carnovale, C., Bryant, G., Shukla, R., and Bansal, V. (2016). Size, shape and surface chemistry of nano-gold dictate its cellular intraction, uptake and toxicity. *Prog. Mater. Sci.*, 83, 152–190.

[5] Centi, M., and Perathoner, S. (2011). Carbon nanotubes for sustainable energy applications. *ChemSusChem.*, 4, 913–925.

[6] Eastlake, A., Hodson, L., Geraci, C. L., and Crawford, C. A. (2012). Critical evaluation of material safety data sheets (MSDS) for engineered nanomaterials. *J. Chem. Health Saf.*, 53, S108–S112.

[7] Guo, L., Lui, X., Sanchez, V., Vaslet, C., Kane, A. B., and Hurt, R. H. (2007). Window of opportunity: designing carbon nanomaterials for environmental safety and health. *Mater. Sci. Forum*, 511–516, 544–545.

[8] Hristozov, D., and Ertel J. (2009). Nanotechnology and sustainability: benefits and risk of nanotechnology for environmental sustainability. *Forum der Forschung.*, 22, 161–168.

[9] Hua, M., Zhang, S., Pan, B., and Zhang, W. (2012). Heavy metal removal from water/wastewater by nano-sized metal oxides: a review. *J. Hazard. Mater.*, 211, 317–331.

[10] Huh, A. J., and Kwon, Y. J. (2011). "Nanoantibiotics": a new paradigm for treating infectious diseases using nanomaterials in the antibiotics resistant era. *Journal of Controlled Release*, 156(2), 128–145.

[11] Hullmann A., and Meyer M. (2003) Publications and patents in nanotechnology. Scientometrics. 58, 507–527.

[12] Hutchison, J. E. (2008). Greener nanoscience: A proactive approach to advancing applications and reducing implications of nanotechnology. *ACS Nano.*, 2, 395–402.

[13] Jayapalan, A. R., Lee, B. Y., and Kurtis, K. E. (2013). Can nanotechnology be green? comparing efficacy of nano and microparticles in cementitious materials. *Cem. Concr. Compos.*, 36, 16–24.

[14] Khan, A. K., Rashid, R., Murtaza, G., and Zahra A. (2014). Gold nanoparticles: synthesis and applications in drug delivery. *Trop. J. Pharm. Res.*, 13, 1169–1177.

[15] Lam, C. W., James, J. T., McCluskey, R., Arepalli S., and Hunter R. L. (2006). A review of carbon nanotube toxicity and assessment of potential occupational and environmental health risks. *Crit. Rev. Toxicol.*, 36, 189–217.

[16] Manuel, C., Toni, U., Allan, P., and Katrin, L. (2018). Challenges in determining the size distribution of nanoparticles in consumer products by asymmetric flow field-flow fractionation coupled to inductively coupled plasma-mass spectrometry: the example of Al2O3, TiO2, and SiO2 nanoparticles in toothpaste. *Separations*, 5, 56.

[17] Munguia, J., and Nizet, V. (2017). Pharmacological targeting of the host–pathogen interaction: alternatives to classical antibiotics to combat drug-resistant superbugs. *Trends in Pharmacological Sciences*, 38(5), 473–488.

[18] Rickerby D. G., and Morrison M. (2006). Nanotechnology and the environment: a European perspective. *Sci. Technol. Adv. Mater.*, 8.19–24.

[19] Schulte P. A., McKernan L. T., Heidel D. S., and Okun A. H. (2013). Occupational safety and health, green chemistry, and sustainability: A review of areas of convergence. *Environ. Health.*, 12, 1186–1476.

[20] Som, C., Wick, P., Krug, H., and Nowack, B. (2011). Environmental and health effects of nanomaterials in nanotextiles and façade coatings. *Environ. Int.*, 37, 1131–1142.

[21] Zou H., Wu S., and Shen J. (2008) Polymer/silica nanocomposites: Preparation, characterization, properties and applications. *Chem. Rev.*, 108, 3893–3957.

Chapter 61

Assessment and Optimisation of Integrated Solid Waste Management Strategies for Sustainable Urban Environments

Rajdip Utane[1], Ujwal Nage[2], Nikhil Manwatakar[2], Gopal Kruahi[2], Prasanna Walde[2], Harsh Diwakar[2], and Naqui Siddiqui[2]

[1]Department of chemistry, Sant Gadge Maharaj Mahavidyalaya Hingna Nagpur, Affiliated to RTM Nagpur University Nagpur, MH, India, Indira Gandhi National Open University (IGNOU), Nagpur

[2]Institute of Science, Nagpur

Abstract: In recent study to comprehensively investigate and propose solutions for improving integrated solid waste management in urban environments, with a focus on sustainability and efficiency. The study will assess current waste management practices, explore environmental implications, evaluate technological innovations, examine community engagement, analyse policies, explore waste minimisation strategies, assess economic considerations and highlight successful case studies. The outcomes will offer valuable insights for policymakers, waste management authorities and urban planners to foster a cleaner, healthier and more environmentally friendly urban environment. The project draws from a wide range of literature, addressing topics such as e-waste, municipal solid waste, composting, recycling and more.

Keywords: Solid waste management, urban environments, municipal solid waste, composting, recycling.

1. Introduction

Integrated waste management (IWM) is a comprehensive approach to managing waste that takes into account all aspects of waste generation, collection, transportation, treatment and disposal. The goal of IWM is to minimise the negative environmental and health impacts of waste while maximising its economic value. In India, IWM holds particular significance due to the country's large population and limited resources. With a population of over 1.3 billion people, India generates a substantial amount of waste on a daily basis, estimated at 62 million tonnes in 2016 alone. This figure is expected to rise to 165 million tonnes by 2030 by Modak *et al.* (2010); PPCB (2010) and Rana *et al.* (2015) while some of this waste can be

DOI: 10.1201/9781003527442-61

recycled or repurposed, a significant portion ends up in landfill sites, where it can release harmful toxins into the environment.

This contamination can affect water sources, soil quality and the well-being of local residents. Therefore, the implementation of IWM is crucial for safeguarding the environment and public health in India. Additionally, IWM has the potential to create job opportunities and stimulate the economy by harnessing waste as a resource for energy generation and other purposes. The Indian Government has been actively working on the development of an IWM strategy since the early 2000s. In 2006, it initiated the National Mission for Clean Ganga (NMCG), which included a focus on waste management as part of its efforts. Various hazardous solid wastes from different sources can be categorised as organic, reusable and recyclable materials suggested by Jha *et al.* (2003); Kansal (2002) and Sridevi *et al.* (2012). These solid wastes primarily originate from villages, agriculture, municipalities, and hospitals.

*1.1. **Village Wastes** (**VWs**)*: Village wastes consist mainly of decomposable and recyclable materials.

1.2. Agriculture Solid Wastes (ASWs)

These wastes can potentially lead to groundwater contamination and soil infertility due to toxic organic materials and metals.

1.3. Municipal Solid Waste (MSW)

MSW accounts for a significant portion of solid waste generation, often amounting to thousands of tons per day. It includes hazardous, non-hazardous and organic waste, with some being compostable. The annual increase in MSW generation per capita is around 1–1.33%. In India, the per capita MSW generation rate ranges from 0.2 to 0.5 kg per day in small towns. MSW can release toxic gases and substances, posing a threat to soil, groundwater and the environment. (Kansal *et al.*, 1998; Srivastava *et al.*, 2014).

1.4. Hospital Solid Wastes (HSWs): HSWs are generated during patient observation, diagnosis, treatment and curative processes in both human and veterinary fields. They also result from the production and testing of biological products. HSW is considered hazardous, with only 5% being non-infectious suggested by Annepu (2012); Dasgupta *et al.* (2013).

The infectious waste component may contain pathogens that can cause diseases in humans and animals, including serious illnesses such as AIDS, hepatitis, tuberculosis and more. Some toxic chemicals, like dioxins and furans, can also be generated from HSW, harming animals and birds. HSW may contain small amounts of radioisotopes used in therapeutic and diagnostic studies.

1.5. Radioactive Solid Waste (RSW): RSW mainly contains uranium and plutonium.

1.6. Electronic Waste (E-Waste)

E-waste is generated at an alarming rate due to globalisation and rapid technological advancements. It includes discarded electrical and electronic equipment and contains toxic elements like acids, polychlorinated biphenyls, hexavalent chromium (PVV), as well as heavy metals such as lead, mercury, cadmium and arsenic. Improper disposal of e-waste can lead to serious health issues, including bronchial diseases, lung cancer, liver and kidney damage and contamination of groundwater.

To mitigate the harmful effects of solid waste on human civilisation, proper solid waste management (SWM) is essential. SWM involves various steps, including waste generation, storage, collection, transfer, transportation, processing and disposal. The management of solid waste is influenced by national income and legal policies, and it requires significant financial resources. Finding suitable locations for waste treatment is increasingly challenging due to the "Not In My Backyard" (NIMBY) attitude in communities.

Efforts such as the "Reduce, Reuse, Recycle" (3Rs) approach and integrated solid waste management, prioritise waste prevention, reduction and recycling over treatment and disposal. Disposal methods for solid waste include landfilling, incineration, pyrolysis and composting. Proper SWM is crucial to minimise the adverse impacts of solid waste on the environment and public health.

2. Review of Literature

The literature review for the assessment and optimisation of integrated solid waste management (ISWM) strategies for sustainable urban environments provides an essential foundation for understanding the key concepts, challenges and solutions in this field. Here's an overview of the relevant topics covered in the literature (; Das *et al.*, 2014; Kaushal *et al.*, 2012).

2.1. Solid Waste Management (SWM) Overview

The literature often begins with an introduction to the significance of SWM in urban areas. It highlights the rapid urbanisation and population growth trends that have led to increased waste generation. This section also emphasises the importance of sustainable SWM practices for mitigating environmental impacts and ensuring public health.

2.2. Components of ISWM

The review typically discusses the core components of ISWM, which encompass waste generation, collection, transportation, treatment and disposal. Researchers delve into the intricacies of each component, highlighting challenges, opportunities and best practices.

2.3. *Waste Minimisation and Source Segregation*

Many studies emphasise waste minimisation strategies including reduction, reuse and recycling. Source segregation of waste at the household level is often explored as an effective means of enhancing recycling rates and reducing the volume of waste sent to landfills.

2.4. *Technological Innovations*

Researchers often investigate the role of technology in SWM, including waste-to-energy conversion processes, advancements in waste sorting and recycling technologies, and the use of data analytics for optimising collection routes.

2.5. *Environmental Impacts*

The literature frequently examines the environmental consequences of improper SWM, including groundwater and soil contamination, air pollution and greenhouse gas emissions. Sustainable practices are explored as solutions to mitigate these impacts.

2.6. *Economic Aspects*

Economic considerations are a crucial aspect of ISWM. Researchers assess the cost effectiveness of different waste management strategies, explore the potential for revenue generation through recycling and waste-to-energy projects and discuss financing mechanisms for SWM infrastructure.

2.7. *Social and Behavioural Factors*

The review often delves into the social dynamics of SWM, including the role of public awareness, community engagement and behavioural change in promoting sustainable waste practices. Cultural factors and waste picker communities may also be discussed.

2.8. *Policy and Regulatory Frameworks*

The literature commonly examines the role of government policies and regulations in shaping SWM practices. Researchers analyse the effectiveness of existing policies, propose regulatory improvements and discuss the importance of stakeholder collaboration.

2.9. *Case Studies and Best Practices*

Numerous studies feature case examples from different urban areas worldwide, showcasing successful ISWM initiatives and highlighting lessons learned. These case studies often serve as practical models for optimising waste management systems.

2.10. *Sustainability and Circular Economy*

The concept of sustainability and the transition to a circular economy are recurring themes. Researchers explore how ISWM aligns with broader sustainability goals, such as resource conservation and reducing environmental degradation.

2.11. *Challenges and Barriers*

Challenges and barriers to implementing ISWM are thoroughly analysed. Common obstacles include inadequate infrastructure, limited financial resources and resistance to change among stakeholders.

2.12. *Future Trends and Research Gaps*

The review concludes by identifying emerging trends in SWM, such as the adoption of smart technologies, increased emphasis on organic waste management and the potential for decentralised waste solutions. Research gaps and areas requiring further investigation are often highlighted.

A comprehensive overview of the multifaceted field of ISWM, offering insights into the challenges, opportunities and strategies for achieving sustainable urban waste management. It serves as a valuable resource for researchers, policymakers and practitioners seeking to optimise waste management systems in urban environments. USEPA (2002) (Pattnaik & Reddy, 2010) and (Agamuthu *et al.*, 2007).

3. Methodology

3.1. *Materials*

Waste can be managed properly by adequate public awareness, effective collection and recycling system, following rules and regulation, fixing responsibilities, adequate research, and generating statistics.

3.2. *Methods*

ISWM hierarchy lays emphasis on source reduction and reuse, recycling, composting, waste to energy and finally landfill.

3.3. *SWM by Sustainability*

It captures various waste management scenarios with different waste generation rates and distribution among recycling, composting, landfilling and energy recovery. The overall sustainability score provides a quantitative measure of how well each scenario aligns with the goal of sustainable urban waste management.

Table 1: Assessing and optimising integrated solid waste management strategies for sustainable urban environments by sustainability.

Scenario	Waste Generation Rate (kg/person/day)	Recycling Rate (%)	Composting Rate (%)	Landfill Rate (%)	Energy Recovery Rate (%)	Overall Sustainability Score
1	1.5	20	30	40	10	75
2	1.8	25	20	30	25	80
3	1.2	30	40	20	10	85

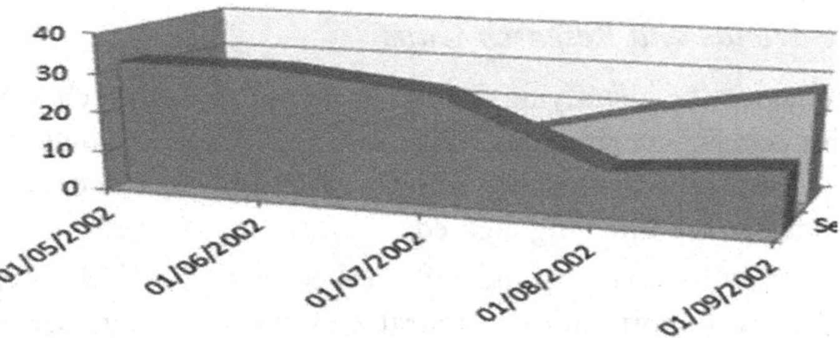

Figure 1: Assessing and optimising integrated solid waste management strategies for sustainable urban environments by sustainability.

3.4. SWM by Component

It includes columns for observation number, strategy type, components of the strategy, implementation cost, environmental impact, social acceptance and effectiveness of the strategy.

Table 2: Assessment and optimisation of integrated solid waste management strategies in sustainable urban environments by component

Strategy Type	Components	Implementation Cost	Environmental Impact	Social Acceptance	Effectiveness
Recycling	Collection, Sorting, Processing	Moderate	Reduced landfill waste, Energy savings	High	Effective in waste diversion
Composting	Collection, Processing	Low	Reduced methane emissions, Soil enrichment	Medium	Effective in organic waste reduction
Waste-to-Energy	Collection, Incineration, Energy Conversion	High	Energy generation, Reduced landfill volume	Medium	Effective in waste volume reduction

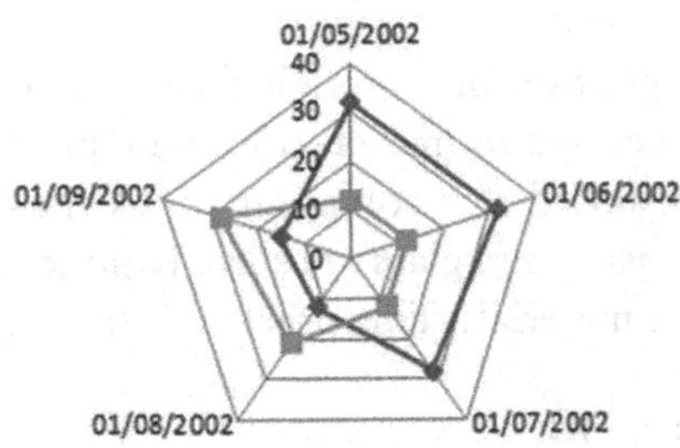

Figure 2: Assessment and optimisation of integrated solid waste management strategies in sustainable urban environments by component.

4. Result and Discussion

In the year 2000, the Ministry of Environment and Forests (MoEF) issued rules for the management and handling of Municipal Solid Waste (MSW) in India. These rules aimed to ensure proper waste management practices were established in the country. A recent update to the draft rules further emphasises this effort. The implementation of these rules has allowed municipal authorities to develop infrastructure for the "collection, storage, segregation, transportation, processing, and disposal" of MSW. Notably, Chandigarh has shown significant progress in waste management compared to other Indian cities, demonstrating improved waste handling practices.

4.1. Segregation

Currently, waste segregation in India lacks an organised and systematic approach, both at the household and community levels. Typically, segregation is carried out by the informal sector and sometimes by waste producers themselves. The conditions and efficiency of these segregation and sorting processes are generally unsafe, unhealthy and ineffective. This is primarily because only valuable waste materials that can yield financial gains when recycled are sought and extracted.

4.2. Solid Waste Collection and Transportation

Crucial components of any Solid Waste Management (SWM) system include waste collection, storage and transportation. However, cities in India face challenges in these aspects. Municipal corporations in India are responsible for waste collection and are often provided with bins for separating biodegradable and inert waste. Despite these efforts, waste is frequently mixed, dumped and openly burned. Improvements in waste collection and transportation processes, as well as enhancements in India's transportation infrastructure, have the potential to create more job opportunities, improve public health and boost tourism. Efforts to address these issues are essential for sustainable waste management practices in India.

4.3. *Reuse/Recycle*

Recycling involves the collection of materials from waste which appear to have economic value and can be used to manufacture new products. At the same time, non-segregated waste is disposed of in community bins, making recycling less likely. However, materials like plastics and glass are sorted and separated by waste pickers who trade these recyclable materials. This practice is quite prevalent.

4.4. *Disposal and Treatment*

According to a UNEP report from 2004, composting is considered one of the more suitable waste treatment methods, especially in developing countries in Asia, as opposed to incineration. In India, composting is practiced at a rate of approximately 10–12%, while countries like Bangladesh, Nepal, Sri Lanka and Pakistan have even lower rates, less than 10%. Among the developing Southeast Asian countries, the final stages of SWM often involve open dumping, accounting for over 50%, with landfilling at 10–30%, composting at less than 15% and incineration at 2–5%.

5. Conclusion

The research project's outcomes will contribute valuable insights to policymakers, waste management authorities and urban planners seeking to enhance waste management practices for a more sustainable and resilient urban future. By identifying best practices and proposing optimisation strategies, the study will play a crucial role in fostering cleaner, healthier and more environmentally friendly urban environment.

Acknowledgment

We would like to thank IGNOU, Nagpur, for providing the necessary support.

Conflicts of Interest

The authors declare no conflict of interest.

References

[1] Bhoyar, R. V., Titus, S. K., Bhide, A. D., and Khanna, P. (1999). Municipal and industrial solid waste management in India. *J IAEM*, 23, 53–64.

[2] Das, S., and Bhattarcharya, B. K. (2014). Estimation of municipal solid waste generation and future trends in greater metropolitan regions of Kolkata, India. *Ind Eng Manage Innov*, 1, 31–38.

[3] Dasgupta, B. Yadav, V. L., and Mondal, M. K. (2013). Seasonal characterization and present status of municipal solid management in Varanasi, India. *Adv Environ Res*, 2, 51–60.

[4] Guria, N., and Tiwar, V. K. (2010). Municipal solid waste management in Bilaspur city (C.G.) India. *National Geographer*, Allahabad, 1, 1–16.

[5] Jha, M. K., Sondhi, O., and Pansare, M. (2003). Solid waste management-a case study. *Indian J Environ. Protect*, 23, 1153–1160.Kansal, A. (2002). Solid waste management strategies for India. *Indian J Environ Protect*, 22, 444–448.

[6] Kansal, A., Prasad, R., and Gupta, S. (1998). Delhi municipal solid waste and environment: an appraisal. *Indian J Environ Protect*, 18, 123–128.

[7] Kaushal, R., Varghese, G. K., and Chabukdhara, M. (2012). Municipal solid waste management in India-current state and future challenges: a review. *International Journal of Engineering Science and Technology*, 4, 1473–89.

[8] Modak, P., Jiemian, Y., Hongyuan, Y. U., and Mohanty, C. R. (2010). Municipal solid waste management: turning waste into resources, in Shanghai manual. *A guide for sustainable urban development in the 21st century, Shanghai, United Nations Department of Economic and Social Affairs* (UNDESA), 1–36.Pattnaik, S., and Reddy, M. V. (2010). Assessment of municipal solid waste management in Puducherry (Pondicherry), India. *Resources, Conservation and Recycling*, 54, 512–520.

[9] PPCB. (2010). Status report on municipal solid waste in Punjab, Punjab Pollution Control Board, Patiala N T, "Municipal solid waste management in India," Waste disposal to recovery of resources? *Waste Manage*, 29, 1163–1166.

[10] Rana, R., Ganguly, R., and Gupta, A. K. (2015). An assessment of solid waste management system in Chandigarh city India. *Electron J Geotech. Eng*, 20, 1547–1572.

[11] Sridevi, P., Modi, M., Akshmi, M. L., and Kesavarao, L. (2012). A review on integrated solid waste management. *Int J Eng Sci Adv Technol*, 2, 1491–1499.

[12] Srivastava, R., Krishna, V., and Sonkar, I. (2014). Characterization and management of municipal solid waste: a case study of Varanasi city, India. *Int J Curr Res Acad Rev*, 2, 10–16.

Chapter 62

Quantum Computing and Simulation – A review

Sanjivani Shastri[1]

[1]Department of Physics, Cummins College of Engineering for Women Hingna, Nagpur, Maharashtra, India

Abstract: One of the emerging technologies is quantum computing and its simulation. Quantum mechanics deals with the particles at atomic or subatomic state. Electrons, photons are the examples of it. Thus, quantum physics plays a very important role in quantum computing. Using simple principals of superposition and entanglement quantum computers can do wonders of fast computation which a classical computer cannot. In 1980 Feynman proposed ideal model for computer-based quantum mechanical system and its simulation. This has revolutionised computer difficulties. Since 1982 in a journey of quantum computing, there were many ups and downs. Numerous advances have been made to uncover quantum mechanics. Nowadays there has been a commercial interest because of the performance of quantum computers over classical computers.

Keywords: Quantum computing, quantum simulation, quantum mechanics.

1. Introduction

1.1. Classical Computers

Classical computers were based on classical mechanics. Binary bits 0 and 1 are used for computing in such computers. The main advantage of such computers is that it can be used at normal temperature. But the computers can have one state at a time that is either 0 or 1 (Kumar *et al.*, 2022). The binary bits can be copied from one to another state in a easy way. Processing of information is also possible subsequently. In such type of computers back tracking is not possible as well as data teleportation is also not possible. The basic of classical mechanics is based on particle nature of electron. The principals of quantum mechanics can help to build novel computers called as quantum computers. Study by ID James *et al.* (2017) reveals this fact. In 1980 Feynman (Feynman, 1982) proposed that logic gates can be built up using quantum mechanical systems. This has brought a revolution in the area of computing research. Gabriel (1994) cited that Peter Shor developed a technique for solving basic prime factorisation problem. His technique can easily

DOI: 10.1201/9781003527442-62

used to solve integer factorisation problem over classical computers as studied by Anders *et al.* (2010). Educating the stakeholders about the quantum computing is necessary to keep pace with the changing world and to meet the global and national needs in this area.

1.2. Quantum Computers

Quantum computers are fast and compact. Chip size is reducing fast as studied by Microsoft and Google team. Computers can do the complex calculations very easily in very short time. This is because it uses quantum bits or qubits. But such computers can work if the temperature is kept below zero-degree kelvin. Qubits can have four various states at the same time but copying of qubit is not possible. Information is in the form of qubit which can be used in quantum logic gates. The back tracking is also possible in such type. Quantum computers use properties of quantum mechanics such as superposition, interference and entanglement. Using entanglement data teleportation is possible. Quantum computers work on the wave particle duality theorem. Here in quantum computers every electron–electron interaction is to be considered. It can optimise the problem and can be used to handle big data, ML, forecasting, simulation of proteins for medicine, etc. Hence it can be used as basic encryption tool in information security (Jazaeri *et al.*, 2019).

2. Quantum Computing System

The system consists of the following characteristics:

a) **Qubit:** Qubit has replaced a bit in quantum computers. The qubit can have ground and excited state at the same time. Eigen states are important here. The state of qubit can be mapped to Eigen state. Polarisation of photon and spin of electron play important role in qubit. Quantum computers differ from classical computers due to two reasons. First and important in quantum superposition and another quantum entanglement.

b) **Quantum superposition:** A classical bit is a data that can have two different states of either 0 or 1. But a qubit (quantum bit) can have a continuous state between 0 and 1. It can be represented as a wave function which is a combination of both. Thus, if a classical computer can take only one state at a time out of four various combinations. In case of quantum computer, quantum two bit can represent any of the above state simultaneously. If the number of qubits increases it leads to exponential increase in a number of states. This is called as superposition.

c) **Entanglement:** If there are two non-interacting atoms, that is if they are separated by some distance they cannot communicate. But due to entanglement it is possible. This is applicable to subatomic level. If the subatomic particles are far apart, they can be correlated and hence their entanglement is possible. Thus, even if the spins of atoms are different they may be mathematically connected (BC Sanctuary, 2004.) The property of entanglement along with superposition helps the quantum computing system to do much faster calculations, which is not possible for classical computer system.

Parallelism: It is the ability of performing large number of operations in parallel. With the use of this concept new quantum microprocessors can be developed (Shailesh *et al.*, 2021). Using parallelism, a computer can search a data from large database and can do several actions simultaneously. Figure 1 shows the characteristics of quantum computing.

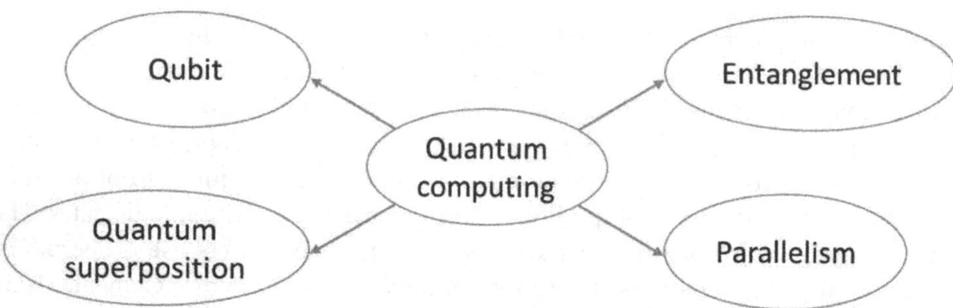

Figure 1: Quantum computing.

2.1. *Quantum Simulation*

Quantum simulation involves quantum properties. Using microscopic properties of particles real-world applications such as biology, industry, nitrogen fixing, material science, logistics and optimisation, information security as well as artificial intelligence can be solved (Ladd *et al.*, 2010). Quantum simulation can offer many issues such as quantum dots, superconducting circuits and atoms within optical lattices (Buluta *et al.*, 2009; Houck *et al.*, 2012).

2.2. *Quantum Simulator*: These are the special types of machine that gives insight about problems in physics (Johnson *et al.*, 2014). Simulators are developed using property of trapped ion, ultra cold atom and superconducting qubits.

a) **Trapped ion simulator:** Such type of simulator consists of tiny crystals that have single plane and have less than one millimetre diameter. Each outermost electron acts as a tiny quantum magnet and can represent a qubit which is equivalent of 0 or 1 in conventional computer. Trapped ion simulator with 2 and 3 spins were studied. (Friedenaur *et al.*, 2008; Joseph, 2012). Quantum simulation up to 18 trapped ion spins that can control the level of spin compensating antiferromagnetic interaction range a new cryogenic ion trapping system was designed (Islam *et al.*, 2013; Joshi *et al.*, 2022).

b) **Ultracold atom simulator:** Many experiments have been done with bosons and fermions in optical lattice. The main aim of such 2008 experiments is low-temperature phases or tracking. Many experiments showed that condensed matter models are difficult for conventional materials.

c) **Superconducting qubits:** There are two main categories. Quantum annealers that determines the ground state of certain Hamiltonians. The other includes many systems that emulate specific Hamiltonian study with their ground state properties. Figure 2 shows the properties of simulators.

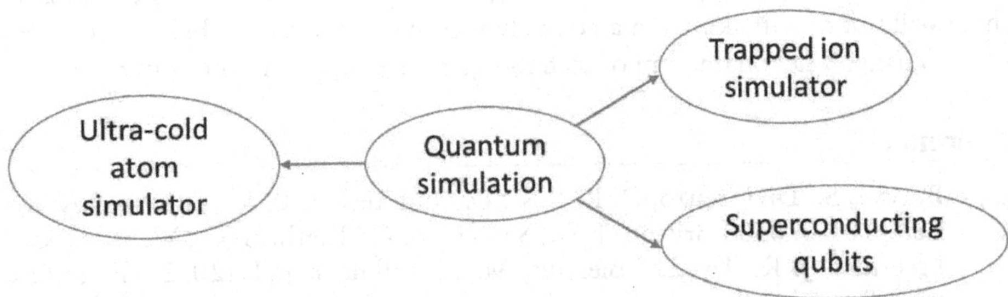

Figure2: *Quantum simulator.*

It is difficult to directly simulate quantum systems on classical computers. It requires a large amount of memory by quantum computers to store data. Quantum states grows exponentially thus causing exponential explosion. To measure some quantity, initial state is prepared and final state can be obtained after evaluation. This is not easy as it includes steps with many resources. Many a times extracting the information from quantum simulation is difficult.

There are two types of simulators Analog and Digital simulators.

Analog quantum simulators: It is device in which system will mimic the evolution of another. Here the Hamiltonian of a system is important Hamiltonian of a system is simulated which can be used for long-time storage, which uses an ion chain upto 44 ions (Joshi *et al.*, 2022) and mapped to Hamiltonian of the simulator. Thus, this type of simulator has limited systems. Such simulators are not accurate as the degree of simulation is less and hence cannot even capture the key features of the real systems, unable to provide meaningful results about the system which is being simulated. Bravy *et al.* (2008) developed several methods, but it is not an easy task. There are many applications stated by Buluta *et al.* (2009). Such simulators can emulate large quantum system than classical computers.

Digital quantum simulator: Using Schrödinger time-independent Hamiltonian which can be represented by many local interactions. But quantum circuits which are having one and two qubit gates used for unitary transformation can efficiently work on many finite dimensional local Hamiltonian. This is the biggest advantage of this digital system.

3. Conclusion

This review article highlights the use of quantum mechanics in the field of quantum computing. Quantum computers are developed using the basic idea of qubits, superposition and entanglement. Quantum simulation involved two major simulators: analog and digital simulators. The simulation requires a lot of research. Analog computers have a lot of impact on science. There are lot of models coming up for analog quantum simulators. A lot of optimisations is required for digital quantum simulators. A hybrid model can be expected in near future which will be accurate and time saving with high programmable ability. India has established the

National Mission for Quantum Technology and Applications (NM-QTA) in 2020. These will not only make India a competent contender in the global quantum race but also usher a new paradigm of technology policymaking in the country.

References

[1] Bravya, S., Divincenzo, D. P., Loss, D., and Terhal, B. M. (2008). *Phy. Rev. Lett*, 101 070503.Britton, J. W., Sawyer, B. C., Keith, A. C., Wang, C. C. J., Freericks, J. K., Uys,H., Biercuk, M. J., Bollinger, J. J. (2012). Engineered two-dimensional using interactions in trapped ion quantum simulator with hundreds of spins. *Nature,* 484(7395), 489–492.

[2] Buluta, I., and Nori, F. (2009). Quantum simulators. *Sci*, 326, 108–111.

[3] Dur, W., and Briegel, M. J. (2008). *Phy Rev*, A78, 052325.

[4] Feynman, R. P. (1982). Simulating physics with computers. *International Journal of Theoretical Physics*, 21(6), 467–488. [online]. Available: https://doi.org/10.1007/BF02650179.

[5] Friedenauer, A., Schmiz, H., Glueckert, J. T., Porras, D., and Schaezt, T. (2008). Simulating a quantum magnet with trapped ions. *Nature Physics*, 4(10), 757–761.

[6] Gabriel, T. (2012). Quantum Vs classical logic: the revisionist approach. *Logos & Episteme*, 3(4), 579–590.

[7] Houck, A. A., Tureci, H. E., and Koch, J. (2012). On chip quantum simulation with superconducting circuits. *Nature Physics*, 8, 292–299.

[8] Islam, R., Senko, C., Campbell, W. C., Korenbit, S., Smith, J., Lee, A., Edwards, E. E., Wang, C. C. J., Freeicks, J. K., and Monroe, C. (May 2013). Emergence and frustration of magnetism with variable range interactions in a quantum simulator.

[9] James, I. D. (2017). A history of microprocessor transistor count 1971 to 2017.

[10] Janet, A., Saroosh, S., Stefanie, H., and Eric, L. (2010). Landauer's principle in quantum domain, Developing in computational model. *Cornell University Library quant -Phy*, 1, 13–18.

[11] Jazaeri, F., Beckers, A., Tajalli, A., Sallese, J. M. (2019). A review on quantum computing: qubits, cryogenic electronics and cryogenic MOSFET Physics. arXiv:1908.02656V1Quant-phy.

[12] Johnson, A., Tomi, H., Stephen, R., and Jaksch, D. (2014). What is a quantum simulator? *EPJ Quantum Technology,* 1(10). arXiv:1405.2831. https://doi.org/10.1140/epjqt10.S2CID 120250321.

[13] Joshi, M. K., Kranzl, F., Schuckert, A., Lovas, I., Maier, C., Blatt, R., Knap, M., Roos, C. F. (2022). Observing emergent hydronomics in a long- range quantum magnet. *Science* 6594(376), 720–724.

[14] Ladd, et al. (2010). Quantum computers. *Nature*, 464, 45–53.

[15] Microsoft and Google quantum architectures and Computation team, January 2014, defining and detecting quantum speedup, center for quantum information science & technology, University of Southern California.

[16] Sanctuary BC, (2004). Quantum correlations between separated particles. McGill University, Canada.

[17] Shailesh, S., Mohmad, Z. K., and Ravendra, S. (2012). Green computing: an Era of Energy Saving Computing of Cloud Resources. *I J Mathematical sciences and computing*, 2, 42–48. https://doi.org/10.5815/ijmsc.

[18] Sumit, K., Anjali, T., and Ankit, M. (2022). The review on quantum computer and their application in future era. *IJERT*, ISSN 2278-0181, 11(12).

Printed in the United States
by Baker & Taylor Publisher Services